# XIX International Coal Preparation Congress

# Congress Proceedings

## Volume I

XIX International Coal Preparation Congress
13 – 15 November, 2019,

New Delhi – India

*Editor*

**Raj K. Sachdev**

**Coal Preparation Society of India**

Woodhead Publishing India Pvt Ltd

New Delhi

Published by Woodhead Publishing India Pvt. Ltd.
Woodhead Publishing India Pvt. Ltd.,
303, Vardaan House, 7/28, Ansari Road,
Daryaganj, New Delhi - 110002, India
www.woodheadpublishingindia.com
Phone: 91-11-23266107, 011-43612145

First published 2019, Woodhead Publishing India Pvt. Ltd.
© Woodhead Publishing India Pvt. Ltd., 2019

Woodhead Publishing India Pvt. Ltd. ISBN: 978-93-88320-19-1
Woodhead Publishing India Pvt. Ltd. e-ISBN: 978-93-88320-20-7

Typeset by Bhumi Graphics, New Delhi
Printed and bound by Replika Press Pvt. Ltd.

# Acknowledgement

I sincerely thank all the 300 plus authors and co-authors who have contributed articles in these two volumes of proceedings. I also thank the Members of the International Organising Committee (IOC) of the XIX International Coal Preparation Congress (ICPC) who took pains in reviewing all 170 plus 'ABSTRACTS' received and shortlisted some 115 abstracts for inclusion in the congress agenda. On behalf of the Coal Preparation Society of India (CPSI), I express deep gratitude to the Ministries of Coal, Power, New and Renewable Energy, Environment, Forest & Climate Change, Steel, Mines, Science & Technology, Earth Sciences and Economic Diplomacy & States Division, External Affairs for extending their full support to this prestigious global event on COAL.

I also thank the editorial team of the publishers Woodhead Publishing India Pvt. Ltd., for their efforts in bringing out this two volume set of 'Congress Proceedings'.

My special thanks to the Saint Petersburg Mining University and the International Competence Centre for Mining-Engineering Education, Saint Petersburg (Russia) under the auspices of UNESCO for their appreciation and recognition of this work.

**Raj K. Sachdev**

# Foreword

India is a country, which has been fortunate to be endowed with very large reserves of coal. As of date, the reserve is over 300 billion tonnes of coal. While the abundance of this fuel is acknowledged,most of this coal has a very high ash content, which makes beneficiation a necessary process to improve its quality and to render its usage as a fuel, which is more environment friendly. The need to wash and improve the quality of the coal has been endorsed by the government of India repeatedly, but the progress in establishing washeries by Coal India Ltd., the main producer of coal, is abysmally poor. Coal Preparation Society of India is the front-runner in promoting the use of washed coal. There are some very critical issues involved in promoting this activity despite clear benefits in respect of preservation of the environment. Any enterprise would welcome any sort of move to get and use good quality coal to improve its business. We in CPSI are constantly engaged in creating forums and encouraging industries to recognize both environmental and commercial benefits of washing.

CPSI is a member of the Organising Committee of the International Congress on Coal Preparation (ICPC) and in that capacity has become the host for the 19thcongress, which is to be held in New Delhi from 13th to 15th November 2019. The first ICPC was held in 1950 in Paris, France. The main objective was to make the coal available and usable from the damaged mines in Europe after the World War. ICPC is coming back to India after 37 years. The 9th ICPC was held in New Delhi in 1982. ICPC is a forum where intensive deliberations take place on varied technical aspects of Coal Preparation reflecting the worldwidestatus with regard to the latest state-of-the-art technologies being developed or deployed in the industry in respect of coal preparation. This congress will provide all stakeholders, which may include coal miners, steel and cement manufacturers, thermal power generating companies, manufacturers of equipment used in coal handling and washing and many others an excellent opportunity to gather information on the latest technologies and also be able to interact with the technology providers. CPSI finds itself in an exceedingly advantageous situation in furthering its objectives as the organization, which is primarily responsible for holding the 19th ICPC.

ICPC is a forum where only technical matters are presented. We in CPSI initiated the process of seeking papers for presentation in the congress more than a year ago. Initially, abstracts were obtained which were scrutinized by the International Organising Committee of ICPC and thereafter the authors of the articles were requested to send us the full text of the articles. We were fortunate enough to receive more than 110 articles for presentationin the Congress. These proceedings contain the 100 articles, which are being presented in the Congress. All the proceedings of ICPC since its initiation have become an invaluable record of scientific advancements that have taken place in the area of coal preparation over the late 70 years. We are sure these proceedings will be of immense value to the industry and the scientists.

**Alok Perti**
*Chairman, Coal Preparation Society of India (CPSI)*
*Chairman, National Organising Committee (NOC)*

# Editorial

Coal has been and will continue as the main source of energy for India for many more years. India's hunger for coal is more than justified because of its meagre known hydrocarbon resources and the country imports nearly 85% of its consumption of oil and natural gas. Furthermore, out of its 1.35 billion population about 25% is yet to be served with clean cooking fuel and electricity. Therefore, continuing use of coal which provides about 75% of India's electricity generation becomes a critical necessity. However, notwithstanding the fact that per capita electricity is only about 1100 Kwhr, country's economy has to grow at a reasonable rate of 8 to 9%, that makes it imperative that India continues to depend on coal based electricity, despite heavy investment on creating non-fossil based generation capacity of 40% of total electricity generation capacity of 800 – 900 GW required by 2030. With country's aim of becoming a Five Trillion Dollar economy by 2025, the requirement of coal for generation of electricity, for making iron & steel, cement for meeting requirement of growing infrastructure like ports, highways, housing, education, health facilities and other basic needs to provide required quality of life to its growing population will also grow commensurately.

With poor generic quality of domestic coal, it is imperative that all coals must be washed to reduce the ash content and improve the heat value so that it burns more efficiently with lesser emissions. With this, as its main objective, the Coal Preparation Society of India (CPSI) has been dedicatedly promoting washing of coal. Fortunately, with persistent efforts, CPSI has been able to convince the government as well all stakeholders about positives of coal washing and the benefits that accrue to the Nation as a whole.

Coal Preparation Society (CPSI) with strong industry support has made a niche for itself as a credible professional body representing the Indian coal industry. It was support of the member companies that prompted CPSI to undertake the onus of hosting the XIX International Coal Preparation Congress & Expo (ICPC).

The 9th ICPC was held in India in 1982 when the world was faced with sudden doubling of oil prices after Iranian revolution and beginning of Iran-Iraq war. However, coal continued to be the King. The XIX ICPC is now being held under the shadow of climate change, fears of melting glaciers, fast changing weather patterns and concerns about floods, droughts and crop losses. Naysayers are predicting death of coal. However, China, India and many other countries having limited choice between renewable and coal, will continue to use coal for power, iron and steel, cement etc for many more years to come. International Energy Agency (IEA) has predicted that in 2045 or so, around 26% (from present 38%) of world electricity will still be coming from coal. But we cannot afford to adopt an ostrich approach. India has accepted the need to wash coal, adopt high efficiency low emission (HELE) technologies for power generation and in coming years CCUS projects are sure to become reality.

In the above background, we are fortunate to have the XIX ICPC at this juncture. With expected participation of around 500 delegates including some 120 subject experts from overseas, this prestigious international event on coal offers a great opportunity to all participants including government policy makers to share experience and exchange knowledge on various topical issues on coal. At the same time, this is a unique opportunity for business community, investors, project developers and technology suppliers to show case their strengths and team up with local entities to establish business in India and elsewhere.

Coming to the subject of this publication, we have over 110 high quality papers from some 15 countries, covering all aspects of coal processing and also touching upon gasification, coal to chemicals and emerging uses of coal through its conversion into high value, low polluting energy and petro substitutes.

While every care is taken to edit these two volumes of 'Congress Proceedings', we will be open to receive suggestions for any omissions or errors. In case any reader desires to get in touch with the author(s) for any clarification, he or she may approach the concerned author directly. CPSI will provide the e mail contact of the concerned author, if requested.

I must record my sincere thanks to all 300 plus authors and co authors, my own CPSI team and the editorial team of Woodhead Publishing (India) who have put their heads together to bring out a two volume set of 'Congress Proceedings' running into over thousand pages.

**Raj K. Sachdev**

*President, Coal Preparation Society of India (CPSI),*
*Chairman, International Organising Committee (IOC),*
*XIX International Coal Preparation Congress & Expo (ICPC),*
*New Delhi, India*

# Messages

The most crucial element required for rapid socio-economic development is energy. In India, despite a serious effort to increase the dependence on renewable energy, **COAL** will remain the most important source for electricity generation for several years. Electricity, which in India is largely coal based is crucial for building modern infrastructure, supporting urbanization and improving the quality of life. The objective is to produce 1.5 billion tonne of coal by 2020 and increase it to 2.5 billion by 2032. With such increase in coal production and consumption, adoption of clean coal technologies for improved efficiency in power generation should be part of our energy security plan.

With every passing day, global climate change concerns are becoming more pronounced. India is committed to fulfill its declared targets in reduction of $CO_2$ emissions. In this effort the government has taken significant policy initiatives and I am hopeful that these efforts will helps in reducing overall $CO_2$ emissions, so that, India can meet its climate priorities.

I extend my heartiest felicitations to the **Coal Preparation Society of India** for spearheading the cause and concept of washing of coal in India. I am happy to note that **19th International Coal Preparation Congress** will be held in 2019 in New Delhi, India after a gap of 37 years. I extend heartiest welcome to the subject experts and delegates both from within the country and overseas who will be participating in this important event on coal. I wish the **Coal Preparation Society of India** all success in their undaunted efforts in making this international conference a great success.

**(M. Venkaiah Naidu)**

**New Delhi**
**04th May, 2018.**

**Shri Narendra Modi**
**Hon'ble Prime Minister of India**

"Due to the use of washed coal, the energy consumed in transportation, handling and milling, is optimized as the inert material from coal is eliminated. This helps in reducing the auxiliary consumption of equipment involved in coal processing because the use of improved quality coal ultimately results in reduction of emission of GHG as compared to conventional coal."

**'CONVENIENT ACTION – continuity for**
**Change' by Narendra Modi.**

प्रल्हाद जोशी
**PRALHAD JOSHI**
ಪ್ರಲ್ಹಾದ ಜೋಶಿ

सत्यमेव जयते

संसदीय कार्य, कोयला तथा खान मंत्री
भारत सरकार
नई दिल्ली
**MINISTER OF PARLIAMENTARY AFFAIRS,
COAL AND MINES
GOVERNMENT OF INDIA
NEW DELHI**

No. M(C/M&PA/Message/2019/        June, 2019

## MESSAGE

Coal is valued as a vital energy and strategic resource essential to the world's sustainable development and energy security objectives.

With 75% of electricity generation, 65% steel production and 90% cement production being dependent of coal as a critical fuel; Coal is and will continue to be a major source of India's commercial needs for many more years to come. Therefore, it is essential that coal must be washed so that it is burnt efficiently and power plant emissions are also reduced. Therefore, washing of coal is an essential subset of coal production-supply chain in India.

It is heartening to note that the Coal Preparation Society of India (CPSI) has been dedicatedly working towards creating awareness about the benefits of using washed coal, among the coal producers as well as the consumers. I convey my best wishes to the Coal Preparation Society of India (CPSI) for its stupendous efforts in promoting coal washing in India.

I am happy to know that the XIX International Coal Preparation Congress (ICPC) will be held at New Delhi during 13-15 November, 2019 New Delhi and Coal Preparation Society of India (CPSI) is the organiser of this International event on Coal. I wish the XIX International Coal Preparation Congress (ICPC) a grand success and I am confident that this International Conference will be an excellent opportunity for exchange of experience, knowledge and expertise amongst subject experts from various coal producing countries attending this important congress on coal.

( PRALHAD JOSHI )

Office : Room No. 15, Parliament House, New Delhi-110001, Tel : 011-23017780, 23017798, 23018729, Fax : 011-23792341
Office : Room No. 504, C Wing, 5th Floor, Shastri Bhawan, Tel : 23070522, 23070524, Fax : 23070529
Residence : 5, G.R.G. Road, New Delhi-110001, Tel : 011-23094650, 23093497
H. No. 122-D, 'Kamitartha' Mayuri Estate, Keshwapur, Hubli-580023 (Karnataka)
Tel. No. : 0836-2251055, 2258955 E-mail : pralhadvjoshi@gmail.com

प्रकाश जावडेकर
Prakash Javadekar

मंत्री
पर्यावरण, वन एवं जलवायु परिवर्तन मंत्रालय,
सूचना एवं प्रसारण मंत्रालय
भारत सरकार
Minister
Ministry of Environment, Forest & Climate Change,
Ministry of Information and Broadcasting
Government of India

## MESSAGE

Coal is the backbone of India's commercial energy supply and long term energy security and sustainable development. Nearly 75% of India's electricity is generated using coal. The Iron & steel, cement and many other industries are also heavily coal dependent. Various projections made by credible institutions including NITI Ayog, indicate that COAL will continue to be a major source of our country's energy needs for few decades.

India is endowed with abundant coal resources but its oil and gas resources are very limited. Due to its very generic nature, the quality of our domestic coal is poor due to its high ash content. Since washing of coal reduces its ash content and its heat value also improves, it is essential that coal must be washed so that it is burnt efficiently and power plant emissions are also reduced, therefore, coal has to be washed for its efficient burning and with less environmental pollution. We must ensure that our coal is washed as it will help to a significant extent in meeting India's commitments made at the Paris Climate Treaty.

It is gratifying to note that the **Coal Preparation Society of India (CPSI)** has been dedicatedly spearheading the concept and cause coal washing by creating awareness about the benefits of using washed coal, among all stake holders.

I am delighted to learn that the **XIX International Coal Preparation Congress (ICPC)** will be held during 13-15 November 2019 at New Delhi and Coal Preparation Society of India(CPSI) is the organiser of this international event on COAL. It is further heartening to know that this prestigious international congress on coal is being held in India after 37 years. Last one the IX ICPC was held in 1982.

I convey my best wishes to **Coal Preparation Society of India (CPSI)** for its stupendous efforts in promoting coal washing and clean coal technologies in India.

I wish the **XIX International Coal Preparation Congress (ICPC)** a grand success. I am sure that this international conference will afford an excellent opportunity for exchange of knowledge and sharing amongst experts from abroad who will be participating in this prestigious international event on coal.

Date: 22.08.2019

**(Prakash Javadekar)**

Paryavaran Bhawan, Jor Bagh Road, New Delhi-110 003
Tel. : 011-24695136, 24695132, Fax : 011-24695329
E-mail : minister-efcc@gov.in

5th Floor, Shastri Bhawan, New Delhi-110 003
Tel. : 011-23384340, Fax : 011-23782118
E-mail : mib.inb@nic.in

अनिल कुमार जैन, भा॰प्र॰से॰
सचिव
**ANIL KUMAR JAIN,** IAS
SECRETARY
Tel.: 23384884 Fax : 23381678
E-mail : secy.moc@nic.in

भारत सरकार
GOVERNMENT OF INDIA
कोयला मंत्रालय
MINISTRY OF COAL
शास्त्री भवन, नई दिल्ली–110 001
SHASTRI BHAWAN, NEW DELHI-110 001
www. coal.gov.in

## MESSAGE

**Coal** is the key stone of India's energy supply. Apart from nearly 75% of India's electricity being generated using coal, 65% of iron & Steel and 90% of cement production is also coal dependent. Besides, there are many other industries using coal either as their feedstock or for generating heat and electricity.

India's coal resources being of 'drift' origin, the ash content is high. This is true both for coking coal as well as thermal variety of coal. Therefore, coal must be washed to reduce the ash content and improve its heat value.

The **Coal Preparation Society of India (CPSI)** is well known for its efforts in spreading awareness among the stakeholders about the techno-economic benefits that accrue to the power plants of using washed coal.

I am very happy that the **Coal Preparation Society of India (CPSI)** is organizing **XIX International Coal Preparation Congress (ICPC)** from 13th to 15th November, 2019 at New Delhi. It is further heartening to know that this prestigious international congress on coal is being held in India after 37 years.

I convey my best wishes to **Coal Preparation Society of India (CPSI)** for its dedicated efforts in promoting coal washing among all stakeholders.

I wish the **XIX International Coal Preparation Congress (ICPC)** a grand success. I am sure that this international conference will afford an excellent opportunity to Indian coal professionals and experts from overseas for exchange of knowledge and sharing experiences.

**(Anil Kumar Jain)**

Date: 03.10.2019

विनय कुमार
सचिव
**Binoy Kumar**
Secretary

सत्यमेव जयते

भारत सरकार
इस्पात मंत्रालय
**GOVERNMENT OF INDIA
MINISTRY OF STEEL**

28th August, 2019

## MESSAGE

The Indian Steel Industry has entered into a new development stage envisaging 250 million tonnes of steel production in coming decade. In 2018 Indian Crude Steel production was 2nd highest in world and this position is poised to be retained..

While Indian Steel Sector is on a growth trajectory, it is required that our steel companies are globally competitive. For achieving this, both availability and the cost of major inputs like iron ore and coking coal are key factors. As the global suppliers of metallurgical grade coking coal are limited in number, our endeavour must be to maximize the use of domestic coking coal. Since around 12% of total coal resources are known be of coking variety but of higher ash content, washing of such coals must be undertaken to obtain clean coal of desired quality to suit the requirement of steel mills. This would naturally require appropriate designing and setting up of coal washeries based on coal specific washing technologies. More importantly, additional sourcing of coking coal from domestic sources will reduce our import dependence to a significant extent.

It is really commendable that the **Coal Preparation Society of India (CPSI)** has been dedicated to promoting awareness among the stakeholders about the techno-economic benefits of utilizing domestic coking coals after proper washing. Towards this endeavour I am happy that the CPSI is organizing **XIX International Coal Preparation Congress (ICPC)** from 13th to 15th November, 2019 at New Delhi. I wish this prestigious international event on **COAL** all success.

Binoy Kumar
(Binoy Kumar)

Room No. 291, Udyog Bhawan, New Delhi-110 011, Tel. : +91 11 23063912, Fax : +91 11 23063489
E-mail : secy-steel@nic.in

सी.के.मिश्रा
C.K.Mishra

सचिव
भारत सरकार
पर्यावरण, वन एवं जलवायु परिवर्तन मंत्रालय
SECRETARY
GOVERNMENT OF INDIA
MINISTRY OF ENVIRONMENT, FOREST AND CLIMATE CHANGE

## MESSAGE

Energy is an essential requirement for any economy and coal is the major supplier of India's commercial energy supply and a critical fuel for our country's long term energy security and sustainable development.

Nearly 75% of India's electricity is generated using coal. The iron & steel, cement and many other industries are also heavily coal dependent.

While India has abundant coal resources, its known hydrocarbon oil and gas resources are rather limited. However, the quality of our domestic coal is poor due to its high ash content. Since washing of coal reduces its ash content and its heat value also improves, it is essential that coal must be washed so that it is burnt efficiently with lower emissions.

Therefore, we must ensure that our coal is washed as it will help to a significant extent in meeting India's commitments made at the Paris Climate Treaty.

It is well known that the **Coal Preparation Society of India (CPSI)** has been dedicatedly spreading awareness about the benefits of using washed coal among coal producers as well as coal consumers.

I am very happy that the **XIX International Coal Preparation Congress (ICPC)** will be held during 13 -15 November, 2019 at New Delhi and Coal Preparation Society of India (CPSI) is the organiser of this international event on COAL. It is further heartening to know that this prestigious international congress on coal is being held in India after 37 years. Last one, the IX ICPC was held in 1982.

I convey my best wishes to **Coal Preparation Society of India (CPSI)** for its stupendous efforts in promoting coal washing among all stakeholders.

I wish the **XIX International Coal Preparation Congress (ICPC)** a grand success. I am sure that this international conference will be a good opportunity for exchange and sharing of knowledge amongst Indian coal professionals and experts from overseas participating in this prestigious international event on coal.

[C. K. Mishra]

Dated: 5th August, 2019
Place:  New Delhi

इंदिरा पर्यावरण भवन, जोर बाग रोड, नई दिल्ली-110 003 फोन : (011) 24695262, 24695265, फैक्स : (011) 24695270

INDIRA PARYAVARAN BHAWAN, JOR BAGH ROAD, NEW DELHI-110 003 Ph. : (011) 24695262, 2465265,  Fax :  (011) 24695270
E-mail : secy-moef@nic.in, Website : moef.gov.in

United Nations · International
Educational, Scientific and · Competence Centre for
Cultural Organization · Mining-Engineering Education

The Saint Petersburg Mining University and the International Competence
Centre for Mining-Engineering Education in Saint Petersburg (Russia) under
the auspices of  UNESCO, recognise excellent efforts put in by Mr. Raj K
Sachdev, President Coal Preparation Society of India in editing these **Congress
Proceedings**, and appreciate his contribution to the coal preparation industry.

Vice-Rector for Scientific and Innovative Activities                V. Bazhin
of Saint Petersburg Mining University

# Contents

# About International Coal Preparation Congress (ICPC)

The **International Coal Preparation Congress (ICPC)** was an offshoot of the Allied Coal Commission which was constituted as part of the Marshall Plan post World War II to initiate the process of reconstruction of Europe. The first ICPC was held in 1950 in France and subsequent Congresses were held in Germany, Belgium, UK, USA, Poland, Australia, Ukraine, Russia, India, Canada, Japan, South Africa, Turkey and China. ICPC is now held every three years in a country selected by the IOC through ballot.

The International Organizing Committee (IOC) of the ICPC is a body which has representatives from 15 countries. Representation on IOC is by a non-government organization which deals in their respective country with the issues relating to coal preparation.

The goals of the ICPC inter alia, are:

- to ensure the growth and development of the coal preparation industry, support and promotion of scientific and technical cooperation among all countries;
- to facilitate the exchange of information about the state-of-art technologies in the field of coal mining, preparation and transportation of coal. Also promote new and emerging technologies in the area of environment friendly and low emission coal utilisation.
- to bring together experts from different countries in order to form professional contacts, exchange experimental results, and promote international cooperation.

History of International Coal Preparation Congress

| | | | | | |
|---|---|---|---|---|---|
| I. | | 1950 - France, Paris | XI | | 1990 - Japan, Tokyo |
| II | | 1954 - Germany, Essen | XII | | 1994 - Poland, Krakow |
| III | | 1958 - Belgium, Liege | XIII | | 1998 - Australia, Brisbane |
| IV | | 1962 - UK, Harrogate | XIV | | 2002 - Republic of South Africa, Johannesburg |
| V | | 1966 - USA, Pittsburgh | XV | | 2006 - Peoples Republic of China, Beijing |
| VI | | 1973 - France, Paris | XVI | | 2010 - USA, Lexington |
| VII | | 1976 - Australia, Sydney | XVII | | 2013 - Turkey, Istanbul |
| VIII | | 1979 - Ukraine, Donetsk | XVIII | | 2016 - Russian Federation, St. Petersburg |
| IX | | 1982 - India, New Delhi | XIX | | 2019 - India, New Delhi (13-15 November 2019) |
| X | | 1986 - Canada, Edmonton | | | |

# About Coal Preparation Society of India (CPSI)

Representing India's commitment to the use of Clean Coal to the world, Coal Preparation Society of India (CPSI) is a not - for - profit, non-government professional body of coal, power, iron and steel sectors and their allied industries. CPSI's mission is to promote washing of high ash domestic coal to improve its quality and enhance the calorific value, making it more environment friendly and suitable for use in High Efficiency Low Emission (HELE) power generating systems as well as for gasification and conversion into chemicals, CTL and other high value petroleum substitutes. It will, therefore, be a step, which will facilitate fulfilling the country's commitment made at the Paris Climate Treaty.

**Main Objectives of CPSI inter alia are:**
- To facilitate policy formulation in coal beneficiation and preparation;
- To provide an effective network amongst coal producers, consumers, R & D organisations etc;
- To provide an independent platform for deliberating technological, operational, financial, commercial and policy aspects of the Indian Coal Preparation Industry.
- To promote and encourage the exchange of technical information relevant to the Indian coal industry.
- Meeting India's commitment to the Climate Change Treaty

CPSI is registered under the Societies Registration Act, XXI of 1860 and its head office is located in New Delhi. CPSI is a member of the International Organizing Committee (IOC) of the International Coal Preparation Congress (ICPC). CPSI is a member of IOC representing India.

**Domestic and Global Network:**

CPSI is an Associate Member of the World Coal Association - a global industry association formed of major international coal producers and stakeholders and has a bilateral relationship with IEA Clean Coal Centre, UK for promoting clean coal technologies for use in High Efficiency Low Emission (HELE) power generating systems.

CPSI has bilateral arrangements with China National Coal Association (CNCA), Coal Preparation Society of Ukraine and similar organisations of other IOC member countries for technical knowledge sharing and exchange of experts and trainees.

CPSI is a Member of ASSOCHAM and also an Associate Member of the PHD Chamber of Commerce and Industry.

Major industry bodies like Federation of Indian Mineral Industries (FIMI), Sponge Iron Manufacturers Association (SIMA), Association of Power Producers (APP), ASSOCHAM of India, World Coal Association (WCA), IEA Clean Coal Centre (UK) are associating and supporting CPSI's efforts towards cleaning of coal and introduction of clean coal technologies so that India can meet its emissions reduction target committed at Paris Climate Treaty.

# List of IOC Members

1. Raj K. Sachdev, India, Chairman, IOC
2. Alok Perti, India, Chairman, NOC
3. Andrew Swanson, Australia
4. Zhang Shaoqiang, P. R. China
5. Xie Wenbo, P. R. China
6. Zhou Shaolei, P. R. China
7. Maria Holuzko, Canada
8. Dieter Ziaja, Germany
9. Ireneusz Baic, Poland
10. Leonid A. Vaisberg, Russian Federation
11. Vladimir Yu Bazhin, Russian Federation
12. Quentin Campbell, South Africa
13. Neil Jenkinson, United Kingdom
14. Olexandr Yegernov, Ukraine
15. Ihor Shemelov, Ukraine
16. Barbara J. Arnold, United States of America

**Corresponding Members**

1. George Anastasakis, Greece
2. Bokanyi Ljudmilla, Hungary
3. ............., Kazakhstan

# 1

# The economic value of coal in India

Paul Baruya

*IEA Clean Coal Centre, London, UK*

**Abstract:** The macroeconomic impacts of coal are complex, for example, direct employment creates wealth, which filters to the surrounding local economies. Importantly, for export producers coal is a valuable source of wealth creation for the national economy through tax receipts, royalties and foreign currency earnings. These aspects of the value of industries to local and national economies would be discussed. Other macroeconomic issues might include the value of coal-fired power to economies. In many non-OECD (Organisation for Economic Co-operation and Development) nations in Asia, coal is the lowest cost form of reliable electricity, and some studies suggest that economic growth in low and middle-income countries have been possible due to low-cost coal power.

This presentation focuses on the Sasan power plant, one of India's ultra-mega power plant projects built by Reliance and assesses the value of the project with regards to employment and electricity supplies to the local and national economy. This presentation accompanies a report published by the International Energy Agency Clean Coal Center (IEA CCC) investigating the macroeconomic and social value of coal in major world economies.

**Keywords:** Economic value, employment, mining, electricity, macroeconomic

## 1. Introduction

The negative impacts of coal use are well documented by environmental lobby and political groups, but less is known about the positive feedback associated with coal plants, particularly in economies that are lacking electricity supplies. In many developed economies in western Europe and North America, coal markets have contracted from a variety of regulatory and market pressures making the case for unabated coal less compelling compared to natural gas and renewables. However, in many regions in developing (and developed) economies a low carbon pathway is not simple, and economic growth and employment is still heavily dependent on the existence of a commercial and thriving coal sector. This study will examine the strategic value of coal to a large proportion of the world's population and examines why the move away from coal is not as easy as first thought on a macroeconomic level. On a micro-economic level, the local impacts of any coal project are a broad and complex list of socio-economic and environmental impacts.

The analysis is based on published research examining the impact of coal and power related to the creation of gross domestic product (GDP), incomes and employment. These positive impacts are often directly related to coal activities, while further economic impacts are felt indirectly by the wider economy from the spending of wealth created in the coal and power sector. Many academic and consultancy papers have examined the economic impacts of coal and power activities across the world, and nearly all use the same methodology by measuring the impacts in three ways: direct impacts, indirect impacts and induced impacts.

Assessing the economic value of *direct impacts* can be represented simply by the level of employment and/or the income generated by the business of operatinga coal mine and power generation plant. Any large infrastructure project creates wealth, which filters to the surrounding local economies. A simple example is to estimate the incomes paid to the workforce of a coal mine, but also measure the expenditure on components and materials, which generates incomes to those component and material supply companies, these are *indirect impacts*. While coal activities create full time employment, part time positions are converted into full-time equivalent. To avoid exaggerating the employment data, two people working part-time shifts might be considered as a single full-time role. These incomes will then be further disseminated across the local economy as households spend on goods, services and leisure (*induced impacts*).

The operation of a power plant will similarly generate direct and indirect wealth, firstly by creating incomes for its workforce, but also by creating cashflow for companies responsible for supplying the fuel and (fuel) transportation during the operation of the plant. New projects will attract an influx of new labour in the form of construction and engineering jobs, as well as other opportunities in administration, management and catering. Public wealth is also boosted through tax receipts and royalties during both the construction and operation of a mine or power plant. Reliable power supplies are considered vital for economic growth in regions.

## 2. Macroeconomic value of coal fired power as a driver of GDP

In many non-OECD nations in Asia, coal is the most affordable form of baseload and flexible electricity. Some studies suggest that economic growth in low and middle income countries have been possible due to the development of coal-fired power. Furthermore, the contribution of coal to an emerging economy facing a complex energy trilemma (security, affordability and sustainability) might be considered as a positive feature that benefits the energy economy.

The relationship between the cost of energy and its impact on the economy is well understood. For example, the effects of the price of oil on inflation are linked. Between 2000 and 2015, analysis by Choi et al. (2018) showed that for every 10% increase in global oil prices, domestic inflation increased by an average 0.4 percentage points across 72 countries.

With regards to the impacts of energy prices and wider economic impacts, the consensus among economists suggests that the electricity price elasticity coefficient of employment is negative, in other words, a rise in the price of electricity leads to a fall in employment.

The accompanying report published earlier this year examines the socio-economic impacts of coal power and/or mining in several OECD and non-OECD countries. India is an important country in terms of its coal economy and several aspects of the value of coal.

## 3. Case study: Sasan coal power plant

In the case study for India, the socio-economic impacts of a power station project was reviewed with material drawn from Russo et al. (2014). The evidence was provided for one of India's newest power stations, the Sasan supercritical coal power plant. Sasan is one of India's ultra-mega class of power plant and has a capacity of 3960 MW, comprising $6 \times 660$ MWe (gross). The plant is equipped with ESP particulate control systems. According to the plant's parent company, Reliance, Sasan is capable of generating power for 17.5 million people across seven states. The project was initially estimated to cost $4 billion, of which 75% was debt (Reliance, 2018).

### 3.1 Employment created by Sasan

A detailed socio-economic impact of the plant was carried out by Russo et al. (2014) and included a range of direct, indirect and induced impacts as follows:

- Direct: employment during plant construction (4 years);
- Indirect: employment and benefits associated from setting up the coal supply infrastructure such as local coal mining, transportation and fuel treatment plants (25 years);
- Induced: lower electricity prices and expenditure on environmental improvements;
- Societal impacts: taxation revenue and energy independence.

The economic benefits arising from the four-year construction period and 25-year operation of the plant for three of the economic impacts: direct, indirect and induced impacts (see Table 1). The evaluation showed a total positive impact of $54.54 billion. Around a fifth of the economic benefits arise during construction, while four fifths of the benefits come from the operation (and maintenance, or O&M) of the power plant over 25 years (Fig. 1).

**Figure 1:** Sasan ultra mega power project (UMPP)
(Image courtesy of Reliance, 2018)

**Table 1:** Direct and indirect economic impacts due to construction and operation (Russo et al., 2014)

|  | Construction (USD billion) | O&M (USD billion) |
|---|---|---|
| Direct economic impact | 2.40 | 9.21 |
| Indirect economic impact | 3.51 | 11.29 |
| Induced economic impact | 6.24 | 21.88 |
| Total impact | 12.15 | 42.39 |
| Total economic impact | 54.54[a] | |

[a] Cumulative benefits accrued during the construction phase of our years and the operating lifetime of 25 years.

Around 41,100 jobs were expected to be created during construction and operation (see Table 2), half being created during construction and half being sustained over the 25 years of operation. With regards to the Sasan power plant the construction of the plant created 20,950 jobs over the four-year period prior to operation. This comprises 5000 directly employed people annually over a four-year period. Indirect jobs (mining, coal transport and coal preparation) and induced jobs added a further 16,000 jobs during the construction period.

**Table 2:** Sasan UMPP employment impacts (Russo et al., 2014)

|  | Construction | O&M | Total |
|---|---|---|---|
| Years of employment (years) | 4 | 25 | |
| Direct jobs | 5000 | 639 | 5639 |
| Indirect jobs | 3700 | 3970 | 7670 |
| Induced jobs | 12,250 | 15,532 | 27,782 |
| Total jobs created | 20,950 for 4 years | 20,141 for 25 years | 41,091 |

The operation of the plant requires considerably fewer people in terms of direct employment. Some 639 people are required for the O&M of the plant. This employment level is sustained over the 25 years life of the plant. Indirect jobs include mining, transportation, communication, spares and consumable supplies, financial services and other ancillary essentials ensure the operation of the power plant is supported with adequate mining and transportation infrastructure coal preparation plant. Reliance Sasan has three captive coal mines associated with the power project and are quite labour-intensive operations. A range of skilled, semi-skilled and unskilled jobs have been created, and estimates suggest every household in Sasan would have at least one member working directly or indirectly for the Sasan power project.

The direct and indirect jobs that are created by the Sasan plant created higher disposable incomes for individuals and organisations. The income spreads in the form of higher expenditure in food, consumer durables, leisure activities and general spending. This wealth generation is called the multiplier effect (Russo et al., 2014). These induced jobs are seen across the agricultural sector (due to rising demand for food) and retail and commercial sectors. In total 27,782 induce jobs would be created by the Sasan UMPP (see Table 2).

## 3.2 Social benefits

Some anti-coal lobby groups raised issues regarding environmental, human rights and worker conditions which should be addressed where valid (Banktrack, 2016; Carbon Market Watch, 2014), however the reports did not report the positive economic impacts of the Sasan project. At full capacity, it is estimated that the plants would fulfil the power requirements of more than 17.5 million people across seven states and enable 22 million people to get access to safe water supplies.

Electricity from the Sasan power plant is expected to electrify 12,000 schools increasing the enrolment of education by more than 96,000 students. Some 400,000 households will benefit from better street lighting which will improve well-being and security of many communities. The healthcare sector will see an additional 10,000 beds in rural areas. Affordable electricity supplied by coal-fired grid power will also create savings of $19 billion in urban healthcare centres, allowing expenditure to be diverted to patient care.

Other impacts include the electrification of the Indian railways system. Despite a massive growth in personal transportation, rail services are still an important and cost-effective means of transporting large numbers of people across large distances. Electrification of the railways are an effective means of saving costs and lowering the air quality impacts in densely populated areas by switching away from diesel powered trains, especially for shorter routes.

## 3.3 Macroeconomic impact due to the construction and operation of the plant

The Sasan power plant was expected to have a net present value of $55 billion (2014) over the construction and operation period of the plant. The LCOE was calculated as 2 cts/kWh, a low figure by world standards partly due to the economies of scale of such a large project as well as the low cost of coal (Russo et al., 2014). The main plant components were sourced from China, although the finance came from a broad syndicate of Indian, Chinese, Japanese and US funding sources. The boiler, steam and electricity generating turbine systems were supplied by the Shanghai Power Corporation, while financing came mainly from Indian and US financial sources.

The plant and associated mine development was intended to cost $4 billion over the four-year construction period, however the project experienced a cost overrun of $1.45 billion (Thorum, 2015). Some 60% of the materials was sourced within India. Some 40% of the investment will be for materials and equipment from abroad, such as China. One of the major recipients of the funding will be the plant supplier Shanghai Electric Group Co Ltd of China which is supplying the boiler, steam turbine system and the power generator (Platt's, 2018).

## 3.4 Better access to reliable electricity

Increased access to electricity will aid industry. According to Russo et al. (2014), the agricultural sector could see a GDP impact of $2.5 billion, compared with national agricultural GDP of $320–350 billion. Agriculture and associated services account for 18% of the nation's GDP but just 5% of energy use (IEA Statistics, 2019).

Improvements resulting from better access to electricity include improvements in productivity through the use of pumped irrigation improving crop yields by 10%. Better farming practice and the availability of affordable electricity permits a greater use of irrigation allowing farmers to grow a more diverse cash crop and potentially improves incomes by up to five times. Some 90% of electricity used in the agriculture sector is used for water pumping and replacing diesel pumps with electric systems lowering the costs considerably (Russo et al., 2014).

Industry accounts for 29% of India's national GDP and accounts for 34% of total final electricity consumption (2017) the industrial sector is one of the most electricity intensive sectors in the economy. Stable electricity supplies are essential to all sectors of the Indian economy, especially to small to medium enterprises (SMEs) which represent 45% of industrial output in India and employs 60 million people and growing by 1.3 million jobs every year (EISBC, 2019). An inconsistent supply of electricity lead to a number of problems for SMEs due to:

- the reduction in productivity, due to power outages and unplanned stoppages in production;
- an increase in input costs as industries switch to backup diesel power costing $80/h;
- cost of installation of backup power, sometimes $40–85,000, a large outlay for SMEs;
- maintenance costs of backup diesel generators;
- additional cost of fuel inventories for diesel generators (Russo et al., 2014).

Distributed power generation in rural communities has proven to be a valuable addition to off-grid communities, however, industrialisation and larger grid supplies are necessary for large scale energy usage. Evidence suggests that a large-scale power plant like Sasan will offer numerous benefits to modern business and commerce, such as:

- help alleviate grid outages that range between 1 and 40 h/week in India;
- help alleviate the monetary losses accrued by SMEs due to power outages, typically ranging $15–650/day;
- total losses incurred by SMEs of $1.5 billion/year due to power losses.

## 4. Conclusion

India's programme of UMPP provide large amounts of controllable and reliable electricity. The contribution that a coal power station can make to a local economy cannot be understated. Employment opportunities are immense and can be sustained over several decades due to the operation of the plant and local coal mines and transportation infrastructure. The impact wider community benefits from the provision of affordable electricity can greatly enhance the life expectancy of the community. Improvements to healthcare, education through providing electricity can have profound impacts on several thousands of people. Unreliable or the poor provision of electricity impairs business and adds costs due to the reliance on diesel backup. The plant has given the Sasan region identity and is an important contributor to the economy.

## 5. References

[1] Banktrack (2016) Sasan ultra mega coal power projects (UMPP). Available at: https://www.banktrack.org/project/sasan_ultra_mega_coal_power_project_umpp_/pdf; 5 pp (update 7 Oct 2016).

[2] Carbon Market Watch (2014) The Sasan Coal-Fired Power Plant, in India a Clean Development Mechanism project supporting human rights, environment, and labor violations. Available at: https://carbonmarketwatch.org/wp-content/uploads/2013/07/Sasan-report-final-text.pdf (Dec 2014).

[3] Choi S, Furceri D, Loungani P, Mishra S, Poplawski-Ribeiro M (2018) Oil prices and inflation dynamics: evidence from advanced and developing economies. Journal of Intenrational Money and Finance. Available at: https://www.sciencedirect.com/science/article/pii/S0261560617302541; p 71–96 (Apr 2018).

[4] EISBC (2019) Definition of Indian SMEs. Europe-India SME Business Council, Mumbai, India. Available at: http://www.eisbc.org/definition_of_indian_smes.aspx (2019).

[5] IEA (2019) Energy balances. IEA Data Services, IEA/OECD, Paris, France (accessed Jan 2019).

[6] Platt's (2018) World Electric Power Plant Database. Platts/McGraw Hill (version Sep 2018).

[7] Reliance (2018) Coal based projects. Available at: https://www.reliancepower.co.in/coal-based-projects (2018).

[8] Reliance (2019) Coal based projects. Image courtesy of Reliance. Available at: https://www.reliancepower.co.in/coal-based-projects#menu (downloaded 16 Jan 2019).

[9] Russo C, Thomas M, Hinkamper R (2014) The socio-economic impacts of advanced technology coal-fuelled power stations. Available at: https://www.iea.org/ciab/The%20Socio-economic%20Impacts%20of%20Advanced%20Technology%20Coal-Fuelled%20Power%20Stations_FINAL.pdf (2014).

[10] Thorum M (2015) Report on the project financing of Sasan Power Limited. OIG-INS-15-02, Office of Inspector General Export-Import Bank of the United States. Available at: https://www.exim.gov/sites/default/files/oig/reports/Final%20Sasan%20Report%20-%20Redacted.pdf (28 Sep 2015).

# 2

# The status and development of coal preparation and processing in China

Zhang Shaoqiang

*President, Coal Preparation Branch,*
*China National Coal Association, China*

**Abstract:** The remarkable characteristics of China's energy and natural resources are abundant with coal and lack of oil and gas, so coal is the basic energy in China. In order to avoid and overcome the negative impact of coal on ecological environment, the Chinese government has been vigorously promoting the development of clean and high efficient utilization of coal, including coal preparation, efficient combustion, and clean emissions, and coal conversion. The technology of coal preparation and processing has been developed rapidly in recent years. The coal preparation and processing and clean utilization still have potential development in a certain period in China.
**Keywords:** China, Coal preparation, Clean utilization

## 1. Coal is basic energy in China

### 1.1 China coal production in the past five years

China is abundant with coal and lack of oil and gas, which determines the leading role of coal in primary energy production and consumption. Since the beginning of the 21st century, the Chinese government has vigorously developed new and renewable energy and tried to adjust energy consumption structure, but coal is still the main energy due to the rapid economic and social development and continuous increase of energy consumption. At present, coal accounts for about 70% of energy production and 60% of energy consumption in China. In 2018, China produced 3.68 billion tons of coal.

**Table 1:** Raw coal production in China since 2014

| Year | 2014 | 2015 | 2016 | 2017 | 2018 |
|---|---|---|---|---|---|
| Output/Mt | 3870 | 3750 | 3410 | 3520 | 3680 |

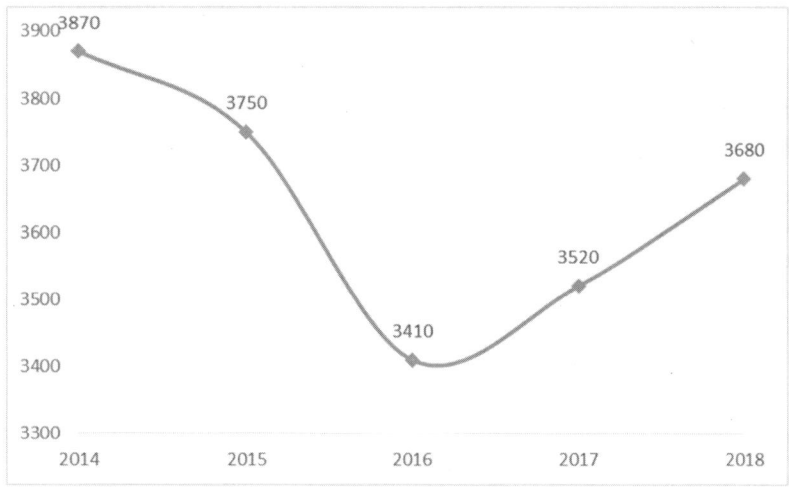

**Fig. 1:** Variation of raw coal production in China since 2014

## 1.2 Coal resource in China

(1) Although China is rich in coal resources, it has a large population, and the per capita share of coal resources is low, which is about 50% of the world average.

(2) Coal resources are unevenly distributed. The main coal mining areas are in the north and northwest regions, which produce about 86% of the total in China. But the main coal consumption areas are concentrated in the southeast region. This situation results in long-distance coal transportation and higher freight.

(3) The geological conditions of coal resources are complex, and difficult to be mined. Most of the coal mining is underground, and some coal mines are in-depth of over 1000 meters. Only a small part of coal mining is an open-pit with mainly lignite.

(4) Coal resources and water resources are inversely distributed. The northwest region with abundant coal resources is in arid and semi-arid conditions. The shortage of water resources has brought serious constraints to the deep processing and clean conversion of coal.

## 1.3 Coal import and export of China

Since 2009, China has become a net coal importer, mainly from Indonesia, Australia, Russia, Mongolia, accounting for 90% of total coal import.

China imported 327 million tons of coal in 2013. With the energy-saving and efficiency improvement and industrial restructuring measures, the total imported coal declined in 2014, but rebounded in 2016, and reached 281 million tons in 2018. On the other hand, China exported less than 10 million tons of coal each year.

**Table 2:** China coal import since 2014

| Year | 2014 | 2015 | 2016 | 2017 | 2018 |
|------|------|------|------|------|------|
| Mt   | 290  | 200  | 260  | 271  | 281  |

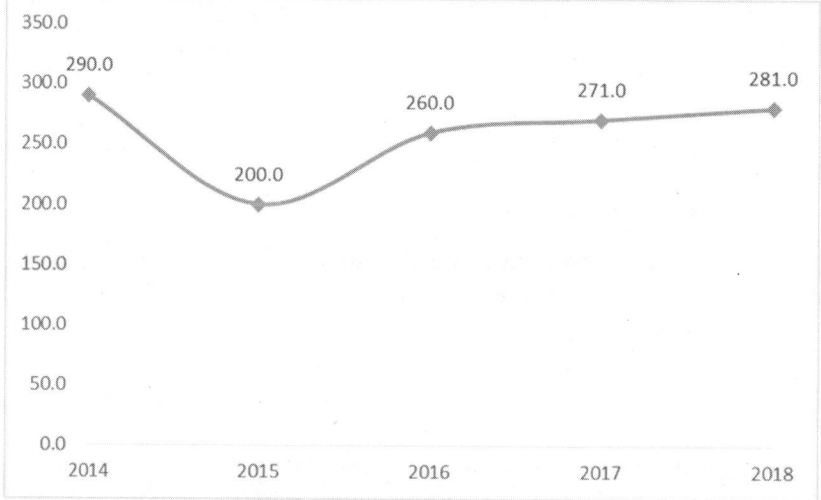

**Fig. 2:** Variation of China coal import since 2014

**Table 3:** China coal export since 2014

| Year | 2014 | 2015 | 2016 | 2017 | 2018 |
|------|------|------|------|------|------|
| Thousand Tons | 5740 | 5330 | 8780 | 8170 | 4930 |

China's domestic coal supply and demand made basically balance, but the loose and tight situation coexist in some areas due to local transportation. The imported coal in southeastern coastal areas is competitive in the advantage of shipping.

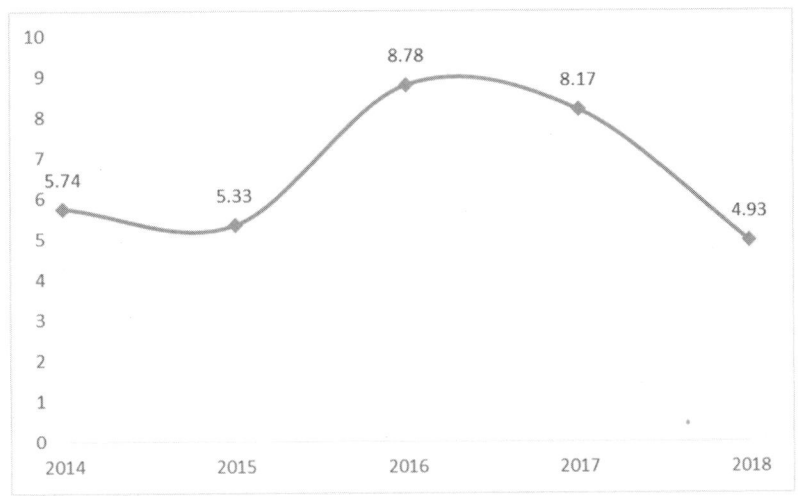

**Fig. 3:** Variation of China coal export since 2014

**Table 4:** China coal consumption since 2014

| Year | 2014 | 2015 | 2016 | 2017 | 2018 |
|---|---|---|---|---|---|
| **Consumption/Mt** | 4120 | 3970 | 3780 | 3860 | 3900 |

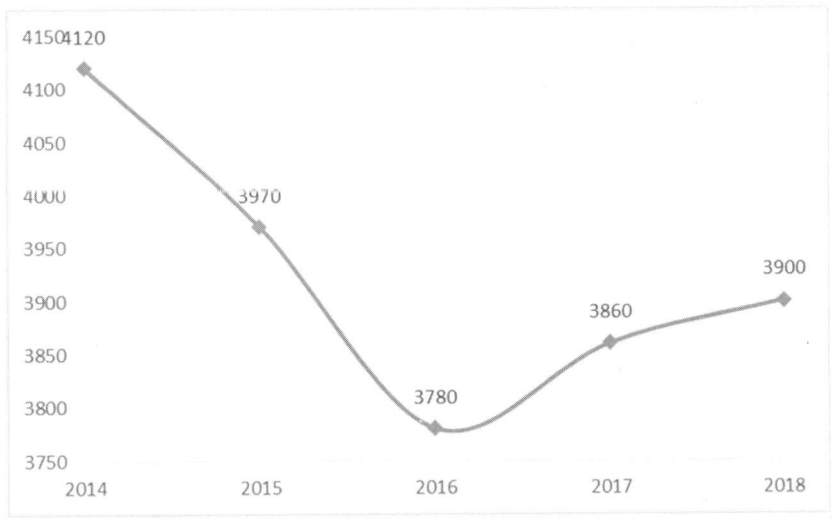

**Fig. 4:** Variation of China coal consumption since 2014

## 2. Development of coal preparation in China

With raw coal preparation and processing, Coal quality and utilization efficiency can be improved, transportation reduced, energy saved and emissions of coal-fired pollutants reduced. Over 2.85 billion tons of coal preparation capacity has been built and put into production. The rate of raw coal washed has increased from 62.5% in 2014 to 71.8% in 2018. The amount of raw coal washed has increased from 2.42 billion tons to 2.642 billion tons.

### 2.1 The amount and rate of raw coal washed has been improved rapidly

**Table 5:** China raw coal production and preparation since 2014

| Year | 2014 | 2015 | 2016 | 2017 | 2018 |
|---|---|---|---|---|---|
| Raw coal output (Bt) | 3.87 | 3.75 | 3.4 | 3.52 | 3.68 |
| Coal washed (Bt) | 2.42 | 2.47 | 2.34 | 2.47 | 2.64 |
| Coal washed rate (%) | 62.53 | 65.9 | 68.9 | 70.17 | 71.79 |

### 2.2 The number, scale and concentration of coal preparation plants have been greatly increased

There are more than 2300 coal preparation plants in China. In 2018, the coal preparation capacity exceeded 2.85 billion tons, and the average capacity of a single plant reached 1.23 million tons per year. In recent years, the capacity of a new coking coal preparation plant is more than 1.2 million tons per year, and the capacity of a new steam coal preparation plant is over 3 million tons per year. There are 80 super-large coal preparation plants with an annual capacity exceeding 12 million tons, accounting for 42% of total capacity in China. Among them, 12 coking coal preparation plants have 160 million tons capacity, and 68 steam coal preparation plants have about 1040 million tons capacity. At present, the largest coking coal preparation plant has an annual capacity of 17.5 million tons, and the largest steam coal preparation plant has an annual capacity of 35 million tons.

### 2.3 Coal preparation method and process

Various coal preparation methods and processes have been adopted in China. Since the beginning of the 21st century, dense medium coal preparation has become the main coal preparation method in China, especially for the newly built coal preparation plants in recent five years. The characteristics of dense medium coal preparation are high separation efficiency, a wide range of separation size, strong handling capacity and easy to realize automatic control. Large dense medium cyclone is mainly used in coking coal preparation plant, and dense medium shallow slot separator is mainly used in steam coal preparation plant. Flotation technology has been set up in all coking coal preparation plants and most coal preparation plants with injection coal. Flotation links have also been set up experimentally in several steam coal preparation plants.

**Table 6:** Proportion of different coal preparation method

| Jigging | Dense Medium | Floatation | Other |
|---------|--------------|------------|-------|
| 10% | 75% | 10% | 5% |

With the moving of China's coal producing areas from east to west, the dry coal preparation technology has been developed rapidly, nearly 200 million tons of dry coal preparation capacity has been built and put into operation.

## 3. An outlook of China's coal preparation industry

### 3.1 Planning target of China's coal preparation and processing

The annual capacity of raw coal preparation will reach over 3.2 billion tons in 2020, and the preparation rate reaches 75%, among which the steam coal preparation rate exceeds 70%. Average CV for commercial coal exceeds 4500 kcal/kg.

### 3.2 Sub-objectives for China's coal preparation and processing

(1) Realizing clean production, controlling the influence of gangue and slime to the environment, increasing waste resource recovery rate, fundamental realizing the reduction and recycling of solid waste including gangue and slime in the whole production process.

(2) Carrying out intelligent upgrade and reconstruction, energy-saving and consumption reduction in construction and operation of coal preparation plants, and reducing the consumption of energy and other auxiliary materials.

(3) Promoting the construction and technical transformation of steam coal preparation plant, increasing preparation rate, and gradually improving the quality of commercial steam coal.

According to the plan, China will increase 350 million tons of coal preparation capacity per year in 2019 and 2020, newly constructed 180 million tons capacity per year.

## 3.3 R&D Achievements and future development

### (1) Achievements

### Ultra large-scale dry separation complete equipment

To meet the demand of steam coal preparation with over 10 million tons, Shenzhou Manufacturing Group develops 13-80mm ultra large compound dry separation equipment, Tianjin Meiteng Technology Company develops 80-200mm TDS (Telligent Dry Separator )for lump coal, and Beijing Great Dragon Rising Electromechanical Tech Company develop 80-200mm γ-ray lump coal separation equipment.

### Large-scale dry deep screening equipment

Aury (Tianjin) Industrial Technology Co., Ltd and Qinhuangdao EuroCMA Industrial Technology Co., Ltd has developed large-scale screening equipment with efficiency, maximum screen width of 4.9 meters, +6mm to +3mm lower limit of sorting, the longer work life of sieve plate, maximum capacity of 1500Tph.

### Large efficient floatation concentrate dewatering equipment

Jingjin Environmental Protection Co., Ltd developed large intelligent dewatering filter press with the moisture of clean coal under 16%, lower material consumption and working strength, continuous good performance and easy maintenance,non-manual operation. The equipment can be a substitute for pressure filter press.

### (2) Development trend

### Intelligent construction of coal preparation plant

Intelligent coal preparation plant is supported by Internet of Things technology, constantly improving data system, forming independent large data system application, achieving intelligent management by means of artificial intelligence and intelligent equipment,realizing intelligent decision-making of production tasks, intelligent setting of process parameters and intelligent adjustment of production system, which can greatly reduce the labor intensity of employees, stabilize product quality and reduce operation costs,enhance production efficiency, and ultimately achieve intelligent preparation and improve the competitiveness of coal preparation enterprises in an all-round way.

### Promoting large-scale and intelligent development of key coal preparation technology and equipment

We will strive to promote the development of key technology and equipment of large-scale coal preparation plant, research and develop supporting equipment for coal preparation plant with capacity of 10 million tons ,including ultra large dense medium separator, flotation machine,screen, centrifugal dehydrator,filter, etc., research on matching equipment of single system for coal preparation plant with capacity of 10 million tons, achieve the fault self-diagnosis of large key equipment linked in intelligent operation and management system.

### Full size fraction dry preparation technology

We will strengthen basic theory research on dry preparation technology, promote scientific and technological innovation, increase the separation accuracy of dry separation equipment, continuously improve the full size fraction dry preparation technology, and ultimately achieve the lower limit of effective dry separation to + 3mm.

## 4. Design of current coal preparation plants in China

### 4.1 Typical technology process

The current leading coal preparation process for steam coal preparation plant is screening and classification of 13mm(25mm) crushed raw coal, dense medium shallow slot separation of lump coal, and dense medium cyclone separation of pulverized coal based on coal quality.

The current leading coal preparation process for coking coal preparation plant is dense medium cyclone separation for 50mm crushed raw coal(desliming or non-desliming), coarse slime is separated in coal slime dense medium cyclone, TBS or hydrocyclone, fine slime is separated by flotation equipment.

### 4.2 New trend and improvement of design and construction

(1)  Intelligent development of lump coal and gangue separation. Practical application in recent years has proved that lump coal

separation equipment has an excellent effect in replacing manual gangue separation, which greatly reduces labor intensity and play a role in impurity removal. Now it been widely promoted in the design of coal preparation plant, and some mines with increased mining depth directly discharge gangue underground. In some steam coal preparation plants, the gangue is pre-discharged and then the products are re-sorted, which will reduce the pressure of subsequent coal preparation and the loss of coal preparation equipment.

(2) Compound dry preparation equipment is under a large-scale application. With greatly improved processing capacity, efficiency, precision, and reliability, good effect of 13-80mm sized coal separation, the compound dry preparation equipment can meet the demand for steam coal customers.

(3) Companies actively promote the intelligent construction and upgrading, combining big data, cloud computing, IoT and information technology to the design and operation of the preparation plant, and pushing on digital control, automated production, remote operation of preparation equipment.

(4) The dry flip-flop screening process is gradually generalized in new coal preparation plants, +6mm efficient classification means increase of lump coal, decrease of fine coal preparation, lower pressure of coal slurry system and less coal slime output, which can improve economic efficiency of enterprises, reduce cost of waste disposal, and also protect ecological environment.

(5) The process design of coal preparation plant is gradually simplified. With continuous improvement of single equipment capacity and preparation efficiency, the single system and process has replaced the past multi-system process, plant pipeline design is no longer complicated, construction cycle is shortened, cost is greatly reduced, problem of uneven distribution of raw material is resolved, the amount of equipment maintenance is reduced, the intensity of work is reduced, and the health of workers is further guaranteed.

## 5. Coal deep processing and clean transformation are developing rapidly in China

China has made breakthrough in ultra low emission in coal fired power generation, ultra low emission in industrial heating boiler and kilns, clean and efficient transformation of modern coal chemical, green upgrading and ultra low emission of conventional coal coking, clean and efficient utilization of low rank coal with pyrolysis technology at medium - and - low temperature, substitution of clean coal-based fuel for civil bulk coal and special cleaning furnaces. The emission from various coal-consuming users has reached the national emission standard of natural gas, with smoke and dust emission concentration $\leq 5$ mg/Nm$^3$, SO2 $\leq 35$ mg/Nm$^3$, NOx $\leq 50$ mg / Nm$^3$. The demercuration and desulfuration are also being carried out.

## 5.1 Coal-fired power generation

China has large coal-fired power plants with 1.26 billion kilowatt, and over 75% units have achieved ultra low emission by the end of 2018.

## 5.2 Modern coal chemical

China has achieved demonstration and commercialization of coal liquefaction, coal gasification, coal to olefin, coal to aromatics, coal to methanol, coal to ethanol, coal to ethylene glycol, coal to dimethyl ether, coal to methanol to gasoline, coal tar hydrogenation to oil and so on. Annual productivity of coal liquefaction products reach 15 million tons, 20 million tons for chemical raw materials including the coal to olefin, 15 billion cubic meters of coal gasification, 50 million tons of coal synthesis ammonia to urea, and 100 million tons of low rank coal pyrolysis quality-based utilization, which basically realize substitution of oil gas and oil-gas chemicals.

## 5.3 Coal-fired heating and civil utilization

The annual coal consumption for heating and civil utilization is near 1 billion tons due to the vast territory and population in China. In order to reduce air pollution of bulk coal, Chinese government enacted strict regulations for to harsh the control of bulk coal, including the substitution of natural gas and electric energy for small heating boilers under 35t/h, the transformation of ultra-low emission for industrial boilers over 35t/h, using gas and electricity to replace civil bulk coal, supplemented with coal based clean fuel and special stove (including semi-coke, sulfur-fixing and dust-suppressing briquettes) , which have been carried out for three years, with remarkable results, replacing more than 100 million tons of bulk coal.

## 5.4 Steel and coking industry

The capacity for steel and coking in China are 1.1 billion tons and 600 million tons respectively. In 2018, China produced 438 million tons of coking. Chinese government required the metallurgy and coking industry to carry out ultra low emission transformation, which is a significant air pollution control action in China.

## 5.5 Building materials and ceramics industry

Annual coal consumption of building materials and ceramics industry in China is 500 million tons, and the industry is also under ultra low emission transformation and upgrading.

**3**

# Coal policy in an era of renewables

Jitendra Roychoudhury

*King Abdullah Petroleum Studies and Research Center,*
*Riyadh, Saudi Arabia*

**Abstract:** For the states in India, the past few years have seen a massive focus on renewables – primarily solar and wind. This focus on renewables has caused states to come up with renewables policies, and these are being implemented as per the states. However, the focus on renewables brings in major challenges for resource-rich states. For states which are rich in fossil fuel resources, it means developing new policies and balancing the needs of its citizens and its energy economies. Not all states are blessed with abundant solar insolation and wind, and these states will have to focus on enhancing fossil fuel resources to sustain their economies.

This paper intends to analyze the challenges for developing a resource policy in such a time of transition.

**Keywords:** Fossil fuel, Energy mix, Coal, Natural gas, Renewables

Fossil fuel, which has so far been the energy mainstay of all economies globally, is facing its biggest challenge yet as more stringent emission regulations, carbon taxes and ever cheaper renewables are rolled out. Strong growth in renewables, especially wind and solar, over the past few years as well as regulatory and environmental headwinds point toward a change in the energy status quo that is already starting to manifest. These changes reflect the growing concerns of policymakers as they try to find meaningful solutions to the challenges of energy affordability, accessibility, and sustainability while trying to meet climate change obligations.

Countries like the United States, China, and India that historically were majorly dependent on fossil fuels are making gradual moves to transition to an energy mix with lower carbon levels. Policy measures are being developed globally to tackle the ubiquitous presence of carbon following combustion by taxing it, and ensuring that emissions are reduced through the introduction of stringent control mechanisms. However, these policy measures have yet to drastically reduce the world's dependence on fossil fuels or significantly curb carbon emissions. The United States Energy Information Administration's outlook to 2040 (Sieminski, 2016) forecasts that fossil fuels will continue to be a major source of world energy, contributing more than three-fourths. Other forecasts including the World Energy Council (World Energy Council, 2013) also mirror such views. Despite the world's dependence on fossil fuels in the foreseeable future, it is still critical to understand the impact that renewables will have on coal policy and demand scenarios as policymakers push for clean energy to have a greater renewables share in the energy mix.

Fossil fuels output had grown at a brisk pace over the past few years spurred on by brisk demand in coal, oil and natural gas. Investments in the sector aided with technological advancements led to the development of new supplies, shale oil being the oft-repeated example. Major fossil fuel users such as the United States, China, India, and Japan have increased domestic production, but as a group, they also expect to continue using fossil fuels for the foreseeable future.

Renewables are proving to be a paradox in that they are disrupting existing energy markets with the promise of cheap and sustainable energy, which sit within a system where the costs of building spare capacity to cater to their intermittency are predominantly based on fossil fuels. Advancements in storage technology could provide solutions to intermittency. However, the overall costs would probably be higher. The challenge of renewables storage technology is still in its infancy, and as such, it is still too early for policymakers to be able to provide a concrete direction where fossil fuels are no longer a part of the energy mix. The experience of countries including Germany and Australia illustrate the difficulties inherent in such energy transitions.

These issues raise pertinent questions for Indian states. The recent push towards increasing renewables penetration into the Indian energy mix makes renewables a critical part of India's energy security along with domestic coal. However, while the growth of renewables has been strong in like Gujarat, Maharashtra, and Karnataka, they haven't matched the potential in other states like Jharkhand and Odisha. For coal-rich states like Jharkhand, Chhattisgarh, and Odisha, stranded coal will be a major loss of an economic mineral. They need to come up with solutions along with the central government as a key stakeholder to ensure sustainability of the coal industry.

## 1. Role of coal

Coal has been a key part of the Indian energy sector and its growth over the previous decades has been a key driver of increasing energy consumption. Being a critical part of the energy mix, the role of coal trickles down into several aspects of the economy –both at the central and state level. While at the central level, coal provides energy security and affordability, at the state level, the coal economy generates employment opportunities and revenue for the state governments. Coal's presence is felt across logistics chains, commercial factories and employment for a range of skilled and unskilled labor.

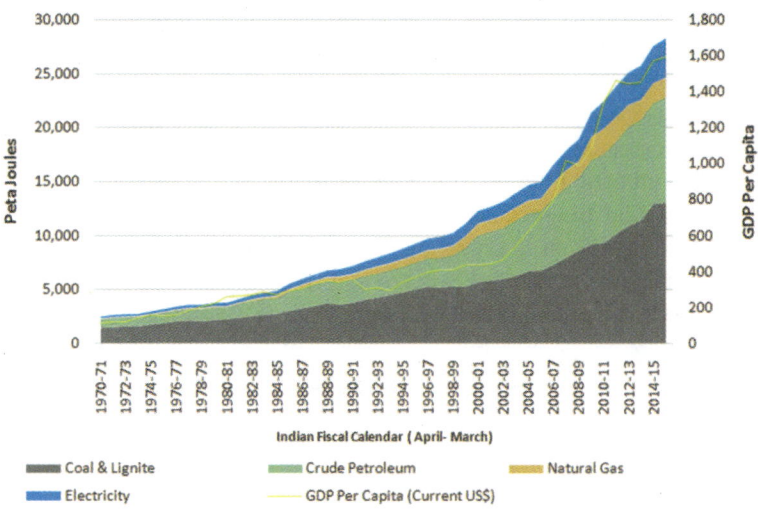

*Source: MOSPI, KAPSARC analysis*

**Figure 1:** Consumption of primary sources of conventional energy in India

However, coal faces new challenges which seek to change its position as the key energy resource for India. Greater penetration of renewables, supported by technology changes making solar and renewables, extremely competitive commercially and a massive push for a cleaner and more sustainable environment has meant that the policymakers are increasingly under pressure to curtail and limit the role of coal in the Indian energy economy.

The inflection point which will mark the peak of coal consumption in India is still some years away in the future. Demand in steel, cement and power sector for coal continues to be strong and growing. However, the changes which have occurred in the past portend that the inflection point will be upon the industry much sooner than anticipated. The oil shock of the 1970s helped to invigorate the coal industry in India and the past few decades since then have seen an increasing entrenchment of the coal sector. Increasing economic growth in India will require a cheap and accessible energy source. Renewables still hasn't made a play for that role as yet in India, but increasingly the demand for cleaner air and sustainability could as easily help to ensure that the role of coal is diluted over some time. This period of the energy transition is a critical phase for Indian policymakers and also states in general. Any change in the energy mix will have wide-ranging ramifications which could impact the livelihood of millions and impact Indian economic growth – current and future. In India, coal is present in several other sectors like cement, brick kilns, steel and fertilizer to name a few.

Transitioning away from all these sectors to a completely carbon free is going to be extremely challenging and economically expensive. The transition from a coal-centric electricity sector to a renewable one will be a very difficult transition, especially if policymakers rush to implement low carbon pathways without (a) adequately identifying stakeholders and (b) without preparing and enabling the transition. Observing how Germany, a vastly developed country with a highly industrialized economy and with heavily funded social services infrastructure in place is delaying the decarbonization of its electricity sector, holds important lessons for policymakers. It is not going to be an easy process, and it is better to be deliberate and sure than to rush headlong into space which is fraught with policy missteps. This is difficult in today's world where environmental activists would like the policymakers to implement climate-friendly solutions immediately. The pressure from the activists to transition at a faster pace is only going to increase in the future.

## 2. Policy influences on energy infrastructure

The headwinds that coal faces relate most to emissions and their impact on the earth's atmosphere. Unabated carbon, be it in the form of any fossil fuel like coal, oil or gas has a commercial appeal that is going to be increasingly under pressure from international climate agreements on carbon emissions. To ensure that the carbon footprint is made more climate-friendly, policymakers have to increasingly focus on newer and better technologies, which increase efficiencies and reduce emissions.

This is much easier said than done as the legacy depreciated infrastructure costs and the cheap economics of coal are what makes it a viable fuel. Infusing newer and better technology will increase energy costs to consumers exacerbating the shift to cleaner energy options. Already the shutting down of old state-owned coal-fired power plants has meant that states are now forced to procure higher priced electricity from newer, more efficient power plants. Their Renewable Purchase Obligations (RPO) have also increased over the past few years, increasing energy costs for their industries. The economic pressures from the transition are already being felt. This when the penetration of renewables like solar and wind are still very low as the figure below illustrates. However, it can be easily observed that it is coal-fired electricity generation that is absorbing the seasonality and intermittency of renewables.

Policy makers should increasingly focus on the transport links as these are the ones which will impact other sectors of the economy. Upgrading electricity transmission infrastructure to be able to handle the intermittency of renewables will be of benefit to the coal sector also as it will enable the coal-fired power plants to play a much more flexible role to ensure balancing. Incorporating larger load balancing areas as might be the case with increased renewable penetration will help efficient coal-fired power plants in these balancing areas to be used more with the benefits of higher PLF and efficient boiler fleet impacting the environment positively. The larger balancing areas could also potentially benefit pithead coal-fired power plants, which could potentially help to reduce pollution in the cities and decongest railway networks. Some states in the east like West Bengal have opted to go slow on their renewable energy plans as they believe that with their portfolio of coal-fired power plants and low solar insolation, it is much more financially feasible for them to buy renewable RPOs and continue to operate their coal-fired fleet at a higher PLF. States have to cater to their own regional imperative and ensure that their financial investments are not impacted negatively. States can also explore policies to shave the developing evening peak as much as possible. This can ensure that the idling of the thermal power assets is optimized. Newer investments could also be reviewed in light of the projected availability of renewable-based power. Policymakers will have to be extremely cautious while proceeding with future investments to reduce stranded assets – both in the renewable (due to curtailment) and coal sectors.

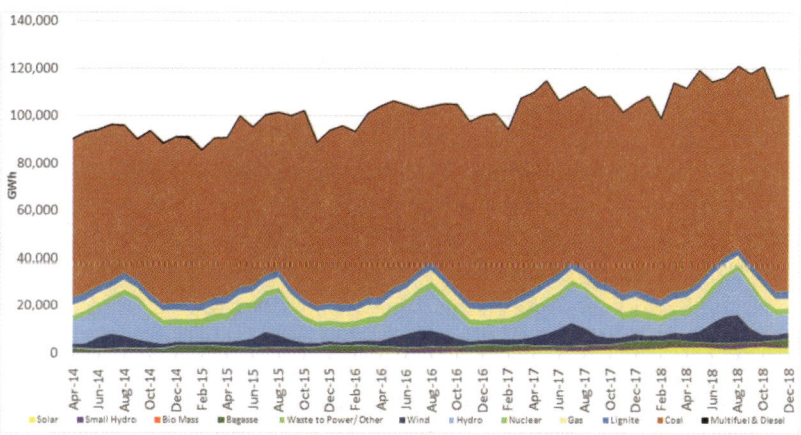

*Source: CEA, KAPSARC analysis*

**Figure 2:** Monthly generation of electricity

## 3. Policy influences on future coal investments

To ensure that future investments do not turn into stranded assets, policymakers will have to be extremely careful to ensure the longevity of the project regarding both financial feasibility and carbon emissions. Policymakers will have to explore projects, which ensure that future coal-based projects ensure carbon abatement through carbon capture and sequestration (CCS). Developments in CCS technology could help to ensure future coal consumption without impacting the environment negatively. For this, states have to ensure that they help incubate technology developers and provide a mechanism for innovators to work with industry and ensure that carbon emissions are kept within norms or reduced. For this to crystallize,investments in academic research institutes and encouraging industry to work with academia through policy support would greatly help to build an environment conducive to help develop such technology.

## 4. Conclusion

The transition that is underway in the global energy mix is an ongoing one, and one which will accelerate as storage prices decline and renewable technologies become more efficient. The push for energy consumption from a growing population in India will impose severe stresses on existing energy frameworks. To ensure that the transition is as smooth as possible, it will become inherent for policymakers to ensure the following:

- Understand in-depth the penetration of coal in energy systems across value chains in diverse energy sectors. Low carbon pathways have been drawn and are helpful, but as the recent issue with petroleum coke consumption has illustrated, there is a severe lack of on the ground real-time data availability regarding consumption patterns and changes occurring in those consumption patterns. These can only be rectified through enhanced data collection and reporting. Ensuring transparent availability of these consumption data would enable energy researchers to ensure that policymakers are provided with workable models and the assumptions are reduced as much as possible. All stakeholders should work together to ensure that the availability of data is as transparent and accessible as possible. This will involve close coordination with producers, consumers, and transporters. Putting together a workable framework will ensure that policymakers are aware of the impact of the transition occurring on the ground and can tweak policy to ensure that it reflects the pace of actual scenario and not an assumption based one.

- Communication is a critical part of change management, and each state should work with all stakeholders to ensure that policy impacts are not limited to pockets of the population but are broad-based enough to be acceptable to all. For this continuous communication strategies should be adopted to ensure that the population is aware of the changes required and the pathway that the policymakers are following to avoid a chaotic transition. We have seen from the Brexit movement that transitions are not easy and even with adequate planning, issues could crop up that could derail previously planned pathways. For this sharing of transition efforts across states would help to increase acceptability and also ensure greater debate.

- The greatest impact from the transition to a low carbon environment would be on communities living close to coal mines. For them, their access to coal would be of continued economic interest, and as previous efforts at Jharia has shown, it is not an easy task to move communities in India. States will need to plan far ahead to ensure that the economic impact of lowered coal consumption is as low as possible. This will involve working with social workers and building adequate infrastructure – physical and social to enable communities to move away. The US election of 2016 illustrated that coal retains a huge emotional connect and it will be difficult to manage the transition without impacting communities negatively. The key will be to manage expectations to ensure that the transition in energy systems mirrors the one on social systems.

The advantage that policymakers in India and other countries still consuming coal is to be able to observe the impact of the transition in countries withdrawing from coal and ensure that the policy learnings are absorbed, updated and deployed using local contexts. Like Britain, the United States and China work on transitioning away from coal, a treasure trove of policy learnings will be available for researchers, and this will enable future policymakers in other transitioning countries to ensure that their energy transition is more orderly and nuanced.

## 5. References

[1] Sieminski, Adam. 2016. "International Energy Outlook 2016." *Energy Information Administration.* May 11. http://www.eia.gov/pressroom/presentations/sieminski_05112016.pdf.

[2] World Energy Council. 2013. "World Energy Scenarios." *World Energy Council.* September. https://www.worldenergy.org/wp-content/uploads/2013/09/World-Energy-Scenarios_Composing-energy-futures-to-2050_Full-report.pdf.

# 4

# Russian coal industry innovations and perspectives

Vladimir Litvinenko[1] and Bernd Meyer[2]

[1]*Saint-Petersburg Mining University, 2, 21st Line, St Petersburg, Russia*
[2]*Institute of Energy Production and Chemical Technology, Freiberg Mining Academy,*
*Akademiestraße 6, Freiberg, Germany*

**Abstract:** A significant percentage of the coal deposits in Russia contains high-ash coals, characterized by a lower specific heating value. The selection of the most appropriate conversion pathway for coal utilization depends on the local site condition and the prospective markets for the coal-derived products. The present study focuses on the evaluation of economically feasible application options for high-ash coals from Russian coal district, which are located far away from large industrial centres. Two Russian coals are selected as examples, which represent the given broad range of coal characteristics: low-ash coal from the Taldinskaya mine and high-ash coal from the Komsomets mine (Kuzbass coal region). The key technology considered for coal utilization is syngas production and conversion. Based on the detailed analysis, two gasification technologies and two syngas conversion routes are chosen as the most feasible ones that fits the general framework of Russian coal industry. Among the syngas utilization products, methanol and synthetic gasoline are selected. Despite the low gasification temperature, virtually complete carbon conversion can be achieved using COORVED gasification (typically 1.2 wt% C content in the bottom ash). Calculations show that coal-based methanol production could be economically feasible under the boundary conditions of the Russian coal industry.

**Keywords:** Russian coal industry, methanol, syngas, gasification, COORVED

## 1. Introduction

Coal is the fossil fuel with the most uniform distribution worldwide. But its quality differs significantly depending on the region. The selection of the most appropriate coal conversion pathway depends on the local site conditions and the prospective markets for the products considered. Russia helds second position worldwide in terms of coal reserves [1,2]. The present study focuses on the evaluation of economically feasible application for high-ash coals from Russian coal district. For such low-quality coals, the non-energetic utilization could be economically more feasible than selling the raw coal on the world market. The most promising use for such coals is the production of chemicals with high added value. The key technology in this respect is syngas production and utilization. Today, 65% of all the syngas produced is converted into chemicals (methanol and its non-fuel-derivatives, olefins, oxo-alcohols, ammonia, urea, hydrogen, etc.), 18% into liquid fuels (diesel, kerosene, naphtha, etc.), 10% into gaseous fuels (synthetic natural gas), and 7% into power (IGCC power plant, polygeneration) [2,3]. The gasification of Russian coal is the most suitable way to maximize the added value by the refining of these natural resources. To select the coal gasifier, the most important task is to choose a gasifier that fits the respective coal. Also one of the main challenges today is minimizing the plant capital expenditure (CAPEX). In contrast of oil-based process chains, coal utilization requires expensive facilities for coal preparation, syngas production and cleaning as well as wastewater processing.

## 2. Experimental material

Two coals were selected as examples which represent the given broad range of coal characteristics: low-ash coal from the Taldinskaya mine and high-ash coal from the Komsomolets mine. The most important coal parameters are given in Table 1.

**Table 1:** Coal properties

| Parameter, unit | Taldinskaya | Komsomolets |
|---|---|---|
| Immediate analysis | | |
| Fixed carbon, wt% (wf) | 55.43 | 33.81 |
| Volatile matter, wt% (wf) | 30.97 | 30.66 |
| Ash content, wt% (wf) | 13.61 | 35.53 |
| Water content, wt%[a] | 11.80 | 9.65 |
| Elemental analysis | | |
| C, wt% (wf) | 70.60 | 53.47 |
| H, wt% (wf) | 3.97 | 3.82 |
| O, wt% (wf) | 9.31 | 4.96 |
| N, wt% (wf) | 2.15 | 1.59 |
| S, wt% (wf) | 0.22 | 0.55 |
| Cl + P + As wt% (wf) | 0.15 | 0.10 |
| Heating value | | |
| Lower heating value, MJ/kga | 23.85 | 19.71 |
| Higher heating value, MJ/kga | 24.90 | 20.69 |
| Ash melting behavior | | |
| Softening temperature, °C | 1440 | 1310 |
| Hemisphere temperature, °C | 1470 | 1360 |
| Flow temperature, °C | 1500 | 1410 |

*Contd...*

| Parameter, unit | Taldinskaya | Komsomolets |
|---|---|---|
| Ash composition | | |
| $SiO_2$, wt% | 60.90 | 42.66 |
| $Al_2O_3$, wt% | 23.40 | 16.75 |
| $Fe_2O_3$, wt% | 5.00 | 10.13 |
| CaO, wt% | 2.90 | 10.58 |
| MgO, wt% | 2.00 | 3.86 |
| $TiO_2$, wt% | 0.90 | 0.89 |
| $MnO_2$, wt% | 0.03 | 0.03 |
| $P_2O_5$, wt% | 0.40 | 0.20 |
| $SO_3$, wt% | 1.30 | 9.06 |
| $Na_2O$, wt% | 0.8 | 2.18 |
| $K_2O$, wt% | 2.40 | 2.63 |

[a]Values on the "as received coal" base.

## 3. Gasification

The most promising gasification technologies are identified as the Siemens entrained-flow gasification and the new COORVED staged gasification, comprising a fixed bed at the bottom and jetting fluidized bed above. Among the syngas utilisation products, methanol and synthetic gasoline were selected. An overview of the overall process chains and modelled cases considered is given in Fig. 1.

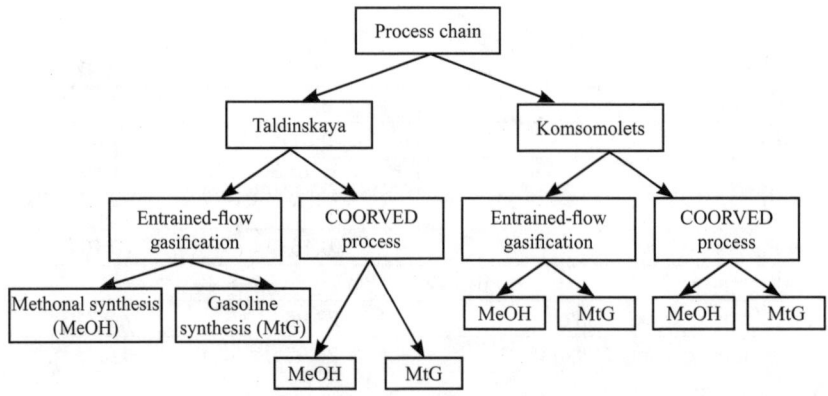

**Figure 1:** Overview of the processes cases

Both gasification technologies considered the use of raw coal, crushed to the required size range. The COORVED gasifier utilizes a broad coal fraction in the range of 0–6 mm. Entrained-flow gasification requires coal fines <250 μm. The Siemens entrained-flow gasifier uses dense coal fed with $CO_2$ provided as an off-gas from the acid gas removal unit. The COORVED gasifier works with a gravitational solids feeding system.

The entrained-flow gasifier developed by Siemens is shown in Fig. 2. The gasifier is equipped with an integrated cooling screen inside the gasifier, which is used to generate saturated intermediate-pressure steam [4]. The syngas produced is directed with a high gas velocity from the gasifier into the water quench chamber placed directly below the gasifier. Preheated water is sprayed into the hot gas stream to suddenly cool the gas down to the water steam dew point. Liquid slag drops into the water bath and solidifies. The granulated slag is discharged at the bottom of the quench chamber. Water-saturated raw gas leaves the quench chamber at a temperature of 200–250°C.

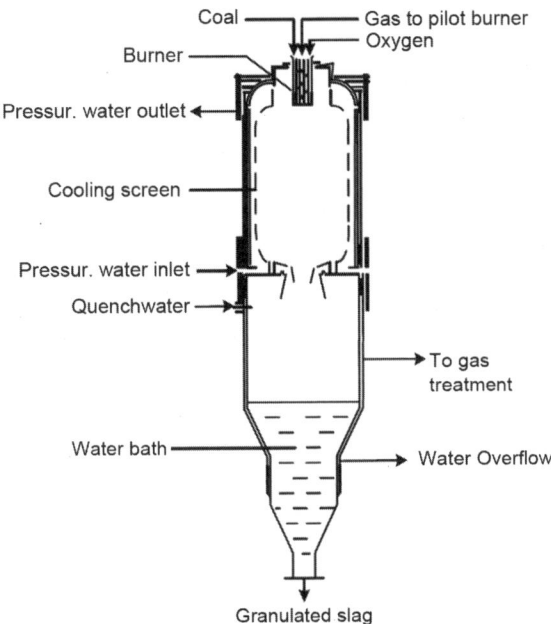

**Figure 2:** Siemens gasifier

The COORVED gasifier is described on Fig. 3. It is characterized by increasing gas velocities from bottom to top. Zone 1 is a fixed bed at the bottom of the reaction chamber, where char agglomerates are contacted with a secondary gasification agent (mixture of steam and oxygen) [5]. In the second zone, which is placed above the fixed bed, particles are separated according to their weight. Larger particles move downwards and fall onto the bottom fixed bed. Smaller particles are directed upwards into the main reaction zones – a jetting fluidized bed and fast fluidized bed. The fresh feedstock enters the reactor in the area of the jetting fluidized bed and comes into contact with the primary gasification agent, which leads to exothermic reactions. A flame-like hot zone of more than 2000°C is formed. If particles reach a certain carbon conversion, their ash surface will begin to partially soften or melt due to the high temperatures. In this way, particle surfaces become sticky, which allows agglomerates to form. Along with large feedstock particles, they fall through the bubbling bed and form the fixed bed at the bottom. In the main reaction zone, particles are dried and devolatized. Because the temperatures are comparably high, devolatilization gases such as CH4 and higher hydrocarbons are destroyed. Thus, the produced gas quality is comparable to an entrained-flow gasifier.

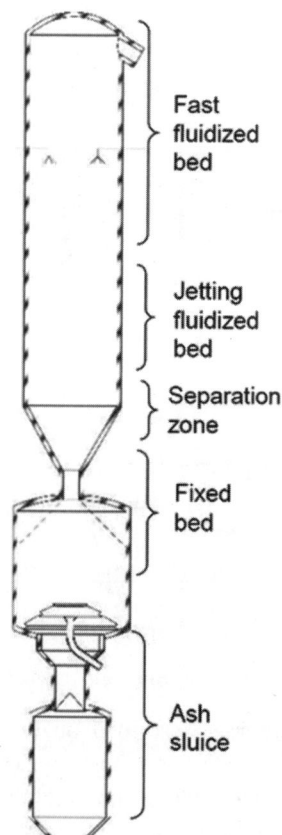

Fast fluidized bed

Jetting fluidized bed

Separation zone

Fixed bed

Ash sluice

**Figure 3:** COORVED gasifier

The comparison of the gasification processes itself is the starting point for the comparison of the overall process chain. The basic characteristics of both investigated coals, the process parameters selected for modelling and the calculated key performance indicators are listed in Table 2.

**Table 2:** Input data and calculation results for the gasification process

| Parameter, unit | Taldinskaya | | Komsomolets | |
|---|---|---|---|---|
| Gasification technology | Siemens | COORVED | Siemens | COORVED |
| Softening temperature, °C | 1440 | 1440 | 1310 | 1310 |
| Flow temperature, °C | 1500 | 1500 | 1410 | 1410 |
| Temperature at reactor outlet, °C | 1650 | 1200 | 1550 | 1200 |
| Pressure, bar | 40 | 40 | 40 | 40 |
| $O_2$ demand, m3/kg coal | 0.711 | 0.583 | 0.774 | 0.606 |
| Steam demand, kg/kg coal | <0.01 | 0.361 | <0.01 | 0.379 |
| Cold gas efficiency, % | 77.7 | 84.9 | 75.8 | 84.2 |
| Carbon conversion, % | 99.5 | 99.8 | 99.5 | 99.2 |
| Syngas yield, $m^3$ $H_2$+CO/kg coal | 2.02 | 2.21 | 2.13 | 2.37 |
| $H_2$/CO, mole/mole | 0.315 | 0.649 | 0.362 | 0.747 |
| Raw gas composition ex gasifier (vol.%) | | | | |
| $H_2O$ | 9.4 | 10.0 | 11.4 | 10.4 |
| $H_2$ | 21.8 | 35.9 | 23.8 | 39.3 |
| CO | 69.3 | 55.4 | 65.8 | 52.6 |
| $CO_2$ | 7.5 | 7.2 | 8.9 | 6.6 |
| $CH_4$ | 0.0 | 0.2 | 0.0 | 0.3 |

Simulation verifies the numerous advantages of the COORVED process in comparison with the conventional Siemens technology. It is characterized by a comparably low energy demand for the required sensible heat. The oxygen demand for COORVED gasification is lower than for Siemens gasification by 18% for Taldinskaya coal and 22% for Komsomolets coal. Besides the better syngas yield, the COORVED process is favored compared to the Siemens entrained-flow technology because of the H2/CO ratio in the syngas produced, which is twice as high. The cold gas efficiency for the ash-rich coal from the Komsomolets mine is lower than for the low-ash coal from the Taldinskaya mine.

Using the modelling data for the overall process chains, an economic assessment of the process chains was performed. The estimation of CAPEX costs is based on the capacity rating method [6,7]. Product costs were calculated as an average over the whole payback period presumed as 20 years. Averaged product costs are listed in Table 3 (methanol production) and Table 4 (gasoline production).

**Table 3:** Averaged product cost in Euro for the process chains based on methanol synthesis

| Parameter | Taldinskaya | | Komsomolets | |
|---|---|---|---|---|
| | Siemens | COORVED | Siemens | COORVED |
| CAPEX | 34.30 | 27.01 | 38.47 | 29.44 |
| Fuel costs | 7.70 | 6.88 | 7.64 | 6.69 |
| OPEX including service | 19.99 | 16.47 | 22.27 | 18.26 |
| Annual costs | 62.00 | 50.36 | 68.38 | 54.39 |
| Revenue for the by-products | -5.77 | −3.97 | -6.07 | −3.99 |
| Product costs | | | | |
| In Euro/MWh of product | 56.24 | 46.39 | 62.31 | 50.40 |
| In Euro/t of product | 310.86 | 256.45 | 344.44 | 278.62 |
| In Euro/L of product | 0.25 | 0.20 | 0.27 | 0.22 |

**Table 4.4:** Averaged product cost in Euro for the process chains based on gasoline production

| Parameter | Taldinskaya | | Komsomolets | |
|---|---|---|---|---|
| | Siemens | COORVED | Siemens | COORVED |
| CAPEX | 40.31 | 31.73 | 44.84 | 34.17 |
| Fuel costs | 8.47 | 7.46 | 8.38 | 7.22 |
| OPEX including service | 22.66 | 17.71 | 25.01 | 17.79 |
| Annual costs | 71.44 | 56.90 | 78.23 | 59.18 |
| Revenue for the by-products | −13.15 | −10.91 | −13.38 | −10.86 |
| Product costs | | | | |
| In Euro/MWh of product | 58.29 | 45.99 | 64.85 | 48.32 |
| In Euro/t of product | 697.17 | 550.01 | 775.63 | 577.93 |
| In Euro/L of product | 0.54 | 0.43 | 0.60 | 0.45 |

# 4. Conclusion

The application potential of different gasification technologies for the non-energetic utilization of Russian coals of varied quality (low-ash coals and ash-rich coal) was evaluated in this study. In both cases the COORVED process has clear advantages compared to the commercial entrained-flow

Siemens process. An assessment of process chains based on the utilization of the innovative COORVED gasification reveals product costs for methanol of around 250 Euro/t for the low-ash coal and approximately 280 Euro/t for the ash-rich coal.

## 5. References

[1] BP Statistical review of world energy 2015. http://www.bp.com/content/dam/bp/pdf/energyeconomics/statistical-review-2015/bp-statisticalreview-of-world-energy-2015-full-report.pdf.

[2] Higman C. State of the gasification industry: worldwide gasification database 2015 update. Colorado Springs, 2015.

[3] Grabner M. Industrial coal gasification technologies covering baseline and high-ash coal. Weinheim, Wiley-VCH, 2014.

[4] Grabner M, Meyer B. Performance and exergy analysis of the current developments in coal gasification technology. Fuel 2014, 116, 910–920.

[5] Laugwitz A, Meyer B. New frontiers and challenges in gasification technologies. In: Nikrityuk PA, Meyer B, editors. Gasification processes: modelling and simulation. Hoboken, Wiley, 2014.

[6] Remel A, Wasserscheid P, Baldauf M, Hammer T. Techno-economic analysis for the synthesis of liquid and gaseous fuels based on hydrogen production via electrolysis. Int. J. Hydrogen Energy. 2015, 40, 11457–11464.

[7] TU Bergakademie Freiberg, internal cost calculations. 2016.

# 5

# Coal mining and coal preparation in Vietnam

KWaldemar Mijał[1], Nhu Thi Kim Dung[2], Nguyen Ngoc Phu[2], Nguyen Hoang Son[2]

[1]*AGH University of Science and Technology, Kraków, Poland*
[2]*Hanoi University of Mining and Geology, Hanoi, Vietnam*

**Abstract:** Vietnam is rich in mineral resources, from which coal is one of the most important minerals of Vietnam due to its abundance and its role in the national economy. Quang Ninh Coal Basin is the most important resource of coal as most of the coal exploitation and preparatory works are being run in this area. Red River Delta Coal Basin is the second large coal basin, which is still under development stage. VINACOMIN Group is the biggest coal producer in the country which is 100% dependent on state policy.

The article deals with issues related to the hard coal mining sector and coal processing in Vietnam. The final part of the article will discuss a simple preliminary coal enrichment system at mining plants and coal processing plants with schemes in Vietnam. The summary will be the description of future plans for the coal mining industry and enrichment systems.

**Keywords:** Coal preparation, coal resources, coal basin, coal industry, coal processing, coal preparation flowsheet, Vietnam coal industry

## 1. Introduction

Vietnam is one of the most important anthracite producers in the world. VINACOMIN is the leading company in Vietnam coal market. Coal industry plays an important role in the fast growing economy of Vietnam, particularly in the energy sector. By the mid of the 1990s, total annual coal production was about 6–7 Mt/y, two thirds of produced coal were mined by surface mining while the rest by underground method and most of the coal were not washed, however, the picture is now turned in reverse due to the harder government pressure on environmental protection. Current annual coal production is kept at a level of 45 Mt/y and about 30% of coal is washed at large coal preparation plants utilizing both jigging and dense medium separation.

## 2. Coal resources and coal basins in vietnam

Currently, available data shows that coal reserves in Vietnam area about 49.8 billion tons. Coal resources are classified into a few categories: measured & indicated reserves (categories A+B+$C_1$) are 33%, inferred ($C_2$) is 39%, and prognostic resources (P) are 28%. In Vietnam appeared all types of coal: anthracite (67% and already mined), bituminous coal (insignificant), sub-bituminous coal (26%), lignite coals (2%, rarely mined), and peat (5%).

Most important coal basins are located in Quang Ninh, Red River Delta, Thai Nguyen, Backan, North Path, Da River, Ca River, Na Duong, Nong Song, Ba River, and Mekong River Delta (Fig. 1). Vietnam has one of the biggest resources of the world anthracite. Quang Ninh coal basin plays the key role as most of the coal mines and coal preparation plants are operated in this area. Quality parameters of raw coal in Vietnam from a few coal mines are described in Table 1.

**Table 1:** Parameters of coal from opencast mines in Vietnam (Buiand Drebenstedt, 2004)

| Quality factor | Coc Sau | Cao Son | Deo Nai | Ha Tu | Nui Beo |
|---|---|---|---|---|---|
| Ash (%) | 2.24–40.0 | 8.08–11.27 | 2.49–39.28 | 2.64–28.33 | 2.52–30.6 |
| Inherent moisture (%) | 1.42–4.92 | 0.35–3.5 | 0.99–3.56 | 1.88–13.83 | 1.82–14.2 |
| Volatile matter (%) | 2.82–9.86 | 7.0 | 4.2–24.51 | 4.48–13.83 | 4.52–13.96 |
| Sulfur (natural) (%) | 0.1–1.64 | 0.01–2.72 | 0.07–0.6 | 0.24–0.33 | 0.23–0.35 |
| Heat content (kJ/kg) | 25.79–36.26 | 28.45–35.56 | 31.17–39.31 | 29.31–35.77 | 29.1–34.9 |
| Density (t/m³) | 1.39–1.46 | 1.38–1.46 | 1.39–1.44 | 1.38–1.45 | 1.37–1.5 |

Quang Ninh basin, located in the northeast part of the country, occupies the area of about 5900 km² of which 2800 km² is forestland and 510 km² is agricultural. Coalfields in this area are located very close to the coast to form good conditions for mining and trading. Quang Ninh coalfield got 8.7 billion tons of coal resources (anthracite). Most important coal deposits in Quang Ninh basin: Mao Khe, Trang Bach, Nam Mao, Vang Danh, Uong Thuong, Dong Vong, Nga Hai, Khe Tam, Giap Khau, Nui Bao, etc. The main coal areas of Quang Ninh Basin are Uong Bi, Hon Gai, and Cam Pha. The coal has been mined since 1839 and the mining is being expanded continuously.

Red River Delta coal basin was discovered in 1960 during the search for oil and gas. More than 39.4 billion tons of sub-bituminous coals lie beneath Red River Delta over the area of 2000 km². The coal seams are located at 300–2500 m. VINACOMIN attempted to develop here projects for few coal mines like Binh Minh, Khoai Chau I, Khoai Chau II, however, the projects are delayed due to the dense population and the main rice field of Vietnam (Chuan, 2011a, b; Binh, 2015; Strzałkowska and Strzałkowski, 2011; Hoa, 2010; Paul, 2010; Bui and Drebenstedt, 2004) (Table 2).

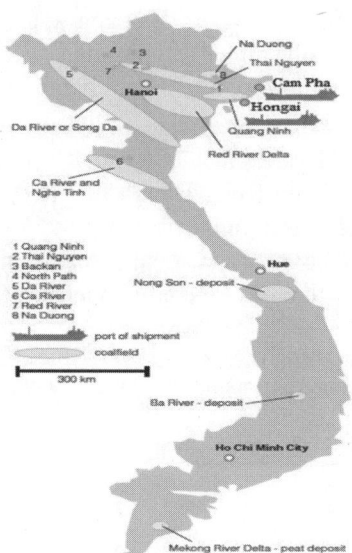

**Figure 1:** Map of Vietnam coal resources (Paul, 2010)

**Table 2:** Total coal resources in Vietnam (Mt) (Buiand Le 2011)

| Mine areas | Total resources (A+B+C$_1$+C$_2$+P) | Measured resources (A+B) | Indicated resources (C$_1$) | Inferred resources (C$_2$) | Prognostic resources (P) |
|---|---|---|---|---|---|
| Total | 49.777 | 285 | 2.220 | 2.928 | 44.344 |
| 1. North-East basin | 9.904 | 236 | 1.468 | 2.234 | 5.966 |
| + VINACOMIN | 3.826 | 234 | 1.325 | 1.663 | 604 |
| + Government | 6.078 | 2 | 143 | 571 | 5.362 |
| 2. Other investor basins | 191 | 49 | 99 | 24 | 19 |
| 3. Peat basins | 332 | | 129 | 107 | 96 |
| 4. Red River Delta Basin | 39.352 | | 525 | 564 | 38.263 |
| + Khoai Chau area (80 km²) | 1.581 | | 525 | 564 | 492 |
| + Phu Cu-Tien Hai area 2000 km² | 37.771 | | | | 37.771 |

# 3. Vietnam coal industry policy

The Ministry of Industry and Trade (MOIT) is responsible for the state management of all energy industries as it is not only to determine first-line policy but also has supervisory responsibilities for state owned companies VINACOMIN and Electricite de Vietnam (EVN). The Ministry is also responsible for master plans for electricity, coal, oil and natural gas exploitation and supply.

The coal industry has a strategic position in the Vietnam economy. Government still accepts new plans for new coal-fired power projects. It will need more coal supply from coal mines. These decisions will change the policy of Vietnam and maybe even stop the export of anthracite coal. Government now focuses on developing and increasing coal industry practice and productivity. VINACOMIN has been merged from Vinacoal and Vietnam Mineral Corp in 2005. VINACOMIN Holding Corporation Ltd. with 54 coal mines, it is the biggest coal mining company in Vietnam. It operates five big open pit mines, 15 smaller pits, and 30 underground coalmines. This is an economic corporation with 100% owned by the state. Ninety-five percent of coal production in Vietnam is from VINACOMIN Company. It also runs the most important washing plants such as Hon Gai (2 Mt/year), Cua Ong (10 Mt/year), and Vang Danh (3 Mt/year).

According to the master plan of coal industry development in Vietnam by 2020, with perspective to 2030, the total coal output will reach 60 million tons (in 2020), 65–70 million tons (in 2025), and 65–75 million tons (in 2030).

The master plan is also presented for the coal mining industry in Table 3 (Viet, 1994; Chuan, 2011a, b; Binh, 2015; Hoa, 2010; Paul, 2010; Giang, 2014; Dung, 2014).

**Table 3:** Master plan for development of Vietnam coal industry (Mt) (Bui and Drebenstedt, 2004)

| Coal area | 2020 | 2025 | 2030 |
|---|---|---|---|
| Total run-off-mine coal | 92,430 | 119,250 | 120,732 |
| I. North-East basin | 72,330 | 85,050 | 83,282 |
| 1.1. VINACOMIN | 64,530 | 67,150 | 59,782 |
| In which: banned area & coal bearer | 6200 | 7400 | 7300 |
| 1.1.1. Uong Bi coal field | 19,280 | 20,550 | 20,950 |
| 1.1.2. Hon Gai coal field | 9,350 | 9,800 | 8,800 |
| 1.1.3. Cam Pha coal field | 35,900 | 36,800 | 30,032 |
| 1.2. New coal mines | 7,800 | 17,900 | 23,500 |
| II. Other interior basins | 3,050 | 2,650 | 2,700 |
| III. Out of VINACOMIN | 3,550 | 6,550 | 9,750 |
| IV. Red River Delta basin | 13,500 | 25,000 | 25,000 |

Vietmindo is the only joint venture with a foreign country (Indonesia). It is operating in Vang Danh-Uong Bi area of Quang Ninh province. The concession is for 30 years and can be renewed for another period if the company still needs more time to achieve longer-term production goals (Chuan, 2011a, b; Binh, 2015; Hoa, 2010; Paul, 2010; Giang, 2014; Dung, 2014).

## 4. Coal mining and coal preparation status

Surface mining currently is dominated by truck and shovel methods. Some of the major equipment includes Russian EKG hydraulic shovels, dump trucks of 30–80 ton payload from BalAZ, Komatsu, and Caterpillar 769C.

Typical surface mine is showed in Fig. 2 (Coc Sau mine). Large surface mines with production more than 2 million tons/year are Cao Son, Coc Sau, Deo Nai, Ha Tu, and Nui Beo, from which Ha Tu and Nui Beo are converted into underground mines. There are a number of average and small-scale surface coal mines with a total year capacity between 0.5 and 2 million tons/year.

Rocks in surface coal mines are mainly conglomerates of sandstone or mudstone leading to the use of drilling and blasting. Surface mining major problem is environmental protection (such as how to locate waste dumps, ways of dumping, the stability of waste dumps, wastewater draining, and the processing of water before it flows into rivers and seas). Surface coal mines are now facing huge pressures from the local governments and communities due to the destruction of the landscape, deforestation, and pollutions. There is a trend to close surface coal mines and convert existing surface mines into the underground.

**Figure 2:** General view of opencast coal mine Coc Sau
(Strzałkowska and Strzałkowski, 2011)

Currently, the depth of underground mining in Vietnam is on 300 m. Under VINACOMIN control, there are 30 underground coal mines, nine of which have capacity of more than 1 million tonnes/year (Mao Khe 1.6 Mt, Nam Mau 1.5 Mt, Vang Danh 3.1 Mt, Ha Lam 1.8 Mt, Duong Huy 2.0 Mt, Thong Nhat 1.6 Mt, Mong Duong 1.5 Mt, etc.). Rest of coal mines have an annual capacity of less than 1 million tonnes.

Coal seams of Quang Ninh Coal Basin have complex geological structures with many faults and folds and most of the coal seams need to be mined by selective mining technologies to ensure the required factors of loss and dilution. Complicated geological structures of coal seams decrease the efficiency of longwall mining, decrease the mining face production, while the high cost of coal makes underground production less competitive and there are many safety issues related to complex geology, mine pressure, gas, and ventilation.

Coal preparation in Vietnam is carried out in two alternatives. The first is to implement the run-of-mine (ROM) coal pre-treatment system by handpicking, screening, grinding or blending. All coal mining companies have their pre-treatment systems. The second alternative is typical coal washing by jigs and dense medium separation (DMS) at central coal preparation plants. About 30% of ROM coal can be cleaned by coal preparation plants. Management system in Vietnam coal preparation industry is showe in Fig. 3.

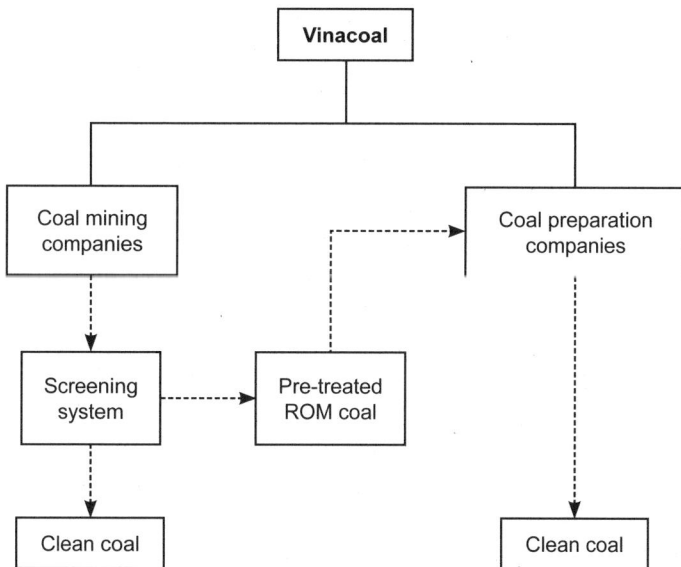

**Figure 3:** Coal preparation management in Vietnam
(Viet Bach and Gheewala, 2008)

ROM coal is prepared at coal mines by simple technologies as described by Fig. 4.

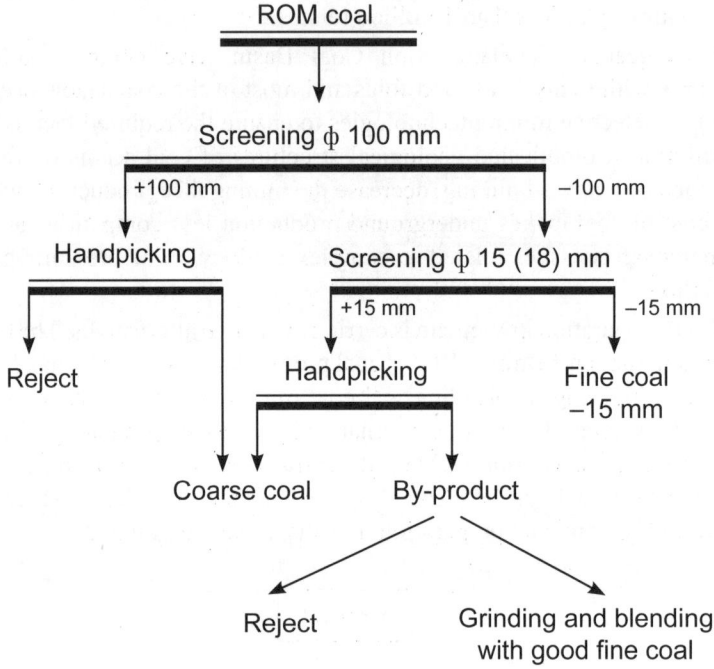

**Figure 4:** Simple treatment system for ROM coal at coal mines
(Viet Bach and Gheewala, 2008)

Coal preparation plants use different cleaning methods including jigging, DMS, spiral separation, cyclones, and flotation. Typical coal preparation flowsheet and technology can be described from the examples of Cua Ong Coal Preparation Company and Vang Danh Coal Mining Co.

The Cua Ong plant No. 2 flowsheet is shown in Fig. 5. First coal preparation plant Cua Ong CPP II was designed and built by Polish engineers in the 1970s. Production was started in 1980. First using capacity was 3.2 Mt/y and used equipment was dense medium separators DISA. In the 1990s the Cua Ong Coal Preparation Plant No. 2 was remodified by Australian Technology and the used DISA was replaced by a line of jigs and DMS cyclones.

Vang Danh Coal Preparation Plant No. 1 was first designed by Russian in 1972 with designed capacity of 0.6 Mt/y. Then the plant modified to adopt the capacity of 2.7 Mt/y. Current flowsheet is shown in Fig. 6.

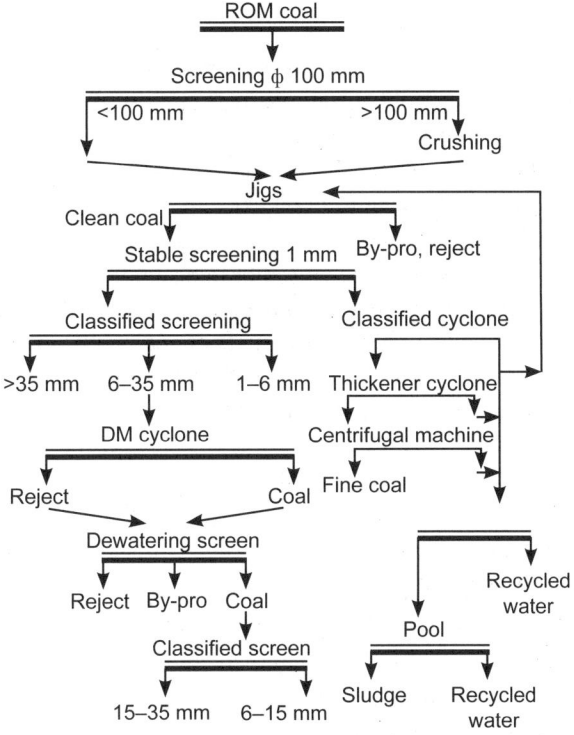

**Figure 5:** Coal preparation plant Cua Ong No. 2 (Viet Bach and Gheewala, 2008)

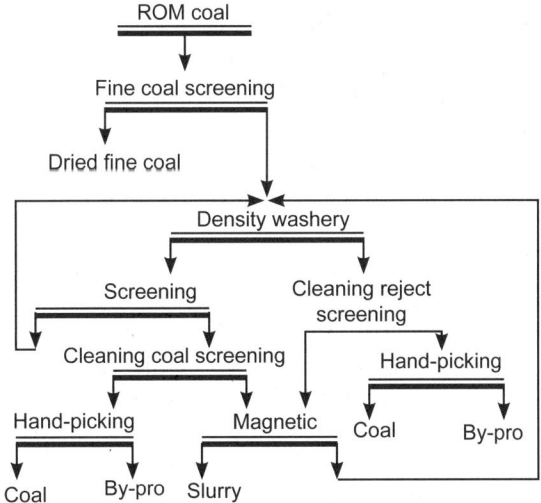

**Figure 6:** Flowsheet of Vang Danh coal processing plant
(Viet Bach and Gheewala, 2008)

## 5. Conclusion

Coal industry of Vietnam plays an important role in the national economy as the country is rich in coal resources, particularly in anthracite. Vietnam is one of the key anthracite producers of the world. The industry produces approximately 45–50 Mt of ROM coal/annum, and one-third of this volume is cleaned in central coal preparation plants.

Environmental impacts are the major challenges for the coal industry. Many surface coal mines are to be closed or to be converted into underground mines due to the government's increasing pressure for environmental protection. Complicated geological structures of coal seams prevent the use of long wall mining and decrease the mining face production; high cost of coal makes the underground production uncompetitive; and many safety issues related to complex geology, mine pressure, gas and ventilation. Coals fed to the preparation plants are more unpredictable as more fines and slimes in the feed.

To overcome the difficulties and challenges, the coal mining industry of Vietnam should concentrate on the followings:

- To invest in modern mining and preparation technologies;
- To exclude ineffective old equipment and outdated technologies;
- To reuse more recycle water at preparation plants to reduce fresh water consumption (Zero-freshwater policy);
- To increase the percentage of coal cleaned by preparation plants to increase the coal value and its efficiency;
- VINACOMIN and coal miners should actively cooperate with foreign countries such as Russia, Poland, Australia, Indonesia, India, etc. for better coal market and technologies;
- Establish more effective environmental management systems.

## 6. References

[1] Baruya Paul – Prospects for coal and clean coal technologies in Vietnam, Copyright IEA Clean Coal Centre 2010, ISBN 978-92-9029-484-9.

[2] Dung Le Viet – Problems of beneficiation in coal preparation plants in Vietnam. New Trends in Coal Preparation Technologies and Equipment: Proceedings of the 12th International Coal Preparation Congress, Cracow, Poland 1994, p. 553–555. ISBN 2-88449-139-2.

[3] Le Ba Viet Bach, Gheewala S. H. –Cleaner production in the coal preparation industry of Vietnam: necessity and opportunities. Asian J. Energy Environ., 9, 2008, Issue 1 and 2, p. 65–100.

[4]  Le Ba VietBach, Gheewala S.H. – Cleaner production options at a coal preparation facility in Vietnam.J.Sustain. Energy Environ., 1, 2010, p. 41–46.

[5]  LeMinh Chuan–Current status of coal demand and supply in Vietnam and plan of VINACOMIN in the coming time, 2011, available at www.jcoal.org.jp.

[6]  LeMinh Chuan–Perspective development of Vietnam coal industry. Clean Coal Day, Japan 2011, available at www.jcoal.org.jp.

[7]  LeVan Dung–Ứng dụng một số tiến bộ KHCN trong chế biến và sử dụng than, 2014, available on www.vampro.vn.

[8]  Nguyen Binh–Vietnam coal potential and development orientation.APEC Coal Supply Security Tokyo, Japan 2015, available at www.jcoal.org.jp.

[9]  Pham Huu Giang–Tình hình nghiên cứu tuyển than mỡ ở Việt Nam, 2014, available on www.vampro.vn.

[10]  Strzałkowska E., Strzałkowski P. –Coal mining in Vietnam – chosen information, Górnic.Geol. 2011, 2, p. 203–209.

[11]  Tran Xuan Hoa–Coal export and the future in Vietnam. Clean Coal Day, Tokyo, Japan 2010, available at www.jcoal.org.jp.

[12]  Xuan Nam Bui, Drebenstedt Carsten – The situation of the surface coal mines in Vietnam and their future development, World of Mining – Surface and Underground 2004, Vol. 56, p. 210–216, ISSN 1613-2408.

[13]  Xuan Nam Bui, Qui Thao Le – Coal mining industry in Vietnam, Proceedings of International Symposium on Earth Science and Technology 2011, 6–7 December, Fukuoka, Japan 2011, p. 155–158, ISBN 978-4-9902356-1-1.

# 6

# When to blend raw coal with clean coal from full-wash thermal coal preparation plants – case studies from Indian and US coals

Swadhin Saurabh[1], Rainer Stephenson[1], Barbara Arnold[2]

*[1]Millcreek Engineering Company, Salt Lake City, UT, USA*

*[2]PrepTech Inc., Apollo, PA, USA*

**Abstract:** Full-wash thermal coal preparation plant (CPP) circuits in which typical +1 mm raw coal is processed in heavy media vessel (HMV) and/or cyclone (HMC), and 1 mm × 0.15 mm raw coal is processed in an intermediate water-only gravity separation circuit, such as spirals, water-only-cyclone/spirals combination, teeter-bed separators, or Reflux® classifiers, are commonplace in coal preparation plants in the United States. Similar cleaning circuits are also becoming popular in thermal CPPs in India in which the equal incremental ash method is applied to determine the cut-points ($\rho_{50}$) that can produce the maximum amount of product while meeting the overall product specifications. However, due to washability characteristics, equipment performance limitations, and/or product specification requirements, there are situations in which these full-wash thermal CPPs alone cannot meet the overall clean coal specifications and produce slightly lower ash, higher moisture, and higher thermal values. The economics of the CPP are optimized if the overall product specifications are as close to the required specifications as possible.

This paper presents several case studies from United States and Indian coals in which the full-wash thermal CPPs did not appear to produce the products that met the required specification. Simulation case study results have shown that if the CPP products are blended with the raw coal, the overall product specifications can be met, and an increase in the overall yield (including blending) of the facility can be achieved. The effect of blending raw coal with the coal preparation plant product on various product specification parameters, such as ash, moisture, and gross calorific value, will also be discussed.

**Keywords:** Coal preparation plant optimization, Reflux® classifier, water-only-cyclone, full-wash plant, spiral, coal blending

## 1. Introduction

Thermal coal preparation plant (CPP) plays a big role in the coal mining cycle. The CPP yield is the second most sensitive factor (after transportation) affecting the economics of coal production (Norton, 1979). These CPPs process the raw coal to meet the desired clean coal or product specifications of ash, sulfur, and gross calorific value (GCV) for thermal power plants. To meet these specifications, the coal preparation plants normally go through size-based separations typically using screens and/or hydrocyclones to form specific raw coal size fractions suitable for different processing circuits that are mainly gravity-based. These gravity-based separation technologies perform efficient separation within certain size fractions (Saurabh et al., 2017). For the last 20 years, the optimization of a CPP consisting of multi gravity-based separation technologies normally has followed the *equal incremental ash* method (Bethell et al., 2006; Watters et al., 2010; Saurabh et al., 2015, 2017) in North America. This methodology of CPP flowsheet selection, optimization, and design has also been accepted more recently in the Indian coal industry (Wang et al., 2017).

North American thermal power plants typically require around 12.0–13.0% ash on an as-received (AR) basis, >6100 kcal/kg GCV (AR-basis), <2.0% sulfur (AR-basis), and <11.0% total moisture (Saurabh et al., 2017). The CPPs meet these specifications using multi gravity-based separation circuits. Indian thermal power plants normally require product specification of around <34.0% ash and 8.0–12.0% total moisture. The full-wash CPPs are now being pursued even in the Indian coal industry (Wang et al., 2017). To meet Indian clean coal specifications, full-wash plants consisting of multi gravity-based separation technologies are recommended for coal having a head-ash of >40.0% (Saurabh et al., 2015).

Several case studies demonstrate that in full-wash plants sometimes the size distribution and washability characteristics of coal can reduce the product ash to several percentage points lower than the desired product ash specification. The authors have also seen similar characteristics in North American coal. Such CPP circuits lead to lower than required ash specifications and higher than required GCV specifications. According to the Coal Preparation Directory and Handbook 2014 (a CPSA publication):

- The supply of high quality coal may be inadequate to meet customer demand.
- Coal blending can be used to extend those supplies by mixing-in lower quality coal and creating a blended product that meets, but does not exceed, customer specification.
- Marginal quality coal may be hard to sell, but it can be blended with higher quality coal to create a product that is more easily marketable.
- Coal blending can be used to substantially improve the profitability of an operation by increasing the revenue or reducing the cost from each transaction.

This paper provides several simulation case studies of Indian as well as North American coals that could produce lower ash product than the desired specifications using the *equal incremental ash* method because of the gravity-based separation equipment performance limitations, for example heavy media (HM) cyclone can separate at maximum 1.90 ρ50 while spirals can separate at maximum 2.20 ρ50. The CPP products were further blended with raw coal to meet the desired ash specifications. The effect of blending raw coal with clean coal on total moisture and GCV is also discussed.

## 2. Coal preparation plant optimization

Two distinct raw coal plant feed data sets, one from India and the other from North America, were used in this case study. Table 1 shows the raw coal characteristics of these coals. The same coal data was used by Saurabh et al. (2015) to present a case study for Indian thermal coal. Table 2 shows the desired clean coal specifications. Table 3 shows the CPP circuits used for simulations. In this case study, the plant throughput was assumed to be 500 t/h air dry (AD) basis. All plant simulations were performed using LIMN® flowsheet software and the *equal incremental ash* method.

**Table 1:** Raw coal characteristics

| Sample location | | Size distribution | | | Head ash (%) | Moisture | | |
|---|---|---|---|---|---|---|---|---|
| | | 50 mm × 1 mm | 1 mm × 0.15 mm | 0.15 mm × 0 mm | Dry basis | Inherent (%) | Surface (%) | Total (%) |
| India | Wt. (%) | 79.00 | 11.00 | 10.00 | 43.70 | 4.00% | 5.00% | 8.80% |
| | Ash (%) | 43.50 | 43.50 | 45.50 | – | – | – | – |
| North America | Wt. (%) | 81.63 | 11.04 | 7.33 | 21.98 | 7.00 | 2.15 | 9.00 |
| | Ash (%) | 20.60 | 21.30 | 38.70 | – | – | – | – |

**Table 2:** Clean coal specifications

| Sample location | Product specification | | |
|---|---|---|---|
| | Ash | kcal/kg | Total moist. |
| India | 33.0% (AD-basis) | – | <12.00% |
| North America | 13.00% (AR-basis) | >6115 | <11.00% |

**Table 3:** CPP flowsheet circuits used for simulations

| North American coal – CPP option 1 | | | |
|---|---|---|---|
| **Size fraction** | **Processing equipment** | **Dewatering equipment** | **Product?** |
| 50 mm × 12.7 mm | Heavy media (HM) cyclone | Centrifuge dryer | Y |
| 12.7 mm × 0 mm | None | None | Y |
| North American coal – CPP option 2 | | | |
| Size fraction | Processing equipment | Dewatering equipment | Product? |
| 50 mm × 12.7 mm | HM vessel | Centrifuge dryer | Y |
| 12.7 mm × 1.0 mm | HM cyclone | Centrifuge dryer | Y |
| 1.0 mm × 0.15 mm | Spiral | Screen bowl centrifuge | Y |
| 0.15 mm × 0 mm | None | Belt press | N |
| North American coal – CPP option 3 | | | |
| **Size fraction** | **Processing equipment** | **Dewatering equipment** | **Product?** |
| 50 mm × 12.7 mm | HM vessel | Centrifuge dryer | Y |
| 12.7 mm × 1.0 mm | HM cyclone | Centrifuge dryer | Y |
| 1.0 mm × 0.15 mm | None | Screen bowl centrifuge | Y |
| 0.15 mm × 0 mm | None | Belt press | N |
| North American coal – CPP option 4 | | | |
| **Size fraction** | **Processing equipment** | **Dewatering equipment** | **Product?** |
| 50 mm × 1.0 mm | HM cyclone | Centrifuge dryer | Y |
| 1.0 mm × 0.15 mm | Spiral | Screen bowl centrifuge | Y |
| 0.15 mm × 0 mm | None | Belt press | N |
| North American coal – CPP option 5 | | | |
| Size fraction | Processing equipment | Dewatering equipment | Product? |
| 50 mm × 1.0 mm | HM cyclone | Centrifuge dryer | Y |
| 1.0 mm × 0.15 mm | None | None | Y |
| 0.15 mm × 0 mm | None | Belt press | N |

*Contd...*

*Contd...*

| Indian coal – CPP option 6 | | | |
|---|---|---|---|
| **Size fraction** | **Processing equipment** | **Dewatering equipment** | **Product?** |
| 50 mm × 1.0 mm | HM cyclone | Centrifuge dryer | Y |
| 1.0 mm × 0.15 mm | Spiral | Screen bowl centrifuge | Y |
| 0.15 mm × 0 mm | None | Belt press | N |
| **Indian coal – CPP option 7** | | | |
| **Size fraction** | **Processing equipment** | **Dewatering equipment** | **Product?** |
| 50 mm × 10 mm | HM cyclone | Centrifuge dryer | Y |
| 10 mm × 0 mm | None | None | Y |

The North American coal was used in the simulation of five different flowsheet options (Options 1–5), while the Indian coal was used in two flowsheet simulations (Options 6 and 7).

Options 1 and 5 are HM cyclone partial-wash plants. The bottom size of HM cyclone feed were 12.7 mm for Option 1 and 1 mm for Option 5. While all the fines bypass, that is 12.7 mm × 0 mm raw coal, reports to the clean coalin Option 1, because of product moisture limitations, only 1 mm × 0.15 mm raw coal gets dewatered and reports to the clean coal stream in Option 5. Option 3 is also a partial wash plant consisting of HM vessel and HM cyclone that is fed with 1 mm bottom size raw coal. Option 7 is also a typical HM cyclone partial-wash plant operated in India. All of these partial-wash plants have major limitations of optimization because the bypass stream dictates the $\rho50$ of the coarse circuit(s). In these kinds of circuits, the *equal incremental ash* could not be applied.Options 2, 4, and 6 are full-wash plant circuits in which the *equal incremental ash* was utilized.

The CPP simulation results of both the coals produced lower than the desired ash specification product; therefore, the CPP simulation result was further extended to see the behaviour of the overall specification after blending of raw coal with the product.

# 3. Results and discussions

## 3.1 North American coal

Figure 1 shows the overall CPP simulation results for the North American coal. The CPP options (Options 1–5) showed that the target ash of 13.0%

(AR-basis) could not be achieved through any of the CPP options. Options 2–5 produced product of ash value <13.0%, while Option 1 produced product of ash value >13.0%. The overall product from Options 2–5 were then blended with raw coal to increase the ash, decrease the total moisture, and decrease the higher than desired GCV values. Figure 2 shows the effect of blending raw coal to clean coal based on ash value.

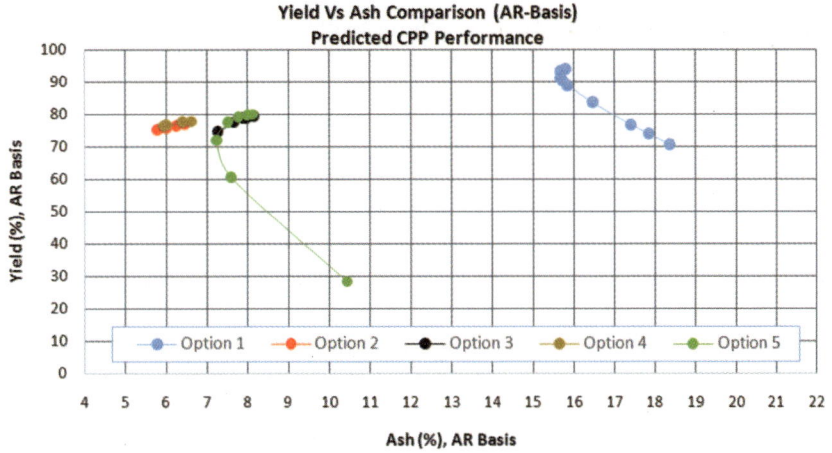

**Figure 1:** Predicted CPP performance for North American coal

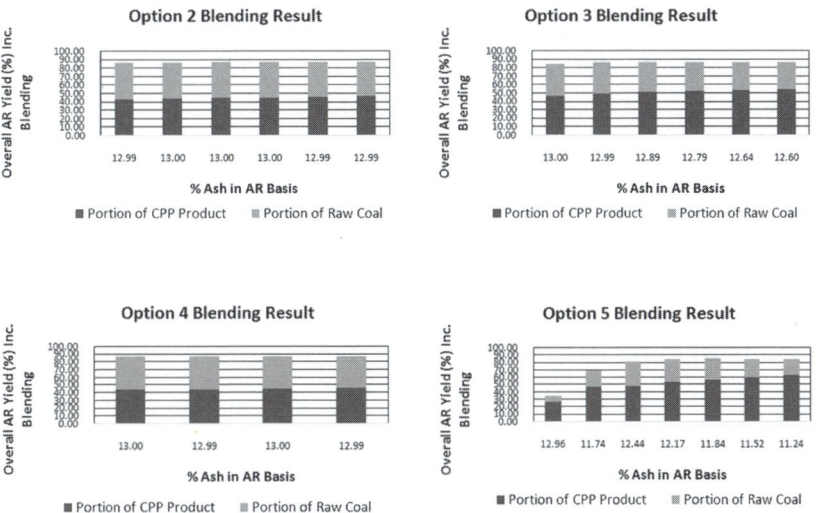

**Figure 2:** Overall yield of North American coal after blending raw coal with clean coal

The various bars in each of the graphs in Fig. 2 represent different CPP product ash values that are also shown in Fig. 1. For each option, the overall product of various ash values were blended with raw coal until any of the specifications (ash, GCV, or moisture) dropped lower than the desired specifications, for example, higher than specified ash, lower than specified GCV, or higher than specified total moisture. This determined the amount of raw coal to be used in the blend. Please note that the Option 1 CPP product was not used in the blending simulation since this option could not produce CPP product that was less than 13.0% ash (AR-basis). The overall yield of the facility (including blending) was calculated using the formula:

$$Y = \frac{(C + B)}{(R + B)} \times 100 \qquad (1)$$

where, $C$ is the product from the CPP in t/h; $R$ is the raw coal in t/h feeding the CPP; $B$ is the raw coal in t/h blended with the clean coal; and $Y$ is the overall yield of the facility in percentage.

After blending clean coal and raw coal, we found that Option 2 provided an overall maximum yield (including blending) of 86.59% @ 13.0% ash (AR-basis), >6115 kcal/kg (AR-basis), and 10.08% total moisture, while Option 4 provided an overall maximum yield (including blending) of 87.04% @ 13.0% ash (AR-basis), >6115 kcal/kg (AR-basis), and 10.41% total moisture.

Option 3 provided an overall maximum yield (including blending) of 86.18% @ 12.89% ash (AR-basis), >6115 kcal/kg (AR-basis), and 10.24% total moisture. Most of the data points of Option 3 could not produce 13.0% ash product even after blending because the overall GCV value was the limiting factor.

Option 5 provided an overall maximum yield (including blending) of 84.85% @ 11.84% ash (AR-basis), >6115 kcal/kg (AR-basis), and 10.87% total moisture. Most of the data points of Option 5 could not produce even 12.0% ash product even after blending because the overall GCV value was the limiting factor.

## 3.2 Indian coal

Figure 3 shows the overall CPP simulation results for the Indian coal. The CPP showed that the target ash of 33.0% (AD-basis) could not be achieved using Option 6. However, Option7 could provide a product of ash value of 33.0% (AD-basis). Please note that the Indian coal size distribution and washability data did not provide any GCV value, so overall product GCV analysis was not done in this study. The overall product (<33.0% ash) from Options 6 and 7 were then blended with raw coal to increase the ash and meet the specification. Figure 4 shows the effect of blending raw coal with clean coal based on the ash value.

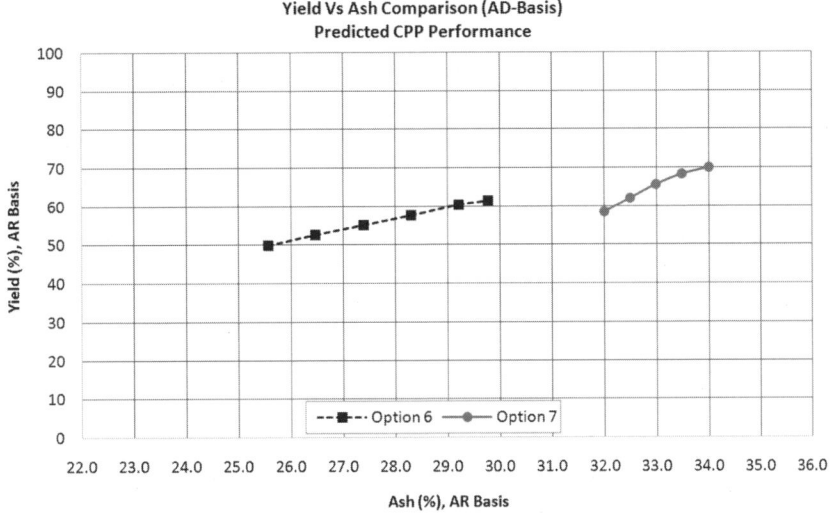

**Figure 3:** Predicted CPP performance for North American coal

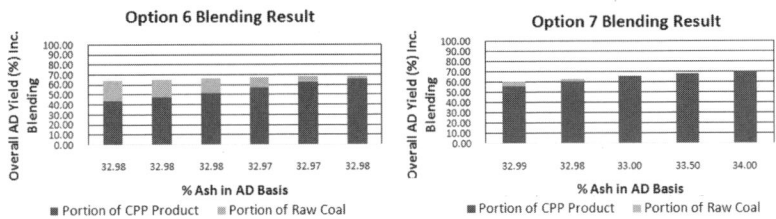

**Figure 4:** Overall yield Indian coal after blending raw coal with clean coal

After blending clean coal and raw coal, we found that Option 6 provided an overall maximum yield (including blending) of 68.07% @ 33.0% ash (AD-basis), while Option 7 could provide an overall maximum yield (including blending) of 65.6% @ 33.0% ash (AD-basis). At the same ash specification, the yield difference of 2.47 percentage point is considered quite significant in the area of coal preparation.

Figures 2 and 4 clearly demonstrated that a full-wash plant consisting HM cyclone–spiral CPP product combined with raw coal blending can produce the maximum overall yield.

Major benefits from the blending raw coal with clean coal to meet the product specification in low CPP product ash are listed below:

- It increases the overall product ash, bringing it closer to the desired specification.
- It decreases the overall product moisture, bringing it closer to the desired specification.
- It decreases the overall GCV value, bringing it closer to the desired specification.
- It helps in increasing the overall clean coal yield and profitability of the facility.

Although the Indian raw coal data did not have the GCV value, the effects of blending raw coal to the clean coal mentioned above is also expected with the Indian coal.

## 4. Conclusion

The case studies both on Indian and North American coal show that if the coal preparation plant alone cannot reach the overall product ash specifications and produces several percentage points lower than the required ash product, raw coal blending with the clean coal can be seriously considered for the overall process optimization. This will not only assist in meeting the overall desired specification, but also assist in increasing the overall yield of the facility, thus it will help in improving the economics of the coal production cycle. It was also found that although a partial-wash plant may produce the desired product specification, a full-wash combined with blending can provide a higher overall product yield.

## 5. Acknowledgement

The authors would like to express gratitude to David Carter (Vice President, Millcreek Engineering Company, USA) for his valuable suggestions and support in writing and reviewing this paper.

## 6. References

[1] Bethell, P.J., Luttrell, G.H., and Firth, B., 2006. *Comparison of Preparation Plant Design Practices for Metallurgical and Stream Coal Applications*, XV International Coal Preparation Conference and Exhibition, Beijing, China, October 17–20, 98–110.

[2] Norton, G., 1979. *The economic role of coal preparation in the production and utilization of coal for the market*, Internal publication, Norton-Hambleton Consultants, Inc., Ann Arbor, MI, 24 pp.

[3] Saurabh, S., Wang, Z., and Kumar, V., 2015. *New optimized flowsheet for Indian thermal coal preparation plants*, Coal Preparation Society of India (CPSI) Journal, 7(17), 89–98.

[4] Saurabh S., Arnold B., and Carter D., 2017. *Future Coal Processing Strategies for India*, Conference Proceedings of Coal Washing, Coal Preparation Society of India (CPSI) Journal, 9(26), 28–33.

[5] Wang, Z., Nielson, J., Arnold, B., Saurabh, S., and Chatterjee, A.B., 2017. *Full Wash Coal Washery Design and Implementation for Indian Thermal Coal Production*, Conference Proceedings of Coal Washing, Coal Preparation Society of India (CPSI) Journal, 9(26), 38–43.

[6] Watters, L.A., Dynys, A., Keles, S., Ali, Z., and Luttrell, G., 2010. *Identification of Optimum Strategies for Process of Fine Coal Streams*, XVI International Coal Preparation Congress 2010 Conference Proceedings, Lexington, USA, April 25–29, 30–36.

# 7

# Recommended practices for design and engineering of heavy media based coal preparation plants

Anath Bandhu Chatterjee, Rakesh Parmar, Akash Prabhakar, Amol Bhosale, Haresh Mayatra, Maulin Shah, Vishal Patel, Ahesanali Momin

*Adani Enterprises Limited, Ahmedabad, India*

**Abstract:** Coal preparation plant design and engineering is not a new buzz. However, reiterating certain norms, standards and guidelines from lessons learnt from live projects will be very useful for project proponents, designers and other stakeholders in design, construction and operation of new plants. The technological choices are discussed here for designing most optimum plant with less time and cost. Adopting best engineering practice helps minimise the teething troubles during initial months, execute the project as per schedule and operate the plant with minimum trouble.

Modern plant design guidelines, layout design, process design criteria, equipment selection considering best/worst cases, interchangeability, technological choices, engineered chutes, stand-by norms, sumps/tanks design, liner philosophy, piping norms and practices, retention times, latest civil, structural, electrical and instrumentation guidelines, etc. are some of the focused areas of this paper. Process design philosophy is the backbone of washery project from concept to its whole life cycle. It has to be decided very sensibly and judiciously.

Also, best engineering practices using latest technology of civil-structure, mechanical, utility, E&I, etc. helps building most useful, trouble free and efficient plant all together. Latest construction/erection methodology helps to control project cost and reduce construction time. Automation advancements help to operate plant in different modes with minimum manpower and in a more controlled way. Further use of technologies can help us install sensors on equipment and capture-running data, proper analytics of these data can be useful in predictive maintenance of equipment and management information system for the whole plant.

The author's paper strongly believes that these modern engineering practices can ensure smooth and systematic execution as well as trouble free efficient operation of a new coal preparation plant.

**Keywords:** Flowsheet, Heavy media, Optimisation, Gravity Control, Design criteria, Yield

## 1. Introduction

Coal preparation technique in India has witnessed continuous and steady changes over past few years. Designing and construction of coal preparation plant in India is not an easy job owing to the inadequate coal characteristics data and lack of common benchmark for designing and construction of all types of washeries. Due to the heterogeneous nature of the coal, designing of coal washeries involve large number of variables, sometimes making the coal circuits very complex. Though there has been a significant development in the coal processing technology, yet the information on the coal processing units is very scanty.

## 2. Proper coal test results

The likely particle size distribution and wash ability test results of the feed to the plant along with the hard grove grind ability index (HGI) and fines degradation from drop shatter and wet tumbling are the most important information required for designing a full wash coal preparation plant. The general practice in India is furnishing particle size distribution and corresponding wash ability analysis based on drop shatter data and dry screening, which is different from practices in United States and Australia where wet pre-treatment data is also furnished. The use of the provided sizing and wash ability data without appropriate size adjustments to simulate the additional breakage that will occur in the coal handling and washing process underestimates the amount of coal fines that will be processed in the plant, and will result in insufficient fines processing capacity being allocated in the washing process. There will also be impacts on the selected process to meet the stringent process guarantee requirements with respect to overall product yield, moisture and equipment performance parameters. An example of dry screening after drop shatter versus wet screening after wet tumbling of coal seam of North Galilee Coal Basin, Australia is plotted in Fig. 1.

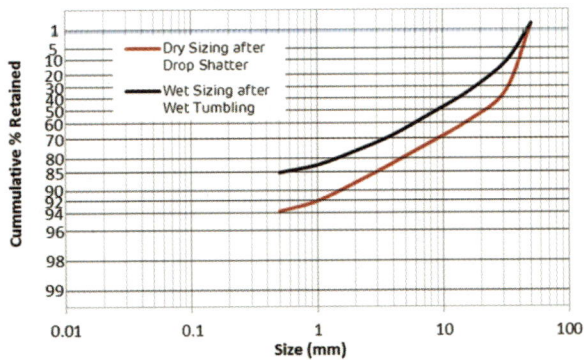

**Figure 1:** Rosin Rammler plot: dry sizing versus wet sizing of coal
(*Source*: Wang et al., 2017)

Thus, it become very important to take into account all the necessary data based on proper coal testing before going on to simulated flowsheet design and mass balance and equipment sizing.

## 3. Standardisation of design criteria and minimum technical requirements

Some essential technical requirements or the design criteria for the process, mechanical, electrical and automation and civil and structural have been given below. The intent of these criteria is to encapsulate the standard design criteria and practices that may be used during the design and construction of a new coal handling and processing plant. Specific design data will be developed during detailed design to support equipment procurement and construction specifications. Thus, it is not the purpose of this paper to furnish the detailed design information for each component and system, but rather to summarise the standards and general criteria that may be used.

## 4. Process design criteria

### 4.1 Technology selection

Proper care should be taken before selection of the washing technology for a coal processing plant. Various selection criteria such as feed size, capacity, achievable cut density, equipment performance criteria, footprint, capital and operating expenditure based on current standards of proven and reliable equipment design must be evaluated before deciding technology. Suitable account must be taken for near gravity materials and ease of washing for all coal types.

| Processes | Probable error in separation $(E_p)$ | Organic efficiency (%) | NGM | Remarks |
|---|---|---|---|---|
| DM drums/baths | 0.01–0.025 | 93–95–97 (99) | High | Range of values essentially dependent on the expertise of the manufacturers vis-à-vis the feed coal characteristics |
| DM cyclones | 0.02–0.07 | 90–96 (98) | High | |
| Modern jigs | ≥0.045 | 88–91 (92) | Low | |
| Baum jigs | ≥0.09 | 80–85–87 | Low | |
| Barrel washers | 0.10–0.25 | ≥80 | Low | |

The dense media washers are the most efficient separators. Unfortunately, most of the plant in India use inferior washers with low efficiency. Due to their high separation efficiency and ability to handle feed coal size up to 50–0.5 mm, DMCs are the choice of washing equipment for washing difficult washing coal. They are also used in the case where the price of coal dictates the highest possible yield to be obtained.

## 4.2 Dense media cyclone (DMC) plant design

### 4.2.1 Dense media cyclone dimension and configuration

Nowadays, large diameter DMCs are commonly used to process higher capacity and large sized feed particles. Most of the available DMCs have the feed, vortex finder and spigot diameter as per the DSM recommendations. But, to meet the demand of higher capacity, larger feed size and ability to handle high amounts of floats and sinks due to variation in the feed coal quality, many manufacturers are supplying DMC with feed, overflow and underflow diameter slightly larger than the DSM recommendations to avoid the overloading of the vortex finder and spigot as the overloading of Vortex Finder and Spigot may cause high misplacement of materials. Adequate openings of overflow and underflow also help to avoid the retention of large middling's particles or surging inside the DMC.

Likewise, it is necessary to size the DMC so as adequate medium flows through the vortex finder in order to carry out the coal particles. If the flow of medium to the overflow is too low, then the excess clean coal cannot be carried through the vortex finder and will instead report to refuse.

### 4.2.2 Media

The PSD and viscosity of the dense media in DMC circuit is of utmost importance. As per DSM Handbook guidelines, for low media specific gravity between 1.3 and 1.5 the grain size of magnetite should be 95% less than 40 µm while for high media specific gravity from 1.4 to 2.0, it should be 95% less than 50 µm.

Inappropriate medium rheology and inadequate mixing of return media severely affect the DMC circuit operation resulting in yield loss. The contamination of the medium by ultrafine slimes increases the viscosity and decreases the separation efficiency. Hence, a fraction of circulating media should be constantly bled to the dilute media circuit for removal of non-magnetic materials and thus keeping the media clean. Bleeding is also needed to control the medium density as increase in media density is related to reduction in the volume of the media in the correct media/over-dense media tanks.

### 4.2.3 DMC feed tank

It is very necessary for coal and medium to mix properly before being fed to the DMC. A good DMC circuit may include a combination of correct media tank (feeding to correct media header) and wing tank or mixing tank for DMC feeding instead of directly feeding of DMC from the correct media tank. The first line from the CM header is mixed with feed coal while an additional line of CM is fed inside the mixing tank. The second medium flow maintains a constant level in the mixing tank and to prevent at the same time an overflow of the products from the mixing tank and thus a constant head on the cyclones is obtained.

### 4.2.4 General equipment design criteria

| Description | Target | Limit |
|---|---|---|
| Drain and rinse screen wash water | 1.0 m³/h/tph of O/F solids | 0.75 m³/h/tph of O/F solids |
| Medium to coal volume ratio | N/A | 3:1 (min) |
| Magnetic separator efficiency | 99.5% | 99% (min) |
| Magnetic separator feed solid concentration | 10–20% (w/w) | N/A |
| Classifying cyclone feed percent solids | 10–12% | 15% |
| Coarse coal centrifuge discharge free moisture | 5% | 6% |
| Thickener feed solids flux | 100–150 kg/m²/h | 300 kg/m²/h |
| Thickener underflow percent solids | 35% (w/w) | 30% (w/w) (min) |
| Pump selection | Duty point to be on the left or adjacent right to the line of BE curve on the pump curve | N/A |
| Pump motor size | Installed power = absorbed power + 20% | N/A |
| Variable frequency drives | Cyclone feed pumps, thickener underflow pumps, vibrating feeders, BP filters | Thickener underflow pumps, vibrating feeders, BP filters |

### 4.2.5 Sumps/tanks design criteria

| Criteria | Specifications |
|---|---|
| Design | Conical or conical + cylindrical<br>Conical tanks to have minimum 45° wall angle |
| Retention time (allowance for excess volume above operating level to capture run-back) | Medium and coarse particles slurry tanks: 1.5 min<br>Fine coal slurry tanks: 1.5–2 min |

Contd...

*Contd...*

| Criteria | Specifications |
|---|---|
| Liners | Dense medium slurry tanks and wing tank: cast basalt of 20–30 mm thickness |
| | Dilute medium and fine coal slurry tanks: cast basalt of 20 mm thickness |
| | All catch pans, kill boxes, distributors, flume boxes, etc. can be lined with 10–12 mm thick high alumina liners |

### 4.2.6 Piping design criteria

Pipe diameters are selected to ensure slurry velocity is above minimum settling velocity. All slurry pipes are provided with minimum slopes to ensure in case of power failure or sudden stoppage of plant, slurry in the system will report to nearest equipment/tanks and there will not be any accumulation of slurry within pipe. Where due to layout constraints it is not feasible to ensure degree of slopes, flushing connections are provided. Drains are provided in case a portion of pipe needs to be drained out.

Three-dimensional model of complete piping system including equipment, structures, electrical cable routing prepared for all type of services (coal slurry + magnetite, coal slurry, water) to check interfaces. Clashes identified at drawing review stage saved substantial time of execution.

Suction pipes are kept as short as possible. Some recommendations for design criteria and piping material (though other alternatives are also available) are as follows:

| Duty/description | Velocity (m/s) | | Material |
|---|---|---|---|
| | Minimum | Maximum | |
| Water only | N/A | 3.5 | Carbon steel (IS1239/3589) |
| Coal slurry | 1.5 | 3.5 | Carbon steel (Sch 80/XS) |
| Correct media | 1.5 | 3.5 | Carbon steel with 20 mm Cast basalt lining |
| Dilute media | 2.0 | 3.5 | Carbon steel (Sch 80/XS) |
| Thickener underflow | 1.5 | 3.5 | Carbon steel with 20 mm Cast basalt lining |
| Flocculant | N/A | 3.5 | PVC |
| Plant and instrument air | N/A | 10 | Carbon steel (IS1239) |

## 5. Mechanical design criteria

Coal handling system is the vital part of any port, power, mines as well as coal washery plant, therefore the coal handling system has to be appropriately planned, designed, with proper factor of safety and safety in design (SID) during basic and detail engineering.

To keep in mind the best design and advanced material handling engineering, some of the modern technologies in coal handling system are discussed in this section.

## 5.1 Chutes and skirt boards

In chutes and skirt boards, a combination of mother plate and suitable liners are conventionally used. When the average hardness of coal is lower than that of the liner plate, the plate is exposed to only moderate wear. However, when the hardness of the material is higher than that of the liner plate, a wearing-down process begins. After complete worn out of liner plate mother plate will start wearing. Therefore, to protect the mother plate in every periodic and predictive shutdown period, new liner plates needed to be fixed. To overcome this problem, only liner plates can be used without mother plate to form a chute, up to the life of liner plate. This is the best method to maintain and replace any damaged face of the chute, which is connected to each other by bolted joints. Only 20 mm sail hard liners can be considered for chutes and skirt board. Major advantages like more chute life, avoid belt damage due to worn out liners and easy in operation and maintenance has been envisaged.

## 5.2 Discrete element (DEM) analysis for chutes

A useful tool in the design of optimum transfer chutes are computer-simulation programs based on the discrete element method (DEM) in connection with computer aided design (CAD). All chutes shall be engineered chutes and analyzed in DEM; a wide range of material properties (friction coefficients, moisture content, cohesive properties, etc.) are modeled and analyzed. Various size distributions and even completely different material types have been simulated. Trouble areas are spotted, and the design is modified. This is major advantage over the historical practice of trial and error.

## 5.3 Tramp detection and removal

In coal washery plant, it is very important to remove the tramp (ferrous and non-ferrous material) before feeding to washery equipment. The question is how and where to mount the tramp removal auxiliaries (i.e. metal detector and magnet). There are number of beliefs of installation of these auxiliaries like, one of it is "first metal detector then magnet and again then metal detector", other one is "metal detector then magnet" and last one is "first magnet then metal detector". More or less all of such combinations are installed in some or the other plant, so it is very difficult to select best combination. After lots of deliberations, analysis of the requirement, application of the system and recommendations, it is recommended that first magnet shall be installed followed by metal detector.

A ferrous tramp received with run of mine is removed by first installed magnet. In case if any ferrous tramp is missed to get removed by magnet and also a non-ferrous material shall be detected by metal detector installed after magnet and thus conveyor will be stopped and protected. This arrangement is the preferred combination of the pair and the best techno-economical solution.

## 5.4 Telescopic chute

If coal will be stacked via elevated conveyor and stockpiling from height, to reduce the material impact and pollution, telescopic chutes can be used to reduce generation of fugitive dust. This is the best practice when stacking coal over stockpile as dust liberation from the material is prevented as chute travels towards the loading point and released at the lowest possible elevation at every step for loading bulk material onto stockpile to ensure pollution free unloading.

## 5.5 Standardization

Inventory control and cost reduction techniques is an important part of materials management which helps to reduce overall cost of inventory control by applying various basic as well as innovative techniques such as inventory turnover ratio, standardization and codification, value engineering and value analysis. To reduce the varieties and inventory cost of conveyor, various components like belts, idlers, pulleys and drives, standardization can be done within the plant as well as with existing plant if any. This is definitely cost economical practice.

## 6. Electrical and instrumentation design criteria

### 6.1 Substation building (cable cellar instead of trench)

Cable cellar with required clearances for cable tray installation, cable laying and men movement is preferred over providing cable trench. This gives advantage over conventional design of trench from water logging, ease of cable laying and maintenance activities.

## 6.2 LED lights over conventional light fixtures

LED lights are preferred for all the area lighting of the plant instead of conventional light fixture. LED lights gives batter lumens, long life, environment safety and energy saving. LED lights use about 50% less electricity than traditional incandescent, fluorescent, sodium vapor, sodium mercury, and metal halide and halogen options, resulting in substantial energy cost savings. LED lights can also reduce the labor cost of replacing bulbs and other component.

## 6.3 Sandwich type compact busduct over huge conventional busduct

Sandwich type cast resin busduct occupies the lesser space over the conventional air insulated busduct. It is totally maintenance free. Erection time is lesser and simpler than other bus ducts. It gives low voltage drop as well as high short circuit capacity.

## 6.4 LHS cable for fire detection

Linear heat sensing (LHS) cables for all the conveyors of the plant for detection of fire due to coal fire or friction of idlers. These cables are further segmented in to specific zone for early identification and remedial measures for each conveyor based on the length of the conveyors. LHS cables are also installed to detect fire in cable cellar.

## 6.5 PLC permissive for better monitoring

Plant is designed for master control with PLC. PLC will provide local permissive to local control station even for starting the motors locally. This will ensure that LOTO (lock out tag out) and maintenance notification to plant control room and then only maintenance is allowed with consent of control room operator.

## 6.6 Standardization of switchgears for fewer inventories

Low voltage switchgear items i.e. MCB, MCCB, Relay, Lamps, etc. are standardized for all the panels of the plant to ensure common make and inventory management. This is applicable even for mechanical package panels supplied by OEM's for Crane, Hoists, Magnets, etc.

## 6.7 Internal Arc class for HV switchboards

IEC 60694/62271 – common specifications for switchgear offers with internal arc class with 1 s/0.5 s/0.1s. Generally, manufacturer provides IAC with 0.1 s for all MV/Hv switchboards, if client specification is silent on the same. We prefer IAC with 1 s. Additional arc venting duct provides with the switchboards to meet the IAC for 1 s. Ducting offered on top of the switchboard, provides for a safe method of venting of arc flash pressure, and hot gasses. It also provides metallic segregation of each compartment.

## 6.8 CCTV monitoring of critical equipment

For critical equipment of DMC washery, CCTV system with monitoring facility at central control room is provided for actual site condition monitoring in addition to electrical and process data shared via PLC for better control and monitoring of the washery plant.

## 6.9 Separate transformer for VFDs

Converter duty transformer is used to feed power to the low voltage VFDs of different rating. It provides isolation between the VFDs and the feeding upstream network. It suppresses the harmonic generated by VFDs, thus protecting the upstream network from harmonic contamination. It also provides protection against radio-frequency interference produced by the rapidly commutating semiconductors.

## 6.10 Cable voltage sensor

Voltage presence indicators can be installed at incomer of HV/MV switchboards. It provides the permanent monitoring of all three phases. It connects with capacitive detecting insulator and voltage presence indicator. It gives the indication for presence of voltage at incoming cable of switchboards, which provides the safety to persons during maintenance activity.

## 7. Technological IOT based solutions

Entire industrial world has started journey to adopt *Industry 4.0* solutions for optimal utilization of men, machine, and materials. Out of many solutions available by OEM's and system integrators in industry and best practices followed by global mining companies; following solutions are adopted for coal handling and beneficiation plants and moving towards complete integrated IOT enabled plant.

## 7.1 IOT based belt-conditioning monitoring

This will facilitate real time condition monitoring of the critical conveyor belts with help of sensors, i.e. vibration, belt slip, belt rip, temperature, thickness, etc. These real time data are even available to field operators for predictive analysis and any rectification majors to be taken.

## 7.2 IOT based coal stack monitoring

IOT enabled thermal imagery for coal stack monitoring system for continuous assessment of likelihood of fire and which is interconnected with fire-suppression systems and audiovisual alerts.

## 7.3 Asset management tool

Having historian data of all the critical equipment (sensor data, operation data and event data) at one place at plant level and at enterprise level enables *advanced asset management* tools for measuring and predicting likely hood of failure of important assets which will lead to best planned maintenance practices and it allows to follow production as scheduled by preventing breakdowns.

## 7.4 KPI Based Management information system

Key performance index (KPI) based management information system (MIS), both web and mobile enabled IOT platform for operational as well strategic reporting. This allow easy and real time access and reporting of field level KPI to plant operators, operational KPI to plant managers and enterprise KPI to business managers. This system is configurable for access and authorization based alerts for respective levels, which help to eliminate manual reporting systems and highlight the relevant issues to relevant personnel.

## 7.5 IOT network security

To enable all IOT solutions mentioned above and for overall security of the IT network and OT network, aspects of network security is taken care by adoption of firewall, network segmentation and patch updation system for Microsoft licenses and Antivirus licenses. In addition to this group level, security solution is implemented which takes care of the intrusion detection and intrusion prevention system for securing IT and OT network and give real time traffic monitoring.

## 8. Civil and structural

For any washery and material-handling project, civil and structural design is very much important because it directly impacts the safety and durability of structures. Efficient structural design will ensure building is safe during the most probable loads like dead load, live load, equipment loads, wind load and earthquake loads and at the same time, it results in economical design solution. Any deficiency in civil/structure design may lead to ill functioning of building (i.e. floor vibration, improper access to equipment, improper head room) and may lead to failure of structure and in turn stoppage of plant.

For RCC frame structures like thickener, substations, water tank and pump houses, reinforcement of grade Fe500 D shall be used for good behavior in earthquake condition.

Substation building (i.e. control room) shall be connected to washery building by an elevated passage for man movements. This arrangement shall be useful with respect to operation point of view.

The layout of the building shall be made in a way to provide free access to man and material for operation and maintenance point of view. Minimum 1200 mm wide walkway shall be maintained around all equipment. Minimum width of staircase shall be 1000 mm for proper access of various floors.

For conveyors, deck type gallery proposed instead of conventional box type gallery, this type of gallery configuration will reduce the wind load on structure and leads to economical design.

For slime pond construction, side slopes shall be provided with sufficient thickness of RCC lining so it will not get damaged during operation. RCC lining shall increase life and reduce operation cost.

For transfer tower/crusher/washery building floors where heavy rotating equipment (i.e. drives, crusher, screens, etc.) are located, RCC floors shall be preferred to control vibration. For RCC floors construction, metal decking sheet with shear studs shall be used as a permanent formwork. For other floors, chequered plate/grating may be provided.

Equipment arrangement in buildings shall be done in such a way that heavy equipment are located near to the ground and other equipment are located at higher floors.

## 9. Plant utility services

### 9.1 Dust extraction system

Dust extraction system is provided with self-cleaning duct instead of providing horizontal ducting. Self-cleaning duct ensures no material accumulation during operation. Providing hoods near sources of dust generation ensures dust is captured immediately after its generation. Flange connections are provided for ensuring only a part of duct can be easily removed during maintenance of material handling equipment.

## 9.2 Water filtration system

Slurry pumps are heart of DMC washery and to ensure filtered water is supplied for gland sealing and to provide filtered water for dust suppression system (to avoid nozzle chocking) proper filtration plant chemical based caustic soda (NaOH), poly aluminium chloride (Al2Cl(OH)5) and hydrochloric acid (HCL)) is recommended. Plants can have provision for by pass also when good quality of water is available.

## 9.3 Heating ventilation and air conditioning system

Instead of providing multiple split AC units, package AC system with 50% redundancy are provided. Separate packaged air conditioning room is provided instead of mounting IDUs under roof which require frequent inspection and maintenance. Switch gear rooms are provided with effective wet type pressurized ventilation system (air washer unit) to ensure proper heat removal.

## 9.4 Compressed air system

Experience dictates that centralized compressed air system is preferable. Separate compressor room with proper ventilation system is provided to ensure the risk of heating. It also ensures less maintenance and increased life of equipment.

## 9.5 Firefighting system

Dedicated fire network system is provided as per prevailing norms (TAC/BIS). A combination of fire hydrants (internal/external) and water monitors with complete ring main circuit is provided to combat fire. It is ensured that complete plant is covered with firefighting network and no part if left out. Fire extinguishers are also provided along with conveyors to combat fire.

## 10. Conclusion

This is quite an exciting time for coal preparation in India; several new projects are either under construction or at various stages of planning and design. So correctly defining the scope and standard design criteria would ensure the participation of majority stakeholders and implementation by the participants of best practices arising out of standardization. These practices would bring modernity in coal preparation plant design, performance enhancement, quality assurance, better process and quality control and consequently improving the plant economics.

## 11. References

[1] Designing the coal preparation plant of the future, SME, Littleton, 2007, Barbara J Arnold, M S Klima and P J Bethell.

[2] The DSM handbook for heavy media cyclone plant design – Anonymous, Netherlands: DSM.

[3] Benchmarking the only way forward for coal preparation in India – Prof. S Bhattacharya 2012.

[4] Full wash coal washery design & implementation for Indian Thermal Coal Production – Zhigang Wang, J Nielson, B Arnold, A B Chatterjee CPSI 2017.

[5] Need for training and skill development in coal preparation – Prof. Nikkam Suresh CPSI 2012.

[6] Engineering design documents and personal communications received from EPC Contractors and Consultants for various Adani CHPP Projects.

# 8

# Mongolia's first gravity fed 3-product heavy medium cyclone coal washing plant

G. Davaatseren, M. Bazarragchaa, Ts. Tsegmid, G. Ochir-Bat, B. Serdamba, David Jiang,
Zhaomin Duan

*Mongolyn Alt (MAK) LLC, Ulaanbaatar, Mongolia*
*Beijing Guohua Technology Group Ltd., Beijing, China*

**Abstract:** This paper represents the design and construction basis of the Naryn Sukhait 170 TPH Modular CHPP Project in Mongolia. Based on the plant raw feed coal characteristics analysis, gravity-fed 3-product heavy medium cyclone process was chosen. Technical and economic indicators are collected after the CHPP entered into full production stage for comparison against the trial washing results, conducted at the jig coal preparation plant. The comparison result demonstrates the advantage of efficient and simplified heavy medium process in regards to maintaining of the product stability, hence improving the clean coal yield.

**Keywords:** Gravity fed 3-product heavy medium cyclone, modular plant, flotation, rotary breaker, crusher

## 1. Summary

MAK is the private Mongolian company which owns and operates the Naryn Sukhait coal mine in South Gobi Province, Mongolia. Naryn Sukhait sub-bituminous coal property is located in the territory of Omnogovi aimag and annually 6–8 million tons of coal is exported, making MAK one of the leading company within Mongolia's mining sector. MAK experts have been actively pursuing exploration and test works in order to improve exporting coal quality. As a result, the most advanced technology selection is made to implement coal handling and preparation washing modular plant (CHPP) with capacity of 1 million tons annually. The project entered into full production in July 2018.

In order to reduce the construction duration and amount of earth-work, modular equipment and materials are selected. Beijing Guohua technology Group (BGTG) from China developed the engineering design and supplied the technology and the major equipment. Moreover, Jiangsu Yangzhou Hengyuan Co Ltd from China specialized in industrial construction carried out the plant erection at the site within 9 months.

The plant technological process flow sheet and its design basis are based on Naryn Sukhait mine's coal washability, lab research and pilot plant testing results. Hence, the most advanced of state of art technologies and technological process flow sheet are selected (3-product heavy medium cyclone + flotation). The chosen suitable technology to process mine coal with varying quality (high ash, oxidized and fine coal) is hybrid technological flow sheet. The project also implemented full automatic production software.

The entire plant equipment start/stop command, control of critical production parameters as per technological requirement (density, level, feed conveyor speed, rotation of pump, etc.) can be automatically performed from dispatcher's room. In addition, to evaluate amount of clean coal and to calculate their balances, the product transportation conveyors are equipped with electric scaler/balance and the parameters such as ash, moisture and calorific contents of raw coal and washed products are analyzed online, allowing real time process control as per technological requirement (Table 1).

**Table 1:** Project design basis

| No. | Description | Unit | Value | Description |
|-----|-------------|------|-------|-------------|
| 1 | Capacity | t/h | 170 | Feed coal |
| 2 | Error ($Ep$) | | $E_{p1}$ = 0.03–0.05 $E_{p2}$ = 0.06–0.08 | |
| 3 | Clean coal misplacement in middlings | % | 3–5 | $E_{p1}$-HMC 1st stage; $E_{p2}$-HMC 2nd stage |
| 4 | Middlings misplacement in reject | % | ≤3 | Particle + 0.5 mm |
| | Product quality | | | Particle + 0.5 mm |
| 5 | Clean coal ash | %, ad | ≤8 | |
| 6 | Clean coal surface moisture | %, ar | ≤10 | |
| 7 | Operation power consumption per ton of raw coal | kwh | ≤6 | |
| 8 | Magnetite consumption per ton of raw coal | kg | ≤1 | |
| 9 | Water consumption per ton of raw coal | m³ | ≤0.1 | |

Therefore, practical operation, technical and economic indicators of CHPP with annual capacity of 1 million tones becomes the basis to making of technological selections for Naryn Sukhait coal washing plant with larger capacity in the near future.

## 1.1 Coal quality and washabilty

Within the project scope, development of CHPP's technological process flow sheet is based on number of research and pilot plant tests carried out at MBE-CMT, Germany, SGS research center, China and at Mongolyn Alt (MAK) corporation's Technological research center. As a result, the Naryn Sukhait's coal characteristics and washability were determined (Tables 2 and 3).

**Table 2:** Coal particle size

| Particle size (mm) | Yield (%) | Ash (%) | Sulphur (%) |
|---|---|---|---|
| 50–10 | 27.60 | 12.03 | 0.63 |
| 10–1 | 52.20 | 30.70 | 1.17 |
| 1–0.5 | 8.87 | 30.14 | 1.10 |
| 0.5–0.15 | 8.27 | 26.80 | 1.33 |
| 0.15–0.1 | 1.06 | 25.18 | 1.65 |
| <0.1 | 2.00 | 24.80 | 1.77 |
| Total | 100.00 | 25.00 | 1.05 |

**Table 3:** Float-sink test result

| Density (kg/l) | Quality | | Float | | Sink | |
|---|---|---|---|---|---|---|
| | Yield (%) | Ash (%) | Yield (%) | Ash (%) | Yield (%) | Ash (%) |
| –1.3 | 49.66 | 3.63 | 49.66 | 3.63 | 100.00 | 24.83 |
| 1.3–1.35 | 15.83 | 5.96 | 65.49 | 4.19 | 50.34 | 45.76 |
| 1.35–1.40 | 4.56 | 9.40 | 70.05 | 4.53 | 34.51 | 64.01 |
| 1.40–1.45 | 1.95 | 12.89 | 71.99 | 4.76 | 29.95 | 72.31 |
| 1.45–1.50 | 1.21 | 16.02 | 73.20 | 4.94 | 28.01 | 76.44 |
| 1.50–1.60 | 1.04 | 21.52 | 74.24 | 5.17 | 26.80 | 79.16 |
| 1.60–1.70 | 0.48 | 27.57 | 74.72 | 5.32 | 25.76 | 81.49 |
| 1.70–1.80 | 0.40 | 37.74 | 75.11 | 5.49 | 25.28 | 82.51 |
| 1.80–1.90 | 0.46 | 45.03 | 75.58 | 5.73 | 24.89 | 83.23 |
| >1.90 | 24.42 | 83.95 | 100.00 | 24.83 | 24.42 | 83.95 |
| Total | 100.00 | 24.83 | | | | |

The washability determination research showed that the Naryn Sukhait coal belongs to category of "easy to wash" and possible to have yield of >70–75% with ash content of <8%. These figures also match to that of coal washing plant technological parameters. Therefore, it can be concluded that the time and financing spent on research and test works do contribute to minimizing of the project risks (Fig. 1).

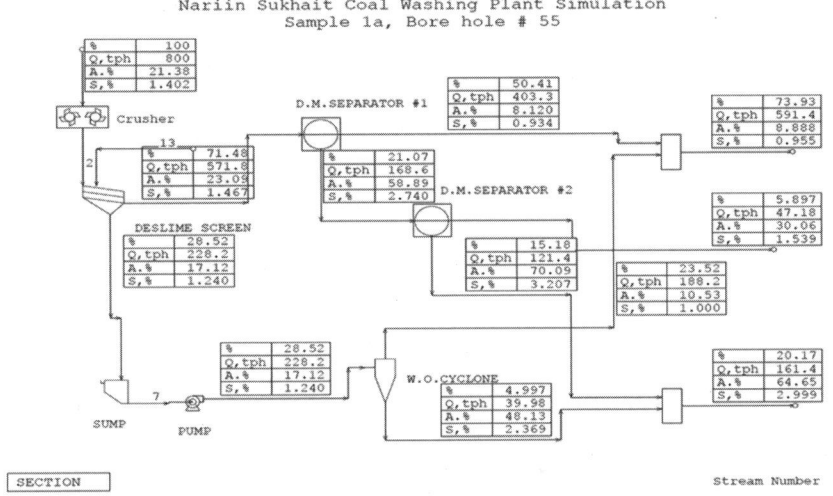

**Figure 1:** METSIM simulation result

## 1.2 Technological process flow sheet and equipment

Naryn Sukhait mine consists of main 5th layer and multi layers of stone and clay. Coal washability quality of these layers is thoroughly studied and future coal to be washed from the mine was taken into account during technological process flow sheet development.

During project phase, oxidized and high slam coal from multilayers were tested at the Qinhua's coal washing plant in China and the results were used in developing and finalizing of flexible hybrid technological process flow sheet to allow varying quality feed coal from mine.

Therefore, as a result of various tests, researches and surveying, the most advanced state of art equipment and technological process flow sheet were selected and finalized (gravity fed 3-product heavy medium cyclone + flotation). The Naryn Sukhait CHPP's technological process flow sheet is shown below (Fig. 2).

The main characteristic of the above shown flow sheet is dual operation mode (system 1 and 2) that can be changed depending on coal-oxidized level, original slam content and the ash content. For example:

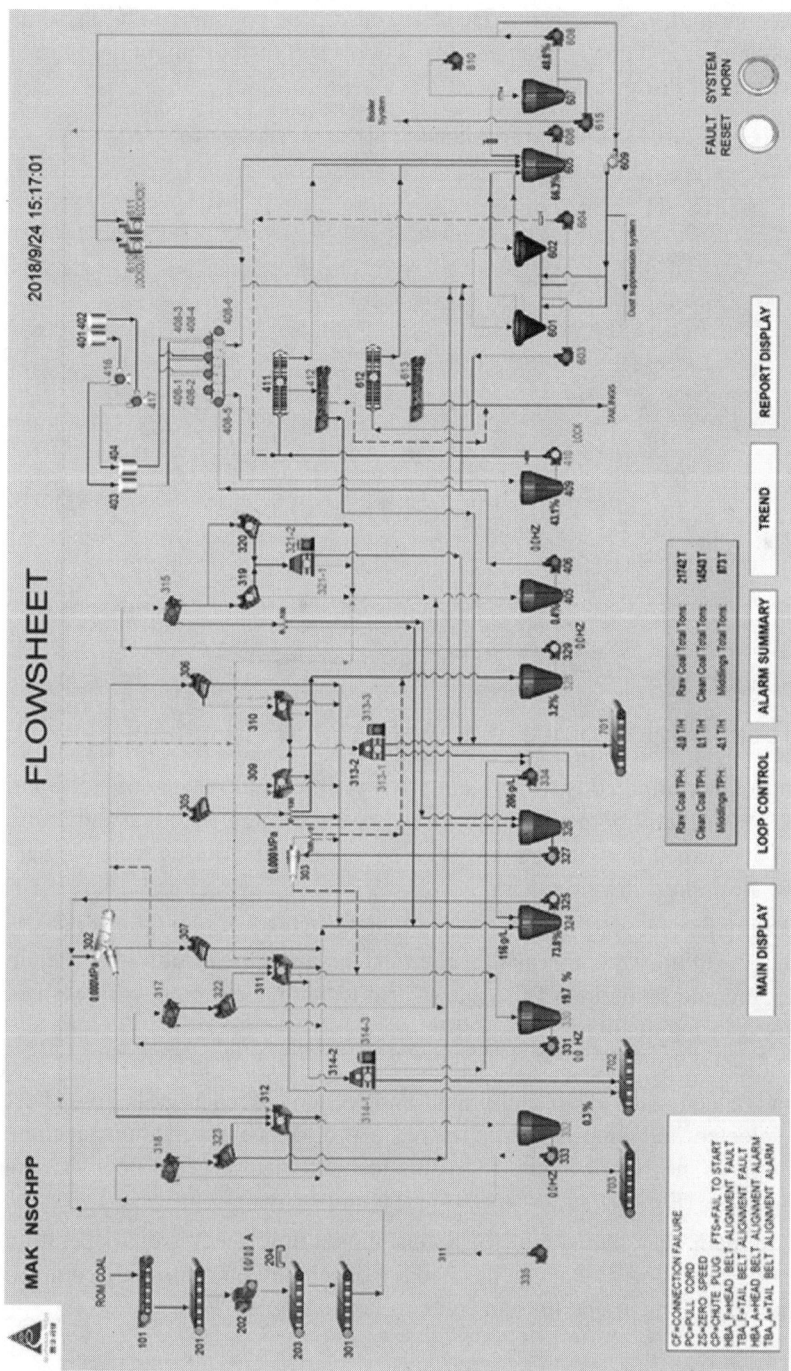

**Figure 2:** Naryn Sukhait CHPP technological process flowsheet

- Powder coal's hydro cyclone system (mode 1) to be used for oxidized high slam content coal. During this mode, the flotation line should be stopped in order to utilize the thickener and the press filter that all the slam tailings gets processed.
- To process coal with less slam content and with stone, the flotation line is brought online. In this case, the flotation clean coal and the tailings should be thickened and filtered separately.

The main feature is the flexibility of changing the systems and change is made automatically via the dispatcher. In addition, depending on the feed coal ash content, if necessary it is possible that the middling brought to stop and hence the corresponding equipment.

Based on the last 2 months of plant operation history and its production parameters, the chosen technological flow sheet is suitable to wash coal with varying quality. This had also been proven through test works before.

In regards to the equipment, it involves selective operation rotary breaker, high capacity 3-product hydro cyclone, banana type D&R screener; high-speed deep cone thickener, automatic pressurized filter and other most advanced equipment are integrated in the technological process flow sheet.

In particular, high efficiency deep cone thickener is located up stream of the pressurized filter to de-water the powder slam with maximum efficiency, which is proven.

## 1.3 Automatic control system

The CHPP is equipped with automation control system from Allen Bradley, which allows control and optimization of the product yield/quality ratio depending on varying feed coal quality. In addition, the automated central control system enables starting of the plant equipment in required sequence, optimization of process parameters according to technological need and the product yield/quality ratio, which can be seen that these functions are proven realistic by looking at the last 2 months of production history parameters (Fig. 3).

Furthermore, the automatic control system allows starting/stopping of equipment as per safety-interlock sequence. For example, hydro cyclone cut-point density, sump level, rotation speed of pumps and feed conveyor efficiency is controlled directly from the central control room.

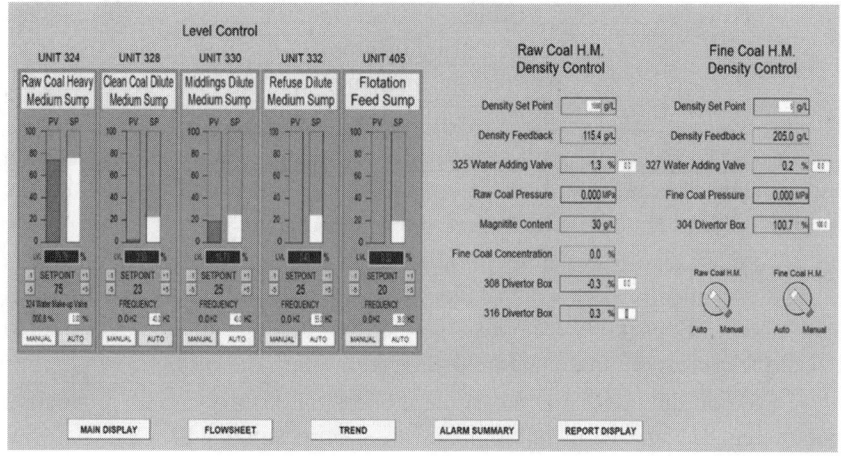

**Figure 3:** Automatic control system screen

The control of process parameters enables optimization of product yield and quality based on feed coal quality. In order to determine the product quantity and the amount and their material balance, the raw feed coal, clean coal and the middling product transportation conveyors are equipped with electronic balance. The production report of particular shift and daily reports are shown below in Fig. 4.

| MAK NSCHPP | REPORT | 2018/9/24 15:18:20 |
| --- | --- | --- |

**TOTAL PER MONTH**

| RAW COAL | 9952.0 TONS |
| --- | --- |
| CLEAN COAL | 6775.0 TONS |
| MIDDLINGS | 378.0 TONS |

RESET

**SHIFT SETTING**

| | START AT | ENDING AT |
| --- | --- | --- |
| SHIFT1 | 7:00 | 19:00 |
| SHIFT2 | 19:00 | 7:00 |

YESTERDAY REPORT    PRINT

**TODAY REPORT**

| | Starting Time | ENDING TIME | Raw Coal Total Tons | Clean Coal Total Tons | Middlings Total Tons | Run Time | Clean Coal Yield | Midding Yield | Availability | Clean Coal Average TPH | Midding Average TPH |
| --- | --- | --- | --- | --- | --- | --- | --- | --- | --- | --- | --- |
| Shift1 | 07:00 | 19:00 | 0 T | 0 T | 0 T | 0H 0M | 0.0% | 0.0% | 0% | 0.0 | 0.0 |
| Shift2 | 19:00 | 07:00 | 0 T | 0 T | 0 T | 0H 0M | 0.0% | 0.0% | 0% | 0.0 | 0.0 |

**Figure 4:** Daily production report

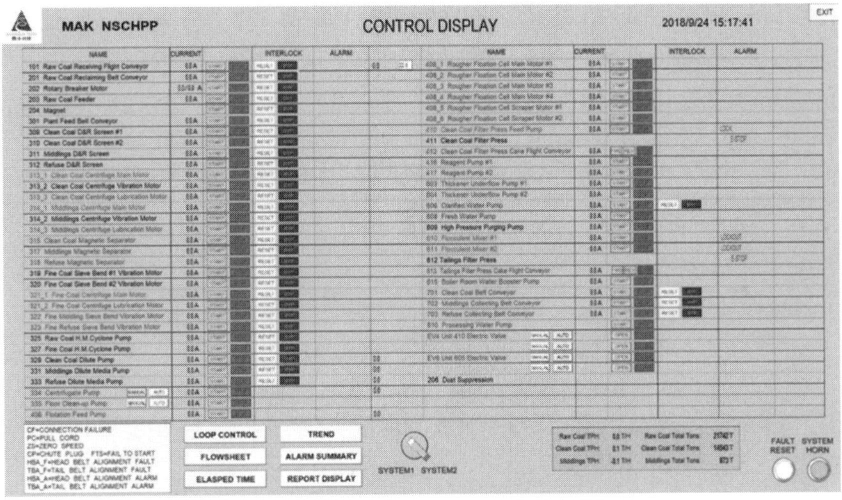

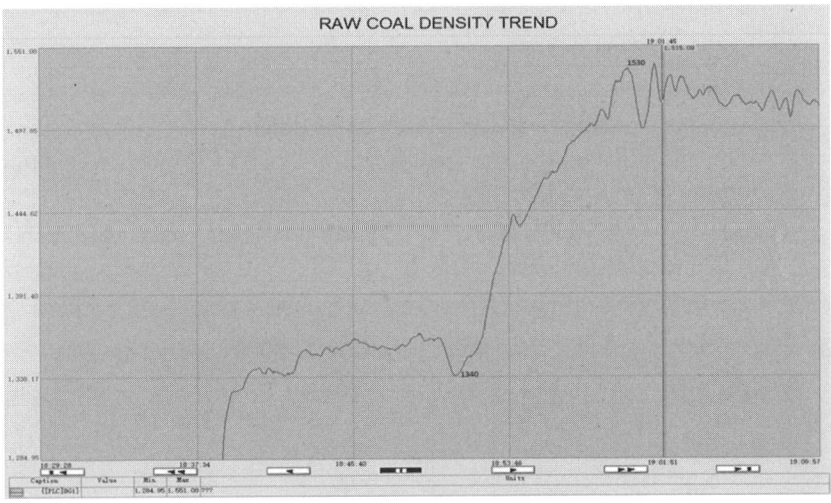

**Figure 5:** Equipment status

Figure 5 above illustrates CHPP's entire mechanical equipment status, their start/stop and automatic shift screen of the technological process flow sheet. Figure 6 below illustrates the point trend of cut-point density optimization on the automatic control system.

**Figure 6:** Density trend

## 1.4 CHPP's first production results

Between August and October of 2018, six different types of coal (mainly different in ash and sulfur content) were considered in the CHPP's test plan and the tests were carried out to fine-tune the production and the process. The table below illustrates the average test results for normal and high ash content coal washing (Table 4).

**Table 4**: Coal washing results

| Product | Parameters | | A (%) | S (%) | Description |
|---|---|---|---|---|---|
| | Quantity | | | | |
| | t/h | Clean coal yield (%) | | | |
| Test-1 | | | | | |
| Feed | 143.3 | 100 | 27.9 | 1.38 | Density: 1480–1530 g/l |
| Clean coal | 106.3 | 73.5 | 8.5 | 1.19 | |
| Middlings | 4.3 | 3.3 | 36.3 | 1.81 | |
| Tailings | 32.7 | 23.2 | 88.1 | 1.92 | |
| Test-2 | | | | | |
| Feed | 139 | 100 | 37.5 | 1.16 | Density: 1380–1520 g/l |
| Clean coal | 84.4 | 58.7 | 9.3 | 1.19 | |
| Middlings | 2.8 | 3.4 | 30.3 | 1.65 | |
| Tailings | 51.8 | 37.9 | 81.8 | 1.07 | |

The product yield amounts were based on readings given on the electronic conveyor balance. In addition, the yield is recalculated using balance equation by considering the ash and the sulfur.

The less product yield difference between conveyor electronic balance and mass balance calculation signifies that the sampling and chemical analysis errors are minimum.

Expression of the balance equation in form of matrix:

$$\begin{bmatrix} 1 & 1 & 1 \\ A_c & A_m & A_t \\ S_c & S_m & S_t \end{bmatrix} \times \begin{bmatrix} \gamma_c \\ \gamma_m \\ \gamma_t \end{bmatrix} = \begin{bmatrix} 1 \\ A_f \\ S_f \end{bmatrix} \tag{1}$$

Herewith: $A_c$, $A_m$, $A_t$ and $A_f$ are the clean coal, middling, tailings and feed ash content, %; $S_c$, $S_m$, $S_t$, $S_f$ are the clean coal, middling, tailings and feed sulfur content, %; $\gamma_c$, $\gamma_m$, $\gamma_t$, are the clean coal, middling and tailings yield, %.

The yield results shown (Б.т**) in Table 4 were calculated from material balance calculation using Eq. (1). The yield results shown in the table for comparison demonstrates that in most cases it matches well with technological balance.

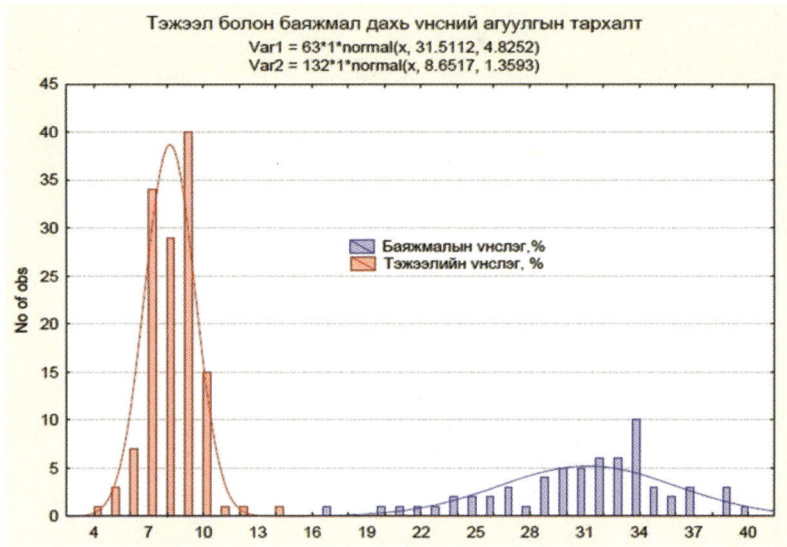

**Figure 7:** Feed and clean coal ash distribution

The ash distribution of the feed coal and the clean coal processed at the CHPP during test period is illustrated in Fig. 7 above. The main parameters that affect the technological process are determined by conducting mathematical statistical analysis on the test results.

The correlation of the feed ash content on clean coal yield is shown below in Fig. 8. The main parameter for heavy medium clean coal technology is the density and effect of cut-point on clean coal yield (Fig. 9).

The clean coal yield that is the important parameter to chosen technology and to the economic feasibility is determined based on multi variable regression equation that also includes the original feed ash and hydro cyclone density. Such regression equation is valid when clean coal quality is <10%.

$$\gamma_c = 0.09711\,\rho_s - 0.6862\,A_f^c - 57.2243 \qquad (2)$$

Herewith: is the clean coal yield, %; is the density, g/l; is the feed ash content,%.

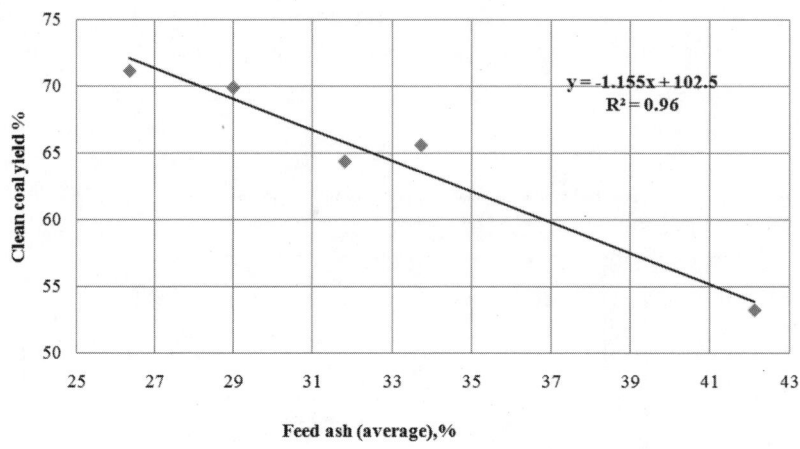

**Figure 8:** Clean coal yield dependence on feed ash

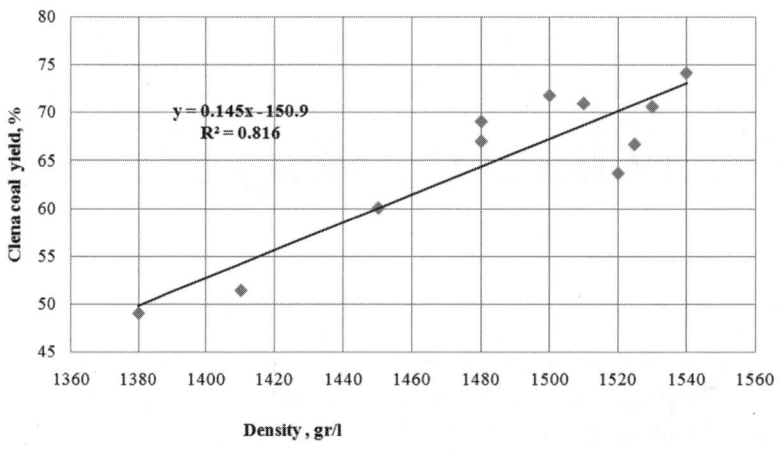

**Figure 9:** Cut-point density dependence on clean coal yield

## 2. Conclusion

1. Heavy medium modular coal washing plant with annual capacity of 1 million tons of coal project has been successfully implemented at the Naryn Sukhait coal mine within short period of time and has been successfully brought to full production.

2. The final chosen technological hybrid process flow sheet allows production of high quality clean coal when the coal washing plant feed is of varying quality (oxidized and has slam and stone content).

3. The important parameter that affects the production is determined through pilot plant testing, which the results were further used in fine-tuning of the automatic control software.

4. The pilot plant testing result shows that the coal washing plant technology is able to give clean coal with high quality ($A = 6–10\%$, $G = 75–85$) when the original feed coal variation (ash 20–60%). The water medium gravitation process is not able to give yield with such parameters (jig, 3HMC, WOC, etc.).

## 3. References

[1] Naryn Sukhait multi-layer coal washability report, 2009.

[2] Feasibility Study Report (FSR) & Front End Engineering Design (FEED) Report, 2017.

[3] Washing plant comparison test report, 2018.

# 9

# What efficiency parameters are crucial to washery operations?

Matthew Swanson and Andrew Swanson

*QCC Resources Pty Ltd, Hexham, Australia*

**Abstract:** The use of efficiency parameters, such as $E_p$ and D50, is well established within coal preparation practice, and the measurement and benchmarking of such parameters are widely used by plant operators for plant improvement studies. However, the impacts of efficiency parameters on overall plant performance vary widely and can depend on the nature of the coal and the operation. So, when considering plant improvement projects, the consequences of the various efficiency parameters need to be understood. This is of appreciable relevance to plant designers and constructors who are often asked to provide quite stringent performance guarantees for individual beneficiation equipment, without any consideration of the true value of such guarantees. This paper will attempt to quantify the impacts of various efficiency parameters on overall plant outcomes with the aim of enlightening all parties involved in the washery projects to ensure that the nominated performance guarantees are measurable and relevant.

Desktop Limn simulations have been used to model the impact of varying the $E_p$, D50, $T_1$ and $T_0$ guarantees on the dense medium cyclone (DMC) and spirals circuits of a single product coal preparation plant. Two coal types were considered; a coal with better washability and limited middlings material and a coal with poorer washability and significant middlings material.

**Keywords:** Process efficiency, simulations, dense medium cyclone, spirals, process guarantees

## 1. Introduction

Operators of coal preparation plants strive to maximise process efficiency. Most coal washing processes are based on density separations and the performance of such equipment is typically determined by producing partition curves that represent the probability of a particle of a certain density reporting to the product stream. These curves, and thus equipment performance, are typically quantified by the following performance variables:

- D50 the density at which the particle has an equal probability of reporting to either the product or reject streams. Typically referred to as the cutpoint and often referred to as ρ50 or RD50.

- $E_{p75-25}$ Ecart probable a measure of separation efficiency that is calculated from the equation (ρ75–ρ25)/2. For simplicity the $E_{p75-25}$ will be referred to as $E^p$ from this point onward.

- $T_0$ low density bypass, the percentage of low density material that will not be separated and report to the reject stream.
- $T_1$ high density bypass, the percentage of high density material that will be separated by the partition curve. $1 - T_1$ is the percentage of high density material that will report to reject.

## 1.1 The Whiten equation

The Whiten curve is most commonly applied to density separations (Atkinson, 2015) and the basic form of the Whiten equation is presented in Eq. (1).

$$PN(\rho_d) = \left(1 - \frac{1}{1 + e^{1.099 \times \frac{D_{50} - \rho_d}{Ep}}}\right) \tag{1}$$

Where PN is the partition number at particle density $\rho_d$; $\rho_d$ is the particle density; D50 is the particle density which has an equal probability of reporting to either the product or the reject streams and $E_p$ is the $E_{cart}$ probable separation efficiency at the selected size range.

Curves derived from this equation are symmetrical around a PN value of 0.50 equal to D50 and asymptotes at 0 and 1. The value of $(\rho_{75} - \rho_{25})/2$ is equal to the $E_p$ value.

This model does not allow for "tails" or bypass of material at either extreme of high or low density material to the product or the reject streams. As a result, the modified Whiten equation is often used and is of the form shown in Eq. (2). The form of the curve can be found in Fig. 1.

$$PN(\rho_d) = 1 - T_1 + (T_1 - T_0) \times \left(1 - \frac{1}{1 + e^{1.099 \times \frac{a - \rho d)}{b}}}\right) \tag{2}$$

Where $T_0$ is the low density bypass i.e. the higher asymptote of the probability curve; $T_1$ is the high density bypass i.e. the lower asymptote of the probability curve; $a$ is a curve fitting number, equal to D50 when $T_0$ and $T_1$ are 0 and 1, respectively and $b$ is a curve fitting number, equal to $E_p$ when $T_0$ and $T_1$ are 0 and 1, respectively.

It is important to note that for the modified Whiten equation, the $a$ and $b$ terms are only curve fitting parameters and the D50 and Ep must be read off the curve. Changing the $T1$ or $T0$ values, for a constant $a$ or $b$, will result in different Ep and D50 values.

## 2. Study approach

To establish the importance of the efficiency parameters, it is essential to determine the impact of each variable on the total plant yield, not just the unit operation. In this study, the impact of each variable was evaluated by undertaking a series of simulations with only that variable changing.

### 2.1 Efficiency parameters

Efficiency parameters are generally specified for a number of size fractions. For the DMC, the usual size fractions are +4 mm and -4 + 2 mm, while for the spirals the size fractions are $-2 + 0.25$ mm and $-0.25 + 0.125$ mm. Some typical guarantees for the $E^p$, $T_1$ and $T_0$ values are shown in Table 1, and these were used as the basis of this study.

**Table 1:** Typical guaranteed performance parameters

| Efficiency parameter | DMC (+4 mm) | DMC (−4 + 2 mm) | Spirals (−2 + 0.250 mm) | Spirals (−0.250 + 0.125 mm) |
|---|---|---|---|---|
| Ep | 0.03 | 0.06 | 0.165 | 0.35 |
| T1 | 0.98 | 0.95 | 0.95 | 0.90 |
| T0 | 0.02 | 0.05 | 0.03 | 0.06 |

### 2.2 Process simulation software

For this exercise, process modelling was undertaken in Limn: The Flowsheet Processor, a Microsoft excel based flowsheet program which is considered a coal industry standard tool for process engineering. Limn simulations allow for practical inefficiencies and resultant misplacement of material, to determine yield-ash relationships and process stream flow rates.

A typical dense medium cyclone (DMC) and spirals plant has been used as the basis for the Limn simulations. The DMC treats the $-50 + 1.4$ w/w mm material and the spiral treats the $-1.4$ w/w $+ 0.125$ mm material. From this point onwards in the paper, 1.4 w/w mm will be referred to as 2 mm for simplicity.

### 2.3 Plant feeds

To investigate the impact of process efficiencies, two different types of raw coal feed were utilised. Coal A has more fines, better overall feed washability and limited middlings material, while Coal B has less fines, poorer overall feed washability and more middlings material. The following Figs. (1–3) demonstrate the feed characteristics, specifically sizing, yield/ash distribution and the material density distribution.

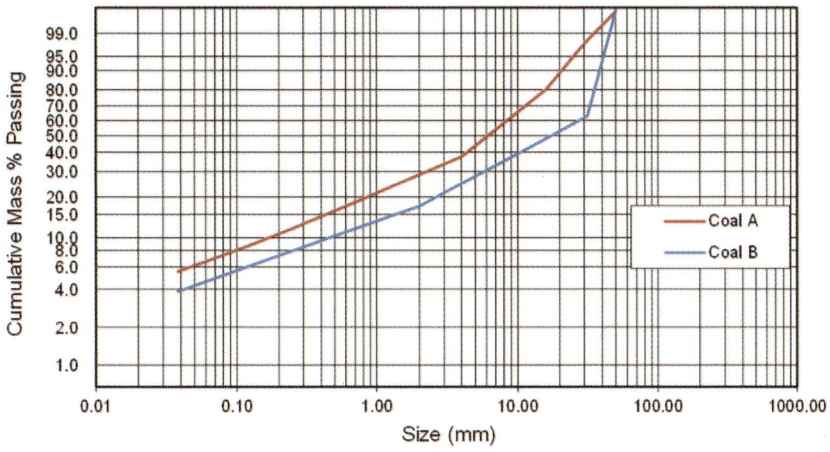

**Figure 1:** Sizing distribution

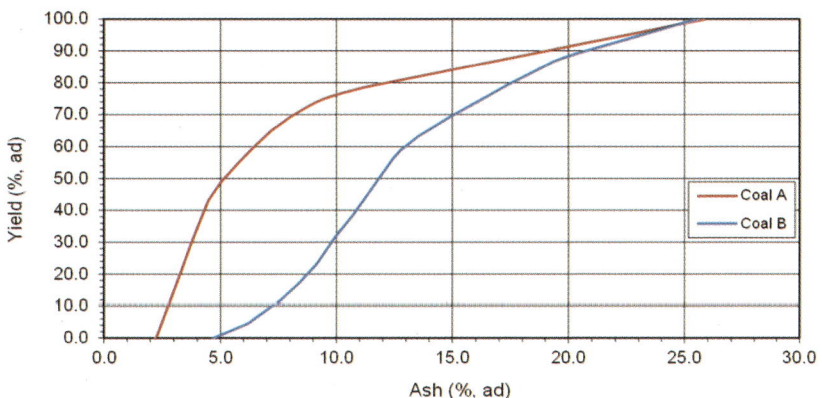

**Figure 2:** Yield/ash distribution

Because Coal A is significantly finer than Coal B, there less material in the DMC circuit and more material in the spirals circuits. Both samples have a similar raw ash but very different density distribution. For Coal A, the mass is principally in the F1.30 RD fraction and quickly tails off with little material above 1.55 RD. For Coal B, the majority of the material is still in the lower density, low ash fractions but considerable material can still be found in the F1.50 to F1.80 RD fractions.

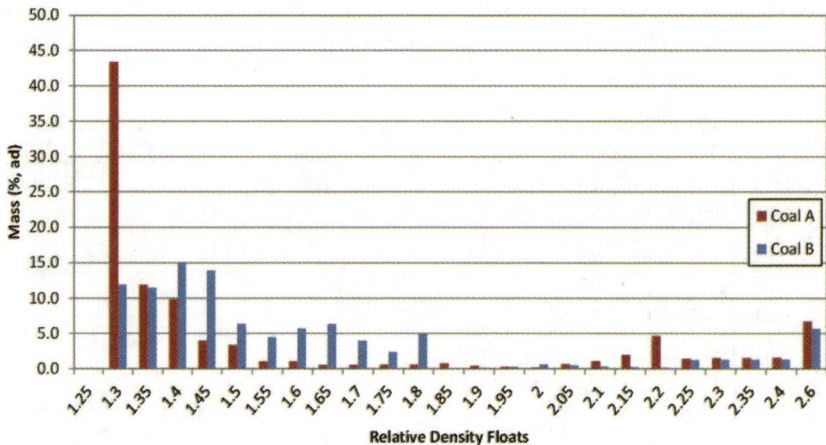

**Figure 3:** Material density distribution (overall +0 mm washability)

## 2.4 Procedure

An initial baseline for each sample was established by carrying out simulations using the standard performance parameters presented in Table 1. The DMC was modelled at cutpoints of 1.40 RD and 1.60 RD while the spirals were modelled with a constant cutpoint of 1.70 RD. The impact of each variable was then established by modifying it in isolation. Three alternative values considered, one better than the typical guarantee and two slightly worse. In each case the DMC cutpoint was adjusted to maintain the same total product ash as the baseline to allow proper comparison of plant performance.

Note: the $E^{ps}$ discussed within this paper are true $Ep$'s calculated directly from the curve (i.e. (RD75−RD25)/2) not the "$a$" and "$b$" Whiten with bypass equation inputs. In each scenario the "$a$" and "$b$" inputs have been selected to generate a partition curve with the required $E_p$. As the $T_1$ and $T_0$ vary, "$a$" and "$b$" are varied to maintain the target $E_p$.

## 3. Simulation outputs

### 3.1 Impact of varying DMC efficiencies

Figures 4–7 contain the yields resulting from modifying the DMC performance variables. It can be seen that Fig. 5 is missing data at a coarse $Ep$ of 0.04 and a coarse $T1$ of 0.97 and 0.96; this is because the impact of these variables on the plant product ash is so significant that at the DMC minimum cutpoint of 1.35 RD the total product ash is still greater than the target baseline.

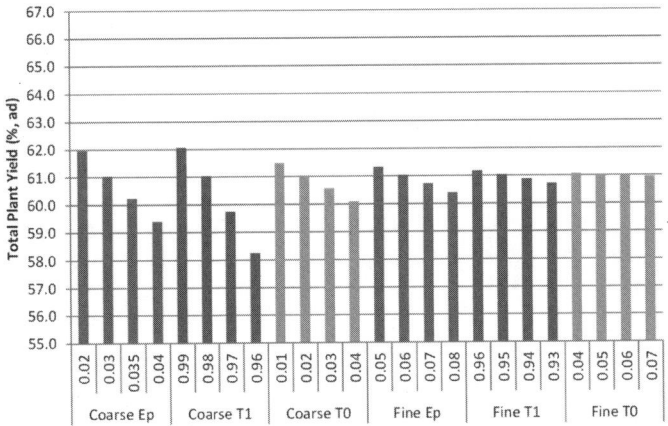

**Figure 4:** Coal A results for DMC cutting at 1.40 RD

Generally, the greatest impact on the plant yield for both coal feeds is the coarse (+4 mm) $T_1$ value, followed by the coarse $E_p$. The low density bypass values ($T_0$) have less of an impact on the plant yield than the $T_1$ values and the fine (−4 + 2 mm) variables appear to have similarly small impact on the yield.

The magnitude of the change in yield is much greater for Coal B than Coal A. When there is a higher quantity of middlings materials, any degradation of separation efficiency will include more of the higher density material in the product. As Coal B has significantly more material in the F1.45 to F1.80 RD fractions the impact of any change in performance variables will be greater.

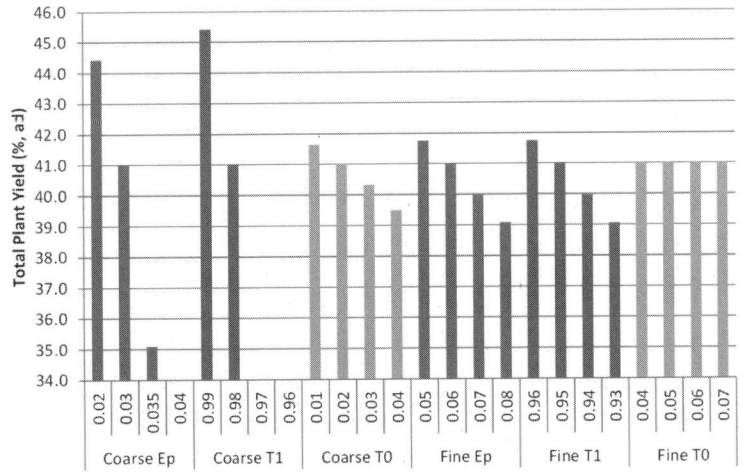

**Figure 5:** Coal B results for DMC cutting at 1.40 RD

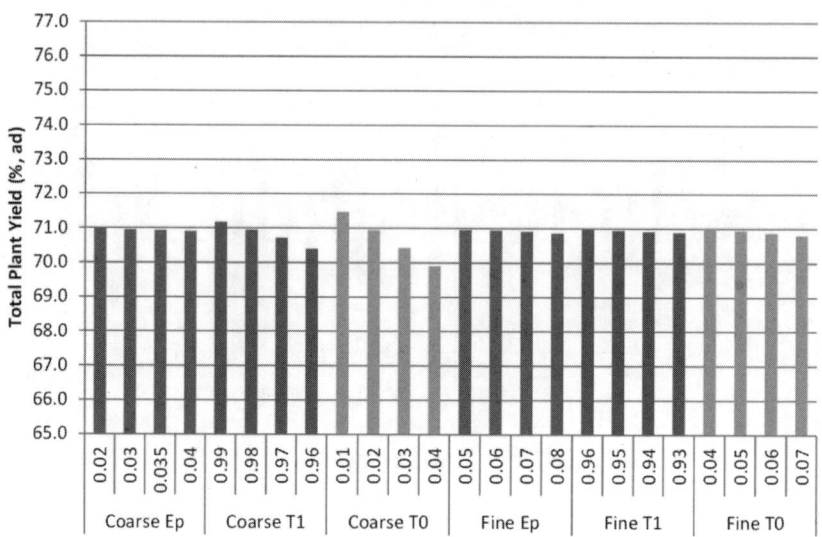

**Figure 6:** Coal A results for DMC cutting at 1.60 RD

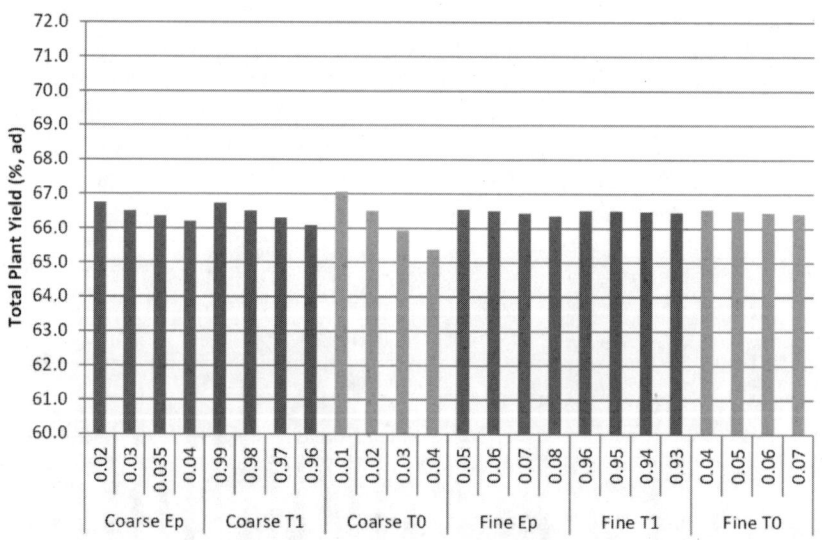

**Figure 7:** Coal B results for DMC cutting at 1.60 RD

The impact on yield by changing the performance variables at a cutpoint of 1.60 RD is shown in Figs. 6 and 7. The impacts on yield at 1.60 RD cutpoint are generally less than those observed at 1.40 RD, with the exception of the coarse $T_0$ values. It is interesting to note that, despite the large differences observed in the yield impact of the performance variables on the two coals at a cutpoint of 1.40 RD, the results for the two coals are very similar at a cutpoint of 1.60 RD. This is due to the density distribution of the two samples, in both samples the majority of the material is around 1.40 RD, so minor changes of cutpoint or separation efficiency will affect significant quantities of material. At a cutpoint of 1.60 RD, there is less material around the cutpoint and thus changes in cutpoint or efficiency will have a smaller impact on the overall yield.

Because there is a relatively small mass of material in the −4 + 2 mm fraction, the impacts are smaller; however, they show the same trends as the results for the coarser material.

## 3.2 Spirals' efficiency effects

To study the spiral performance impacts, the same methodology, as for the DMC case, was applied, with the spirals cutpoint kept constant and the DMC cutpoint being adjusted to maintain the total plant product ash. The results for a DMC cutpoint of 1.40 RD can be found in the Figs. 4.8 and 4.9, while the results at 1.60 RD can be found in Figs. 10 and 11.

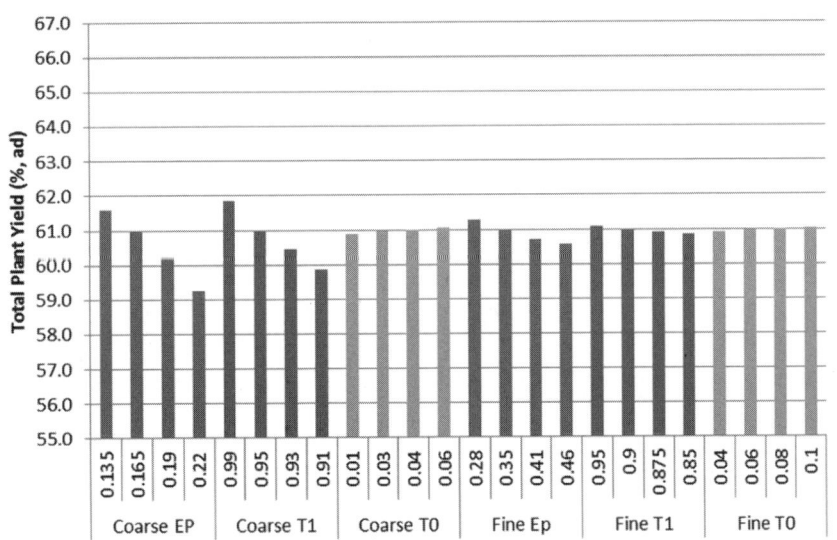

**Figure 8:** Coal A spiral results for DMC cutting at 1.40 RD

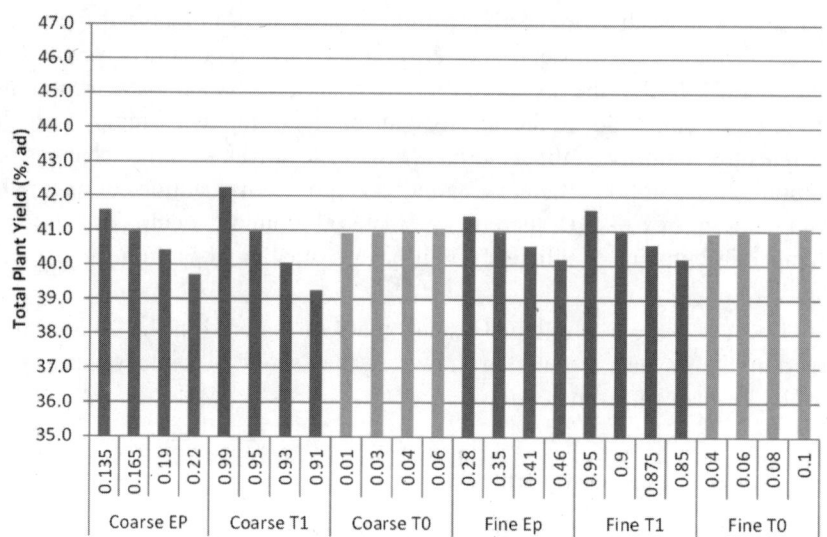

**Figure 9:** Coal B spiral results for DMC cutting at 1.40 RD

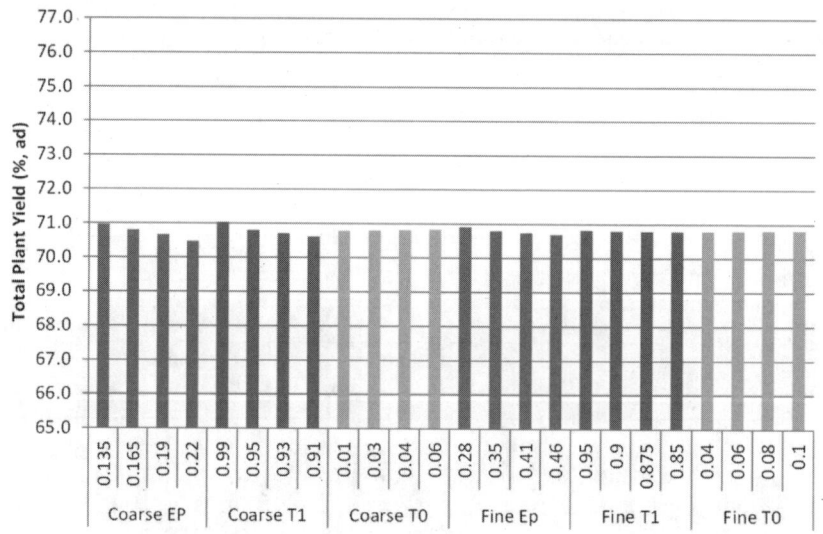

**Figure 10:** Coal A spiral results for DMC cutting at 1.60 RD

For the spiral results, there was a similar observation to the DMC cases in that the impact of changing the performance variables is more significant at a DMC cutpoint of 1.40 RD compared to 1.60 RD. This is due to the quantity of material around 1.40 RD, a change in cutpoint to maintain the product ash will result in a larger change in total yield than at 1.60 RD where there is less material.

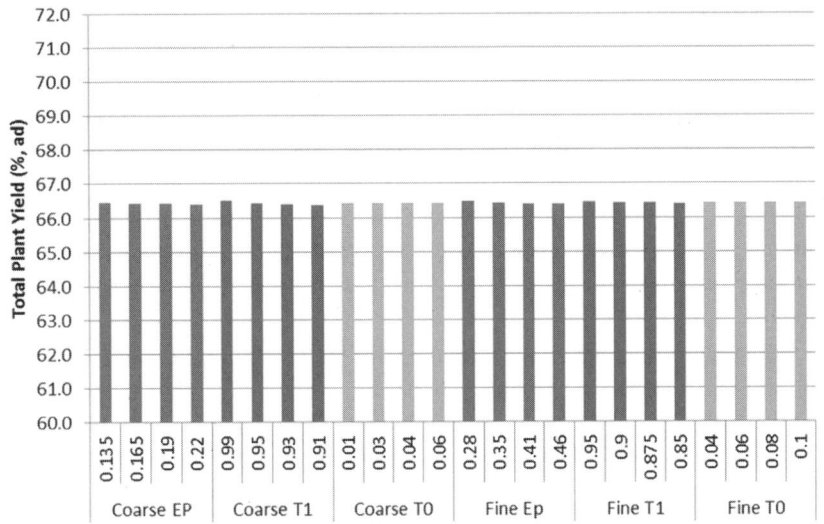

**Figure 11:** Coal B spiral results for DMC cutting at 1.60 RD

The spiral efficiency impacts on total plant yield are greatest for the coarse $(-2 + 0.25$ mm) $Ep$ and coarse $T_1$ values; the $T_0$ value has minimal impact. It is interesting to note in both cases, at a cutpoint of 1.40 RD, the impact of the spiral performance variables is greater than the relative masses would have suggested. This is because at a lower DMC cutpoint the yield decreases and as yield decreases, there is a greater contribution from the spirals circuit to total product therefore the impact of spiral efficiency is relatively more. For Coal B the spiral product makes up only 9% of the total product at a DMC cutpoint of 1.60 RD compared to 14% at 1.40 RD.

When considering spiral efficiency, it must be noted that fine coal float-sink results are notorious for poor reproducibility, and the top size of the fraction tested has a significant impact on the results (Atkinson et al., 2016). Any parameters that can be defined by sampling the $-0.25 + 0.125$ mm fraction should be considered in the light of potentially large error bands. However, the outcomes from this size fraction typically have a minimal impact on the total plant product. The apparent impact for coal B, washed at a low cutpoint, is unlikely to be manifested under normal operating conditions as coal B would almost certainly be washed at a higher cutpoint.

## 4. Concluding remarks

Process simulations have been carried out to determine the impact of performance variables on overall plant yield for dense medium cyclone and spiral circuits. The parameters considered were $E^{cart}$ probable separation efficiency ($E^{p75-25}$), high density bypass ($T_1$) and low density bypass ($T_0$). More details of this work can be found in Swanson et al. (2018).

The DMC cutpoint was shown to be an important factor in the relative impact of these process efficiency factors. Because the majority of coal material is found at lower densities, small changes in performance variables cause large changes in total plant yield when DMC cutpoint is low. As the cutpoint increases away from the bulk of the mass, the impacts are lessened. The presence of middlings materials increases the impact of any performance changes.

The greatest impact on performance, for the DMC circuit at low densities, was shown to be the high density bypass, however, at higher cutpoints, the low density bypass became more significant. The $-4 + 2$ mm fraction shows similar trends to the $+4$ mm fraction but the magnitude is smaller as a result of the reduced quantity of mass in this fraction. At a cutpoint of 1.60 RD, the impact of performance variable changes is reduced with the exception of the low density bypass term where the impact increased.

The impact of the process efficiencies for the spirals showed similar results to the DMC, with the coarse fraction ($-2 + 0.125$ mm) high density bypass ($T_1$) and $E_p$ having the greatest impact at a DMC cutpoint of 1.40 RD, but negligible impact at a cutpoint of 1.60 RD. The impact of the spirals variables was less than that of the DMC, but was more than would have been expected at a 1.40 cutpoint due to the relative masses treated by the two circuits. This is because the DMC product is reduced at low cutpoints so the spirals product is more significant and the higher product ash of the spirals magnifies its impact on the total yield.

In general, this study has shown that the importance of individual performance parameters is highly dependent on coal type and separation cutpoint. For this reason, coal industry stakeholders need to be aware that, when setting goals for efficiency parameters, their materiality must be considered and the goals, and related penalties, adjusted accordingly.

## 5. References

[1]  Atkinson B., 2015, "ACARP Project C23043: Product Coal Loss due to Inappropriate Focus on $E_p$".

[2] Atkinson B., O'Brien G., and Swanson A., 2016, "An Update on Washability Prediction using CGA", Proc. Sixteenth Australian Coal Preparation Conference (Mathewson D. and Eschebach D.), Wollongong, NSW, pp 265–285.

[3] Swanson M., Mackinnon W., and Swanson A., 2018, "Quantifying the Materiality of Efficiency Parameters", Proc. Seventeenth Australian Coal Preparation Conference (Vince A.), Brisbane, Qld, pp 303–318.

# 10

# The intelligent coal preparation plant

Michael O'Brien[1], Nerrida Scott[2], Shenggen Hu[1]

*[1]Queensland Centre for Advanced Technologies, Pullenvale, Queensland, Australia*

*[2]Indooroopilly Centre, Indooroopilly, Queensland, Australia*

**Abstract:** The Australian Coal Association Research Program (ACARP) has supported some projects over the last ten years in CSIRO with the ultimate aim of producing an "Intelligent Coal Preparation Plant". An Intelligent Plant is one that recognises operating issues and takes steps to avert poor operating conditions based on the measurement of certain parameters. The dense medium cyclone (DMC) circuit, for instance, involves the monitoring of the underflow and overflow medium densities using unique non-nucleonic online instrumentation developed by the CSIRO and the JKMRC. These instruments were installed in a coal preparation plant and the data remotely accessed and recorded along with the current plant operating conditions. This type of data has previously not been available to operators.

This paper describes these instruments and how the data can be used to optimise the operation of the dense medium cyclone. A scenario based on the data collected is examined, and the dynamic relationship with this data and plant process changes such as changes in feed medium density is explored. In the case of feed medium densities, the data has shown that in some circumstances changing the feed medium density using a "wrong" method can lead to unstable cyclone operation, higher cut points and thus loss of saleable coal.

**Keywords:** Coal, coal preparation, dense medium cyclones

## 1. Introduction

The idea of a metallurgist/engineer in an apartment by the beach, remotely controlling and operating a coal preparation plant has long been a dream of operators worldwide for many years, while remote operating centres are now common, there is still a need for an in-situ metallurgist/engineer to insure optimum operation of cleaning circuits.

Dense medium cyclone's (DMC's) have changed little since the invention of the Dutch State Mines (DSM) dense medium cyclone first patented in 1942. Despite considerable research in this area, it is unlikely that another separation device as reliable and efficient in capacity, energy and in the separation of coal, will be "discovered" in the near future.

Instrumentation around the DMC circuits is typically kept to bare a minimum due to incorrect cost assumptions and the lack of suitable cost-effective instruments to measure the DMC separation process. Consequently, there is limited knowledge of what is happening in the DMC separation process in near real time. Typically, the cyclone pressure, medium to coal ratio and correct medium density are the only online instrumentation used to control the DMC. The separation density is determined from the washability of the coal being processed, and the ash value of the product coal required by the client is adjusted based on laboratory analysed two hourly ash value determinations. In some cases, an ash gauge may be used with some success on the product coal.

The widely used quote sometimes attributed to Peter Drucker "If you can't measure it you can't manage it" holds true in this case, but we should also add:

- You cannot control it.
- You cannot understand it.
- You cannot improve it.
- You lose money.

## 2. The intelligent coal preparation plant

Coal preparation research in CSIRO has, with support from the Australian Coal Association Research Program completed a number of projects under "The Intelligent Plant" umbrella (Firth, 2008). These projects have progressed over the last 10 years and have looked at the factors that affect the efficient operation of the dense medium cyclone and the required measurements needed to understand, control and optimise the operation of the DMC. They have led to new instrumentation to monitor the DMC operation in near real time and to observe and measure the dynamic changes that may occur with changing feed types, changes in the correct medium density and other plant operations such as the operation of the correct medium bleed valve to the magnetic separators and water and magnetite addition.

The Intelligent Coal Preparation plant concept leads into a wider challenge project which looks at the complete integration of the whole mine with instrumentation and tracking of coal quality from the pit or coal face to the port.

### 2.1 Online dense medium cyclone yield

CSIRO has developed and commercialised a portable screen motion analyser (O'Brien et al., 2006), which uses accelerometers to measure the displacement of the screen in three dimensions, formed the basis for online DMC yield measurements. Accelerometers were used to measure the displacement of a desliming screen (feed to the DMC) the DMC product screen and the DMC rejects screens. Displacement is proportional to the mass on the screen, and with calibration, these instruments were used to provide for the first time online yield of the DMC circuit. Figure 1 shows the output from the mass meters over 24 h and the calculated yield.

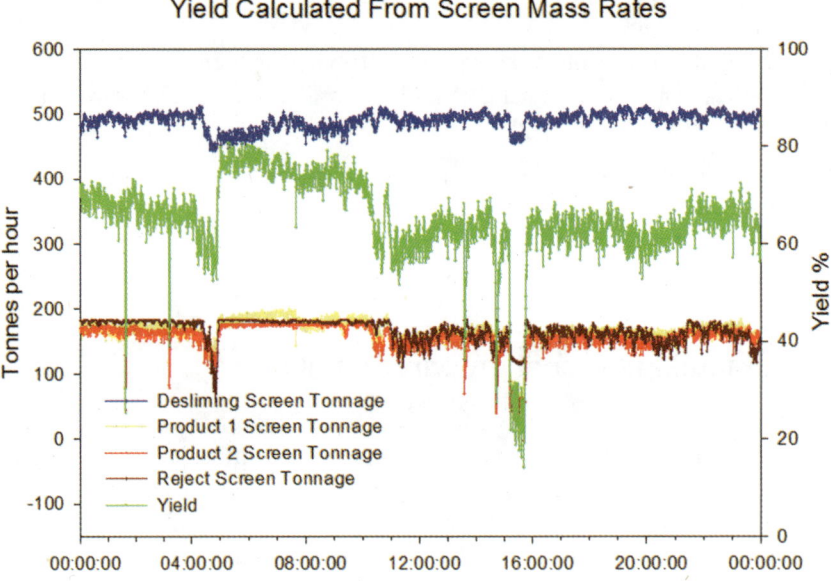

**Figure 1:** Outputs from the mass meters on the desliming and drain and rinse screens showing the tonnages on the screens and the calculated DMC yield in real time

## 2.2 The medium

Stability of the medium is an important factor in the DMC in relation to separation size and efficiency. The size distribution of the magnetite used is important for stable DMC operation as at low densities of separation the magnetite can excessively separate leading to unstable DMC operation. The use of ultra-fine magnetite is recommended in these cases (Crowden et al., 2013). The amount of clays present can also affect the stability of the medium, while normally thought of as contaminants at high densities of separation, at low densities of separations the clays present can aide in stabilising the fine magnetite particles (O'Brien and Firth, 2008). At very high clay concentrations viscosity issues could prevent efficient coal separation leading to coal loss to the reject. Fine coal in the medium can also have a similar effect as the clay however the lighter density of this fine coal reduces its effectiveness. Figure 2 shows the results of tracer tests for a magnetite only medium and medium with 20% by weight of clay added. This experimental work was on a 150 mm diameter DMC at CSIRO's pilot plant in Queensland.

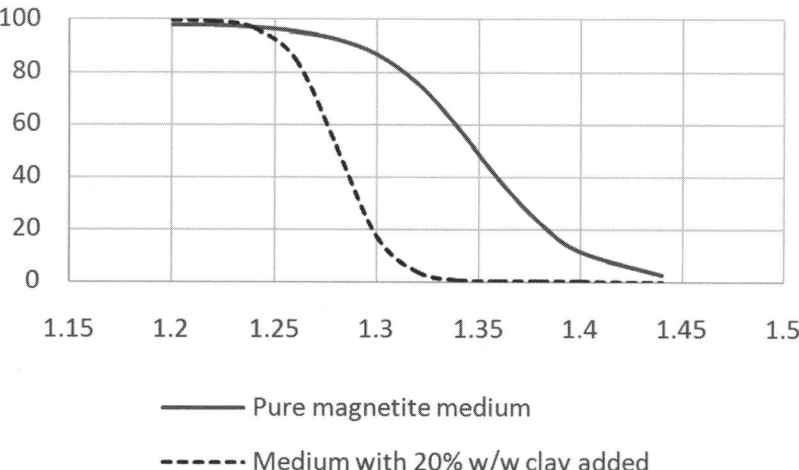

# Effect of the addition of clay into the medium at low separation densities

——— Pure magnetite medium

- - - - -· Medium with 20% w/w clay added

**Figure 2:** Plot showing the partition curves obtained from experiments at low densities of separations and different concentrations of clays in the medium

The DMC medium differential that is, the difference in density between the underflow medium and the overflow medium is an effective indicator of stable DMC operation. As a general rule for good cyclone operation, this differential should remain between 0.2 and 0.4 RD units. However, there are a number of documented cases where the differential is outside of the optimum range with the operation of the DMC uncompromised. Figure 3 shows modeled data based on experimental work at CSIRO's pilot plant using a 150 mm diameter DMC configures to Dutch State Mines specifications.

The figure shows how the amount of non-magnetic material in the correct medium solids effects the density of separation (RD 50) and the differential and indicates that a minimum of 15–20% non-magnetics should be present at low correct medium densities (RD 1.4 and below).

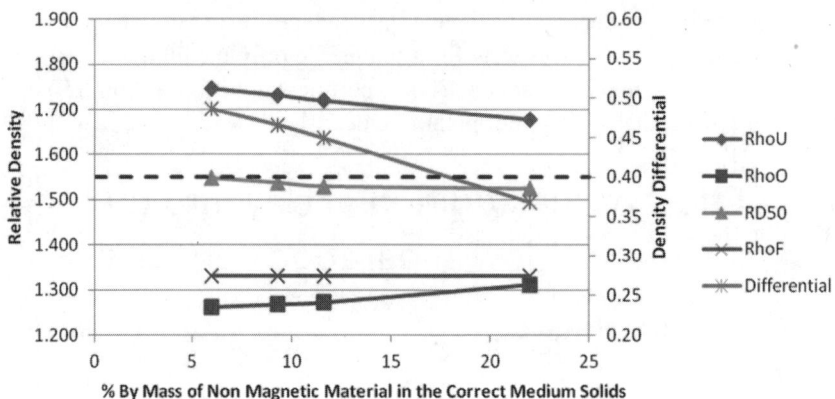

**Figure 3:** Calculated modelled data for the D50 for the different flow streams and the differential

## 2.3 Online measurement of DMC underflow and overflow medium

The measurement of the overflow and underflow medium densities is not currently measured online in the coal industry, samples are normally taken, and the density determined using a Marcy Gauge or other similar methods however this occurs on an infrequent basis. Online measurement of the DMC differential is used in the operation of DMC in the diamond and iron ore industries where a high, medium viscosity, as indicated by a low differential is detrimental to DMC yield. The diamond and iron ore industry have used Ramsey Type coil, and De-Beers developed the DebTech dense medium controller, these instruments like nucleonic gauges need a full pipe for accurate measurement and therefore require new pipe work to be installed for a proper operation which can be a challenge. In the coal industry, a build-up of clays can result in a viscosity which is too high for optimum yields. However this is not as common as the unstable operation due to a high differential particular at low densities of separation. Researchers have used nucleonic gauges online in the overflow and underflow medium to optimize the operation of DMC circuits. Addison (2010), in his master's thesis used nucleonic gages on the overflow and underflow medium to improve the density control, while the gauges were in place for the duration of his thesis work they were not integrated into the plant operating system and were removed at the completion of the test work. Restarick and Krnic (1991), set up a small cyclone circuit and carried out experiments from sampling the underflow and overflow densities. Peter Holtham in 2003 started developing a sensor to measure the overflow density as a feed into control models for DMC operation and at about the same time the CSIRO started developing instrumentation to measure the density of the underflow and overflow mediums. These projects resulted in a series of joint projects, Firth et al. (2010) which lead to the development of the CSIRO electrical impedance spectrometer.

The electrical impedance spectrometer (EIS) measures the impedance of the slurry over a spectrum of frequencies (Hu and Firth, 2007). The resulting impedance spectrum is sensitive to the amount of magnetite in the slurry and is not sensitive to changes in conductivity and temperature normally found in the coal preparation plants medium circuits. Thus for the first time in an Australian plant, the densities of the underflow and overflow mediums could be tracked in real time and related to dynamic changes that may be occurring within the plant.

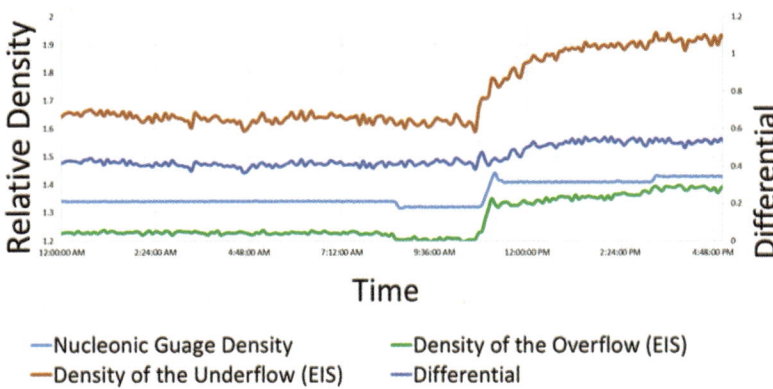

**Figuro 4:** Onlino data from tho EIS inotrumonto in a plant ohowing a situation where the differential is high due to low non-magnetics

Figure 4 shows the data collected from the EIS density instruments in the underflow and overflow medium streams in a plant and the output from the plant's nucleonic gauge. The differential was calculated and plotted as the dark blue line. About 11 am on this day the plant swapped to a different seam and changed the density, at the time the correct medium sump was nearly full and as the magnetite is added as a slurry with significant amounts of water the operators decided to remove water from the circuit by opening the bleed valve, this would reduce the level of the correct medium sump and increase the density. While the method worked and prevented a spillage, it caused the underflow density to increase significantly for a long period (orange plot), with the increase in underflow density came an increase in the differential and the possible loss of saleable coal due to an unstable medium. The increase in the differential was due to loss of clays as a result of the excessive opening of the bleed valve; the clays followed the water resulting in unstable medium conditions. Without these instruments, in place, this situation would have gone unnoticed by the operators.

### 2.3.1 Measurement of non-magnetics in the correct medium

As previously mentioned the non-magnetic material present in the correct medium can help stabilise the medium and prevent poor DMC operation. To manage the non-magnetics, we first needed a method to measure on-line the amounts of non-magnetic material present. While EIS is capable of this measurement, more research and a different spectrometer with an extended range would be required before confidant estimates can be made. For this test work, a Ramsey Coil type instrument was employed in the correct medium circuit to measure the magnetite only concentration. This instrument consisted of a measurement coil through which the medium passed and a reference coil to help reduce the noise in the system, and to minimise the effect of temperature. The output from the plant's nucleonic gauge was used to provide an overall density including any non-magnetics that may be present and these outputs used to calculate the percent non-magnetics in the medium solids (Eqs. 1 and 2).

$$VFNM = -\frac{\left(1-\left(\dfrac{D_{mag}-1}{DMS-1}\right)-VFNM\right)-Dnu-DMS\left(\dfrac{D_{mag}-1}{DMS-1}\right)}{DNMS} \quad (1)$$

Which solves to:

$$VFNM = \frac{D_{nu}-D_{mag}}{DNMS-1} \quad (2)$$

Where VFNM is the volume fraction of non-magnetics in the medium solids, $D_{mag}$ is the density of the magnetite slurry in from the Ramsey Coil device, DMS is the density of the magnetic solids, $D_{nu}$ is the density of the medium from the nucleonic gauge and the DNMS is the density of the non-magnetic solids present (clay).

## 2.4 Prediction of the partition curve and washability

The use of the online measurement for the underflow and overflow mediums together with the plant correct medium density from the nucleonic through models can be used to calculate the separation size (D50) and Ep at which the cyclone is operating (Wood, 1990), the settling model from Firth (2009) and Hu and Firth (2010). Figure 5 shows the partition curves obtained from float sink analysis of samples and that calculated from the medium densities.

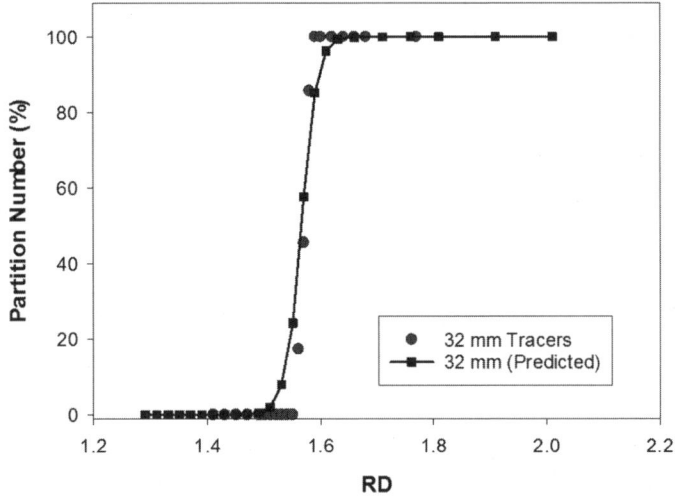

**Figure 5:** Comparison between partition curves determined from float sink analysis of samples and that predicted using the medium densities (Hu and Firth, 2010)

When the yield of the DMC is known (from mass rate instrumentation or weightometers) and there are perturbations in the yield (can be obtained from adjusting the correct medium density by 0.06 RD for a short period of time), then the washability of the coal being processed can be calculated using Eq. (3):

$$Y = \sum_i w_i \, \text{Pn}_i \tag{3}$$

where $w_i$ is the weight percentage of coal feed in the $i$-th density increment (i.e., the washability) and $\text{Pn}_i$ is the partition number of $i$-th density increment.

## 2.5 Online control of the dense medium cyclone circuit

A plant located at west of Brisbane had the mass monitors installed on the product reject and desliming screens. EIS was installed in the drain sections of the overflow and underflow medium streams, a Ramsey Coil type instrument on the correct medium before the nucleonic gauge. A control circuit was installed where the tailings thickener underflow could be returned to the correct medium using a control valve under software control. The software control was designed with a PI controller that would open or shut the valve from the thickener underflow to maintain a density differential between 0.2 and 0.4 and a non-magnetics concentration of 20% by mass as calculated from the inputs from the Ramsey Coil type meter and the nucleonic gauge. Should the differential fall below 0.2 the bleed valve could be opened to remove the excess build up of clays. Figure 6 shows the user interface developed for this test work.

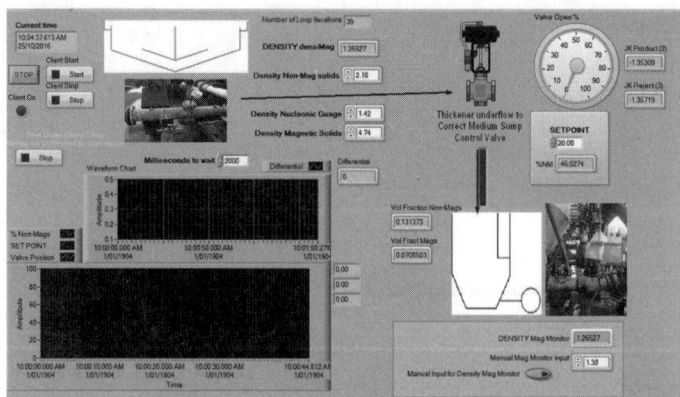

**Figure 6:** User interface for the control system to control the addition on non-magnetics

The system was operated for 12 h with approximately 4 h on and 4 h off repeated. Figure 7 shows the data from the day.

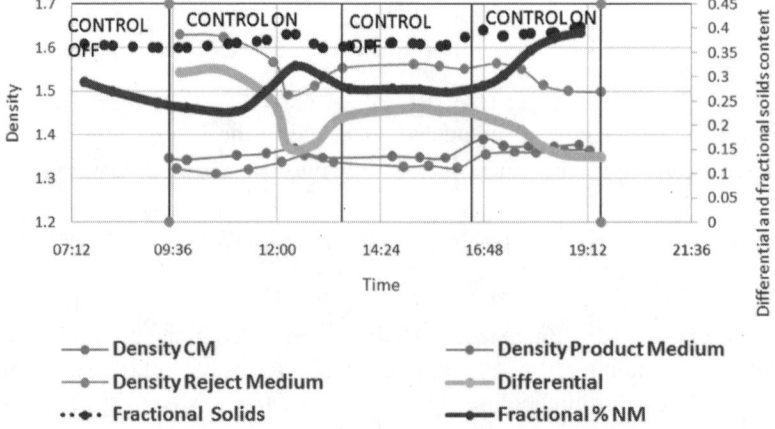

**Figure 7:** Non-magnetic concentration in the medium and the differential over a 12 h period with the control system on and off

When control was switched on the percent non-magnetics in the medium as shown by the green line in Fig. 7 rose from approximately 25% to approximately 34% the set point for this period for the non-magnetics concentration was 33%. A high figure was chosen as the non-magnetics were already high on this day due to the previous coal type processed over the night with higher medium density. The chart clearly shows that as the "control on period" progressed the non-magnetic concentration rose as shown by the green line and the differential dropped (yellow line). When the control system was turned off the differential, and the non-magnetic concentration remained stable, and when the control was again turned on, and the set point for the differential increased to 0.4, the non-magnetic concentration rose, and the differential dropped.

## 3. Conclusion

By using instruments to measure the yield and the underflow and overflow medium densities this paper shows that the concept of remotely operating a DMC circuit is now becoming a reality. These instruments have allowed the operators to dynamically monitor the stability of the DMC and take corrective action. The first step towards automation has been taken with a control loop that maintains the medium stability through the use of thickener underflow addition and opens the bleed valve should the medium be too viscous. Some added control will need to be applied with the bleed valve under automatic control so that water balance and sump levels can remain stable. The use of these instruments in multi-loop circuits will also provide the operators with the D50 from each of the loops allowing corrections for offsets to be applied.

## 4. References

[1] Addison, C. B. (2010). Development of a Multi-stream Monitoring and Control System for Dense Medium Cyclones. Master of Science, Virginia Polytechnic Institute and State University.

[2] Crowden, H., B. Atkinson, C. Clarkson, B. Firth, M. O'Brien, D. Wiseman and C. Wood(2013). Dense Medium Cyclone Handbook. Newcastle, NSW, Australian Coal Preparation Society.

[3] Firth, B. (2008). The Intelligent Plant ACARP Project C1109. Brisbane, ACARP.

[4] Firth, B. A. (2009). Dense Medium Cyclone Control – A Reconsideration. International Journal of Coal Preparation and Utilization 29 (3): 112–129.

[5] Firth, B., P. Holtham, M. O'Brien, S. Hu, R. Dixon, A. Burger and G. Sheridan (2010). ACARP Project C17037 Joint Evaluation of Monitoring Instruments for Dense Medium Cyclones. Brisbane, ACARP.

[6] Hu, S. and B. Firth (2007). On-line Monitoring and Control of DMC Separation Density and Efficiency. Brisbane: 49.

[7] Hu, S. and B. Firth (2010). Prediction of Operating Performance of Dense Medium Cyclones from Medium Densities. Proceedings of 13th Australian Coal Preparation Conference, Paper 4C. B. S. Atkinson. Mackay, Australian Coal Preparation Society.

[8] O'Brien, M., I. Hutchinson and M. Kochanek (2006). Application of the Portable Screen Motion Analyser to Measure Plant Performance. ACARP. Brisbane, CSIRO.

[9] O'Brien, M. and B. Firth (2008). ACARP Project C15053 Quality of Dense Medium.

[10] Restarick, C. J. and Z. Krnic (1991). The Effect of Underflow/Overflow Ratio on Dense Medium Cyclone Operation. Minerals Engineering **4**(3/4): 263–270.

[11] Wood, C. J. (1990). A Performance Model for Coal-washing Dense Medium Cyclones, Ph.D. Thesis, University of Queensland. Brisbane.

# 11

# The study and modeling of coals and cokes behavior with applying thermal analysis technique

Felix Yu. Sharikov and Vladimir Yu. Bazhin

*Saint Petersburg Mining University, Saint Petersburg, Russia*

**Abstract:** The methodology and results of coals and cokes analysis and oxidation kinetic study for energy production application and storing-transportation aspects are presented. Three different coals – two from Kuznetsk basin (Russia) and one from RSA – were tested. DSC (differential scanning calorimetry) and thermal analysis techniques were applied as the main kinetic research technique. Traditional characterization of coals and cokes for their calorific value (heat of combustion), volatiles, fixed carbon ratio, moisture and ash content is far not enough to propose an optimal scheme of utilization or even the best combustion mode in a definite furnace. It was concluded that various coals may be even more unlike at burning them in the same furnace under similar conditions than it might be supposed basing only upon their traditional analysis data. One should also take into account oxidation kinetics for a definite coal batch that in turn may also be a function of storing conditions and coal distribution for pieces size and powder. Kinetic modeling of the combustion process and further simulation of the furnace operation are necessary to determine the optimal conditions for running a definite furnace or gasification unit and finally to increase the efficiency of coal utilization.

**Keywords:** Coal and coke analysis, proximate analysis, thermal analysis, oxidation stability, kinetic modeling, coal storing and transportation

## 1. Introduction

Proximate analysis and direct measuring the combustion heat for coals, cokes, and other carbon materials with applying thermal analysis technique become more popular nowadays. It includes the determination of moisture, volatile compounds, fixed carbon and ash in coals, cokes and other carbon materials. Calorific value of a coal can also be estimated in some cases. It is used to establish or prove the rank of a coal or other carbon material, to show the ratio of combustible to incombustible constituents, to provide the information for a buying/selling procedure, and otherwise to evaluate coal for various purposes. Thus, it is important to both the suppliers and users of coal to have a rapid, accurate and reliable instrumental procedure for obtaining the proximate analysis. Traditionally, the various proximate analysis determinations are strictly fixed and involve heating the sample to constant weight under

ASTM (or other national/international standard) specified conditions [1,2]. These determinations, however, are time consuming and require a significant amount of laboratory equipment and personnel. An alternative method for coal and coke proximate analysis is thermogravimetric analysis (TGA) [3]. This thermal analysis technique also involves heating the coal sample to constant weight under specified conditions, but it does substantially reduce the proximate analysis time and the equipment necessary for testing because of a smaller sample size and rapid temperature and atmosphere control. And application of thermal analyzers has become legal, at least under ASTM regulation [3–5]. One can get the basic characteristics of a coal (moisture, volatile compounds, fixed carbon, and ash) in one run, with one small sample. At the same time this approach may possess some potential uncertainty related to that small sample size and high temperature scanning rate that is usually recommended to reduce the analysis duration time to a maximum possible value. These characteristics should be optimized and it is strictly recommended to perform exactly the same conditions at analyzing a definite batch of samples. And from our experience we do not recommend to save time due to applying a high scanning rate. Moreover, we advise to perform at least two runs for one sample at various atmospheres as in this case we can estimate the coal or coke calorific value as well.

Generally thermal analysis technique for coals and carbon materials is not restricted with only proximate analysis traditional determinations. It is a powerful tool for a more profound study of a definite coal including a kinetic study of coal transformation process. One can mention coal coke production from anthracite or semi-anthracite, coke stability estimation at higher temperature both in an inert and in oxidizing atmosphere, oxidation kinetics study that is necessary, e.g., for coal gasification process optimization. Kinetic modeling with applying TG–DTG (thermogravimetry–derivative thermogravimetry) or DSC (differential scanning calorimetry) experimental data makes it possible to develop a corresponding mathematical model and select optimal process conditions for a definite coal transformation.

## 2. Literature overview

Coal is a complicated natural object. A number of articles and reports are devoted to coals and cokes study with applying TG–DTG and DSC techniques [9–12]. Proximate analysis of coal, coke or shale with applying various thermal analyzers has become common at last [3–5]. It should be emphasized that these instrumental techniques can successfully substitute the traditional analytical procedures in most cases, but they can also supplement them and make it possible to better characterize this object in those aspects where the well established procedures are not working. First of all it is referred to kinetic study and modeling of oxidation/combustion process, to the study of anthracite calcination to obtain coal coke, to the thermal safety evaluation of coal and coal dust storing conditions. Qualitative and quantitative identification of a coal is also possible on the base of its TG–DSC "fingerprint" under specified conditions. And the development of a modern process of low quality coal utilization based on its prior gasification – also needs the exact experimental data and corresponding mathematical models of the gasification reactions [6–8].

The question is still open with the coal calorific value exact determination according to ASTM, ISO, GOST standards, where bomb calorimetry seems to retain its positions up to the moment. The problem is that specified experimental conditions that are run in a calorimetric bomb under oxygen pressure usually cannot be exactly reproduced with a common DSC instrument. But there is a great progress in DSC sensors and accessories development, and now the experiment can be directly done under oxygen pressure up to 100 bar with a special HP DSC version (Mettler TOLEDO, TA Instruments, some others), or with corresponding HP crucibles. Anyway combination of DSC and bomb calorimetry techniques or application of standard samples makes it possible to get the exact results in an optimal way.

## 3. Materials and methods

TG–DSC and DSC techniques were applied for the proximate analysis and oxidation/combustion reaction study for various coals. Kinetic modeling of a combustion reaction with applying formal model was done.

### 3.1 Reagents

Initial coals were from the Kuznetsk region basin (Russia) and the Waterberg Coalfield basin (Limpopo, RSA). The following initial coals were studied:

(1)   coal 1 (cannel coal, low-ash), Kuznetsk, Russia;

(2)   coal 2 (fiery coal, low ash), Kuznetsk, Russia;

(3)   coal 3 (lean coal, high-ash), RSA.

### 3.2 Thermal analysis

Analytical and kinetic measurements were performed using a Labsys evo thermal analyzer and DSC-131 evo differential scanning calorimeter

(SETARAM Instrumentation). Linear heating mode was applied with the heating rates $\beta$ = 10.0, 3.0 and 2.0°C min$^{-1}$. $N_2$ dynamic atmosphere (flow rate 20 mL min$^{-1}$) and air static and dynamic atmospheres were applied.

## 3.3 Data acquisition and primary processing

*Calisto* program package (v 1.086, AKTS AG) was applied for running the experiment, data collection, and initial processing. *Experiment* program package (v 4.31, CISP Ltd) was applied for data preparation for kinetic analysis.

## 3.4 Kinetic analysis of the experimental data and modeling

*ReactOp Cascade* program package (v 3.20, CISP Ltd) was applied for data kinetic analysis.

## 4. Results and discussion

### 4.1 Experimental study results

A number of commercial coals of various kinds (see Section 3.1) were studied with applying TG–DSC technique. The first two are commercial low-ash coals for energy production, but they are very unlike for their characteristics. The third is a high-ash lean coal that cannot directly compete with the first two for energy production but it is an appropriate model sample to compare various ways of its possible utilization. At first the selected coals were compared for their oxidative stability in air under comparable conditions with applying TG/DTG–DSC technique. Temperature intervals of oxidation were found and ash quantity was established. Experimental results are given in Figs. 1–4. At first we recommend the runs in air or oxygen (see Figs. 1, 3 and 4) where we get the information of moisture, combustible fraction, and ash. Peculiarities of every coal oxidation also are fixed at this step. Coal 1 is subject to solid-state oxidation at lower temperature prior to its combustion – see weight increasing up to 102.3% of the starting value after moisture removing to 98.5%. Composition and calorific value of this coal may be changed at storing to a greater extent than that for other coals. And a special attention should be paid to its transportation and dust storing conditions. Coal 2 has a considerable amount of moisture and light volatiles, but it is not oxidized in solid state at low temperature as coal 1. Its composition and calorific value may also be not constant at prolonged storing and transportation, but for another reason. Coal 3 seems to have less problems with its storing and transportation for the reasons mentioned above, but it is obviously a high-ash coal and its calorific value per total mass is rather low.

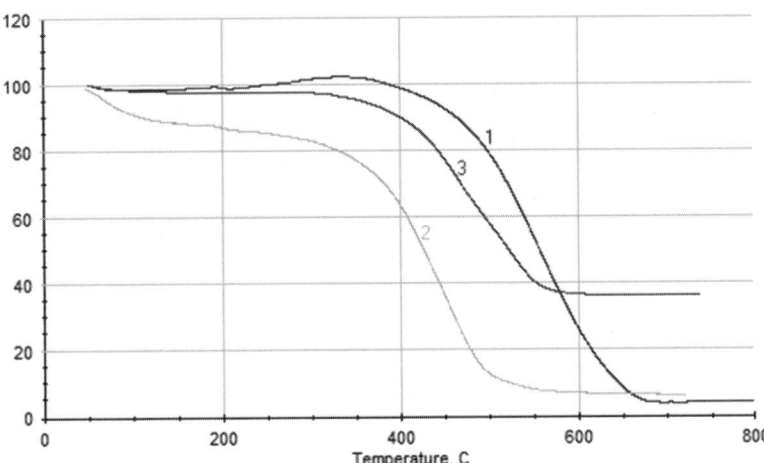

**Figure 1:** Oxidation in air. Weight loss (TG, wt.%) as a function of T (°C) for the cannel coal (1), fiery coal (2), and lean coal (3) under similar conditions. Linear heating, β = 10°C/min, static air. Moisture: 1.5% (1), 8.5% (2), 2.0% (3); combustible fraction: 94.0% (1); 85.2% (2); 62.1% (3); ash: 4.5% (1); 6.3% (2); 35,9% (3)

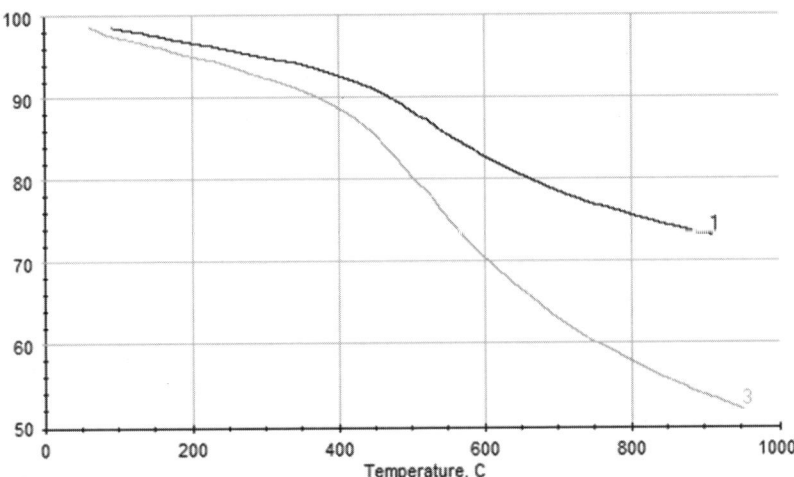

**Figure 2:** Calcination of coals in an inert atmosphere. Weight loss (TG, wt.%) as a function of T (°C) for the cannel coal (1) and lean coal (3) under similar conditions. Linear heating, β = 10 °C min⁻¹, N₂ flow (20 mL min⁻¹). Moisture: 1.5% (1); 2.0% (3). Volatiles: 26.5% (1); 45.9% (3)

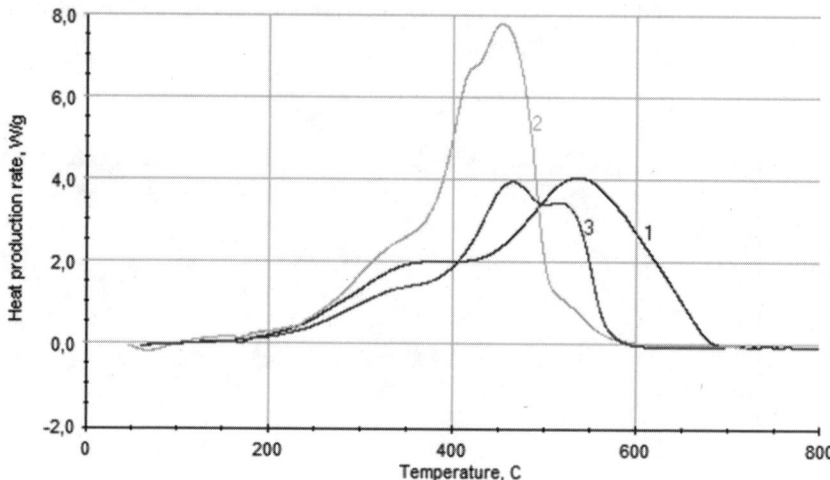

**Figure 3:** Oxidation in air. Heat production rate as a function of T (°C) for the cannel coal (1), fiery coal (2) and lean coal (3) under similar conditions. Linear heating, β = 10 °C/min, static air. Heat is referred to the coal total mass. Relative calorific values in this case are 1.00 (1); 1,075 (2); and 0.711 (3). Calorific value for (3) is 11100 ± 120 kJ kg$^{-1}$

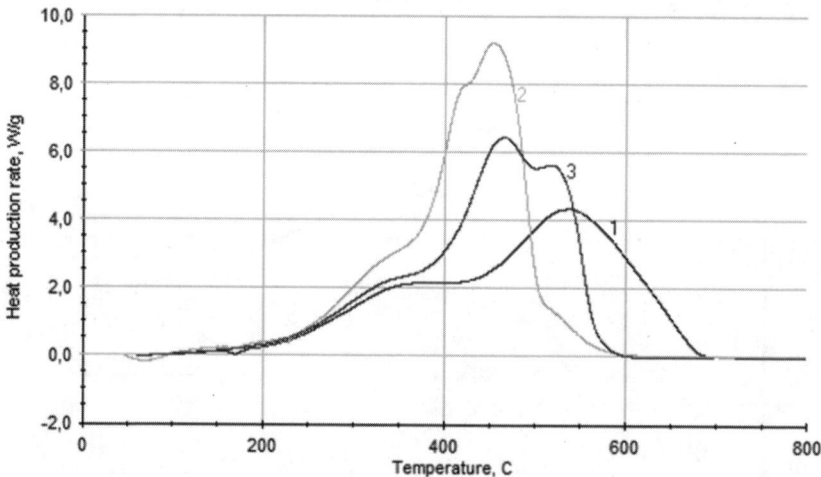

**Figure 4:** Oxidation in air. Heat production rate as a function of T (°C) for the cannel coal (1), fiery coal (2), and lean coal (3) under similar conditions. Heat is referred to the coal combustible fraction. Relative calorific values in this case are 1.00 (1); 1,20 (2); and 1.08 (3). Calorific value for (3) is 17850 ± 200 kJ kg$^{-1}$. Linear heating, β = 10 °C min$^{-1}$, static air

Comparison of various coals with thermal analysis technique should be done strictly under the same conditions and to get the correct interpretation of results one needs to have a deep understanding of the method and of the subject. In Fig. 5 there is the same coal 2, but in a powder form (1) and in "one-piece" form (2). Kinetics and peculiarities of combustion (and even relative calorific values) are so unlike as if there were two different coals. This is typically namely for fiery coal as it contains a lot of gases (methane and light hydrocarbons) and volatiles within it and the sample specific surface value in each run is the key parameter to explain this behavior. Optimal grain-size distribution may be recommended in this case for energy production in order to avoid the invaluable endothermic reactions of coal decomposition in solid state prior to its combustion. Traditional coal characterization techniques usually do not suppose this approach.

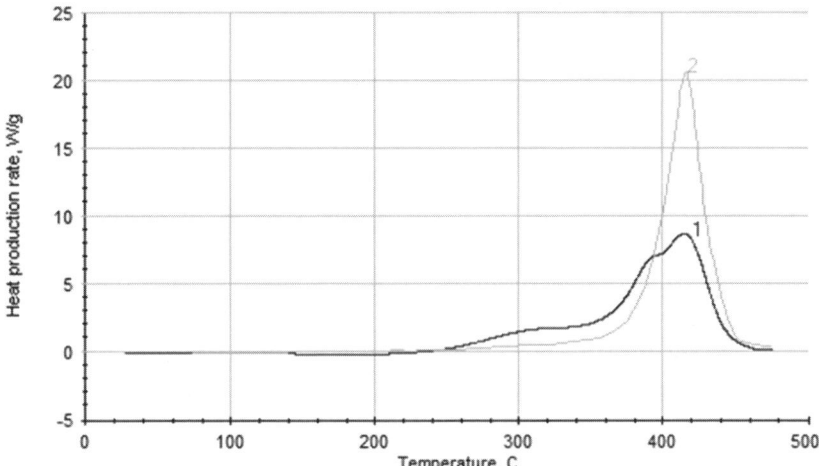

**Figure 5:** Oxidation in air. Heat productions rate as a function of T for two various samples of the same fiery coal (2) under similar conditions: 1, powder sample; 2, one-piece sample. Heat is referred to the coal total mass. Relative calorific values in this case are 1.00 (1) and 1.177 (2). Linear heating, $\beta = 3$ °C min$^{-1}$, static air

Kinetic modeling of heat generation at coal combustion is vital to optimize the mode of an industrial furnace functioning in case of energy production or the mode of a pilot or industrial gasification unit functioning in case of syngas production. This direction of coals utilization – especially for the high-ash and low calorific value coals – becomes more popular at the moment [6,7] as a considerable progress has been reached in gasification unit constructions and in evaluating the possible routes of a definite coal utilization on the base of smart modeling technologies [6].

Heat generation rate curves of powder coal samples combustion (see Figs. 3 and 4) are not uniform. There are several sub-stages (or "shoulders") – from 2 (coal 1) to 4 (coal 2). This is connected mainly with volatiles non-uniform evolving and oxidation in course of combustion. Multi-stage formal models make it possible to find an exact description of experimental data in every case. In case of mainly 1-stage oxidation (as curve 2 in Fig. 5) a 1-stage formal kinetic model may be used to fit the experimental data. An example of applying such a model is given below. The generalized chemical reaction implemented in the model is:

$$Coal + O_2 \rightarrow PROD$$

The reaction is exothermic, coal is presented as 1 generalized component "Coal" that gives 1 generalized combustion products "PROD". The results are given in Fig. 6.

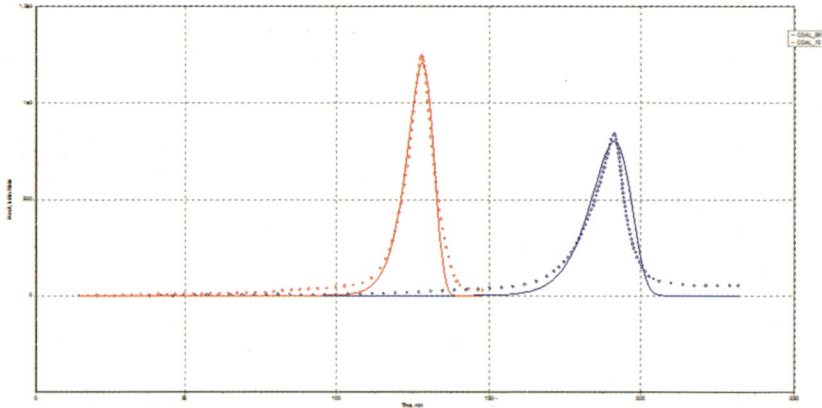

**Figure 6:** The results of kinetic analysis for commercial coal (2) combustion in air dynamic atmosphere. Linear heating, $\beta = 3$ (red) and 2 °C min$^{-1}$, air flow (20 mL min$^{-1}$), fraction of small pieces (2–4 mg).

Experimental (points) and calculated (lines) heat production rate curves (kJ m$^{-3}$ min$^{-1}$) as a function of time (min). 1-stage N-order formal model

Equations are written through conversions (e.g., from 1 to 0 for "Coal"), oxygen is in excess, hence the formal order for it is equal to 0. TG and DSC experimental data are the responses for solving the inverse kinetic task [13,14]. There is an adequate description of the main heat production rate stage.

One or two more stages should be added to this one if we want to describe the whole heat production rate curves at the beginning and in the end perfectly. But their relative inputs will be small enough and the 1-stage model may be used for coal combustion modeling under these conditions. The response for ash formation can also be added as we know the final ash content.

The reaction model developed may be further introduced to the user-specified mathematical model of a definite industrial furnace or gasification unit. In case of a gasification unit modeling one should introduce the oxidizing agent quantity that is necessary for syngas generation.

Chemical equations:

1. Coal + $O_2$ → PROD.

List of the species:

1. C1: Coal.
2. C2: $O_2$.
3. C3: PROD.

Rate of reaction steps:

r1 = K1*C1*C2.

Rate of reaction of species:

R(1) = −r1.
R(2) = −r1.
R(3) = + r1.

| Ea (kJ mol⁻¹) | ln $K_0$ (min⁻¹) | ΔH, (kJ kg⁻¹) | $N_1$ (Coal) | $N_2(O_2)$ |
|---|---|---|---|---|
| 266.11 | 44.85 | 13936.0 | 1.00 | 0 |

## 5. Conclusion

The results obtained in course of the study indicate the effectiveness of applying thermal analysis technique together with mathematical modeling to the task of various coals basic and advanced analysis. The coals may be identified and compared to select an optimal route of their further utilization. Compositional identification of a coal may be done in 1 or 2 runs.

Kinetic modeling of weight loss and heat generation processes in course of coal sample oxidation/combustion makes it possible to select an optimal mode of its combustion in an industrial furnace or estimate an appropriate gasification mode for the coal taking into account its particle-size distribution.

## 6. Acknowledgment

Kinetic measurements and modeling were performed in St. Petersburg Mining University with the equipment and software obtained due to the Development Program "National Research University" (2011–2013) and in course of the Ministry of Science and Higher Education Program "Support of Fundamental Research Works" (2014–2016, 2017–2019).

The authors are grateful to the Mining University Authorities for creating the conditions, support, and attention to the work.

## 7. References

[1] ASTM D3172 Standard Practice for Proximate Analysis of Coal and Coke. Ann. Book of ASTM Stand., Vol. 14.04, ASTM, West Conshohocken, PA.

[2] ISO 1171 as well as BS 1016-104.4. Methods for analysis and testing of coal and coke. Proximate analysis. Determination of ash content.

[3] ASTM E1131 Standard Test method for Compositional Analysis by Thermogravimetry. Ann. Book of ASTM Stand., Vol. 14.02, ASTM, West Conshohocken, PA.

[4] ASTM E1641 Standard Test Method for Decomposition Kinetics by Thermogravimetry. Ann. Book of ASTM Stand., Vol. 14.02, ASTM, West Conshohocken, PA.

[5] Sadek F.S., Herrell A.Y. Proximate analysis of solid fossil fuels by thermogravimetry. American Laboratory, March (1984), P. 75–78.

[6] Litvinenko V., Mayer B. Syngas Production: Status and Potential for Implementation in Russian Industry. Springer, 2018, 161 p.

[7] Higman C., Tam S. Advances in Coal Gasification, Hydrogenation, and Gas Treating for the Production of Chemicals and Fuels. Chem. Rev. 114 (2014), P. 1673–1708.

[8] Xu J., Yang Y., Li Y.W. Recent development in converting coal to clean fuels in China. Fuel, 152 (2015), P. 122–130.

[9] John W. Cumming, Joseph McLaughlin. The thermogravimetric behavior of coal. Thermochim. Acta, 57 (1982), P. 253–272.

[10] Serageldin M.A., Pan W.-P. Thermochim. Acta, 76 (1984), P. 145–160.

[11] Richard L. Fyans. Rapid Characterization of Coal by Thermogravimetric and Scanning Calorimetric Analysis. Presentation at the 28th Pittsburgh Conference in Cleveland, Ohio, March (1977).

[12] Paul Baur. Thermogravimetry speeds up proximate analysis of coal. Power, March (1983), P. 91–93.

[13] Sharikov Y.V., Beloglazov I.N. Modeling of Systems. Part 1. Saint-Petersburg State Mining University. SPb. 2011: P. 43–89.

[14] Sharikov Y.V., Beloglazov I.N. Modeling of Systems. Part 2. Saint-Petersburg State Mining University. SPb. 2012: P. 27–74.

# A study of the effects of thermal shocks on liberation characteristics of high coal ash particles

Soni Jaiswal, A K Mukherjee, A K Bhatnagar

*Tata Steel Ltd., Jamshedpur, India*

**Abstract:** The influence of thermal shocks on the liberation characteristics of a by-product from a coking coal preparation plant, known as coal middlings were investigated. Coal middlings are characterized by high ash content (> 45%), low carbon content (<35%) and presence of highly locked mineral particles. Currently they are used in power plants despite having coking coal fractions. So, to extract the coking coal fractions, improved liberation of these particles is required at optimum size. For this purpose, two kinds of thermal shocks viz. microwave pre-treatment and cryogenic grinding were studied. The microwave pre-treatment was used in both dry and wet modes with varying exposure time. Results showed slight decrease in ash content of treated sample than untreated one. Microscopic analysis confirmed presence of cracks and fractures in treated samples with more pronounced observations in case of wet heating. Among the macerals present in the sample, vitrinite was more affected than inertinite. In terms of recovery, best result was obtained for 6 minutes wet heating with a combustible matter recovery of 23.3% with product ash of 22% and d80 of 2 mm. In case of cryogenic grinding, liquid nitrogen was used and soaking time was varied. Microscopic analysis revealed presence of cracks and fragmentations in both vitrinite and inertinite particles. However, minor reduction in size was observed with d80reaching up to a minimum of 6.5 mm with 60 minutes exposure time. Study reveals that microwave heating is more effective technique than cryogenic cooling for sizing and liberation of ash particles of coal.

**Keywords:** Coal Middlings, Microwave pre-treatment, cryogenic grinding

## 1. Introduction

India is the third largest producer of coal in the world having around 9–10% of total coal reserves of the world. But at the same time India is the second largest importer of coal in the world after China [1]. One of the major reasons behind this situation is the ever-expanding iron and steel industry of the country. The Indian steel industry is broadly classified into three categories based on route wise production viz. BF-BOF, EAF and IF. Among the three routes, the BF-BOF has the largest share with a combined capacity of around 50 MT [2]. Coking coal acts as the source of energy during BF-BOF steel making process. This leads to high demands of coking coal which is not met by indigenous coal production because of the limited resources of coking coal in India as well as their inferior quality. India has a total of 253 billion tonnes of coal reserves out of which only 12.65% is coking coal while the rest is non-coking coal [3]. Moreover, the run-of-mine (ROM) coal is generally high in ash content (ranging from 40% to 50%) and high moisture content (4–20%). This is because the Indian coals belong to the Southern Hemisphere Gondwana coal which is characterised by presence of interbanded mineral sediments. Due to increased open cast mining and production from inherently inferior grades, the ash content of raw coal has been increasing over the past three decades [4]. Hence this worsens the scenario in the country which often leads to situation where as high as 70% of coking coal requirements of India is met by imported coal [5].

A major objective of the National Steel Policy 2017 (NSP 2017) of India aims to increase the domestic availability of washed coking coal and reduce the import of coking coal by 50% by 2030–31 [6]. For this the ministry of steel in co-ordination with the ministry of coal plans to increase the availability of coking coal through various measures such as overseas asset acquisition, establishing of sufficient number of modern coking coal washeries, periodic auction and allotment of coal blocks, exploration and exploitation of deep seated coking coal reserves, etc. However most of the above stated measures require much time and massive capital investment before they start giving results. A long-term solution to this can be increasing the overall yield and process efficiency of the existing coking coal washeries in the country. The overall increment in the yield of coking coal can be achieved by two ways viz. increase the production of raw coal i.e. raw coal throughput to the washery or decrease the losses during production of clean coal. The first option has already been explored by most of coal producers keeping in mind the total reserves of the country as well as sustainability of the mines and preparation process. The second option is a high potential one which can prove to be a game changer if tackled through research and innovation. The term "minimising the losses" refers to more efficient recovery of valuables from raw coal which can increase yield. But despite continuous methods of improvement in the process there are losses which cannot be mitigated. So, the other way to solve this problem is to extract the coal portions lost in the tailings/rejects i.e. by-products/discards of coal washery. In a typical coal washery two or sometimes three types of coal by-products are produced along with the clean coal. In India, a coking coal beneficiation flow-sheet generally consists of two circuits i.e. coarse circuit and fines circuit. The coarse circuit treats coal of the size (+0.5 mm) using gravity separation methods like heavy media cyclone, heavy media bath, etc. whereas the fines circuit employs froth flotation to beneficiate (−0.5 mm) particles. The by-product generated in the coarse circuit is known as coal middlings which represent approximately 22–25% of ROM coal with carbon content of approximately 35%. They are currently used in thermal power plants in place of non-coking coals. This usage is highly inapt as middlings still contain considerable amount of coking coal fractions and if processed further may increase coking coal production by 3–4%. Therefore, as per the current scenario the use of coking coal by-products as a substitute of thermal coal in power plants should be reduced through proper government regulations and their beneficiation to increase the yield of existing coking coal washeries must be initiated urgently.

Coal is a major source of energy all over the world. In the literature, also most of the previous works on consumption of coal by-products is mostly concentrated on their use as fuel after briquetting [7–9] because of their good market demand. Apart from this the other uses of coal discard include use as a construction material [10] or use as sorbent for $SO_2$, $NO_x$, etc. [11]. In addition to this some work to reduce the ash content of the coal discards using different methods like flotation [12], gravity separation [13,14], falcon concentrator [15] and chemical leaching method [16] were found. However, all these works were performed for finer size coal rejects (−0.5 mm or below). Very little work has been done to beneficiate the coarse sized by-products like coal middlings. The main challenge in beneficiating such coarse coal by-products is that to achieve better separation, better liberation of valuables from high ash forming minerals is required at optimum particle size. Furthermore, on considering the economics of the process regular crushing and grinding cannot serve the purpose. In this regard difference in properties of coal and mineral matter may serve the objective through selective grinding achieved by different kinds of pre-treatment methods.

One such method is the microwave pre-treatment of coal. Microwave heating of a material depends on its microwave absorbing properties i.e. dielectric permittivity and loss factor [17]. A material having more than one component with different dielectric properties, undergo differential heating resulting in thermal fracture during microwave treatment. Coal is a highly heterogenous substance with different loss factor values for its organic and inorganic components [18,19]. Past work in literature have shown that coal like other minerals also responds to microwaves and the grindability can be positively affected. The mineral matter gets selectively heated during microwave pre-treatment of coal [20]. This results in weakening of coal–mineral matrix, which increases the rate of breakage and decreases the grinding resistance of coal without altering its fundamental property [21]. Another method of achieving the selective liberation of coal without ultrafine grinding is cryogenic grinding of coal. Cryogenic grinding of coal simply means the pre-treatment of coal with cryogenic liquids like liquid nitrogen followed by its grinding. The basis for this method is that when brought to very low temperature environments coal become very brittle in nature [22]. It occurs due to the creation of cleavage of the physical bond between coal and mineral particles. Interfacial and matrix cracks are generated which increase coal grindability and hence greater brittleness at low temperature environments [23,24] is obtained for coal.

In the present investigation therefore microwave pre-treatment and cryogenic grinding of coal middlings have been explored to liberate more coal particles from the middlings and recover more coking coal from it. The present work involves a detailed characterisation of coal middlings followed by study of microwave pre-treatment and cryogenic grinding on middlings as methods of enhanced liberation for better separation.

## 2. Experimental

## 2.1 Sample collection and characterisation

The coal middlings sample used in the present investigation was collected from a typical coal washery in Jharkhand state, India during a regular working day. The sample was subjected to proximate analysis, size analysis, petrographic classification, liberation studies and washability analysis. The proximate analysis of the sample as on received basis is given in Table 1 which shows that the quality of middlings sample is quite inferior as the ash content is much higher than fixed carbon.

**Table 1:** Proximate analysis of coal sample

| Parameters | Value (%) |
|---|---|
| Moisture (%) | 1.52 |
| Volatile matter (%) | 18 |
| Ash (%) | 46.4 |
| Fixed carbon (%) | 34 |
| Total | 99.92 |

### 2.1.1 Size analysis and size-wise ash analysis

Particle size distribution was carried out in a vibrating sieve shaker and the result is shown in Fig. 1 which shows that the ash content reduces with size which means more liberated coal particles are in finer fractions. However, the difference in ash between coarser fraction (+6 mm) and finer fraction (−6 mm) is only marginal with all the size fractions having ash more than 40%.

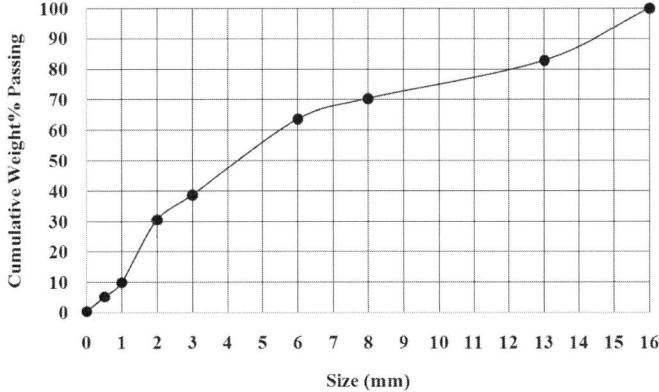

**Figure 1:** Particle size distribution of the coal sample

### 2.1.2 Washability analysis

Sequential sink–float test for coal middling sample was conducted by using organic liquid mixtures in the specific gravity range of 1.5–2.0 at an increment of 0.1. The float products and the sink were washed, dried, weighed and analysed for ash content. The washability of the middlings is given in Table 2. This shows that the combustible matter recovery for middlings is very less and they are very hard to wash coals with low carbon content. The whole process of beneficiating it may become very uneconomical at this low recovery. This is the obvious reason for using it in power plants rather beneficiating. Hence to make it more economical, size reduction with selective liberation of coal particles is required for better recovery.

**Table 2:** Washability analysis of coal middlings sample

| Sp. Gr. | Cumulative ash % | Yield % |
|---------|------------------|---------|
| +1.3 | 10.84 | 0.09 |
| 1.3–1.4 | 15.17 | 1.10 |
| 1.4–1.5 | 27.61 | 10.05 |
| 1.5–1.6 | 31.29 | 41.93 |
| 1.6–1.7 | 34.54 | 63.36 |
| 1.7–1.8 | 36.48 | 72.47 |
| 1.8–1.9 | 37.97 | 78.82 |
| 1.9–2.0 | 40.47 | 84.79 |
| 2.0 sink | 44.72 | 100.00 |

### 2.1.3 Petrography and liberation studies

Petrographic studies were done as per standard methodology [25,26]. Pellets were prepared from the sample with the help of cold setting compounds (resin and hardener) in a crucible and studied under reflected light in oil medium at 500× magnification with the help of petrological microscope. Similarly, for the liberation study, the polished coal pellets of different size fractions were prepared using cold setting material without pressure. The particulate mounts were then studied to determine the locking pattern of mineral matter with organic and distribution pattern of mineral matter.

The maceral analysis of the sample is given in Table 3 which shows that mineral matter is as high as 40%. Inertinite is the main maceral component (>30%) followed by vitrinite (20%).

**Table 3:** Maceral analysis of coal middlings sample

| Component | % Distribution |
|-----------|----------------|
| Vitrinite | 20.1 |
| Liptinite | 5.4 |
| Inertinite | 34.2 |
| Mineral matter | 40.3 |
| Total | 100 |

The microscopic analysis of middling particles shows the minerals present are mostly in the syngenetic form i.e. minerals are mostly in layered and disseminated form within organic matrix (Fig. 2(a) and (b)). Apart from these some large mineral grains are also present (Fig. 2(c)). Layers of macerals are relatively thin in nature.

The liberation studies were carried out for the as received sample (−13+0.5 mm) and for size fractions (−2 mm), (−1 mm) and (−0.5 mm) as shown in Fig. 3. The analysis shows that for the as received sample the particles with less than 20% ash content are very low only 14%. But as the size is reduced it increases and gains a maximum of 32% for both (−1 mm) and (−0.5 mm) due to more liberation.

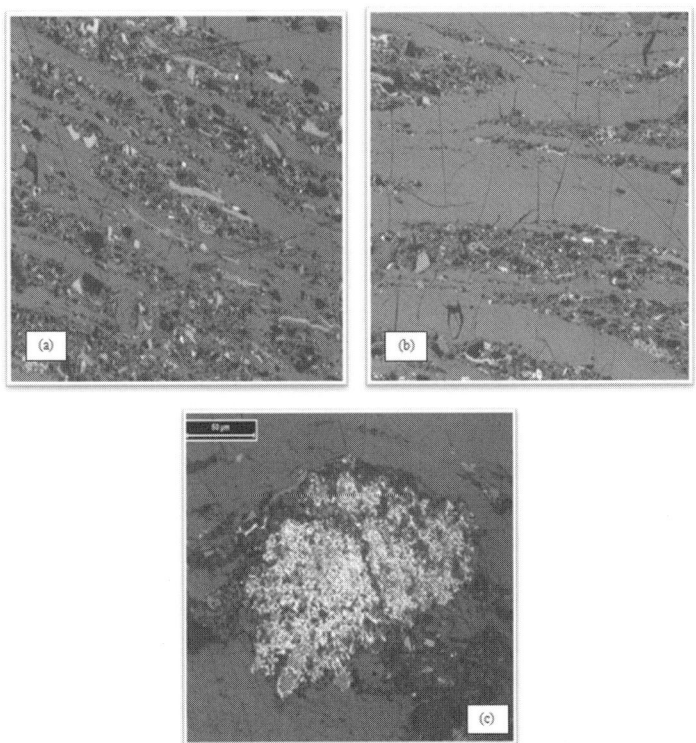

**Figure 2:** Microscopic images of coal middling particles

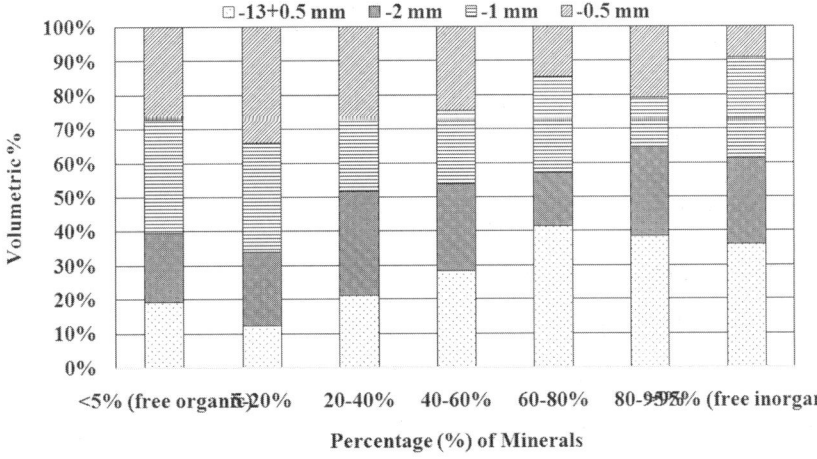

**Figure 3:** Locking pattern of minerals with organic matter in the coal middlings sample

## 2.2 Thermal shocks

### 2.2.1 Microwave pre-treatment: Experimental

The microwave treatment was carried out in a 900 W commercially available LG make microwave oven. The middling sample was uniformly heated in two modes viz. dry mode and wet mode. The time of heating for dry mode was lesser than that for wet mode and operated for durations of 30, 60 and 90 s. Whereas in case of wet heating the soaking time for the sample in water was 24 h and exposure time was 3 and 6 min. After the microwave treatment, the sample was subjected to proximate analysis and microscopic analysis to study the changes in the sample. Finally, the sample was wet ground in a rod mill for 1 h in all the cases keeping other parameters same. The product was then subjected to size analysis and washability analysis.

### 2.2.2 Cryogenic grinding: Experimental

For cryogenic tests, the sample was immersed in liquid nitrogen for different soaking time of 20, 40 and 60 min. After this the sample was removed from liquid nitrogen and air dried. Proximate analysis and microscopic analysis were carried out for all the samples. After this wet grinding as stated above was done in a rod mill for 1 h. Then size analysis and washability analysis of the products were carried out.

## 2.3 Grinding tests: Experimental

Grinding tests were carried out on the coal samples in a laboratory rod mill. The feed to charge ratio was fixed at 1:4 and mill speed was same for all the tests. The quantity of water added during the tests was kept constant. The grinding time of 1 h was fixed for all the samples. After grinding the ground product was taken out from the mill, dried and subjected to size analysis and washability analysis.

## 3. Results and analysis

## 3.1 Microwave treatment

### 3.1.1 Proximate analysis

The proximate analyses of the untreated and treated samples both dry and wet are shown in Table 4. Results show that the ash content of the samples decrease slightly as compared to the original sample during dry heating. But in case of wet heating the reduction is more pronounced. The moisture content of the microwave treated samples is lesser than untreated samples. Likewise, the volatile matter and fixed carbon content of the treated samples were more than the untreated sample. This agrees with the findings of the earlier work done by various authors such as Xia et al. [27], Sahoo et al. [21]. The lowering of the ash content is due to the liberation of the mineral matter present in coal. During microwave heating differential heating takes place because of which the minerals in coal get heated faster and therefore the binding energy between the coal and mineral matter gets weakened [28,29]. Water has a high dielectric constant and thus strong capacity to absorb microwave energy leading to high temperature. So, the moisture content of coal affects the dielectric properties of coal and the results vary from coal to coal [30]. In this case the lowest ash content and correspondingly the highest carbon content were obtained for the 6 min heating of water soaked sample.

Table 4: Proximate analysis of microwave untreated and treated coal samples

| Sample | Proximate analysis | | | | |
|--------|------|------------------|--------------|-------------------|-------|
| | Ash (%) | Volatile matter (%) | Moisture (%) | Fixed carbon (%) | Total |
| Untreated coal middlings | 45.48 | 18 | 1.52 | 35 | 100 |
| Microwave treated (dry, 30 s) | 44.96 | 18.14 | 1.49 | 35.41 | 100 |
| Microwave treated (dry, 60 s) | 44.63 | 18.38 | 1.38 | 35.6 | 99.99 |
| Microwave treated (dry, 90 s) | 44.29 | 18.61 | 1.31 | 35.78 | 99.99 |
| Microwave treated (wet, 3 min) | 44.76 | 18.14 | 1.48 | 35.62 | 100 |
| Microwave treated (wet, 6 min) | 44.21 | 18.45 | 1.32 | 36.02 | 100 |

### 3.1.2 Particle size distribution

The particle size distributions of the untreated and treated microwave samples are given in Fig. 4. It shows the considerable effect of microwave on reducing the size of particles. The $d_{80}$ of the untreated sample is around 10 mm which is brought down to less than 4 mm by dry microwave heating (90 s) and is further reduced in presence of water to 3 mm and 2 mm correspondingly for wet heating of 3 min and 6 min. This shows that the effect of microwave heating is greatly influenced by water and exposure time. The effect of water (or moisture) during microwave grinding is explained by Lester et al. [31] who showed that the rapid expansion of moisture when vapour is formed results in formation of cracks. These cracks in turn are responsible for the huge size reduction in presence of water. The effect of exposure time was studied by various authors [21,32–34] who concluded the same observation and described it in terms of differential heating of different phases in a material. The (+6 mm) fraction for the untreated ground sample is 27% which is reduced to 11% for 90 s dry microwave treated sample. This is even further brought down to only 8% for 6 min of wet microwave heating. The 0.5 mm passing fraction for untreated samples is only 24% which becomes 55.9% by 6 min of wet microwave heating. This shows very substantial reduction in size by wet microwave heating. However, the generation of fines (−150 μm) is quite high (>35%) for the wet microwave heating.

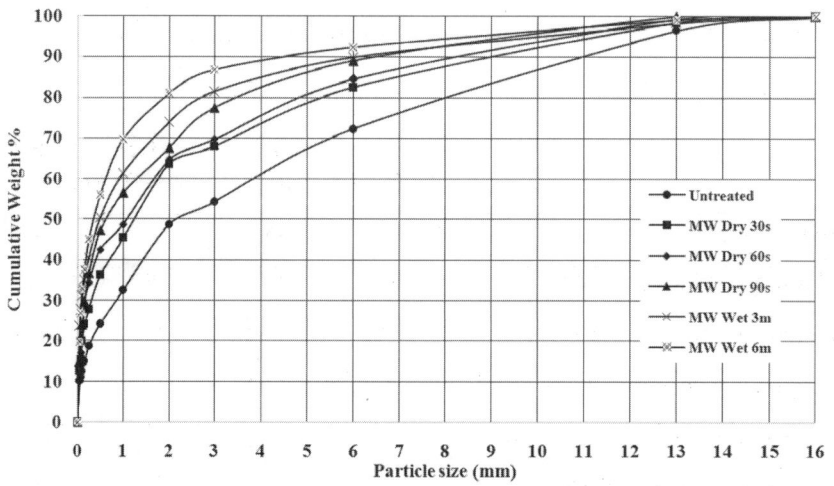

**Figure 4:** Particle size distributions of the ground untreated and microwave treated samples

### 3.1.3 Size-wise ash analysis

Ash distribution of different size fractions for the untreated and microwave treated samples are shown in Fig. 5. The ash distribution of the untreated and treated samples follows the same trend. The coarser fraction viz. +6 mm has higher ash (>50%) while the finer fractions (−6 mm) have lower ash (<45%). This result is quite encouraging for the beneficiation of middlings as it proves that mostly coal particles are broken down to finer sizes. Thus, selective grinding takes place due to microwave treatment of coal samples.

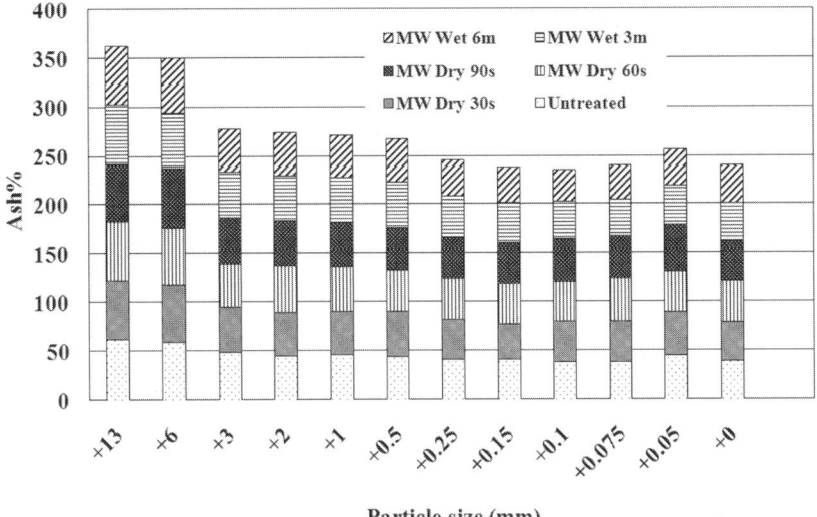

**Figure 5:** Size wise ash distribution of the ground untreated and microwave treated samples

## 3.1.4 Washability studies

Figure 6 shows the washability distributions of the untreated and microwave treated samples. The untreated samples show a combustible matter recovery of 13.8% at product ash of 26% which clearly makes the beneficiation of middlings uneconomical at this stage. However, the use of microwave increases recovery as well as reduces the product ash. The best result is obtained in wet conditions with exposure time of 6 min. It gives a product of ash 22% with a combustible matter recovery of 23.3%. This result is obtained at specific gravity of 1.5. Furthermore, for specific gravity 1.6 or more the ash of the float for all the samples is more than 30%. This abrupt change in float ash confirms the recovery of low ash coal particles at 1.5 specific gravity. Therefore, the substantial reduction in product ash confirms the assumption of the authors in the previous section that selective grinding of coal particles is taking place.

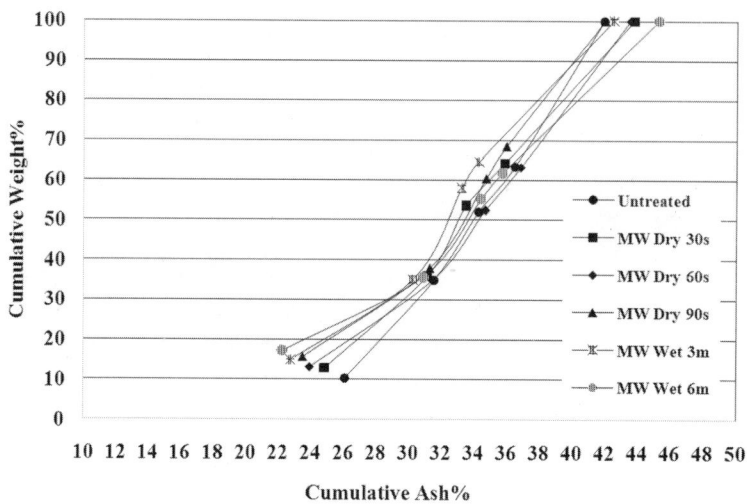

**Figure 6:** Washability analysis of the ground untreated and microwave treated samples

## 3.1.5 Microscopic studies

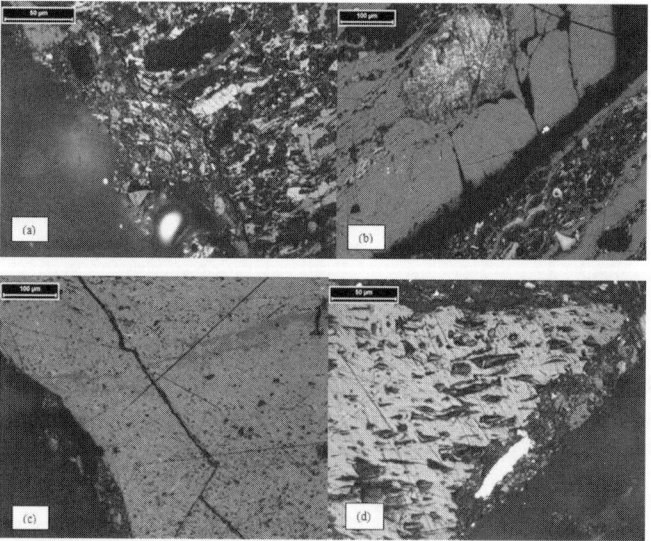

**Figure 7:** Images of the microwave treated samples under the microscope
(a) and (b) dry and (c) and (d) wet

The observations under microscope confirm the presence of cracks and fractures in the microwave treated samples (Fig. 7). More pronounced effect can be seen during wet heating of the samples. The cracks mostly developed at the edges or around the mineral grains during differential heating (Fig. 7(a) and (b)). Fine crack development can be observed within vitrinite (Fig. 7(b) and (c)) whereas inertinite (unreactives) remains unaffected and only crushed grains at the periphery are observed (Fig. 7(d)).

## 3.2 Cryogenic grinding

### 3.2.1 Particle size distribution

The particle size analysis is of the untreated and liquid nitrogen treated samples is shown in Fig. 8. The effect of liquid nitrogen on particle size reduction of coal sample is not as much pronounced as compared to that of microwave treated middling. The decrease in $d_{80}$ is from 10 mm for untreated coal to around 6.5 mm for liquid nitrogen treated (60 m) making the reduction ratio less than 2. In addition to this, the reduction of coarser fraction (+6 mm) is also marginal i.e. from 27% for untreated coal to 20% for 60 min liquid nitrogen sample. The highest 0.5 mm passing fraction is obtained for 20 min liquid nitrogen sample and this decrease when the exposure time is increased. This shows that an exposure time of 20 min is more effective. This is also visible in the analysis of ultrafines (−150 μm) which are produced more (around 35%) for 20 min exposure time and reduces to 24.5% for 60 min.

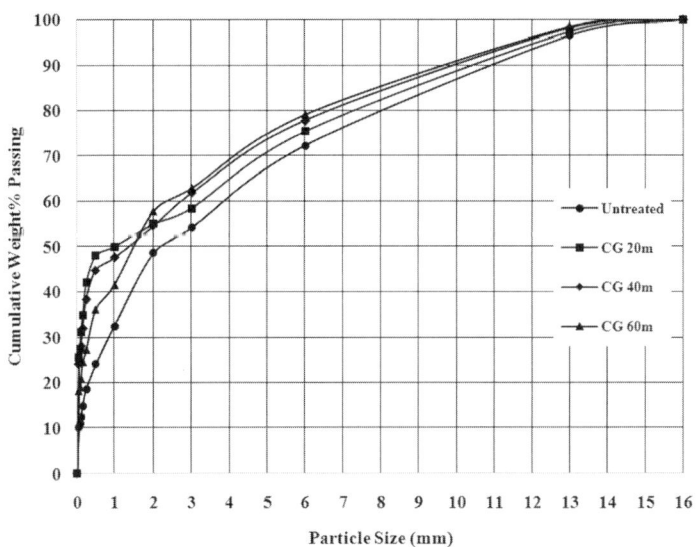

**Figure 8:** Particle size distributions of the untreated and liquid nitrogen treated samples

### 3.2.2 Size-wise ash analysis

As per Fig. 9, the ash distribution of the untreated and liquid nitrogen treated samples follow similar trends with ash continuously decreasing up to 50 μm. Beyond this point there is a slight increment in ash which implies generation of high ash ultrafines and therefore rubs out any possibility of selective grinding. It was also observed that superfine generated during grinding of coal samples treated by cryogenic liquid was very less which can be used for dust control during grinding of such materials.

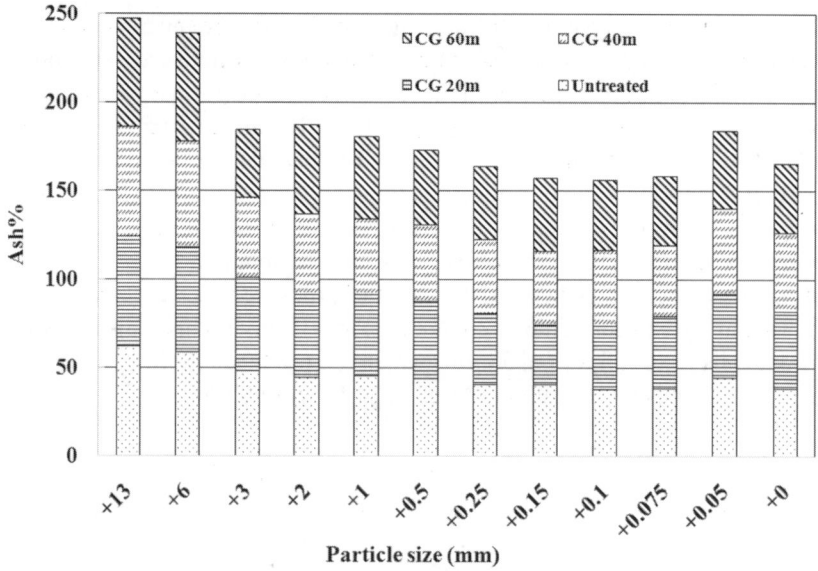

**Figure 9:** Size wise ash distribution of the untreated and liquid nitrogen treated samples

### 3.2.3 Washability studies

The washability analysis in Fig. 10 displays marginal improvement in recovery by use of cryogenic grinding over normal grinding. The best combustible matter recovery of 20.9% is obtained for the sample with 20 min exposure time while the lowest product ash of 22% for sample with 60 min exposure time. However, the recovery at this point is below 20% i.e. 17.6% only.

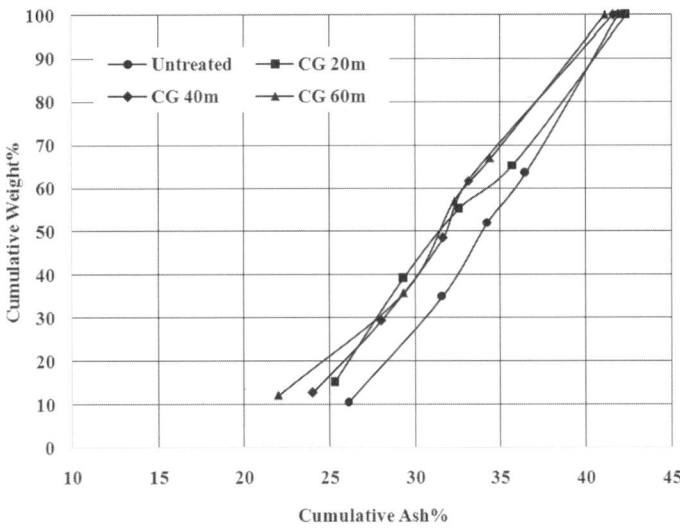

**Figure 10:** Washability analysis of the untreated and liquid nitrogen treated samples

### 3.2.3 Microscopic studies

**Figure 11:** Images of the liquid nitrogen treated samples under the microscope (a) and (b) soaking time 20 min and (c) and (d) soaking time 40 min

The study of the liquid nitrogen treated sample under the microscope reveal the presence of liquid within the intrusions of the coal matrix (Fig. 11(a)). Clear separation of mineral-rich band from maceral band is seen (Fig. 11(b)). Cracks and fragmentation appear in both vitrinite as well as inertinite macerals (Fig. 11(c) and (d)).

## 4. Conclusion

The present study aimed to enhance the liberation characteristics of coal middlings for their beneficiation and better utilisation. The characterisation studies showed that middlings are a hard to wash coal with low carbon content and high ash content mostly in the syngenetic form. Liberation can be improved only by size reduction in a selective manner for coal particles. So, two kinds of thermal shock i.e. microwave pre-treatment and cryogenic grinding were utilised.

Under the microscope presence of cracks and fragmentation was observed for both the methods. The liberation size of the coal sample as found in the characterisation studies must be 0.5 mm or below. Association of ash particles was very complex and neither of the methods could attain the liberated size. The size reduction as well as recovery of coal particles is much better with microwave treatment than compared to that of cryogenic grinding. Presence of water during microwave heating enhanced the results by many folds. However, the unsatisfactory results from cryogenic grinding may because of the absence of any thermal conducting medium like water. Cryogenic liquid treated samples generated lesser number of ultrafine particles which can be explored for further for dust control kinds of applications during grinding and handling of coal.

The lowest product ash achievable from microwave treatment is 22% which is quite higher than the typical ash requirement (i.e. 16% ± 0.5%) of coking coal for coke making. In addition to this the recovery at such product is 23.3% which on raw coal basis for a coal preparation plant producing 20% of its product as coal middlings, comes to around %. This means that the recovery of coking coal can be increased by 4.6% if the coal middlings generated in the plant are treated by the above-mentioned process. This recovery gain is only theoretical and the industrial performance of the pre-treatment method needs to be studied in detail.

Apart from this the economics of the process depends upon the cost of production of the middlings which includes grinding to very finer sizes. The industrial acceptance of the process will be feasible only if profit generated by processing of middlings and generation of coking coal at the desired properties is greater than the current process of selling it as an alternative of non-coking coal.

## 5. Acknowledgement

The authors would like to thank the Tata Steel management and Dr Sanjay Chandra, Chief RD&SS (Tata Steel Limited) for his encouragement during the project and permission to publish the paper. The authors are also thankful to the professionals of the Interface Lab, RD&SS (Tata Steel Limited) for their help in carrying out the characterisation studies.

## 6. References

[1] IEA (International Energy Agency) (2015), India Energy Outlook, World Energy Outlook Special Report 2015, OECD/IEA, Paris.

[2] JPC (Joint Plant Committee) (2014), Annual Statistics 2013/14, JPC, Kolkata, India.

[3] Khadse A, Qayyumi M, Mahajani S, Aghalayam P. Underground coal gasification: A new clean coal utilization technique for India. Energy 2007, 32: 2061–2071.

[4] Chikkatur A P, Sagar A D, Sankar T L. Sustainable development of the Indian coal sector. Energy 2009, 34: 942–953.

[5] Ministry of Coal, 2014. Govt. of India, Provisional coal statistics 2014–2015.

[6] National Steel Policy, 2017. Ministry of Steel, India. http://steel.gov.in/national-steel-policy-nsp-2017; 2018 (accessed 31 January 2018).

[7] Radloff B, Kirsten M, Anderson R. Wallerawang colliery rehabilitation: the coal tailings briquetting process, Miner. Eng. 2004, 17: 153–157.

[8] Borowski G, Hycnar J J. Utilization of fine coal waste as a fuel briquettes. Int. J. Coal Prep. Uti. 2013, 33: 194–204.

[9] Boruk S and Winkler I. Ecologically friendly utilization of coal processing waste as a secondary energy source. In Energy and Environmental Challenges to Security, 2009, 251–259. Springer Publishers.

[10] Okogbue C O and Ezeajugh C L. The potentials of Nigerian coal-reject as a construction material. Eng. Geo., 1991, 30: 337–356.

[11] Lu G Q and Do D D. Preparation of economical sorbents for SO2 and NOx removal using coal washery reject. Carbon 1991, 29 (2): 207–213.

[12] Barraza J, Guerrero J, Piñeres J. Flotation of a refuse tailing fine coal slurry. Fuel Process. Tech., 2013, 106: 498–500.

[13] Ozgen S, Malkoc O, Dogancik C, Sabah E, Sapci F O. Optimization of a multi gravity separator to produce clean coal from Turkish lignite fine coal tailings, Fuel, 2011, 90: 1549–1555.

[14] Rao L S and Bandopadhyay P. Application of a Mozley mineral separator for treatment of coal washery rejects, Int. J. Miner. Process., 1992, 36: 137–150.

[15] Oruç F, Özgen S and Sabah E. An enhanced-gravity method to recover ultra-fine coal from tailings: Falcon concentrator. Fuel, 2010, 89: 2433–2437.

[16] Suresh A, Lingam R K, Sriramoju S K, Bodewar A, Ray T and Dash P S. Pilot scale demineralization study on coal flotation tailings and optimization of the operational parameters with modelling. Int. J. Miner. Process., 2015, 145: 23–31.

[17] Meredith R. "Engineers' Handbook of Industrial Microwave Heating". The Institution of Electrical Engineers: London, 1998: 19–51.

[18] Marland S, Merchant A and Rowson N. Dielectric properties of coal. Fuel, 2001, 80(3): 1839–849.

[19] Marland S, Merchant A and Rowson N. The effect of microwave radiation on coal grindability. Fuel, 2000, 79: 1283–1288.

[20] Meikap B C, Purohit N K and Mahadevan V. Effect of microwave pre-treatment of coal for improvement of rheological characteristics of coal-water slurries, J. Colloid Interface Sci., 2005, 281: 225–235.

[21] Sahoo B K, De S and Meikap B C. Improvement of grinding characteristics of Indian coal by microwave pre-treatment, Fuel Process. Tech., 2011, 92: 1920–1928.

[22] Hartwig G. Advances in cryogenic engineering: Low temperature properties of epoxy resins and composites. Eds K.D. Timmerhaus, R.P. Reed and A.F. Clarke, Plenum press, 1978, 24: 17–36.

[23] Yen S C and Hippo E J. Comminution employing liquid nitrogen pre-treatments. Coal Technology Laboratory first quarterly report, 1989, Southern Illinois University, Carbondale, Coal Research Centre.

[24] Yen S C and Hippo E J. Comminution employing liquid nitrogen pre-treatments. Coal Technology Laboratory second quarterly report, 1989, Southern Illinois University, Carbondale, Coal Research Centre.

[25] ICCP. The new vitrinite classification (ICCP System 1994): Fuel, 1998, 77: 349–358.

[26] ICCP. The new inertinite classification (ICCP System 1994): Fuel, 2001, 80: 459–471.

[27] Xia W, Yang J and Liang C. Effect of microwave pre-treatment on oxidized coal flotation. Powder Tech., 2013, 233: 186–189.

[28] Meikap B C, Purohit N K and Mahadevan V. Effect of microwave pre-treatment of coal for improvement of rheological characteristics of coal-water slurries. J. Colloid Interface Sci., 2005, 281: 225–235.

[29] Bond F C. Crushing and Grinding Calculations Part I, Br. Chem. Eng.,1961: 378–385.

[30] Metaxas A C and Meredith R J. Industrial Microwave Heating. 1983. The Institution of Electrical Engineers, London.

[31] Lester E, Kingman S and Dodds C. Increased coal grindability as a result of microwave pre-treatment at economic energy inputs, Fuel, 2005, 84: 423–427.

[32] Lester E and Kingman S. Effect of microwave heating on the physical and petrographic characteristics of a U.K. Coal Energy Fuel, 2004, 18 (1): 140–147.

[33] Lester E and Kingman S. The effect of microwave pre-heating on five different coals. Fuel, 2004, 83: 1941–1947.

[34] Sahoo B K, De S, Carsky M and Meikap B C. Enhancement of rheological behaviour of Indian high ash coal–water suspension by using microwave pre-treatment. Ind. Eng. Chem. Res., 2010, 49: 3015–3021.

[35] Tang X, Ripepi N and Gilliland E. Isothermal adsorption kinetics properties of carbon dioxide in crushed coal. Greenhouse Gas Sci Technol., 2016, 55 (5): 260–74.

# 13

# Thermal enrichment of high-volatile coals

Dr. Sergey Islamov

*Design Institute for coal Enrichment/SUEK Company, Moscow, Russia*

**Abstract:** The most effective way of deep enrichment of high-volatile coals, both economically and ecologically, is the partial air gasification of coal. In this case, in addition to combustible gas the carbonic residue (thermocoke) is obtained with a calorific value in the range of 27–30 MJ/kg. This technology is realized in standard coal-fired boiler after its minor modification. In the lower part of the boiler, a fluidized bed reactor is installed to partially gasify the coal. The resulting gaseous fuel, dusty with carbonizate (about 10% by mass) meets secondary air above the fluidized bed and completely burns so the boiler produces a nominal amount of heat energy. The solid residue of partial coal gasification comes to the cooler, and then to subsequent processing depending on the purpose. The mineral part of the coal is almost completely capsulated in thermocoke, so the boiler does not have ash-slag waste.

The briquetted high reactive thermocoke with a carbon content of 85–90% is an excellent carbon reductant for electrometallurgy. Also it is recommended as a smokeless household fuel, and in grained form as an effective sorbent for wastewater and gas emissions purification.

**Keywords:** Coal enrichment, partial coal gasification, mid-temperature coke, thermocoke

Traditional coal enrichment methods are designed for the mechanical removal of particles with high ash content in order to increase the calorific value of the final product. We have developed a fundamentally different approach to solving this problem. It applies mainly to high volatile coal, such as lignite and subbituminous coals.

The essence of new technology consists in partial air gasification of coal. As result it is splitting into two parts – gaseous fuel and carbon residue called thermocoke (Fig. 1).

The first one is used for energy production by burning. The positive consequences for the environment of gas combustion instead of coal need no explanation. However, there is one feature that is unexpected for coal use in the context of the Paris Climate Agreement. When burning gas obtained by partial gasification of coal, $CO_2$ emissions per unit of produced useful energy are reduced by 30–40% compared to conventional coal combustion and are comparable to $CO_2$ emissions from the combustion of motor fuels.

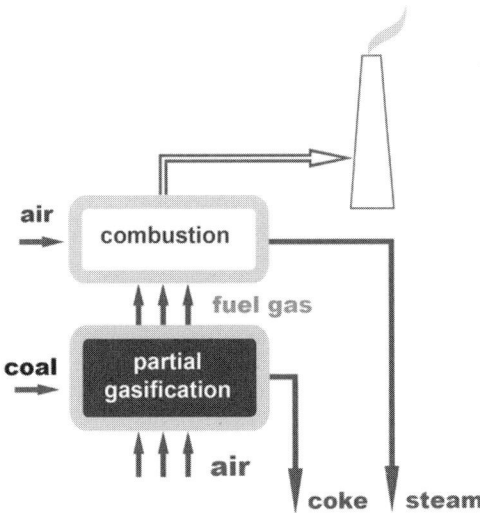

**Figure 1:** Concept of thermal coal enrichment

The second product – small grained thermocoke: We list some areas of its application: briquetted lump coke for electrometallurgy instead of classic coke, smokeless household fuel, pulverized fuel for injection into blast furnaces, high-calorific fuel for clinker burning, etc. It should be noted here that, in contrast to the traditional method of enrichment of coking coal, this technology allows to produce finished metallurgical coke. But the coke made of young coal has a reactivity that is an order of magnitude higher than that of classic coke. Therefore, the most effective application of thermocoke is to use as a carbonaceous reducing agent in the metallurgy of new generation for direct reduction of iron.

The main feature of the new technology lies in the fact that the process of coal conversion is carried out in standard coal-fired boiler with modified fluidized bed chamber. Its schematic diagram is shown in Fig. 2.

Crushed coal (~0–20 mm) is fed into the boiler. Entering in fluidized bed the coal particles are heated rapidly partially fractionized. During the heating they evolve all the water and the most part of volatile matter. Just in this moment to prevent the burning of solid residue the carbonized particles are removed from the bed into coke cooler. Volatile matter released from coal is mixing with hot air and burned inside the bed maintaining its temperature in the range of 800–900 °C. Further, the combustion  products (water vapor and carbon dioxide) enter into gasification reactions with carbon particles, forming the combustible components – carbon monoxide and hydrogen

(CO + H$_2$). Then hot gas mixture with small part of carbonaceous dust is carried out in above the fluidized bed space, were mixing with secondary air stream and burns completely. After such type modernization the boiler provides its nominal power capacity. As a result, the coal-fired boiler is running on gas fuel and instead of ash-slag waste it produces highly enriched coal product – thermocoke. Since gas emissions from a boiler are related to energy production, the process of coke production can be considered as "zero emission" technology. This is an outstanding environmental feature of this technology, which is provided not with the help of sewage treatment plants, but is internally intrinsic to it.

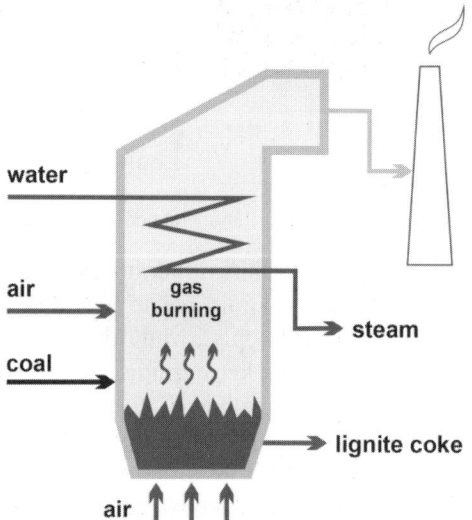

**Figure 2:** Scheme of modified coal-fired boiler for thermal coal enrichment

This type of combined production with additional income from the sale of thermocoke (instead of the cost for processing or storing ash and slag waste in conventional coal combustion) has a very close analogy with the classical power generation cogeneration scheme. However, at approximately the same market price of the second product (thermocoke and electricity), the capital costs for the processing of fine-grained coke are several orders of magnitude less than for the turbine-generator unit of power station. Thus, the technology of thermal coal enrichment **is radically** changing the economic effect of coal utilization.

Finally, for example, compare the quality of the raw brown coal and the product obtained after thermal enrichment of this coal. The relevant data are presented in Table 1.

**Table 1:** Comparative properties of brown coal and its enrichment product

| Parameter | Brown coal, Berezovskoye field, Siberia, Russia | Product of thermal enrichment |
|---|---|---|
| Total moisture (%) | 35.0 | 0.1 |
| Ash (dry%) | 4.7 | 9.3 |
| Volatile matter (daf%) | 45.6 | 7.5 |
| Net calorific value, *as received* (MJ/kg) | 15.66 | 29.76 |

## Conclusion

Thermal enrichment of high volatile coals is undoubtedly a breakthrough in the field of coal processing, which transfers low-grade coal to the category of environmentally friendly fuels and dramatically increases the economic efficiency of coal use.

# 14

# Characterization, beneficiation and carbonization study of some low volatile high rank coals and high volatile low rank coals of Damodar valley basin: Resource identification for augmentation of indigenous coking coal

Zeba Imam[1], Shekhar Saran[1], Dr. H. K. Mishra[2]

*[1]CMPDI, Ranchi, India*
*[2]IIT Bhubaneswar, Bhubaneswar, India*

**Abstract:** Cost of blast furnace (BF) coke accounts for more than 50% in production in hot metal through BF route. This has shown a sharp increase in last few years due to rise in both international and indigenous coking coal prices. A joint CMPDI-SAIL R&D project has been completed for identification of resources for using low volatile high rank (LVHR) which at present are not being utilized for metallurgical purpose and cheaper low rank high volatile Indian coking coal in blends for BF coke making. Detailed characterization of a number of sources of LVHR and low rank high volatile Indian coking coals has been carried out. This paper describes the properties of the different coals investigated and approaches for effective utilization of these cheaper coals in the blend.

The washability, characterization, and carbonization study was carried out jointly by CMPDI and RDCIS, SAIL from various sources of the country and study was carried out on clean coals at 17%, 15%, and 10% ash level. However, two coals one LVHR coal from Muraidih, Jharia coalfield and one low rank high volatile coal from Argada seam of North Urimari, South Karanpura coalfield have been selected for pilot plant washability and pilot oven carbonization study with clean coals at 10%, 15%, and 17% ash levels. This study has shown their usability in coke making.

Use of low ash Muraidih coal in blend produced coke of good quality. More improvement in M10 values was achieved when North Urimari and Muraidih coals used in combination. Lower ash concentrates produced better results. Moreover, under such circumstances, use of higher ash concentrates (up to 17%) did not produce deterioration in M10 values. Thus, use of both high volatile up to 10% and low volatile up to 10% can be used even up to 17% ash level in blends without any deterioration in Micum indices. Low-ash indigenous coals (LVHR and low rank high volatile coals) can be effectively utilized as blends to make coke of required strength through the process of stamp charging.

**Keywords:** Low volatile medium coking coal, low rank high volatile coal, coke blends, stamp charging

## 1. Background

The National Steel Policy, 2017 has envisaged crude steel production around 300 million tonnes (Mt) by the year 2030–31 and forecasts coking coal demand of 161 Mtpa (million tonnes per annum). For the present and even in the foreseeable future conventional blast furnace technology will be the main route for iron and steel making and so will be importance of coking coal as an important raw material. Coking coal is used to produce coke which is used primarily as energy source and a reductant in blast furnaces for the production of steel. The availability of indigenous good coking coal has almost become stagnant. At present about 85% of the current requirement of coking coal is met through imports. Relying on this trend, 35% of the demand during year 2030–31, i.e. 56.35 Mtpa of coal has to be made available from indigenous sources.

As the reserve of good quality coking coal of upper seams are almost depleted there is need to enhance utilization of available resources of inferior indigenous coals as alternate choice. LVMC coal (low volatile medium coking or low volatile high rank coal) and low rank high volatile coals are new sources that can be used in coke-making as blends through processes of suitable beneficiation.

LVMC coals falls into the category of medium coking coal which are also known as non-linked washery coal (NLW) due to ash content greater than 35%. These coals belong to lower seams are more likely to be more mature (Rr > 1.20%) and of low volatile matter content (18–22%), low reactive content (30–45%), and high inertinite content, having lower coking propensities with difficult cleaning characteristics. The laboratory and pilot scale investigations on these (LVMC/NLW) coals by CIMFR/SAIL/CIL in the past have demonstrated that on effective beneficiation these coals show good coking properties and can be blended to produce good quality blast furnace (BF) coke. At present, these coals are being used in power sector. These coals generally occur in the lower seams of Jharia coalfield (combined seam V/VI/VII/VIII and even seam IV, III, and II)) and Karo group of seams (seam VI–XI) in the eastern part of East Bokaro coalfield.

Some low rank high volatile or so called poor coking coals (generally non-coking coals) with less than four free swelling index and vitrinite reflectance within the range of 0.55–0.75%, having more than 10% liptinite content have been experimented for making good quality coke without any ill effects on battery by US Steel makers (Valia, 2009). Contrary to popular belief, not all poor coking coals are bad for coke making. With proper design, up to 30% of poor coking coals can be used in blends. Such coals falling in category of non-coking coals are abundant in the country particularly in SECL, CCL, WCL, and ECL coalfields, which can be tested for their suitability in making blends.

## 2. Geological resources of Indian coking coal and its distribution

India is endowed with vast geological resources of coal. As per the inventory of Geological resources of Indian coal (published by GSI 01.04.2018) the estimated reserve is around 319020.33 Mt up to a maximum depth of 1200 m of which 34522.46 Mt are in coking coal category while the remaining 282910.19 Mt are non-coking coal (Table 1) (GSI, 2018). The coking coal reserves is only about 10.82% of the total coal reserve of the country while 88.68% belongs to non-coking coal grade. Out of the total available coking coal reserve, the share of prime coking coal is 15% (5313.06 Mt), medium coking coal is 80% (27501.88 Mt), and semi-coking coal is 5% (1707.52Mt).

**Table 1:** Inventory of Geological resources of Indian coal (as on 01.04.18) (resources in million tonnes)

| Coal type | Proved | Indicated | Inferred | Total | % Share |
|---|---|---|---|---|---|
| Prime coking | 4648.87 | 664.19 | 0 | 5313.06 | 1.67 |
| Medium coking | 13913.65 | 11708.76 | 1879.42 | 27501.88 | 8.62 |
| Semi-coking | 519.44 | 994.87 | 193.21 | 1707.52 | 0.54 |
| Sub-total coking coal | 19081.96 | 13367.82 | 2072.68 | 34522.46 | 10.32 |
| Non-coking | 129111.66 | 12696.98 | 28101.55 | 282910.19 | 88.68 |
| Tertiary coal | 593.81 | 99.34 | 894.53 | 1587.68 | 0.5 |
| Grand total | 14787.43 | 139164.14 | 31068.76 | 319020.33 | 100 |
| % Share | 46.64 | 43.62 | 9.74 | 100 | |

The geographical distribution of coking coal fields is also uneven. Most of the coking coal resources are available in the Damodar Valley basin. The important coalfields in the basin are Jharia, West Bokaro, East Bokaro, Ramgarh, South Karanpura, and North Karanpura in the state of Jharkhand and Raniganj coalfield in the state of West Bengal. Small resource also found in Madhya Pradesh. The total reserve of coals in Jharia coalfield is 19,430 Mt out of which the entire reserve of prime coking coal i.e. 5313.06 Mt are found here, medium coking coal is 6164.00 Mt (including 295 Mt of high volatile and 5869 Mt low volatile type found in seams VIII and below). Low volatile medium coking coal of about 543 Mt is also available in the Karo group of seams in East Bokaro Coalfield. The total estimated reserve of LVMC type coals of Jharia and East Bokaro coalfield is around 6412 Mt.

## 3. Objective of the study

A joint R&D project was carried out of CMPDI RDCIS-SAIL along with CIMFR to study the characterization, beneficiation, and carbonization characteristics of LVMC/LVHR coals and low rank high volatile coals to justify their suitability to use as blends in coke making for BF. Usage of such indigenous coals as blends will meet the challenges of augmentation of metallurgical coal to greater extent.

## 4. Method of study

The study was carried out in two phases. In the first phase, laboratory washability study and detailed characterization of one sample from seam V/VI/VII Muraidih, Jharia coalfield from low volatile high rank group (Rr 1.35%) and four samples in total from low rank coals (Rr 0.55–0.75%), seam R IV, Jhanjra and R IV/V/VI Sonepur Bazari Raniganj coalfield, North Urimari, and Giddi A from South Karanpura CF to generate clean at around 17%, 15%, and 10% ash level. On the basis of the laboratory scale results samples were selected for second phase of study i.e. pilot scale washability and pilot oven carbonization tests (top charging and stamp charging) with indigenous prime, medium, imported coking coals, and selected low rank high volatile coals and LVHR coals in the blends.

## 4.1 Laboratory scale washability and characterization study

Samples from all the identified sources about a tonne each were collected as per IS 436 (Part I, Sect. I: 1964) (IS, 1964). Representative samples of LVHR (Rr % > 1.35%) and low rank coals (Rr 0.55–0.75%) were crushed to −13 mm size. These sample were subjected to float and sink test with the liquid of 1.5–1.3 sp. gr. (specific gravity) to generate coals at 10%, 15%, and 17% ash level as per IS (Indian Standard), 13810:1993 (IS, 1993). Characterization study of the raw and clean coals was carried out for proximate analysis, ultimate analysis, and tests for coking properties as per IS 1350 (Part I): 1984, IS,

1353:1959 respectively and tests for petrographic properties as per IS 9127 (Part II): 1979 (IS, 1959, 1979, 1984). 9127, (Part III), 1992 (IS, 1992). IS, 9127, (Part V), 1986 (IS, 1986). Based upon the result of Lab washability test and characterization study, samples from Muraidih and North Urimari were selected for pilot scale washability.

## 4.2 Pilot scale beneficiation and generation of cleans at 10%, 15%, and 17% ash content

Clean coal of desired quality was generated through two-stage beneficiation pilot plant operation. The coal crushed of 75 mm was upgraded in a three product jig to produce pre-cleans, middlings and rejects in the 1st stage. The 2nd stage of washing was carried out by heavy medium drum, heavy medium cyclone, and froth flotation accorded to the pre-cleans to generate the cleans. The pre-cleans was screened at 6 mm and deslimed at 0.5 mm for further processing.

## 4.3 Pilot oven carbonization tests

The pilot oven carbonization tests were carried out in two stages. In the first stage evaluation of North Urimari coal was carried out. In second stage, evaluation of Muraidih and Muraidih-North Urimari in combination was carried out. The carbonization conditions for pilot oven tests was done with top charging and stamp charging. Coke produced from the pilot oven carbonization tests were dried and subjected to proximate analysis, Micum indices, and CRI/CSR indices.

## 5. Results and discussions

The washability studies on LVMC/LVHR Muraidih coals and low rank North Urimari coal was carried out with an objective to observe the improvement in coking properties and reactive content in the cleans at 10%, 15%, and 17% ash level along with the recovery percent at each ash level. The characterization studies on raw and cleans indicated significant improvement in the coking properties of both Muraidih and North Urimari cleans, however, the cleans at 10% of both the coals had better coking properties as compared to cleans at 15% and 17% but the yield content at 10% ash was much lower. Table 2 depicts the variation in coking properties with variation in ash content.

**Table 2:** Variation in properties of North Urimari and Muraidih coal with variation in ash content

| Sample details | North Urimari | | | | Muraidih | | | |
| --- | --- | --- | --- | --- | --- | --- | --- | --- |
| | Raw | Clean coal samples | | | Raw | Clean coal samples | | |
| | | 17% | 15% | 10% | | 17% | 15% | 10% |
| Yield (%) | | 32.6 | 27.8 | 17.6 | | 13.9 | 12.5 | 5.0 |
| Ash (%) (air dried basis) | 34.6 | 17.6 | 15.6 | 9.8 | 45.0 | 17.7 | 15.0 | 9.9 |
| Crucible swelling number | 1/2 | 1.5 | 2 | 2.5 | 1 | 3.5 | 4.5 | 6 |
| LTGK coke type | B/C | D | E | F | C | F | G | G4 |
| Maximum fluidity (ddpm) | | 35 | 62 | 203 | | 6 | 12 | 237 |
| Maceral distribution (%) | | | | | | | | |
| Vitrinite | | 50.0 | 63.9 | 73.1 | | 48.0 | 43.1 | 71.2 |
| Liptinite | | 7.9 | 6.8 | 5.3 | | 0.4 | 2.4 | 0.2 |
| Inertinite | | 31.1 | 19.6 | 15.2 | | 40.8 | 45.3 | 22.7 |
| Mineral matter | | 11.0 | 9.7 | 6.4 | | 10.8 | 9.2 | 5.9 |
| Reflectance | | | | | | | | |
| RoR (%) | | 0.64 | 0.64 | 0.59 | | 1.29 | 1.29 | 1.37 |
| MMR (%) | | 0.67 | 0.67 | 0.62 | | 1.36 | 1.36 | 1.46 |

In the first stage pilot oven carbonization tests using top charging process with North Urimari coal at ash content of about 10%, coke of good strength was produced. Increase in ash content resulted in deterioration in M10 values. However, no deterioration in M10 values of North Urimari coal from 10% ash to 17% ash level when using stamp charging method. They could replace existing medium coking coal up to 20% with improvement in M10 values.

In second stage pilot oven carbonization tests, North Urimari coal in association with Muraidih coal combined produced better coke than that produced from only North Urimari. The properties of coke for evaluation under top charging are given at Table 3. Usage of North Urimari up to 10% even at ash level of 17% did not produce any deterioration in M10 value. Table 4 presents the properties of coke sample under stamp charging. Stamp charging produced improvement by about 4–6 units in M10 values. There is significant improvement in CSR values also. Use of both high volatile up to 10% and low volatile up to 10% can be used even up to 17% ash level in blends without any deterioration in Micum indices.

**Table 3:** Properties of coke samples for evaluation of Muraidih and North Urimari coals through top charging process

| Blend no. | B1 | B3 | B5 | B7 | B9 | B11 | B13 | B15 | B17 | B19 |
|---|---|---|---|---|---|---|---|---|---|---|
| | Base blend | Mu10 (5) NU10 (5) | Mu10 (10) NU10 (10) | Mu15 (5) NU15 (5) | Mu15 (10) NU15 (5) | Mu15 (10) NU15 (5) | Mu17 (10) | Mu10 (15) | Mu15 (15) | Mu17 (15) |
| **Micum indices** | | | | | | | | | | |
| M40% | 80.4 | 82.5 | 83.9 | 79.9 | 82.2 | 81.9 | 83.6 | 83.2 | 81.1 | 81.5 |
| M10% | 13.0 | 10.1 | 8.8 | 11.1 | 11.1 | 11.8 | 11.2 | 10.3 | 11.6 | 11.0 |
| **Reactivity of coke** | | | | | | | | | | |
| CRI% | 24.7 | 29.2 | 31.5 | 31.8 | 32.0 | 33.1 | 32.6 | 32.0 | 29.6 | 28.5 |
| CSR% | 56.5 | 59.5 | 57.3 | 54.8 | 64.7 | 49.8 | 49.9 | 64.7 | 73.4 | 63.0 |

Mu, Muraidih; NU, North Urimari coal at different ash level, i.e. Mu10 is Muraidih cleans of 10% ash (figure in parenthesis indicates the blend percentage of Muraidih and North Urimari cleans with the remaining imported and indigenous coking coal)

**Table 4:** Properties of coke samples for evaluation of Muraidih and North Urimari coals through stamp charging process

| Blend no. | B2 | B4 | B6 | B8 | B10 | B12 | B14 | B16 | B18 | B20 |
|---|---|---|---|---|---|---|---|---|---|---|
| | Base blend | Mu10 (5) NU10 (5) | Mu10 (10) NU10 (10) | Mu15 (5) NU10 (10) | Mu15 (10) NU10 (10) | Mu17 (5) NU (5) | Mu17 (10) NU17 (10) | Mu10 (15) | Mu15 (15) | Mu17 (15) |
| **Micum indices** | | | | | | | | | | |
| M40% | 78.6 | 79.6 | 78.4 | 79.9 | 79.6 | 79.2 | 82.8 | 80.4 | 83.3 | 79.8 |
| M10% | 9.2 | 7.0 | 7.2 | 7.3 | 6.8 | 8.0 | 7.4 | 7.0 | 8.1 | 8.5 |
| **Reactivity of coke** | | | | | | | | | | |
| CRI% | 29.1 | 30.5 | 29.0 | 28.5 | 29.1 | 27.8 | 29.5 | 27.2 | 27.7 | 26.6 |
| CSR% | 62.9 | 65.6 | 68.7 | 68.0 | 69.2 | 66.6 | 65.2 | 72.0 | 70.0 | 63.4 |

Mu, Muraidih; NU, North Urimari coal at different ash level, Mu10 is Muraidih cleans of 10% ash (figure in parenthesis indicates the blend percentage of Muraidih and North Urimari cleans with the remaining imported and indigenous coking coal)

# 6. Conclusion

Characterization, beneficiation, and carbonization studies on coal samples from North Urimari and Muraidih coal have proved the suitability of these coals as blends in coke-making, excepting that yield of clean coal of Muraidih is not encouraging.

Pilot oven carbonization studies showed that both North Urimari and Muraidih coals are usable in BF coke making. North Urimari at 10% ash can be used up to 20% with improvement in M10 value, at 15% ash can be used up to 20% with no deterioration in coke properties and beyond 15% ash in cleans produce deterioration.

Use of North Urimari and Muraidih in combination provides more scope for their use in improving coke properties or usability at ash levels.

Stamp charging provides significant scope for improving both M10 and CSR values of coke. It also provided the scope for higher use of soft/medium coking coal in BF coke making.

Out of 34.5 billion tonnes of coking coal reserve in the country, approximately 6.4 billion tonnes fall into NLW/LVMC category and in Jharia Coalfield alone the present resource of NLW/LVMC coals is estimated to be about 5869 million tones and East Bokaro coalfield the reserve is around 543 million tones. The reserve of North Urimari, South Karanpura CF is estimated to be around 200 million tonnes.

The production of NLW/LVMC coals in CIL was around 18.5 million tonnes in the year 2017–18 which at present is not being beneficiated. It is possible to obtain clean coal around 4.0 (Mt) at 18±0.5% ash content assuming the yield around 25% and middling around 9.0 (Mt) (yield 50–55%, ash content below 40%). The clean coals can be used as a blend constituent for coke making in steel plants and the power grade coal can be used in power plants. Utilization of NLW coals after suitable beneficiation can reduce to the tune of around 3–4 million tonnes of import of coking coal, thereby, saving considerable amount of foreign exchange.

# 7. Acknowledgement

The authors wish to thank the management of CMPDI for giving permission to present the paper in the seminar. The authors wish to thank the CIL R&D board for having approved and funded the project No. CIL R&D 13/01/09 (CIL R&D, 2013). The authors are thankful all the members of CMPDI, CIMFR Dhanbad (erstwhile CFRI, Dhanbad) and RDCIS-SAIL Ranchi for their co-operation, support in preparing the R&D report, the findings of which have been presented in this paper.

## 8. References

[1] CIL R&D funded project No. CIL R&D 13/01/09: "Effective utilization of low rank and low volatile high rank Indian coking coals for BF coke making". Feb 2013.

[2] Excerpts from Technical report of 12th Indian Coal Market's Conference, Kolkata 26–28th Nov 2018.

[3] GSI (Geological Survey of India), 2018. An inventory of Geological Resources of Indian Coal.

[4] IS (Indian Standard), 1353, 1959: Methods of test for coal carbonization, caking index, swelling properties and Gray-king assay (L.T.G.K) coke type.

[5] IS (Indian Standard), 436 (Part I, Sect. I), 1964. Methods of test for coal and coke sampling of coal.

[6] IS (Indian Standard), 9127 (Part II), 1979. Methods for petrographic analysis of bituminous coal and anthracite, method of preparing coal sample. (Reaffirmed 2014).

[7] IS (Indian Standard), 1350 (Part I), 1984. Methods of test for coal and coke – Proximate analysis.

[8] IS (Indian Standard), 9127, (Part V), 1986. Methods for petrographic analysis of coal, Microscopic Determination of the Reflectance of Vitrinite. (Reaffirmed 2010).

[9] IS (Indian Standard), 9127, (Part III), 1992. Methods for petrographic analysis of coal, determination of Maceral group composition of bituminous coal and anthracite. (Reaffirmed 2002).

[10] IS (Indian Standard), 13810, 1993. Code of practice for float and sink analysis of coal.

[11] Valia H.S. (2009). Need of the hour – how to cut Met Coal Costs International al conference – coking coals and coke making: Challenges and opportunities ICC-2009 RDCIS, SAIL.

# 15

# Beneficiation and carbonization studies of low volatile coking coals of Jharia for metallurgical uses

T. Gouri Charan, U. S. Chattopadhyay, G. K. Bayen,
S. C. Maji, Manish Kumar, Pradeep Kumar Singh

*CSIR-Central Institute of Mining & Fuel Research,
Digwadih Campus, Dhanbad, India*

**Abstract:** Coking coal is an essential input for production of iron and steel through blast furnace (BF) route. The reserves of coking coal in India is meagre and the good quality coking coals of the upper seams are fast depleting leaving behind the inferior quality lower seam coal. Low volatile coking coal (LVCC), though inferior in qualities but abundantly available in eastern part of the country may be an immediate choice. These coals, being of lower seams are likely to be more matured (Ro ~ 1.30%) than the upper seams and consequently exhibit lower values of volatile matter. The country has a moderate reserve of such coal, amounting to about 50% of the total coking coal reserve. Unfortunately, the washability potential of this coal is so poor that the existing washeries having conventional washing technologies may not able to supply coals of ash 17–18% as desired by indigenous metallurgical industries and cannot stand in competition with foreign coals because of poor yield of clean coal. As such, these coals are being treated as NLW (non-linked washery grade) and are supplied to the thermal power plants, against augmenting the demand of metallurgical coal for coke making thus, wasting the scarce coking coal resources. The present paper highlights some of the beneficiation studies followed by carbonization studies for augmentation of coking coal resources for metallurgical purposes.

**Keywords:** LVMC coal, Near Gravity Material, Coke, Substitution, Washability studies, Jig, Heavy Media

## 1. Introduction

The total reserves of coal in the country are estimated to the order of 315 billion tons [1]. Coking coal, which is merely 14% of the total deposits, is available mainly in eastern part of India. Jharia and Bokaro coalfields constitute the major resources of coking coal in India and the coalfields are presently being operated by Bharat Coking Coal Limited (BCCL) and Central Coalfields Limited (CCL).

The domestic availability of coking coal, a critical raw material required by steel industry is limited and therefore the Indian Steel industry has to depend heavily on imported coking coal to meet its needs. Currently, domestic steel makers meet 80% of their coking coal requirement through imports. The quantum of imports may go up significantly in the near future as steel production in a large number of new projects is likely to be through the BF–BOF (blast furnace–basic oxygen furnace) route and also to meet the requirement of existing steel plants. The cost of the imported coal is always fluctuating and to ensure raw material security and minimize the impact of volatility in coal prices, it is desirable to try for augmentation of indigenous coking coal through various routes and technologies and supply it to steel industries.

Weakly coking coals, though inferior in qualities but abundantly available in eastern part of the country may be an immediate choice. These coals, being of lower seams are likely to be more matured (Ro ~ 1.30%) than the upper seams and consequently exhibit lower values of volatile matter [2]. The country has a moderate reserve of such coal, amounting to about 50% of the total coking coal reserve. Unfortunately, the washability potential of this coal is so poor that the existing washeries having conventional washing technologies may not able to supply coals of ash 17–18% as desired by indigenous metallurgical industries and cannot stand in competition with foreign coals because of poor yield of cleans. As such, these coals are being treated used in thermal power plants, against augmenting the demand of metallurgical coal for coke making thus, wasting the scarce coking coal resources. Beneficiation studies both in laboratory and pilot plant levels were carried out on the low volatile coking coals and the results suggest that after washing they may be used for metallurgical purpose [3,4].

Keeping this in view, an attempt has been made in this paper to focus on washability and pilot beneficiation plant studies carried out on high ash weakly coking coal, for generation of bulk cleans at desired qualities followed by carbonization tests for making of foundry coke for metallurgical uses.

## 2. Experimental

About 100 tons of low volatile coal (LVC) run-of-mine (ROM) was collected from the Western Sector of Jharia Coalfields. The representative raw coal was taken and first screened at 75 mm. The +75 mm fraction was crushed in a single roll crusher to −75 mm. The overall combined fraction of the product below 75 mm was taken for further studies viz., screen analyses and Float & Sink tests.

**Table 1:** Screen analysis of raw coal crushed to 75 mm

| Size (mm) | Wt.% | Ash% |
|-----------|------|------|
| 75–13 | 74.2 | 46.1 |
| 13–6 | 11.4 | 43.6 |
| 6–0.5 | 9.0 | 36.5 |
| −0.5 | 5.4 | 28.4 |
| *Total* | *100.0* | *44.0* |

The size distribution of coal crushed to 75 mm is shown in Table 1. It is seen that about 74.2% of the crushed material was above 13 mm and the ash content being 46.1%. The −0.5 mm fraction is about 5.4%, and its ash percentage being 28.4%. The individual size fractions were subjected to Float & Sink tests, and the relative density range was 1.40–1.80. The washability data of the combined fraction (75–0.5 mm) was generated and were used for plotting the Mayer curve (Fig. 1).

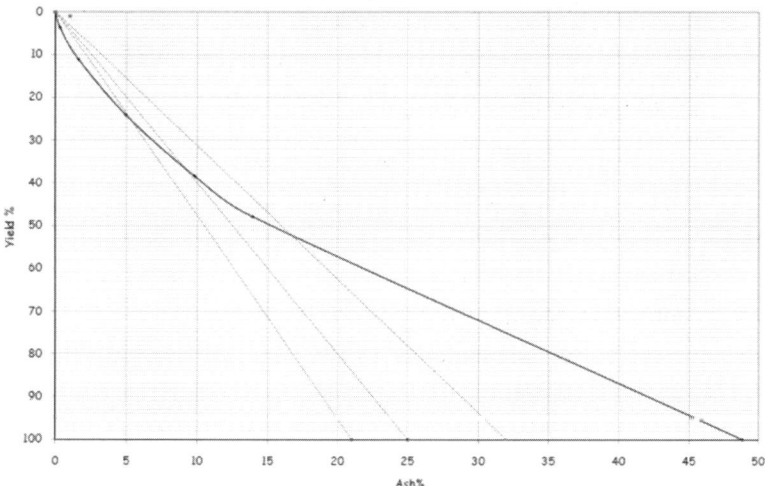

**Figure 1:** Mayer curve

## 2.1. Investigations on pilot plant washing of coals

The main objective of pilot plant washing tests of LVC coals was to get clean coal products of the following specifications:

  (i)   Clean coals of 21% ash level,
  (ii)  Clean coals of 25% ash level,
  (iii) Deshaled coals of about 32% ash level.

## 2.2 Operation of jig for generation of cleans

About 50 tonne of coal drawn from the bunkers and crushed to 75 mm were tested in a three product jig (Fig. 2) to obtain clean coal having 21% ash. The reject discharge rate was controlled in such a way so that about 50% rejects having 60% ash was discarded from the first compartment. Thus, clean coal of 21% ash content with a yield of 24% were obtained as middlings (28%) having about 37% ash were discarded from the second compartment.

**Figure 2:** Three-product Baum jig

The operating conditions of the pilot plant for the above test run are as follows:

- Coal feed rate: 10–12 tones/h.
- Piston stroke per min: 36.
- Average pulsation amplitude: 13.5–20 cm (1st compartment), 15.5–27 cm (2nd compartment).

In the next phase, about 35 tone of ROM coals were similarly crushed to 75 mm and fed to the jig to deliver clean coals of 25% ash. The operating conditions of the jig were varied. The discharge gates openings were controlled and adjusted. The products obtained from the test runs were weighed and the yield and final ash were determined. The deshaled coals were obtained by proportionately blending the cleans and middlings obtained from the 1st test operation as above.

## 2.3 Quality of coals charged to coke oven

The properties of coals as determined by laboratory tests on the representative samples are shown in Table 2.

**Table 2:** The properties of the coal samples

| Particulars | Proximate analysis | | | | CI | SI | LTGK | Gieseler's test | |
|---|---|---|---|---|---|---|---|---|---|
| | Moist % | Ash % | VM % | FC % | | | | Max fluidity (ddpm) | Plastic range (°C) |
| Raw LVC | 1.0 | 42.3 | 15.1 | 41.6 | 8 | 1 | C | 2 | – |
| Deshaled | 0.8 | 33.8 | 17.6 | 47.8 | 7 | 1 | C | 3 | – |
| LVC washed (1) | 1.1 | 24.1 | 17.2 | 57.6 | 11 | 1 | D/E | 7 | 10 |
| LVC washed (2) | 1.1 | 22.2 | 17.9 | 58.8 | 10 | 1.5 | E/F | 11 | 27 |
| Imported | 1.7 | 8.0 | 23.4 | 66.9 | 20 | 7.5 | G7 | 470 | 57 |

It may be seen that high rank, low volatile raw LVC coal with very high ash content (42.3%) has very poor coking properties. After deshaling, although there was appreciable reduction in ash content (42.3–33.8%) there was no improvement in its coking properties. When this coal was washed to ash levels of about 22% and 24%, there was significant improvement in their coking properties. The total reactive and mean reflectance is 34.9% and 1.31%, respectively.

## 2.4 Coking tests in ovens

### 2.4.1 Preparation of test charge

All the samples – raw, deshaled, and washed LVC coals as well as the imported coal were kept in separate heaps. After through mixing requisite quantities of samples were subjected to the coke ovens. Requisite quantities of samples were fed manually to the crusher. Representative samples of crushed coals were drawn from the filled up coal tubs for conducting laboratory tests on blends, samples of individual coals were first weighed, and then mixed thoroughly and tested. The pre-blended coals were fed manually to the crusher for preparing the charge. Representative samples of the crushed coal blends were also drawn for carrying out laboratory tests. The characterization data of the blend samples is shown in Table 3.

### 2.4.2 Carbonization conditions

The straight coal and blends were crushed to a fineness varying from 96.3% to 98.6% through 1.0 mm and the gross moisture of the coal charges was in the range of 2.0–4.4%. The proportion of the material passing through 0.5 mm (bug dust) screen varied from 83.3% to 91.4%. The mean size of the crushed test coal charges was in the range of 0.23–0.31 mm. The temperature (peak) of the coking chamber for all the coking tests, was in the range of

1240–1280°C excepting for the test no. 2 where it was 1100°C. The final coke mass temperature for three coking tests varied between 1010°C and 1060°C while for one test it was 862°C.

**Table 3:** Characterization of blended coal samples

| SI. no. | Particulars | Percent | Proximate analysis (air dried basis) % | | | | CI | SI | LTGK coke type | Gieseler's test | |
|---|---|---|---|---|---|---|---|---|---|---|---|
| | | | M | A | VM | FC | | | | Max fluidity (ddpm) | Plastic range (°C) |
| 1 | LVC washed (1) | 100 | 1.1 | 22.2 | 17.9 | 58.8 | 10 | 1.5 | E/F | 11 | 27 |
| 2 | Raw LVC | 45 | 1.2 | 25.0 | 18.6 | 55.2 | 10 | 2.5 | E/F | 10 | 20 |
| | Imported coal | 55 | | | | | | | | | |
| | Deshaled LVC | 50 | 2.5 | 20.4 | 19.6 | 57.5 | 12 | 2.5 | E | 35 | 46 |
| 3 | Imported coal | 50 | | | | | | | | | |
| 4 | LVC washed (2) | 70 | 1.0 | 20.8 | 19.0 | 59.2 | 15 | 2.5 | F/G | 22 | 42 |
| | Imported coal | 30 | | | | | | | | | |

### 5.2.4.3    Coking tests

Four coking tests (one on straight coals and three on blends) were conducted in the coke ovens. About 0.78–1.0 tonne (on dry basis) representative sample from each of the four test cokes were drawn, for conducting necessary tests. The details of the coking tests are shown in Table 4.

**Table 4:** Characterization of the coke

| SI. no. | Particulars | Percent | Mean size of coal charge (mm) | +25 mm Coke in stabilized coke (%) | Ash in coke (air dried) % | MICUM stabilized coke | | Porosity % | CRI | CSR |
|---|---|---|---|---|---|---|---|---|---|---|
| | | | | | | on 40 mm | through 10 mm | | | |
| 1 | LVC wash (1) | 100 | 0.31 | 94.1 | 26.5 | 89.9 | 6.9 | 36.9 | 22.7 | 62.7 |
| | (22% Ash) | | | | | | | | | |
| 2 | LVC (Raw) | 45 | 0.30 | 92.2 | 30.9 | 87.6 | 10.6 | 40.3 | 26.5 | 44.0 |
| | Imported coal | 55 | | | | | | | | |
| 3 | Deshaled LVC | 50 | 0.30 | 93.9 | 26.2 | 90.6 | 6.7 | 40.9 | 23.6 | 63.9 |
| | Imported coal | 50 | | | | | | | | |
| 4 | LVC wash (2) | 70 | 0.23 | 93.5 | 25.0 | 90.2 | 7.5 | 38.1 | 29.9 | 49.9 |
| | (24% Ash) imported coal | 30 | | | | | | | | |

## 3. Results and discussion

(a) From the pilot plant washing of raw LVC coals, it was observed that a yield of 23.2% cleans was obtained when the ash% of the same was found to be 21.3% from a raw feed coal of 44% ash. Straight carbonization of LVC washed coal (ash% 22.2) produced excellent metallurgical coke although the coal showed inferior/moderate coking property. The $M_{40}$ and $M_{10}$ indices of the stabilized coke were 89.9 and 6.9, respectively. The CRI and CSR values of the coke were 22.7 and 62.7, respectively. The ash air dried coke was 26.5% the proportion of +25 mm coke in the stabilized coke was 94.1%, the porosity of the coke was slightly lower (36.9%).

(b) The raw LVC coal had an ash of 44% after taking representative sample from the feed coal. The coke produced from the blend containing 45% raw LVC and 55% imported coal produced good coke having $M_{40}$, $M_{10}$ CRI, CSR, and porosity values of 87.6, 10.6, 26.5, 44.0, and 40.3, respectively but the ash content of the coke was very high (30.9%). The proportion of +25 mm size fraction in the stabilized coke was 92.2%.

(c) The deshaled coal, after discarding the reject, having yield 51.2% had 30% ash. The blending of 50% deshaled LVC with 50% imported coal produced very strong metallurgical coke ($M_{40}$ 90.6, $M_{10}$ 6.7, CSR 63.9) although the ash content was little higher. The porosity and +25 mm fraction in the stabilized coke were 40.9% and 93.9%, respectively.

(d) The investigation on the pilot plant testing produced the cleans of 31.9% yield having an ash of 25.9%. The 70:30 blends of washed LVC coal (24.1% ash) and imported coal produced good metallurgical coke. The ash content of the coke was 25.0% and the Micum indices were $M_{40}$ 90.2 and $M_{10}$ 49.9, respectively. The porosity and +25 mm fraction in stabilized coke were 38.1% and 93.5%, respectively.

## 4. Conclusion

It was inferred that raw LVC coals of lower seams is washable to the desired specification required for foundry iron industry. The yield of cleans at 21.3% and 25.8% ash were 23.2% and 31.9%, respectively. The deshaled coal at 30% ash showed a yield of 51.2%.

Coking tests on LVC coals at different ash levels and blends with imported coals were conducted in non-recovery ovens to examine the suitability of low volatile medium coking coals for the production of coke of required quality for the Foundry industry. The results tests reveal that:

(1) Straight carbonization of washed LVC coal (ash% 22.2) produced strong metallurgical coke acceptable to Foundry industry if slightly higher ash content in coke may be tolerated.

(2) Blending of 45% raw LVC coal (ash% 42.3) and 55% imported coal produced coke having satisfactory strength but the ash content was high (30.9%). It is expected that much better and stronger coke can be obtained if deshaled LVC is used in place of raw LVC coal. In that case ash content in the coke would not only come down and would be within the acceptable limit but at the same time physical properties of the coke would improve further.

(3) Blending of 50% deshaled LVC and 50% imported coal produced very strong metallurgical coke with slightly higher ash (26.2%) content.

(4) Blending of 70% washed LVC coal (ash% 24.2) with 30% imported coal produced satisfactory metallurgical coke, acceptable to Foundry industry.

## 5. Acknowledgement

Authors are thankful to the Director, Central Institute of Mining & Fuel Research, Dhanbad for giving permission to publish the paper. The authors are also thankful to all the staff members of Coal Preparation Division, CIMFR (Digwadih Campus) for their kind support.

## 6. References

[1] Geological Survey of India (GSI) Report, Government of India. Inventory of Indian Coal Resources, 2009.

[2] K. Sen, Chaudhuri, S.G., Narasimhan, K.S. *"Nature of Low Volatile Coals of Jharia and Their Preparation Aspect International Conference on Energy"*, Asia Energy Vision-2020, New Delhi, 15–18 November, (1996), pp. 67–74.

[3] K. Sen et al. *"Multi Product Beneficiation of Inferior Coals to User Specific Products"*, XIV International Coal Preparation Congress, 2002, pp. 21–28.

[4] T. Gouri Charan et al. *"Washability and Pilot Plant Studies to Generate Bulk Cleans at Desired Qualities from Low Volatile Coking Coal of Jharia Coalfields, BCCL, CIL India"*, International Seminar on Coking Coals and Coke Making: Challenges and Opportunities, 2009, pp. 219–229.

# 16

# Optimisation of dense medium separator performance

Ray Wood and Chris Wood

*Partition Enterprises Pty Ltd, Fig Tree Pocket, Queensland, Australia*

**Abstract:** For some decades, density tracer partitioning data, coupled with model-based understanding of the effects of dense medium cyclone (DMC) dimensions and operating conditions, have allowed
- rapid and precise assessments of partitioning performance, leading to recommendations for adjustments to dimensions and operating conditions,
- prediction of the expected improvements in partition curve parameters for all coal size fractions in DMC feed,
- simulation to estimate the impacts on yield and revenue.

If the recommendations are then implemented, those predictions can be validated by post-adjustment density tracer tests. The coal industry has led these developments.

The use of radio-frequency-identifiable (RFID) density tracers is now routine, allowing real-time generation of a very accurate partition curve by a single operator. This paper reviews those developments and provides case studies.

**Keywords:** Density separators, optimisation, efficiency, Ep, density tracers, RFID

## 1. Introduction

Globally, the most common density separator for coal coarser than 1 mm is the dense medium cyclone, originally developed by the Dutch state Mines (Dreissen, 1945). Rights were acquired by Stamicarbon BV, which produced for licensees a comprehensive guide to dense medium cyclone (DMC) selection, installation and operation. Although subject to copyright, that document was widely distributed and proved very valuable. It laid down clear but restrictive specifications as to DMC configuration and dimensions, medium characteristics, feed rates and medium-to-ore ratio which would be expected to provide reasonably efficient partitioning. However, it was relatively inflexible. If its requirements were to be violated, often for sound financial reasons, it gave little or no indication of how partitioning efficiency may be compromised, or how other conditions may be adjusted to compensate.

To address this matter, mathematical models of DMC performance were developed. They can be used to predict the effects of DMC dimensions and operating conditions on partitioning outcomes.

Density tracers are used to define a partition curve for a density separator, providing a very accurate and low-cost alternative to partitioning performance assessment based on sampling and float/sink analyses. Density tracers in the form of painted ore particles have been used in experimental situations over many decades. Routine production and application of synthetic tracers, manufactured to specific densities, was pioneered at The Julius Kruttschnitt Mineral Research Centre for coal processing and at the De Beers Research Laboratories for diamond processing (Napier-Munn, 1985). Partition Enterprises Pty Ltd has been the principal manufacturer of density tracers, worldwide:

- for coal applications since the early 1980s,
- for processing of diamond and other ores since 2002.

Developments in regard to density tracer properties and automation of the testing process will be discussed.

## 2. The wood model of DMC performance

Perhaps the most enduring DMC performance model has been the so-called Wood Model. It was developed for coal operations, but its principles appear to apply to higher-density operations.

The model was presented as a PhD thesis (Wood, 1990) and is described in Wood (2013). It was based on a large body of test data generated in Australia by the author using DMCs of Dutch State Mines design both at pilot scale and in industrial scale coal preparation plants. The second publication provides guidance as to how the model may be modified to encompass more recent alternative configurations in terms of cyclone size and geometry (inlet shape and area and vortex finder diameter).

The model incorporates nine sub-models which follow a logical sequence (Fig. 1).

## 3. Performance audits using density tracers

### 3.1 Tracer numbers and densities

In Australian coal operations it has become common practice, for statistical confidence, to use 30 tracers at each selected relative density (RD), with intervals between tracers at RD intervals of 0.01 (up to RD 1.60), 0.02 (between RD 1.60 and RD 1.90) and larger intervals for RDs greater than 1.90.

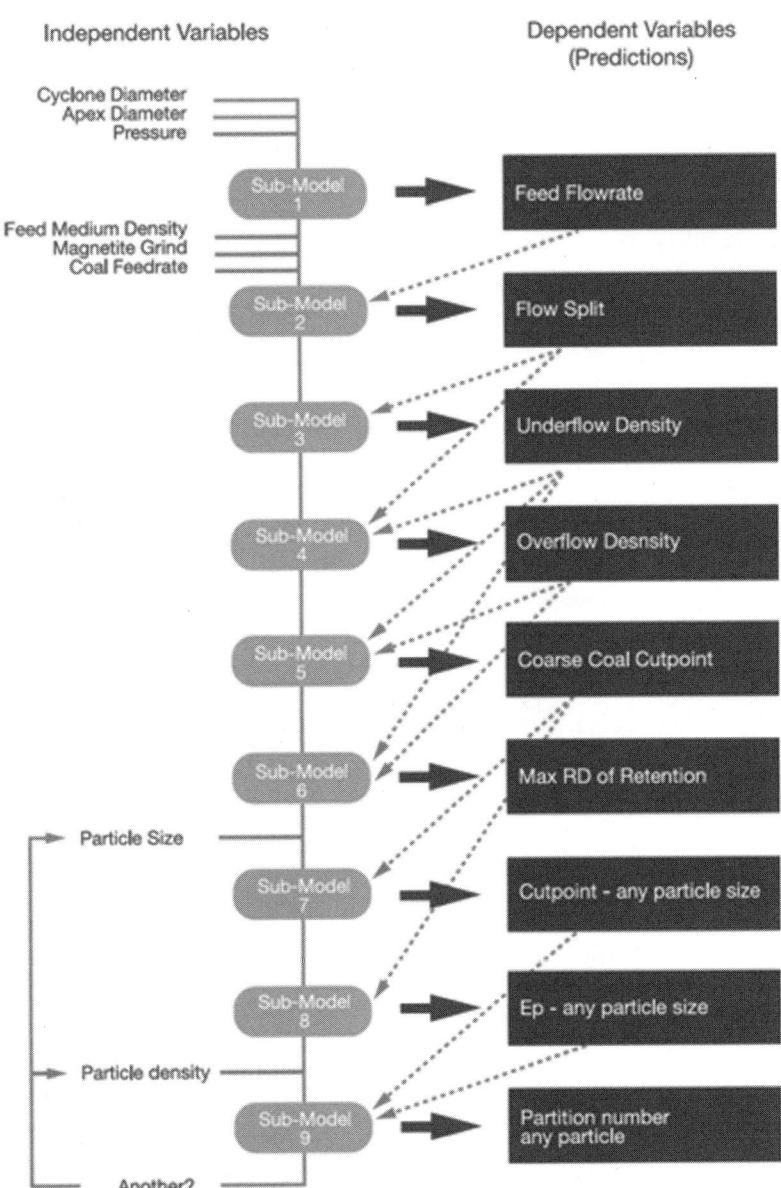

**Figure 1:** Logic flow in application of the Wood model of DMC performance

For diamond operations, density intervals of 0.05 RD are generally considered adequate.

Tracer RD selection should seek to encompass the entire RD range of the partition curve from 0% to underflow to 100% to underflow, unless there is significant bypass, which will be clear from the form of the partition curve.

## 3.2 Density tracer retrieval

Typically, only one or two sizes of density tracers are used. Features of the Wood Model then allow the use of density tracer partition curve parameters, supplemented by determinations of the densities of feed, overflow and underflow media, to estimate the partition coefficient for particles of any size and density. Because small density tracers can be difficult to retrieve manually, diamond plant operators sometimes test with feed off. This can give misleading results, because the partition curve can change dramatically when the circuit is loaded.

Density tracers may be manually retrieved using lightweight nets or scoops (Fig. 2). If the density tracers are ferromagnetic, they may be recovered using purpose-built magnets, but these can be unwieldy.

**Figure 2:** A lightweight net for retrieval of conventional density tracers from floats or sinks streams

## 3.3 Radio-detected density tracers

It was recognised that density tracers could be manufactured with integral radio-frequency-identification (RFID) "tags" which could be encoded with a unique ID which may be referred to a database to determine the tracer RD. Pioneering work in this area was conducted by de Korte (2003) and by Barbee et al. (2008). A commercial service for partition curve determination using RFID tracers was commenced by Partition Enterprises Pty Ltd in 2012.

The preferred configuration now is to detect RFID density tracers (Fig. 3) on the plant product and rejects conveyors. For Partition Enterprises Curve PRO (patented) system, the detectors are permanently installed so that an audit of any circuit in the plant may be conducted at any time. Detections are reported to a control centre which progressively plots the data points and updates the partition curve in real time (Fig. 4). Blue Lias Technologies plc is understood to be researching radio density tracers, but little information is currently available.

**Figure 3:** RFID density tracers of 4 mm, 8 mm, 13 mm and 25 mm edge length (Because of read range limitations the two smaller sizes are currently suitable only for pilot scale operations)

**Figure 4:** A control centre for RFID auditing collates detections and displays the partition curve in real time

## 3.4 Audit procedures

At the time of writing, Australian Standard AS5213 'Density tracer testing for measuring performance of coal density separators' is expected to be published early in 2019. It relates directly to coal processing using DMCs and baths. It could also be used for auditing other types of density separators and in treatment of diamond, iron, lithium, chrome and other ores.

## 3.5 Audit results

A partition curve derived using density tracers indicates considerably higher efficiency (lower $E_{p75/25}$) than one based on sampling and float/sink. This is because density tracers are robust, non-porous and manufactured to precise densities. They therefore provide a reliable measure of the separator performance, unclouded by sampling errors, particle breakage and porosity and wide density intervals between float/sink baths. A well-configured and well-operated DMC partitions 13 mm density tracers with $E_{p75/25}$ of around 0.008 RD or less (Fig. 5).

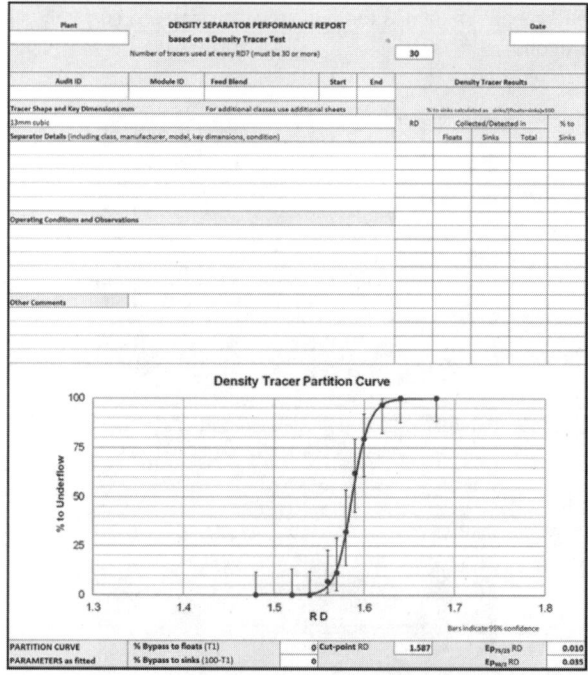

**Figure 5:** Example report from an audit using RFID density tracers (for confidentiality, site-specific details have been omitted)

## 4. Performance optimisation

Key performance indicators (KPIs) for a density separator may include:
- $E_{p75/25}$ (target 0.008 RD or less),
- Percentage misplacement of tracers with RD differing from the cut point by more than 0.1 RD (target 0%),
- Near-equal cut points for separators which operate in parallel and treat the same size fraction (target range of 0.015 RD or less).
- Logical procedures can be devised for correcting KPI violations. For DMCs:
- Differing cut points arise from differences in medium density set points, density gauge accuracies or differing offsets (offset is calculated as cut point minus density of feed medium).
- Serious misplacement arises from overload, or from surging to underflow, triggered by temporary retention of large, near-gravity particles.
- Violation of a target for $E_{p75/25}$ may require reference to a reliable and detailed performance model to determine how it can best be corrected. Possible means, starting with those which are simpler to implement, include adjustments to inlet head, medium bleed rates, magnetite grade, spigot diameter, vortex finder diameter, inlet diameter and complete replacement of the cyclone.

## 5. Partitioning optimisation – Case studies

### 5.1 Reducing $E_{p75/25}$ for an individual DMC

In industrial scale plants, the authors have observed density tracer $E_{p75/25}$ values as low as 0.004 RD (Fig. 6).

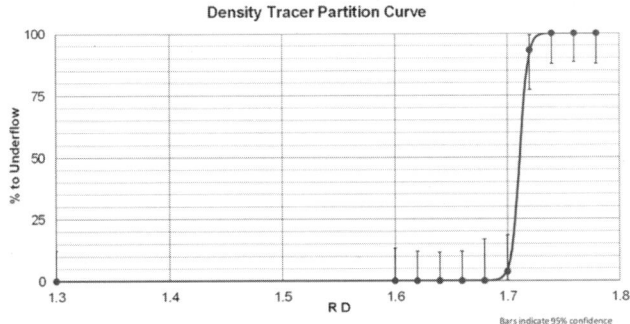

**Figure 6:** Example of an excellent partition curve from a 1500 mm DMC (using 30 tracers per RD)

Figure 7 shows the other extreme ($E_{p75/25}$ of 0.064 with misplacement to sinks exceeding 20%, even for the lowest density particles). Clearly, there is considerable room for improvement of $E_p$, with corresponding very large yield improvement.

In many plants such problems are not well documented, because partition curve audits are rarely conducted. Further the misplacement may not always be present, being triggered by factors such as a change of raw coal type. Finally, plant personnel may not have the knowledge and skills to reliably correct the situation.

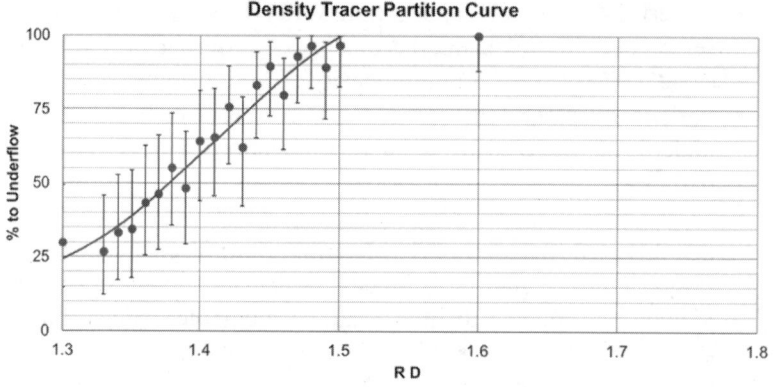

**Figure 7:** Exceedingly poor partitioning

A more typical situation will now be discussed. The first density tracer audit of a circuit in this plant is shown in the first entry in Table 1, and Fig. 8 (red curve). There is no gross misplacement, but $E_{p75/25}$ is poor at 0.024 RD units.

**Table 1:** Partial optimisation of DMC efficiency by progressive adjustments of key dimensions and operating conditions

| Inlet ratio, inlet diameter, cyclone diameter | Inlet head, cyclone diameters | Spigot ratio, spigot diameter, cyclone diameter | Ep75/25, RDU |
|---|---|---|---|
| 0.340 | 10.4 | 0.354 | 0.024 |
| 0.340 | 10.6 | 0.313 | 0.020 |
| 0.311 | 11.7 | 0.313 | 0.012 |
| 0.311 | 12.2 | 0.313 | 0.011 |
| 0.281 | 12.4 | 0.313 | 0.009 |
| 0.281 | 12.1 | 0.313 | 0.009 |

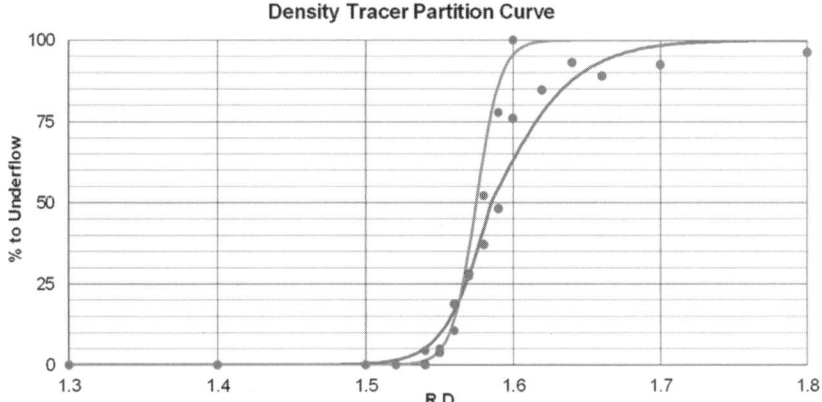

**Figure 8:** Partition curves before and after partial optimisation. $E_{p75/25}$ reduced from 0.024 to 0.009 RD units

The initial result and corresponding observations of DMC dimensions and operating conditions were assessed with the aid of the Wood DMC Performance model. It showed:

- that the prime reason for poor partitioning was the use of oversized orifi, especially the inlet, and
- there was considerable scope for "driving the separation harder" in terms, for example, of increasing the "differential" between the densities of underflow medium and overflow medium.

An optimisation campaign was pursued over three months, and summary audit results are presented in lines 2–6 of Table 1. For simplicity, only three key operating parameters are shown, together with the $E_{p75/25}$ values for RFID density tracers.

The circuit in question utilised a single DMC of nominal diameter 1220 mm operating with cut points in the RD range 1.58–1.60. Feed rates ranged from 760 to 1080 t/h with little effect on $E_p$.

First steps were the simple ones of reducing spigot diameter and increasing inlet head before reducing the inlet area. It was considered that good efficiency could be realised if the inlet ratio were reduced to about 0.22. However, it was recognised that capacity would be reduced, perhaps to an unacceptable level, unless a larger DMC (with appropriate orifice ratios) was installed. The operator decided to proceed with caution, reducing inlet area in small steps.

The first inlet reduction was to a ratio of 0.311. Combined with the changes of spigot diameter and inlet head, this reduced $E_{p75/25}$ to half the original value.

The second inlet reduction was to a ratio of 0.281. Although we consider this still to be oversized, our target $E_{p75/25}$ of 0.008 RD was nearly achieved (lines 5 and 6). Although there was no evidence of overload, for ongoing operation the owner reverted to the inlet ratio of 0.311, with no increase in cyclone body size, and some confidence that there was still spare capacity.

In Fig. 8, the green data present the density tracer partition curve from the last audit in this exercise. Further improvement would have required feed rate reduction, or installation of a larger DMC.

In that regard, it may be noted that in such situations efficiency may be improved without increasing flows and with no changes to ancillary equipment such as pumps and screens. Further, the costs of DMC replacement, including costs of plant downtime and of any plant modifications to fit the larger unit, are trivial when compared with the ongoing revenue benefits of even a small yield increase.

In this case, simulations by the DMC manufacturer put the yield improvement at 1.13% of its initial value. For Australian coking coal circuits with lower operating densities and high levels of near-gravity material, improvements of some percentage points are common.

## 5.2 Matching cut points for parallel circuits

To optimise yield across circuits which operate in parallel, it is important to recognise the "equal incremental ash" principle (Abbott, 1981). For separators of similar efficiencies treating similar coals, conformance with the principle requires that cut points be similar.

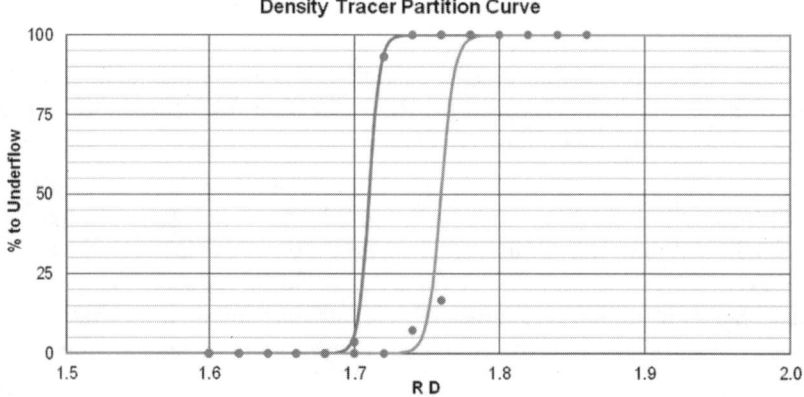

**Figure 9:** Cut point mismatch between parallel circuits with the same density set point

Figure 9 shows partition curves determined on the same day for two efficient DMC circuits operating in parallel. Although feed medium density set points were the same, the first circuit is clearly separating at a lower cut point than is the second circuit. That difference of 0.05 RD units must arise from differences in density gauge calibration errors and/or in the "offset" between cut points and feed medium densities.

To a first approximation, this discrepancy can be eliminated if, temporarily, density control for the second circuit is set for a medium density 0.05 RD units greater than that for the first circuit. To be a little more precise one may make the set point difference greater than 0.05 on the basis that, except at very low medium density, an increase in set point induces a smaller increase in cut point.

## 6. Conclusion

The validity, high precision and utility of density tracer audits have been documented over more than 35 years. The recent advent of reliable systems for detecting and reporting audit results based on RFID density tracers has greatly enhanced the ease of conducting audits and this system is now being adopted by major coal producers, with positive impacts on product yield.

## 7. References

[1] Abbott, J., 1981. The Optimisation of Process Parameters to Maximise the Profitability from a Three-Component Blend. *Proc 1st Aust Coal Prep. Conf.,* Newcastle, April.

[2] Barbee, C., Wood, C., Luttrell, G., 2008. Development of a Transponder – Based Tracer System for Evaluating Dense Medium Separator Performance. *SME annual meeting,* Salt Lake City, UT, USA.

[3] De Korte, G.J., 2003. Comments on the Use of Tracers to Test Dense-Medium Plant Efficiency. South Africa CSIR article.

[4] Dreissen, M.G., 1945. The Use of Centrifugal Force for Cleaning Fine Coal in Heavy Liquids and Suspensions with Special Reference to the Cyclone Washer. *J. Inst. Fuel,* **19**, (105), pp. 33–50.

[5] Napier-Munn, T.J., 1985. The use of density tracers for the determination of the Tromp curve for gravity concentration processes. *Trans. Inst. Min. Met.,* 94, C47–C53.

[6] Wood, C.J. (1990), 'A Performance Model for Coal-Washing Dense Medium Cyclones', PhD thesis, University of Queensland (unpublished).

[7] Wood, C.J. (2013), 'Coal Washing Dense Medium Cyclones', in Mathewson D. and Ryan G. (Eds), *ACARP Dense Medium Cyclone Handbook,* Australian Coal Preparation Society, Australia.

# 17

# Recommissioning of densifiers at export plant

Vutomi Shikwambana, Selby Mphela, Steyn Human

*South 32, Johannesburg, South Africa*

**Abstract:** Export plant is a two-modular plant consisting of double stage wash with a nominal through-put of 2000 t/h. The plant was commissioned in 2010 to produce export and domestic quality fed from open cast mine. Secondary stage requires density range greater than 1.8 t/m3 to achieve the domestic market quality hence the circuit was designed with densifiers. The circuit consists of two 165 mm cyclones with a 45 mm spigot gravity fed from correct medium headbox stationed at a height of 11.55 m and producing a pressure of between 60--160 kPa depending on the feed relative density. The densifiers blocked repeatedly during commissioning in 2010 due to coal misplaced from the mixing box to the correct medium tank. As a result, the densifiers were decommissioned and therefore recommissioned in 2018 because the circuits were using more magnetite from the over-dense. The secondary medium circuit was optimised to reduce the amount of coal misplaced into the circuit by increasing the mixing box pump output and reduce the volume of medium feeding the mixing box. The excess coal in the system was restricted by perforated plate placed on the headbox compartment feeding the densifier and cleaned by operators during plant inspection. This allows free flow of medium into the densifiers to maintain the desired pressure and reduced the use of raw magnetite in the secondary circuit. The weir of the densifying cyclone feed was modified to 0.5 mm perforated plate to prevent coal from flowing through.

**Keywords:** Correct media, Densifying cyclone, Bernoulli equation, Distribution

## 1. Introduction

Export plant is a coal processing plant commissioned in 2010 located in Mpumalanga province of South Africa with an annual production of 14 million tons. The plant consists of a double stage wash mainly primary and secondary. Secondary wash correct medium (CM) circuit normally ranges above 1.8 t/m³ to achieve the product qualities. Production of over-dense stream into the correct medium tank is used during start-ups to increases the rate required to achieve the density. The use of the over-dense stream is the easiest way to increase the density however cost intensive (Fig. 1).

Densifying cyclones are widely used in dense medium circuit to improve density control into the correct medium tank. The combination of magnetic separators and densifiers reduces the time it takes for the correct medium circuit to achieve the required density without using over-dense. The challenge during commissioning was +0.63 mm coal particles blocking densifying cyclones due to overflowing from the mixing box. Mixing box level is controlled by cyclone feed pump pressure and coal to medium ratio addition into the tank. For example, coal circulates in the CM circuit when cyclones lose efficiency over time and coal will be purged from the circuit when the pump efficiency is improved.

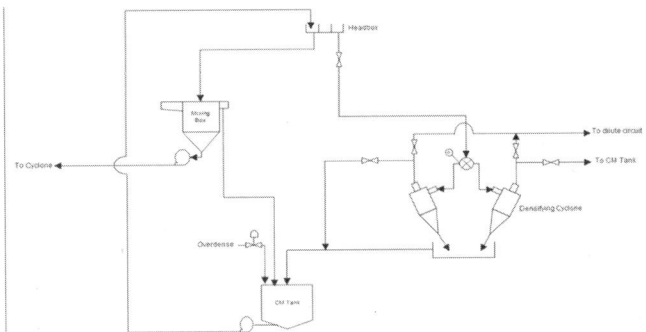

**Figure 1:** Schematic diagram of the densifying cyclones

The headbox was improved because of the coal misplaced into the CM circuit when operating conditions change. Headboxes contain compartment with a weir to control the flowrate of the medium over the weir. The weir is made of steel plates however coal would flow over the weir and flow into the densifying cyclone circuit. The weir of the densifying cyclone feed was modified to 0.5 mm perforated plate to prevent coal from flowing through. However, the perforated plate blocks due to particles that get stuck on the holes. The coal is removed by operators when the control-room technician alerts about the densifying cyclone pressure drop.

## 2. Experimental procedure

Manual sampling on the cyclone feed, overflow and underflow streams was conducted daily and composited weekly. Cyclone pressures were measured during sampling intervals. Apparatus used during sampling was a one litre (L) flask and 20 L collecting bucket. Filter press, Davis tube, oven and screen shaker were used in the laboratory to perform different analysis.

Davis tube was used to determine percentage magnetic in the pulp. Pulp was weighed, filtered, dried in an oven and weighed. Dry sample was split into 10 g using spinning riffler to obtain even distribution. The 10 g sample was introduced into the tube for about 12 min at 90 strokes per minute. Magnetic sample collected by Davis tube was dried and weighed.

Screen shaker with size range of 25 μm, 45 μm, 63 μm, 90 μm, 125 μm, 150 μm and 500 μm was used to obtain particle size distribution. Pulp sample was filtered, dried and weighed. Dry sample was wet screened on the shaker. Different size range collected were dried and weighed.

## 3. Equation

Figure 2 below shows one dimensional Bernoulli equation for viscous fluid.

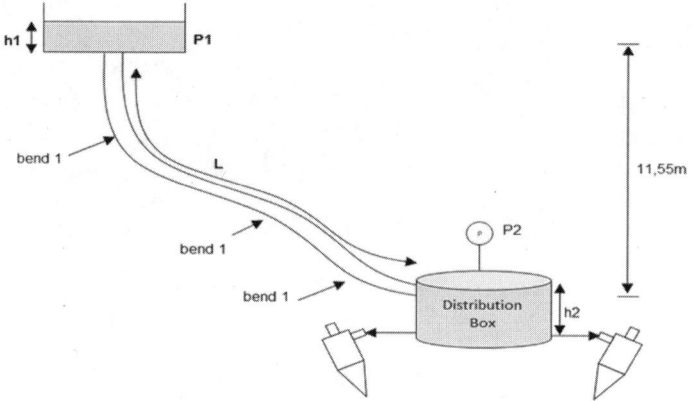

**Figure 2:** Schematic view of waterflow from a reservoir through pipeline

According to Bernoulli's principle:

$$P + \frac{\rho v^2}{2} + \rho gh = \text{constant} \tag{1}$$

Therefore, $\dfrac{P_1}{\rho} + \dfrac{v_1^2}{2} + gh_1 = \dfrac{P_2}{\rho} + \dfrac{v_2^2}{2} + gh_2 + \Delta h_L$ (2)

Hydraulic loss between two cross section along a pipe is equal to the difference of total energy for this cross section:

$$\Delta h_L = H_1 - H_2 \tag{3}$$

$$\therefore \qquad \Delta h_L = \frac{P_1 - P_1}{\rho g} \tag{4}$$

Head loss is expressed by Darcy–Weisbach equation:

$$h_L = f\frac{Lv^2}{D2g} \tag{5}$$

where $P_1$ is the pressure inside tank 1 and $P_2$ is the pressure inside tank 2, $r$ is the density of fluid, $D$ is the pipe diameter, $L$ is the pipe length, $g$ is the gravitation acceleration, $h_1$ is the elevation of tank 1 and $h_2$ is the elevation of tank 2, $v_1$ is the velocity of fluid in tank 1 and $v_2$ is the velocity of fluid in tank 2.

The pressure exerted by the densifying cyclone is:

$$\therefore \qquad P_2 = P_1 + \frac{\Delta v^2 \rho}{2} + \rho g \Delta h - \Delta h_L$$

# 4. Results and discussion

Figure 3 illustrates the relationship between cyclone pressure and change in percentage magnetic. The results show cyclone pressure is direct proportional to the change between feed and underflow percentage magnetics. The percentage upgrade increases with the cyclone pressure and feed RD. Addition of over-dense, densifying cyclone underflow and magnetic separator into the correct medium circuit will improve density control during start-ups. The increase in density is maintained by the densifier after during operation to avoid the use of additional magnetite. Magnetite misplaced to overflow is recovered by magnetic separator and recycled back to the circuit. Secondary circuit medium, magnetic separator and over-dense is used to maintain primary circuit.

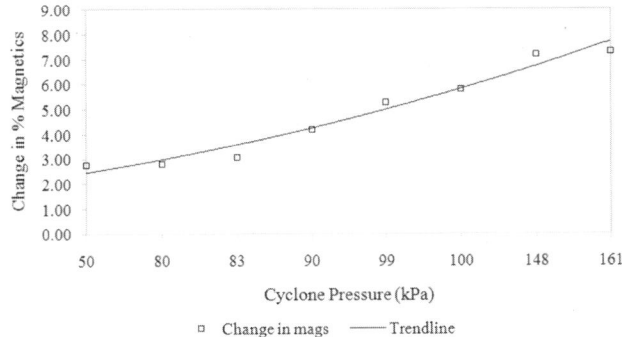

**Figure 3:** Relationship between pressure and percentage magnetic upgrade

The results shown on Fig. 4 illustrate magnetite recovery greater than 90% in the underflow stream with cyclone pressure ranging from 60 kPa to 160 kPa.

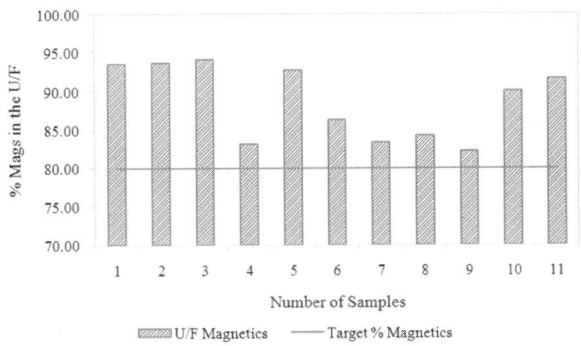

**Figure 4:** Percentage magnetics in the underflow (U/F) per sample

Figure 5 illustrate percentage solids in the feed, underflow and overflow stream. The percentage solids in the feed were ranging between 50% and 55% with an upgrade to the underflow ranging between 60% and 75%. The recovery of solids to the underflow improves as the pressure increase. The pressure decreases as the flow through the perforated deteriorate reduces over time due to non-magnetic particles getting stuck on the holes. When this occurs, operators are required to vibrate the plate so that the particles can float in the medium and inform the control-room to increase the DMS cyclone pressure. Therefore, this will prevent coal from overflowing the mixing box into the correct medium circuit.

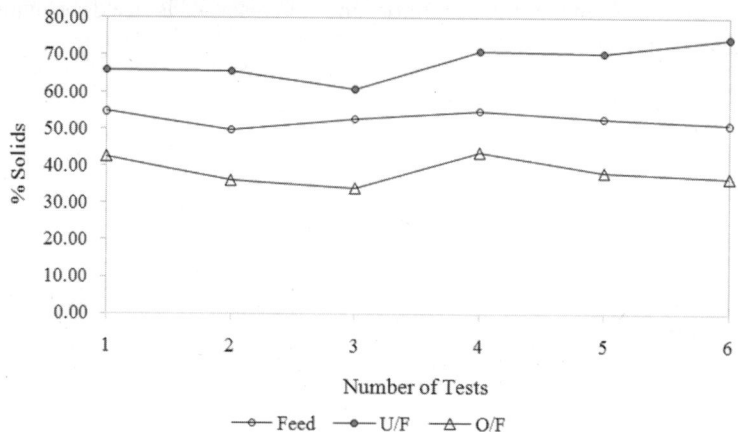

**Figure 5:** Percentage solids in the feed, underflow and overflow

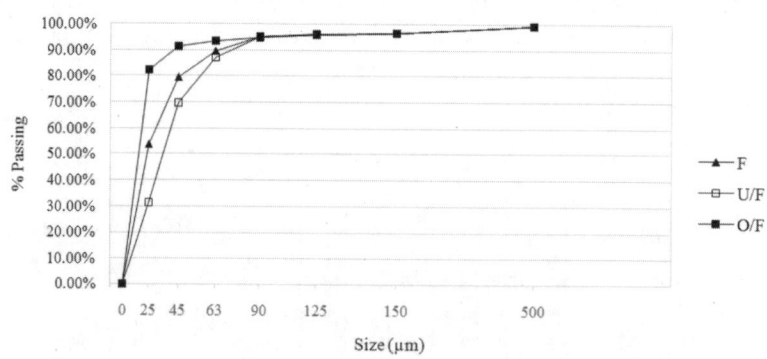

**Figure 6:** Particle size distribution showing percentage passing

Figure 6 shows particle size distribution of the feed, undersize and overflow. Feed, overflow and underflow constituting of 80% particles below

45 µm, 25 µm and 63 µm, respectively. The data shows that there are less than 1% of the 500 µm particles in the circuit. This illustrate that perforated plate is useful in preventing the larger particles into densifying cyclone.

## 5. Conclusion

Densifying cyclones is used to reduce the start-up time required to achieve the set point. Densifiers uses pulp from the correct medium circuit and upgrades the density. Moreover, magnetic separators and densifying cyclones can be used simultaneously to increase the density of the circuit more rapidly. The higher the feed pressure and density, the higher the percentage solids in the underflow. Densifiers are cost effective and cheap to maintain however sensitive to bigger particles that may result in the feed line blocking.

## 6. Acknowledgement

We would like to express our gratitude to South Africa Energy Coal (South32) for giving us this opportunity to conduct the study.

## 7. References

[1] B. R. Munson, D. Y. (1998). *Fundamentals of Fluid Mechanics*. John Wiley and Sons, Inc.

[2] Boucher, Y. N. (1999). *Introduction to Fluid Mechanics*. Butterworth Heinemann.

[3] Cimbala, Y. C. (2006). *Fluid Mechanics*. McGraw Hill.

[4] Yan, D. S., A. a. ( January 2006). *Mineral Processing Design and Operation*. Perth, Australia.

[5] McDonough, J. (2004). *Lectures in Elementary Fluid Dynamics: Physics, Mathematics and Applications*. University of Kentucky: Lexington.

[6] Multotec. (2019, January 21). *Multotec*. Retrieved from Multotec. Website: http://www.multotec.com.au/densifying-cyclones.html.

[7] Multotec. (2019, January 21). *Multotec*. Retrieved from Multotec. Website: https://www.multotec.com/product/mineral-processing-equipment/cyclones/dewatering-cyclones.

[8] WARMAN. (2009). *Slurry Pump Handbook*. Johannesburg: WARMAN.

[9] White, F. M. (1999). *Fluid Mechanics*. McGraw-Hill.

# 18

# A dynamic simulation model for a single stage dense medium cyclone circuit

Nerrida Scott[1], Peter Holtham[2], Michael O'Brien[3]

*[1]Independent Consultant, Brisbane, Australia*
*[2]Independent Consultant, Brisbane, Australia*
*[3]CSIRO, PO Box 883 Kenmore, Queensland 4069, Australia*

**Abstract:** Measurement of dense medium cyclone (DMC) circuits has been enhanced in recent years by the development of novel instrumentation. These instruments, coupled with existing coal handling preparation (CHPP) plant data, have provided new information which has led to better knowledge of DMC circuits and medium behaviour. A DMC circuit model encompassing many years of past empirical work has now been developed in MATLABTM. The dynamic simulation model was based on a single stage DMC circuit and has been created from experimental and electronic data gathered from a thermal coal mine in the Clarence Moreton Basin in Australia. The model was developed as a joint collaboration between CSIRO and The University of Queensland's Julius Kruttschnitt Mineral Research Centre (JKMRC). This paper discusses the dynamic model and important novel aspects of the experimental work. Consideration will also be given to potential future applications of this model.

**Keywords:** Dense medium cyclones, DMC, coal preparation

## 1 Introduction

A dynamic model is essentially comprised of a series of variables that interrelate and change with each increment in time. By comparison, a steady-state model, which is fixed in a point of time and assesses the conditions in a circuit when the circuit is in equilibrium, the dynamic model allows an iterative loop to be created where changes in the variables can occur with time increments. The relationships between variables within the model loop determine the output at each time increment. When an independent variable changes, the output variable will change dependent on controller settings, lags and delays. In a coal dense medium cyclone (DMC) circuit, the various parameters such as pressures, densities, feed washability, medium quality and the level of non-magnetics (medium contamination), size distribution, sump levels and residence times all contribute to variation and hence can influence the resulting outputs from the plant. A dynamic model can be used to simulate the response of a circuit when changes in feed and set points within the circuit occur.

While steady state models are very useful for plant design and for assessing data from plant audits, dynamic models can expose hidden variations within a plant. Coal handling preparation plant (CHPP) models developed in the past include Askew (1983), Wiseman et al. (1987) and Meyer (2010). In the first two models, few empirical models were available for individual unit operations and computer technology was limited. The latter model was developed from a process control perspective and without the use of existing empirical models. The inclusion of empirical work and online plant data from a South Queensland coal mine into a new DMC circuit dynamic model provided an opportunity to take the research further. Instruments and sampling conducted as part of the experimental work at the mine site enabled more comprehensive measurement of streams within the DMC circuit. Empirical models summarised in Crowden et al. (2013) were also included in the model.

## 2. Methodology

Experimental work conducted at the mine CHPP involved numerous density tracer tests, medium sampling and monitoring of the control system. Observations were made during each plant visit and plant operations personnel were also consulted. Additional equipment was added in the underpans of the drain and rinse screens to monitor on-stream medium densities. The DMC overflow density was measured using a JKMRC developed probe (Firth et al., 2010). A standard nucleonic gauge was situated in the correct medium line. Screen motion analysers developed by CSIRO were placed on the screens and were able to measure mass flow rate across the screens. Electrical impedance spectrometers were also developed by CSIRO to measure medium density in the screen underpans (Firth et al., 2013). Medium samples were analysed using a Davis Tube for relative percentages by mass of magnetics and non-magnetics. Travel times through parts of the CHPP circuits were measured using radio frequency identification (RFID) tracers. Ultimately, sufficient data was collated from a series of events which occurred during normal plant operation. These plant events were then used to verify the dynamic model.

Sampling of the medium during various plant events revealed the amount of non-magnetics present and also determined the medium differential. Medium differential was used as a practical plant indicator of medium stability in the model. Research by Firth et al. (2014) found that the prediction of differential by measurement of overflow and underflow densities was linked with stability of the circuit. This also agreed with past work on differential as a predictor of stability by Davis (1987) and O'Brien et al. (2013). While the dynamic model developed considered non-magnetics as a whole, attempts were not made to single out individual clay types or differentiate coal from clay in the medium. It is understood that this is an area where further work could enhance the understanding of medium behaviour in future.

RFID tracers were chosen for determining delays within the dynamic model (Scott et al., 2015). Time lags between different unit operations were accounted for using residence times estimated experimentally using RFID tracers as shown in Table 1 below. The measured residence times showed large standard deviations in cases where the tracers followed the correct medium or dilute circuit suggesting the potential for settling or rafting of particles in the circuits. Smaller standard deviations were found for density tracers that followed the coal pathway through the dense medium circuit, suggesting that there were no route deviations or settling occurring.

**Table 1:** Residence times from plant test work (Scott et al., 2018a)

| Test | Residence time from | Residence time to | Average (mm:ss) | Min | Max | Standard deviation |
|------|---------------------|-------------------|-----------------|-----|-----|--------------------|
| A | Desliming screen | Drain and rinse overall | 01:01 | 00:36 | 02:11 | 0.0002 |
| B | DMC overflow/ underflow | Drain and rinse screen | 00:20 | 00:15 | 00:26 | 0.00003 |
| C, D | Drain underpans | Drain and rinse screen | 02:36 | 00:43 | 29:06 | 0.0029 |
| E | Feed weigher | Drain and rinse screen | 02:25 | 02:00 | 03:27 | 0.0002 |
| F | Magnetic separator concentrate | Drain and rinse screen | 09:50 | 01:10 | 39:36 | 0.0070 |
| G | Deslime water sump | Drain and rinse screen | 08:37 | 02:09 | 35:51 | 0.0072 |
| H | Crusher feed | Drain and rinse screen | 01:55 | 01:36 | 02:25 | 0.0002 |
| I | Wing tank overflow to correct medium sump | Drain and rinse screen | 06:53 | 01:23 | 31:48 | 0.0046 |

Scope of the dynamic model development did not extend to redefining empirical relationships for each unit operation given the wealth of empirical models already available for use as described in Crowden et al. (2013). It was considered that these empirical models provided a reasonable predictor in concert with the experimental data collected through field work to use in the dynamic model. The dynamic model was developed in MATLAB™ software, which offered the ability to sort data into matrices and vectors which greatly simplified the code. The structure of the model involved sequential steps in a repeating loop over a number of iterations equivalent to the plant's run-time. Time increments were in seconds, and the length of the run-time was user-determined (Fig. 1).

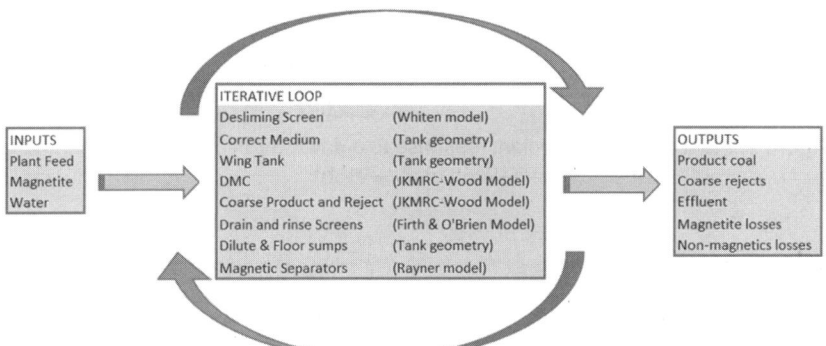

**Figure 1:** The structure of the dynamic model (Scott, 2017)

The iteration loop included all key aspects of the coal dense medium cyclone circuit; i.e. the desliming screen, magnetic separators, correct medium circuit (incorporating the correct sumps, wing tank and DMC), the dilute medium circuit (incorporating the magnetic separators) and the incorporated sumps within the correct and dilute medium circuits.

## 3. Results

The results of the MATLAB outputs were compared with the various events recorded during plant experimental work. While it is not possible to display all events here, a summary of model results used for verification are provided below. The density response shown below (Fig. 2) is a typical plant density response measured from plant instruments. The density drop from 1.35 RD to 1.30 RD a slower model response (5 min vs 12 min) (Fig. 3), however the trend indicates similarities between modelled and actual data. In this particular case, the density controller in the model could have been further optimized, however there are other influences on time taken for a density response, such as magnetite additions and control of the bleed valve.

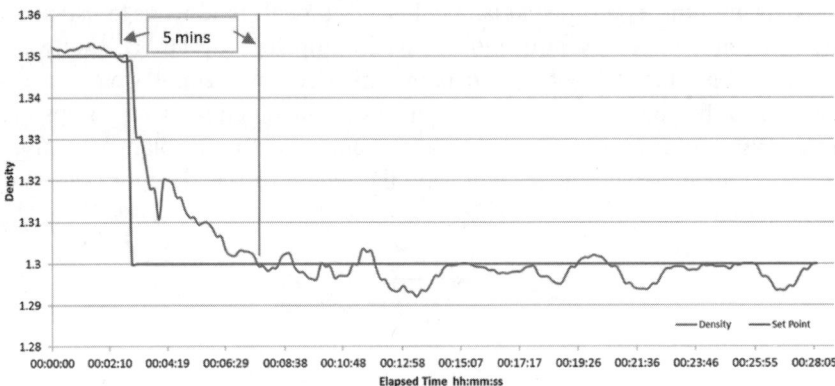

**Figure 2:** Plant data showing plant response to a downwards density set point change (Scott et al., 2018b)

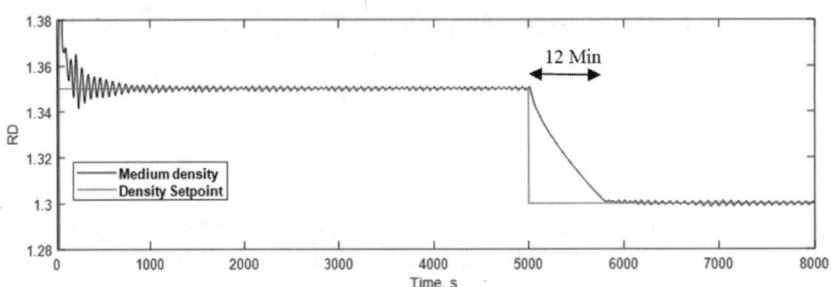

**Figure 3:** Dynamic model was adjusted to drop the density in the plant from 1.35 RD to 1.30 RD (Scott et al., 2018b)

The model also considered the DMC pressure. Figures 4 and 5 below show the plant results compared with the dynamic model result for pressure. Here, the plant feed tonnage was turned off on two occasions, the latter occasion being for a prolonged period of time. The causes of the feed off (coal off) periods in this case were unrelated to the dense medium circuit. Note the plant's pressure response to the short plant feed outage ($t = 20$ min) was considerably more rapid than the response for the longer feed outage ($t = 180$ min). While not noted in the graph below, a corresponding drop in non-magnetics was also noticed during the plant feed-off events.

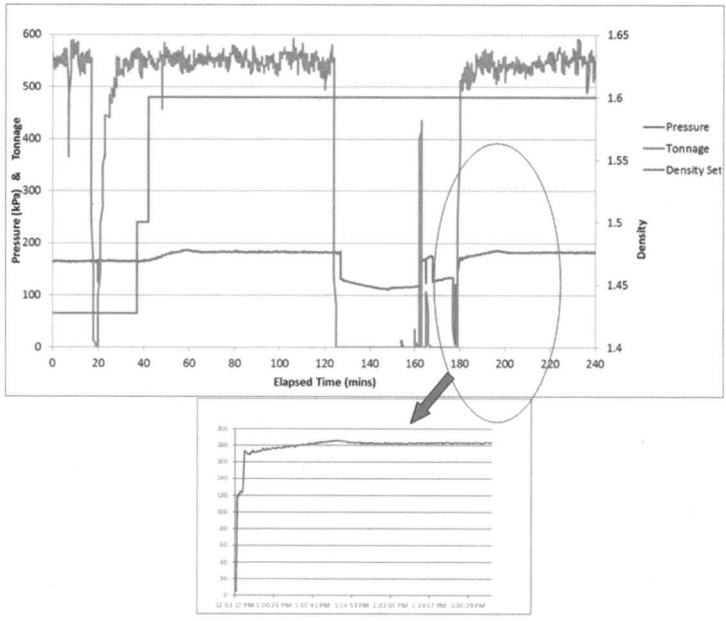

**Figure 4:** Typical plant pressure response – expanded graph of pressure (Scott et al., 2018b)

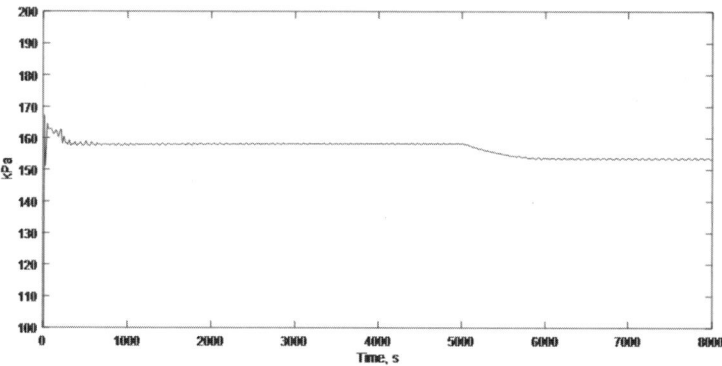

**Figure 5:** A similar pressure curve predicted by the dynamic model. The curve corresponds to a similar plant start up after the feed off events circled in the previous graph at 180 min (Scott et al., 2018b)

Relative proportions (by mass) of non-magnetics in the medium were also considered in the dynamic model. A breakage model was incorporated to simulate the breakdown of clays in the circuit. Figure 6 shows the results of physical measurements taken by in-plant sampling of the medium and the corresponding model response from start-up conditions. The time taken for non-magnetics to rebuild in a circuit was surprisingly long relative to the time it took for the DMC density and pressure to reach a steady state.

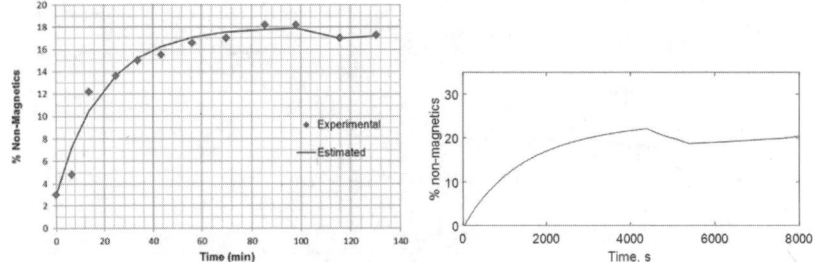

**Figure 6:** (Left) Experimentally determined build-up of percentage non-magnetics from plant start up condition as determined by plant data (Firth et al., 2014). (Right) MATLAB dynamic DMC model prediction of build-up of non-magnetics from a start-up condition ($t = 0$ s) with a density change at 5000 s, bleed opened at 4400 s (Scott et al., 2018b) (Timescale conversions: 20 min = 1200 s, 4500 s = 75 min)

The dynamic model percentage non-magnetics response was able to be altered by varying the amount of breakage estimated in the model. Altering the amount of correct medium bled to the dilute circuit also changed the rate of recovery. Opening the bleed appears to have a strong effect on non-magnetics as the circuit is quite efficient at removing non-magnetics via the magnetic separators. Closing the bleed holds more non-magnetics in the correct medium circuit, however this may have volume implications in some plants. The timing of fresh magnetite additions to the circuit is also important as can be seen in Figs. 7 and 8 below.

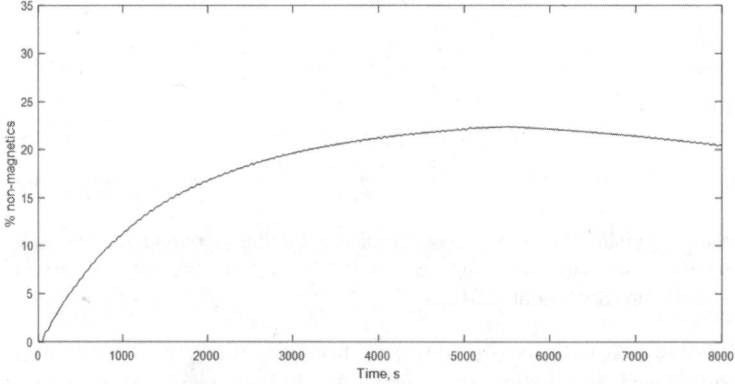

**Figure 7:** Model bleed opened at 5000 s. Non-magnetics in the circuit dropped slightly (Scott et al., 2018b)

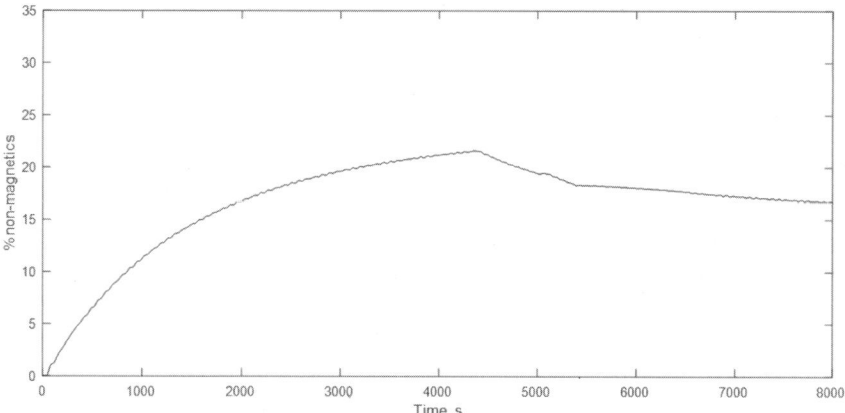

**Figure 8:** Model: Bleed opened at 5000 s with magnetite addition at 6500 s. Significant drop in non-magnetics in the circuit once magnetite is added (Scott et al., 2018b)

The unfortunate reality of current operating plants is that non-magnetics levels are not detected online. Given that, this parameter is invisible to plant operators on a control room screen it is useful to use the dynamic model to better explain what happens to the system response when a change is made to a DMC circuit. The powerful potential of incorporation of non-magnetics into the model's component balance is that dynamic predictions of medium behaviour can now be modelled. This provides useful insights for future plant control. It also enables use as an educational tool to demonstrate the effects of DMC circuit control. If operators are able to see the effects of a change, they will have the ability to better control the circuit and respond appropriately.

## 4. Conclusion

As part of the research project, a model validation process was completed and the trends generated closely resembled real plant operation. With modest adaptations, customising changes to equipment dimensions, control philosophies and plant layouts it is thought that this dynamic model could be applied to other DMC operations and possibly extended to a full plant model. There is much scope to use the dynamic model as a teaching tool, or possibly incorporated as part of a larger full plant model. Further development could also assist with the prediction of DMC losses, assessment of two stage washing and applying the model across other mineral sectors.

## 5. Acknowledgement

The authors would like to recognise the highly significant contributions of the late Dr Bruce Firth to the supervision of this PhD project and his related research work during his time at CSIRO. Dr Firth was instrumental in guiding the course of this research.

The authors would also like to acknowledge the assistance of New Hope Group in enabling and facilitating research work to be conducted at their site, in particular, Robert Rashleigh and the CHPP team. Also we would like to acknowledge the contributions of Dr Christopher Wood from Partition Enterprises also contributed valuable time, expertise and developed specialised RFID density tracers and associated equipment to assist with the project. The authors also greatly appreciate the financial support of ACARP, who funded this research work.

## 6. References

[1] Askew, H. (1983), Dynamic Simulation and Automatic Control of Dense Medium Circuits. PhD Thesis, The University of Queensland, Brisbane, Australia.

[2] Crowden, H., Atkinson, B., Clarkson, C., Firth, B., O'Brien, M., Wiseman, D., and Wood, C. (2013), *"ACARP Dense Medium Cyclone Handbook"*, Eds. Mathewson, D. and Ryan, G. Australian Coal Preparation Society, Newcastle, Australia.

[3] Davis, J.J. (1987), A Study of Coal Washing Dense Medium Cyclones. PhD thesis, The University of Queensland, Australia.

[4] Firth, B., Holtham, P., O'Brien, M., Hu, S., Dixon, R., Burger, A., and Sheridan, G. (2010), *"Joint Evaluation of Monitoring Instruments for Dense medium Cyclones"*, ACARP Project C17037.

[5] Firth, B., Holtham, P., O'Brien, M., Hu, S., Dixon, R., Burger, A., and Scott, N. (2013), *"Linkage of Dynamic Changes in DMC Circuits to Plant Conditions"*, ACARP Report C20050, ACARP.

[6] Firth, B., O'Brien, M., Holtham, P., Scott, N., Hu, S., Dixon, R., and Burger, A. (2014), *"Dynamic Impacts of Plant Feed and Operating practices on a Dense Medium Cyclone (DMC) Circuit"*, 15th Australian Coal Preparation Conference Proceedings, 14–18 Sept Gold Coast, Australia.

[7] Meyer, E.J. (2010), The Development of Dynamic Models for a Dense Medium Separation Circuit in Coal Beneficiation. University of Pretoria, Pretoria.

[8] O'Brien, M., Firth, B., McNally, C., and Taylor, A. (2013), *ACARP Project C20051 – The Effect of Dynamic Changes in Medium Quality in Coal Processing*, February 2013.

[9] Scott, N., Wood, C., Holtham, P., O'Brien, M., and Firth, B. (2015), *"Integration of Plant Residence Time Measurement into a Dynamic Model of a Coal Dense Medium Circuit"*, Coal Prep 2015, Lexington, USA, 27–29 April.

[10] Scott, N. (2017), Dynamic Analysis of Dense Medium Circuits, PhD Thesis, The University of Queensland, Brisbane, Australia.

[11] Scott, N., O'Brien, M., Wood, C., Holtham, P., and Firth, B. (2018a), *A Study of Residence Times and Tracer Partition of Different Sized Particles in a Coal Dense Medium Circuit*, 17th Australian Coal Preparation Conference and Exhibition Proceedings, Sept 2018 Brisbane, Australia.

[12] Scott, N., Holtham, P., Firth, B., and O'Brien, M. (2018b) *The Development of a Dynamic Model for an Australian Dense Medium Cyclone Circuit*, 17th Australian Coal Preparation Conference and Exhibition Proceedings, Sept 2018 Brisbane, Australia.

[13] Wiseman, D.M., McKee, D.J., and Lyman, G.J. (1987) *Dynamic Simulation of Coal Preparation Circuits and Control Systems,* NERD&D Project No. 839, JKMRC, University of Queensland, Sept 1987.

# 19

# Overview of prospective coal processing technologies

Ilia Beloglazov and Vladimir Bazhin

*Saint-Petersburg Mining University, Saint Petersburg, Russia*

**Abstract:** The Russian coal industry is an intensive sector of the economy. Volumes of coal mining have been steadily growing for 7 years and outstripping the growth of oil production. Russia is the second largest country in terms of coal reserves in the world (157 billion tons). The main part of the coal reserves is high-ash (low-grade), which leads to the need to introduce innovative solutions for their processing. The share of coal exports is 40–50%, which is used in particular for gas synthesis in Europe. Perspective technologies of deep coal processing allow using high-ash coal effectively for obtaining gas with optimal parameters for its further synthesis. Thus, the development and implementation of advanced deep coal processing technologies to produce products with high added value (methanol, ammonia, synthetic fuel, and kerosene) for competition with refined products is an urgent task.

In the present article, perspective technologies of deep coal processing, exploited in industrial or pilot industrial scale, are presented. These technologies are implemented on the basis of plasma-fuel systems (PTS), created to increase the efficiency of burning energy coal and reduce harmful emissions.

**Keywords:** Plasma technology, gasification of coals, combined gasifier, plasma-fuel systems

## 1. Introduction

The existing commercial projects connected with the commissioning of new technologies for coal processing, for example, technologies with in-cycle gasification of solid fuels, were realized in the USA, the Netherlands, Italy, Spain, Japan, and China through the construction of demonstration units [1,2–5]. Their main advantages from the position of the second decade of the 21st century are [1,2,4]: high ecological purity exceeding existing and expected European environmental standards (SOx and NOx emissions significantly below 100 mg/N m$^3$, dust less than 5 mg/N m$^3$); the possibility of achieving a wide range of load regulation of power units and providing maneuvering operating modes of the units; availability of waste technologies for complete utilization of liquid and solid waste; the experimentally proven possibility of full utilization of $CO_2$ with less cost for this than with flaring or burning in different modifications of fluidized bed furnace [6].

Taking into account that over the last 5–10 years the degree of readiness of new industrial facilities and installations with in-cycle gasification of coal has increased significantly and exceeded 80% [1,4,5], interest in such technologies is increasing [7]. However, technologies based on plasma coal gasification in the world are not so widely represented. In this situation, there is an interest in analyzing and comparing the effectiveness of introducing various coal gasification technologies and coal properties [8].

## 2. Coal processing technologies

### 2.1 Direct-flow plasma-fuel system

Technology based on plasma thermochemical preparation of coal consists in heating the dust–air mixture (coal dust + air). Plasma-fuel systems (PTS) are pulverized-angle burners equipped with electric arc plasma torches, which are the main element of the PTS (Fig. 1).

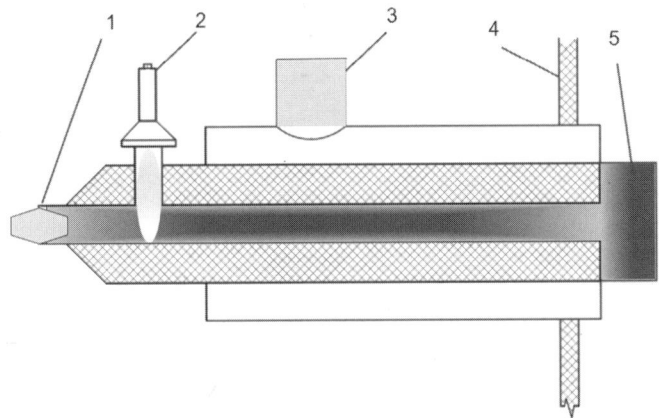

**Figure 1:** Direct-flow plasma-fuel system: 1, air mixture feed; 2, plasma torch; 3, secondary air; 4, furnace wall; and 5, furnace

Heating is performed to the temperature of the volatiles and partial gasification of the coke residue. In this case, the high-reactive two-component fuel (combustible gas + coke residue) is obtained from the initial low-grade coal. When it is mixed with secondary air, the two-component fuel ignites and burns steadily in the combustion chamber. It has been experimentally shown that due to the two-stage combustion of pulverized coal fuel during the operation of the plasma torch in the stabilization mode of the pulverized-coal flare, the NOx concentration at the outlet from the furnace is reduced by half, while reducing the fuel burner by four times [9].

## 2.2 Plasma stabilization

This technology is ready for realization for any types of power furnace, coal-fired torches, and coals. To provide liquid slag removal in furnaces with liquid slag removal, a plasma technology has been created for the oil-free stabilization of the liquid slag outlet. This technology is based on electro thermochemical preparation of fuel for combustion. For its implementation, a superspecific PTS was developed [10], as a replacement for fuel oil supernumerary injectors. The project scheme of the installation on the furnace and the construction of supernodes PTS are shown in Fig. 2.

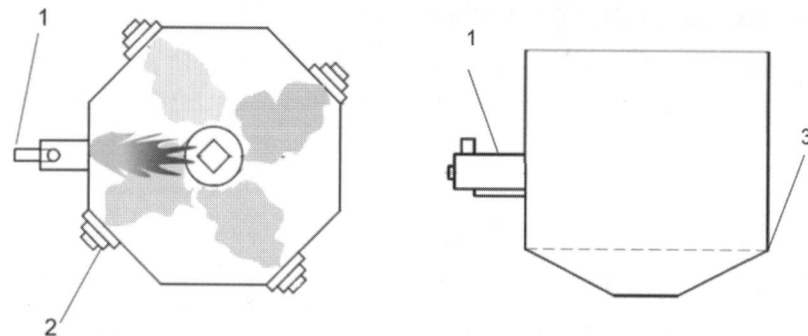

**Figure 2:** Scheme of plasma stabilization of the output of liquid slag in furnaces with liquid slag removal using supernodes PTS: 1, superspecific PTS; 2, main pulverized-coal torch; 3, chamber feed line

Tests were carried out on the furnace «BKZ – 640 of Gusinoozerskaya GRES» at a steam load of 450 t/h furnace, i.e. 70% of the rated steam capacity. The power of the plasma torch is 70 kW, the coal consumption through the plasma–coal burner is 3.7 t/h. The temperature of the torch at the outlet of the plasma–coal burner reached 13,000°C, and the temperature of the liquid slag increased by 70–80°C, which ensured a stable evacuation of the slag in the furnace with liquid ash removal [11].

Thus, it can be concluded that superelevated plasma–coal burners stabilize the output of liquid slag in the oil-free operating mode and are ready for practical use on pulverized-coal furnace with liquid ash removal.

## 2.3 Plasma steam–air gasification

The technology of plasma steam–air gasification of coal for obtaining energy gas and a combined plasma gasifier for its implementation has been developed [12] (Fig. 3). This technology is designed to solve the problem of oil-free combustion of the furnace, illumination of the pulverized-coal flare, stabilization of the liquid slag outlet in the furnaces with liquid slag removal, reduction of nitrogen oxides and sulfur oxides (with addition of dolomite to the initial coal), and expansion of the range of varieties burned in the same furnace coal without reducing its technical, economical, and environmental indicators. Figure 3 shows a combined plasma gasifier used to increase the reactivity of energy coals and the environmental performance of pulverized coal furnaces. On the furnace BKZ-640, a combined plasma gasifier with two plasma stages was installed with a steam output of 640 t/h at the Gusinoozerskaya TPP. The capacity of the gasifier for coal was 32 t/h.

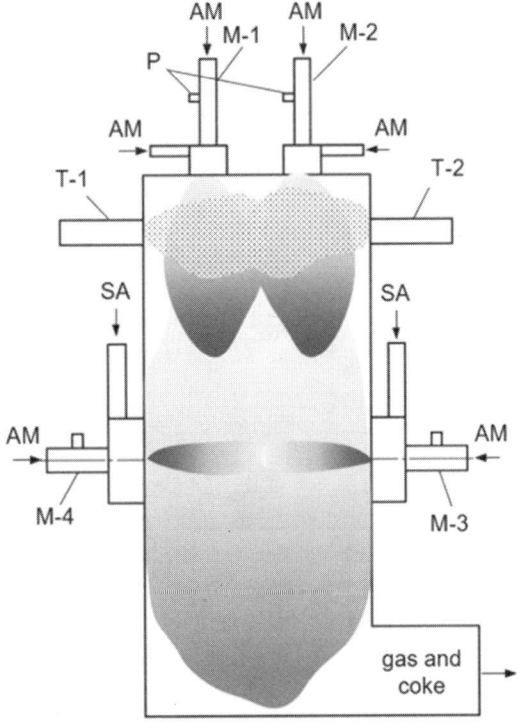

**Figure 3:** Scheme of combined industrial gasifier: AM, air mixture; SA, secondary air; P, plasmatron; M, muffled PTS; T, pulverized coal torch

## 2.4 Complex processing of coal for synthesis gas production

Figure 4 presents a pilot installation for the implementation of the technology of plasma gasification and complex processing of coal for synthesis gas

(CO + H$_2$), hydrogen (H$_2$), and valuable components from the mineral mass of coals, oriented to the development prospects of several industries (power engineering, chemical industry, and metallurgy) [13]. From an ecological point of view, this technology is the most promising. Its essence is the heating of coal dust by an electric arc plasma, which is an oxidizer, to the temperature of full gasification, in which the organic mass of coal is converted into environmentally friendly fuel – synthesis gas, free of ash particles, nitrogen oxides, and sulfur. At the same time, mineral oxides of coal are reduced and valuable components such as technical silicon, ferrosilicon, aluminium, and carbosilicium, as well as trace elements of uranium, molybdenum, vanadium, etc., are formed.

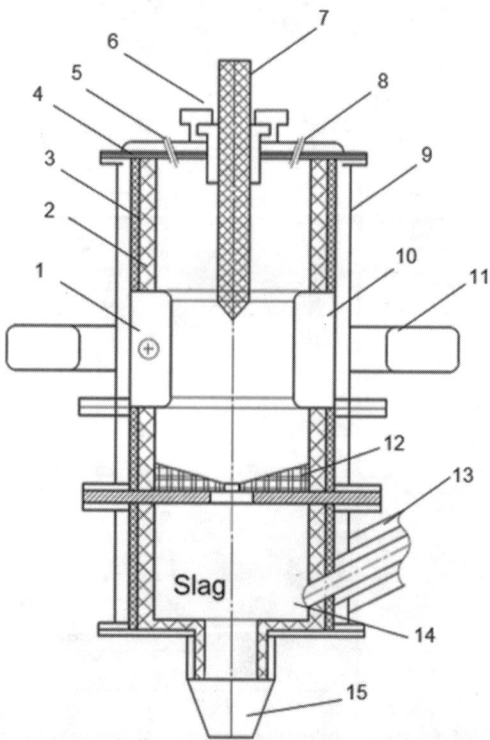

**Figure 4:** Installation of plasma gasification and complex processing of coal for synthesis gas production. 1, electric arc; 2, graphite backfilling; 3, graphite lining; 4, water-cooled cover; 5, a branch pipe of giving of a coal dust; 6, insulator with electrode sealing system; 7, graphite electrode; 8, a branch pipe of giving of steam; 9, reactor cooling water jacket; 10, annular graphite electrode; 11, electromagnetic coil; 12, graphite diaphragm; 13, branch pipe of synthesis gas; 14, gas and slag separation chamber; and 15, outlet pipe of slag

The results of preliminary tests of the gasifier showed that at a current of 1500 A and a voltage of 470 V, the power of the plasma gasifier of 705 kW was achieved. At the same time, the thermal efficiency of the plasma gasifier was 61%, and the useful dedicated power, respectively, was 430 kW.

After heating the reactor for 30 min, a working mixture consisting of coal dust, whose flow rate was 150 kg/h, and water vapor at a temperature of 493 K, was fed into the reaction zone at a flow rate of 75 kg/h. The air sucks were about 30 kg/h. Measuring the wear of the graphite rod electrode showed that its consumption was 0.4 kg/h. In the process of plasma steam gasification of the Kholbolzhinskiy brown coal (ash content – 22%, net calorific value – 18,400 kJ/kg, yield of volatiles – 44%), gas samples were taken from the gasification product removal chamber and after the condensed phase sample experiment in all units of the installation [14]. Synthesis gas was obtained, consisting mainly of CO (41.5%) and $H_2$ (48.7%). Impurities of $CO_2$ (0.4%), $CH_4$ (1.4%), and $N_2$ (8%) were found. As a result of X-ray phase analysis of solid residue samples, the degree of gasification of coal carbon was determined, amounting to 81%. The obtained results allow using plasma technologies of steam gasification and complex coal processing to create a pilot plasma plant for the production of high calorific, environmentally friendly synthesis gas, hydrogen, and valuable components of the mineral mass of coal (ferrosilicon, carbosilicon, technical silicon, etc.). The main part of hydrogen is formed during the decomposition of water vapor by coal, which is very promising for the use of plasma–steam gasification in hydrogen energy.

## 2.5 Energy coals into carbon sorbents

One of the new methods for obtaining active and cheap sorbents is the plasma technology for processing energy coals into sorbents (Fig. 5). The technology of obtaining carbon sorbents from Tugnui coal with the use of plasma thermochemical preparation was tested. The obtained sorbent passed preliminary tests, which confirmed the principle possibility of its production using this technology. On this basis, a technological scheme for the production of carbon sorbents was jointly developed, in which three chambers of thermochemical fuel preparation (TCFP) with plasmatrons are used in different sections of the technological chain. In accordance with the obtained experimental data and numerical studies, an experimental industrial installation with three plasma stages for the production of carbon sorbents from low-grade coals with the productivity of 500 kg of sorbent per hour was developed [15].

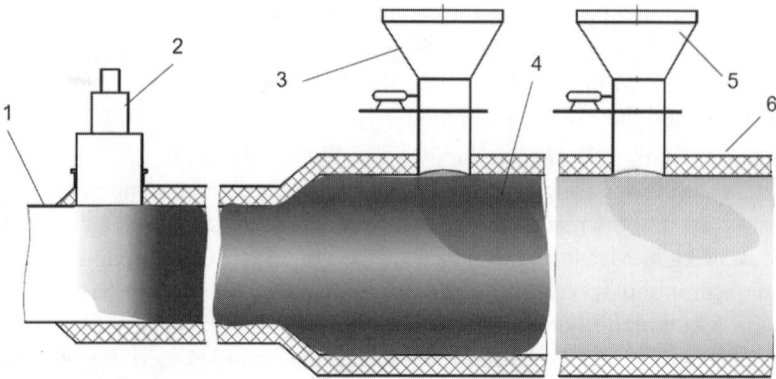

**Figure 5:** Scheme of the plasma plant for the processing of energy coals into carbon sorbents: 1, PTS; 2, plasmatron; 3, coal dust chamber; 4, TCFP chamber; 5, chamber of crushed coal; 6, TCFP chamber of crushed coal

## 2.6 Aluminum industry

Achinsk Alumina Combine has 11 rotating furnaces with a diameter of 5 m and a length of 185 m. The capacity of the furnace for dry alumina is 102 t/h. The thermal capacity of the furnace is 108 MW. The total fuel oil consumption per stove is 17,700 t/year. To replace fuel oil with coal and improve environmental and economic indicators, a plasma technology was developed for the non-oil ignition and stabilization of the combustion of coal dust in inclined rotary kilns. According to this technology, the fuel oil burner that constantly works in the furnace (Fig. 6, left), with a fuel oil consumption of 2 t/h, is replaced by a PTS with a coal consumption of 3 t/h (Fig. 6, right). This technology can be used without significant changes for the free-burning of clinker for cement production.

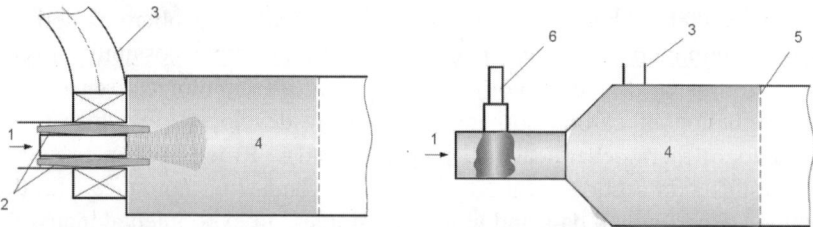

**Figure 6:** Scheme of ignition of an air mixture in an existing kiln (left) and using a PTA (on the right): 1, an aerosol mixture; 2, fuel oil nozzle; 3, secondary air; 4, rotary kiln (combustion zone); 6, PTS with three plasmatrons located at an angle of 120°; 7, end of the combustion zone

With the increase in the depth of oil refining at refineries in Russia, the output of light petroleum products is increasing, the output of fuel oil is reduced, and the mass of heavily utilized deep oil processing waste (DOPW) is increasing. At existing refineries, the output of the DOPW (depending on the quality of oil) can be from 4% to 6%. For example, at the largest refinery in Europe, «Naftohim» Refinery (Burgas, Bulgaria), the DOPW output reaches 180,000t/year with the processing of 12 million tons of crude oil per year.

Traditional technology of utilization of DOPW implies their combustion in special rotating gas furnaces. At the same time, the degree of conversion of the DOPW does not exceed 30–40%, and the specific energy consumption reaches 100–120 kW h/t of recyclable DOPW. To increase the efficiency of the OGPP utilization process, the plasma technology of DOPW utilization was developed and tested, based on the preliminary plasma thermochemical preparation of the DOPW for incineration in a special chamber. At a plasma plant capacity of 1 t/h (Fig. 11), the estimated specific energy consumption was 33 kW h/t, which is 3–4 times lower than existing ones.

Presented plasma technologies provide energy-efficient and environmentally acceptable use of solid fuels in basic industries.

## 3. Conclusion

Presented plasma technologies provide energy-efficient and environmentally acceptable use of solid fuels in basic industries.

## 4. References

[1] Gomez E, Amutha Rani D, Cheeseman CR, Deegan D, Wise M, Boccaccini AR. Thermal plasma technology for the treatment of wastes: a critical review. J Hazard Mater 2009;161(2–3):614–26.

[2] Wang Qin, Yan Jianhua, Tu Xin, Chi Yong, Li Xiaodong, Lu Shengyong, et al. Thermal treatment of municipal solid waste incinerator fly ash using DC double arc argon plasma. Fuel 2009;88(5):955–8.

[3] Messerle VE, Lavrichshev OA, Karpenko EI. Plasma preparation of coal to combustion in power boilers. Fuel Process Technol 2013;107:93–8.

[4] Messerle VE, Ustimenko AB, Lavrichshev OA. Comparative study of coal plasma gasification: simulation and experiment. Fuel 2016;164:172–9.

[5] Shin Dong Hun, Hong Yong Cheol, Lee Sang Ju, Kim Ye Jin, Cho Chang Hyun, Ma Suk Hwal, et al. A pure steam microwave plasma torch: gasification of powdered coal in the plasma. Surf Coat Technol 2013;228:S520–3.

[6] Beloglazov II and Kuskova YV. "Simulation of Roasting Metallurgical Concentrates in Fluidized Bed Using CFD-DEM," in IOP Conference Series: Materials Science and Engineering, 2017, vol. 1, no. 221.

[7] Messerle VE, Karpenko EI, Ustimenko AB. Plasma assisted power coal combustion in the furnace of utility boiler: numerical modeling and full scale test. Fuel 2014;126:294–300.

[8] Bazhin VY, Beloglazov II, Feshchenko RY. Deep conversion and metal content of Russian coals. Eurasian Min 2016;2:28–32.

[9] Gorokhovski M, Karpenko EI, Lockwood FC, Messerle VE, Trusov BG, Ustimenko AB. Plasma technologies for solid fuels: experiment and theory. J Energy Inst 2005;78(4):157–71. DOI: 10.1179/174602205X68261.

[10] Karpenko EI, Messerle VE, Ustimenko AB. Plasma-Fuel Systems for Enhancement Coal Gasification and Combustion, Abstracts of Work-in-Progress Poster Presentations of 30th International Symposium on Combustion, University of Illinois at Chicago, July 25–30, 2004.-5F4-03;-P.423.

[11] Messerle VE, Ustimenko AB, Karpenko EI. Plasma-energy Technologies for Improvement and Economy Indexes of Pulverized Coal Incineration and gasification. The Proceedings of the 28-th International Technical Conference on Coal Utilization and Fuel systems. Clearwater, Florida, USA. Published by U.S. Department of Energy & Coal Technology Association of USA. 2003. P. 255–66.

[12] Rutberg PG, Kuznetsov VA, Serba EO, Popov SD, Surov AV, Nakonechny GV, Nikonov AV. Novel three-phase steam–air plasma torch for gasification of high-caloric waste. Appl Energy 2013;108:505–14.

[13] Fridman A, Gallagher MJ. Plasma reforming for H2-rich synthesis gas. Fuel Cells Technol Fuel Process 2011;8:223–59.

[14] Rutberg PhG, Bratsev AN, Kuznetsov VA, Nakonechny GV, Nikonov AV, PopovVE, Popov SD, Serba EO, Subbotin DI, Surov AV. Production of synthesis gas by conversion of methane in a steam–carbon dioxide plasma. Tech Phys Lett 2014;40(9):725–9.

[15] Messerle VE, Karpenko EI, Ustimenko AB Plasma assisted power coal combustion in the furnace of utility boiler: numerical modelling and full scale test. Fuel. June 15, 2014. V. 126. P. 294–300. DOI information: http://dx.doi.org/10.1016/j.fuel.2014.02.047.

# 20

# Beneficiation strategies for Indian coals

H.L. Sapru

*Monnet Group, New Delhi, India*

**Abstract :** Coal being abundantly available, affordable and reliable energy source in India. All energy sources in India need to be utilised in environment friendly manner. Metallurgical coal is traditionally washed washing of thermal coal is picking up. High ash with poor washability characteristics i.e. high NGM make Indian coal difficult to wash. Keeping Capex and Opex of coal preparation plants within acceptable limits is itself a big challenge.

Therefore, proper design and selection of washing circuits and equipment after in-depth study of comminution/liberation studies and keeping overall organic efficiency of washing plants within acceptable limits is need of the hour. New strategies both for coarse coal and fine coal washing to get maximum yield with minimum wrong migration of clean coal into rejects and vice-versa need to be adopted.

This paper makes an attempt to address these issues, both for thermal and metallurgical coals, and suggest the optimum techniques/flow sheets for Indian coal without blindly adopting the technologies used in western countries as Indian coals are significantly different from that of other countries in their washing characteristics.

**Keywords:** Gondwana, Drift Origin, Fluidity, Demand/supply, Giga Watt, IGCC, Emissions

## 1. Coal resources in India

The coal resources in India are primarily from Gondwana segment occurring mainly in eastern and central part of India. There are also some resources in northeast part of the country mainly in Assam and Sikkim. Based on the information from Geological Survey of India the resources as on 01-04-2018 is given below:

| Category | Qty. (MT) |
|---|---|
| Measured resource | 1,48,787.43 |
| Indicated resource | 1,39,164.14 |
| Inferred resource | 31,068.76 |
| Total | 3,19,020.33 |

The breakup in terms of type of coal is given below:

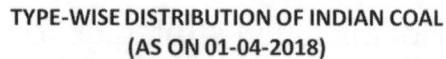

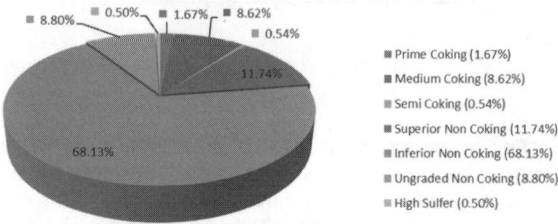

**TYPE-WISE DISTRIBUTION OF INDIAN COAL (AS ON 01-04-2018)**

- Prime Coking (1.67%)
- Medium Coking (8.62%)
- Semi Coking (0.54%)
- Superior Non Coking (11.74%)
- Inferior Non Coking (68.13%)
- Ungraded Non Coking (8.80%)
- High Sulfer (0.50%)

*Source:* Geological Survey of India.

Indian coal resources are mainly of drift origin. Due to their generic nature the ash content is varying from 22% to 50%. At present coal production is predominantly done from opencast mines which further deteriorate the quality of run-of-mine coal due to out of seam dilution. Indian coal is difficult to wash due to high NGM (near gravity material) at required cut gravities. At present washed coking coal is produced at 18% ash level whereas non coking coal i.e. thermal coal is produced at 30–34% ash level at moisture of 9–10%.

## 1.1 Energy mix

The % share of commercial primary energy resources (projected for the year 2031–32) is given below:

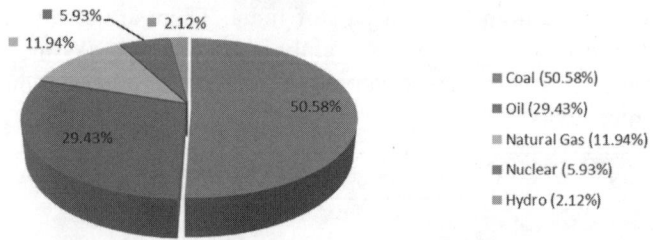

**% Share of Commercial Primary Energy Resources (2031-32)**

- Coal (50.58%)
- Oil (29.43%)
- Natural Gas (11.94%)
- Nuclear (5.93%)
- Hydro (2.12%)

*Source:* Integrated Energy Policy, Planning Commission, Govt. of India.

It is clear that the coal will remain as the primary source of energy for decades to come. As per National Electricity Plan, 2018, the all India coal based generation by the year 2021–22 is estimated to be around 63% of the total energy generated from all sources, which is projected to come down to approx. 51% in 2031–32. Thus, country's power generation over next decade will largely continue to depend on coal.

## 2. Characteristics of Indian coal

In general ash content in Indian coal varies from 22% to 50%. The Indian washed coking coal contains ash content from 15% to 18% which is higher compared that of imported washed coking coal. Besides that Indian coal contains low vitrinite 40–55%, low vitrinite reflectance 0.98–1.3 and low swelling index 5–6 compared to that of imported coal from Australia and elsewhere having vitrinite 55–70%, vitrinite reflectance 1.17–1.55 and swelling index 7.5–9.0, respectively. The coke type is also inferior. But Indian coal has superior fluidity (dial divisions per minute, DDPM) in the range of 772–2400 compared to that of imported coal with 75–1100. Indian washed coal has also higher inertinite content i.e. in the range of 38.1–45.6% compared to that of imported washed coal with 27–42%. The sulphur content in Indian coal is generally below 0.5% except some coal resources in north east i.e. Assam and Sikkim which have higher Sulphur content in the range of 4–5%.

The ash content in non-coking coal is in the range of 34–50% or even more sometimes. The near gravity material (NGM) is also very high. Generally the ash content in the washed thermal/non-coking coal is maintained in range of 30–34% depending upon the requirement of consumers/end users and in compliance to the directive of Ministry of Environment and Forests, Govt. of India.

## 3. Demand and supply of coal and strategies for its augmentation

### 3.1 Demand and supply/consumption scenario as per the coal controller's organization

| | | | | | | (Quantity in million tonnes) | | |
|---|---|---|---|---|---|---|---|---|
| | **2013–14** | **2014–15** | **2015–16** | **2016–17** | | **2017–18** | | |
| **Sector** | **Actual supply/ con- sumption** | **Actual supply/ consump- tion** | **Actual supply/ consump- tion** | **Actual supply/ consump- tion** | **Estimated demand** | **Actual supply/consumption** | | |
| | | | | | | **Indigenous** | **Import** | **Total** |
| I. Coking coal | | | | | | | | |
| 1. Steel/ coke oven and cokeries (indigenous) | 15.49 | 12.02 | 12.37 | 12.50 | 13.05 | 11.33 | – | 11.33 |

*Contd...*

*Contd...*

| | 2013–14 | 2014–15 | 2015–16 | 2016–17 | 2017–18 | | | |
| | | | | | | | | (Quantity in million tonnes) |
| Sector | Actual supply/con-sumption | Actual supply/consump-tion | Actual supply/consump-tion | Actual supply/consump-tion | Estimated demand | Actual supply/consumption | | |
| | | | | | | Indigenous | Import | Total |
|---|---|---|---|---|---|---|---|---|
| 2. Steel (import) | 36.87 | 43.72 | 43.51 | 41.64 | 50.12 | – | 47.00 | 47.00 |
| Sub total | 52.36 | 55.74 | 55.88 | 54.15 | 63.17 | 11.33 | 47.00 | 58.33 |
| II. Non-coking coal | | | | | | | | |
| 3. Power (utilities) | 438.83 | 435.44 | 445.98 | 470.98 | 622.96 | 504.72 | – | 504.72 |
| 4. Power (captive) | 54.42 | 62.26 | 62.27 | 56.28 | 90.34 | 73.35 | – | 73.35 |
| 5. Cement | 11.94 | 11.36 | 8.93 | 6.43 | 22.32 | 7.70 | – | 7.70 |
| 6. Sponge iron/CDI | 18.49 | 17.77 | 7.76 | 5.68 | 24.61 | 8.51 | – | 8.51 |
| 7. BRK and other including fertilizers | 163.30 | 239.57 | 251.57 | 98.74 | 85.00 | 83.46 | – | 83.46 |
| Sub total | 686.98 | 766.40 | 776.40 | 787.41 | 845.23 | 677.74 | 161.27 | 839.01 |
| Total raw coal | 739.34 | 822.14 | 832.39 | 841.56 | 908.40 | 689.07 | 208.27 | 897.34 |

*Source:* Coal Controller's Organization, Ministry of Coal.

## 3.2 Projected demand of coal

National Steel Policy 2017 envisages:

**Steel:**

- Crude steel demand to grow threefold to 255 MT by 2030–31.
- Crude steel capacity to grow to 300 MTPA by 2030–31.

**Coal:**

1. Coking coal – 161 MTPA.
2. Non-coking coal for PCI – 31 MTPA.
3. Non-coking coal for DRI route – 105 MTPA.

National Steel Policy 2017 envisages to bring down the dependence of imported coal from present level of 85% to 65% by 2030–31. Balance need to be met from the domestic resources.

Overall coal demand is estimated to be 900–1000 MTPA by 2020 and 1300–1900 MTPA by 2030. By 2030, out of overall coal demand, thermal coal demand is estimated to be 1150–1750 MTPA and the balance is coking coal demand.

In financial year 2017–18, coal production was 612.6 MT and lignite 48.25 MT. Total electricity generation in 2017–18 was 206 GWh, out of which 76% was coal based. India's estimated power requirement in 2030 is projected to be 750 GW, out of which 300 GW is coal based. According to published data, year 2018 saw robust growth in peak demand by 8% to 177 GW. Since last year 2017–18 India imported 208.272 MT of coal compared to 190.953 MT in 2016–17, which is expected to increase further.

| Demand, dispatch and import of coking coal in India (MT) | | | |
|---|---|---|---|
| Year | Demand | Dispatch | Import |
| 2013–14 | 53.980 | 58.464 | 36.872 |
| 2014–15 | 55.460 | 56.438 | 43.715 |
| 2015–16 | 77.000 | 59.213 | 43.561 |
| 2016–17 | 56.620 | 59.308 | 41.644 |
| 2017–18 | 63.170 | 45.380 | 47.003 |
| Demand, dispatch and import of non-coking coal in India (MT) | | | |
| Year | Demand | Dispatch | Import |
| 2013–14 | 715.710 | 513.596 | 129.985 |
| 2014–15 | 731.570 | 547.334 | 174.068 |
| 2015–16 | 833.000 | 573.229 | 159.388 |
| 2016–17 | 828.250 | 586.670 | 149.309 |
| 2017–18 | 845.230 | 642.451 | 161.269 |

*Source:* Annual Plan, MOC.

## 4. Strategies

1. Keeping in view the huge demand and unavoidable dependence on coal in decades to come, attention need to be given for its use in an environmental friendly manner by focusing on energy efficiency and HELE generation technologies like super critical, ultra-super critical, IGCC, etc. to limit the emission of $CO_2$ into the atmosphere. In recent past in India, $CO_2$ emissions rose about 6.3% which need to be brought down. The other top $CO_2$ emitters like USA and China have also in the same period registered an increase in $CO_2$ emissions of 2.5% and 4.7%, respectively. This calls for global approach to reduce the emissions.

2. Coal production need to be given greater importance by involving private sector and removal of irritants such as minimizing delay in granting statutory and regulatory clearances and other procedural delays.

3. Washing of coal has to be increased judiciously as India is reported to wash only 15–20% of its coal produced. Traditionally all the coking

coal is being washed in India. There has been some reluctance by some consumers for using washed thermal coal due to higher price but slowly and steadily, the consumers reconciled to the fact that there is no way but to wash Indian thermal coal in the overall interest of consumers and producers alike. The major benefits of using washed coal are summed up below:

- Reduction in transportation costs.
- Lower ash content in coal reduces SPM and SOx emissions.
- Significant improvement in thermal efficiency of boilers.
- $CO_2$ emissions reduce by 2–3% results in increase of each percentage point of thermal efficiency.
- Slagging problems in the boiler can be avoided by using washed coal which increases efficiency. As per estimates of IEA, thermal efficiency can be increased up to 10% by switching from unwashed to washed coal in Power Plants.
- The reduction in operation cost in the shape of fly ash generation corresponds to lower ash handling cost. The area required for ash dumps also get reduced by reduction in ash generation by using washed coal.
- The overall Capex (capital expenditure) and Opex (operational expenditure) reduce up to 10% if power plants are designed based on use of washed coal instead of raw coal.

4. Keeping in view the above, there should be an express programme both by the coal producing companies and coal consumers to establish coal washeries at the earliest.

5. For augmentation of coking coal, besides establishing modern coking coal washeries, LVMC (low volatile medium coking) coal should be exclusively supplied to steel plants after washing by restricting its use elsewhere.

6. Expeditious implementation of Jharia Action Plan to improve the production of coking coal is required. Optimum use of deep coal reserve and acquisition of overseas coal assets must be persuaded at faster pace.

7. Coal blocks allotted to private sector by Government of India must be developed immediately with provision of washeries.

8. The Government of India need to finalize national policy for setting up and operation of coal washeries as quickly as possible and impress upon coal producing companies to sell only washed coal based on GCV (gross calorific value) pricing instead of raw coal with ash higher than 34%.

## 5. Present washing techniques in India

In India most of the washeries are using age old washing techniques. Due to change in raw coal characteristics these technologies have become outdated and are not suitable for today's washery feeds. This has resulted in low capacity utilization (20–30%) of our installed washery capacity. Traditionally, in coking coal sector, all the coal feed to the washery is washed by using jigging, heavy media separation, flotation circuits, etc. but the size of the feed is generally 20–100 mm only.

In thermal coal washeries, −13 mm coal is not being washed at present and is being blended directly with washed thermal coal with an intention to keep the cost of washed coal low. This process need to be corrected as the washery feed characteristics have deteriorated considerably in terms of increase in ash content and increase in −13 mm undersize fraction in total feed due to mechanized mining.

In most of the plants centrifugal driers to lower the moisture of the washed coal are not in place which results in increased moisture in washed coal.

The existing capacities of washeries are mostly less than 2.5 MTPA. However, the recent trend is to set up higher capacity washeries up to 10 MTPA in the interest of overall economic viability of washery.

In thermal coal washing, only jigging and heavy media separation systems are being used to make the process simpler and to keep the Capex and Opex of the plant down. However, deterioration of coal quality makes it imperative to go for full washing in case of thermal coal also.

## 6. Optimum washing techniques and emerging technologies in coal beneficiation

*Coking coal*: In coal beneficiation, apart from other things the yield of clean coal at a particular ash and moisture level in the final product assume a greater importance. The circuits are to be designed to ensure that no effluents go outside the premises of the plant and environmental concerns are suitably addressed to.

Endeavor must be made to avoid wrong migration of one product to another, specially reject to clean coal and middling and vice versa. The ash content of the refuse/reject should be as high as possible, in any case it should be more than 60%. Scheme for disposal of rejects should form an important part of washery design/operation. This need to be planned well before setting up of washery.

Optimum design of beneficiation plants need to be done only after comminution studies, washability studies, flotation studies, filtration characteristics, etc. are carried out for raw coal which is to be processed. Data bank need to be created for raw coal characteristics to be processed and other characteristics/parameters of products like clean coal, middlings, rejects, etc.

The sources of raw coal feed to the coal preparation plant need to be fixed beforehand for the entire life of the plant. However, the washery circuits must be designed to take care of minor fluctuations in the raw coal feed characteristics, both in terms of quality and size constituent of the feed.

Based on liberation studies of raw coal the optimum size for different processes need to be arrived at for getting better yields of clean coal.

Considering the requirement of coarser cut at specific gravity of 1.8 and above for deshaling coal, modern electronically controlled jigs perform better but range of size constituents need to be kept as close as possible for better jigging efficiency. The heavy media cyclone washing can be done at lower size ranges i.e. −13 mm/10 mm to 1 mm for sharper separation efficiencies, TBS/reflux classifiers beneficiation for −1 mm to +0.25 mm and floatation for −0.25 mm size fraction of coal.

Normally at present the ash of middling product in both coking coal and non-coking coal washeries is kept around 34% ash level. However, it is strongly recommended to enhance this limit to 40% as this could increase the yield of clean coal. The ash of rejects should be kept above 60% which can be used in FBC boilers. Only the rejects in the ash level of 80% or above should be called as a throw away rejects for possible use as filling material, etc.

In any case, Float & Sink data and other data will only dictate the selection of process, the washing equipment and level of gravity cut to be used for getting desired quality and quantity parameter of products.

For keeping the moisture of clean coal and middling within the accepted limits of 9–10%, centrifuges should be installed in washeries. Equipment like solid bowl centrifuge, horizontal filter press, etc. should be introduced for reducing the moisture.

*Non-coking coal:* Ash content of −13 mm coal fraction has increased to 40–45% in the washery feed and quantity of this fraction has also increased to 35–40% due to machine mining and mining of lower seams. It is imperative to resort to total washing of thermal coal, so that quality of product is maintained and yield of clean coal also increased at a marginal higher cost of washing.

Generally, the raw non-coking coal was subjected to only single cut washing/de-shaling by jigs or heavy media systems, baths or cyclones. Need of the hour is to use double cut washing for non-coking coal also for better yields and to avoid losing precious coal through rejects. The liberation size also needs to be lowered to the extent possible even up to −10 mm.

Centrifuges/driers need to be used extensively for keeping the moisture below 10% as any increase of moisture in the washed coal affects the GCV, which is not acceptable to the consumers and affects overall economic viability.

Recovery of the fine coal by using vacuum disc filter, horizontal filter and belt filter presses need to be resorted to, so that there is complete closed circuit washery operation. Use of additives/chemicals should be encouraged in the fine coal washing for better recovery of fines and ensure that the recovered process water is clean before recycling it to washery as the muddy water directly affects the washing process.

The capacity of the washing plants should be 5–10 MTPA and above, in order to reduce the Capex and Opex of washeries. Instrumentation system for better process control needs to be employed to control the process which will contribute to overall economic viability as well as quality parameters of products. The overall organic efficiency must be kept at 95% or above.

## 6.1 Emerging technologies

1. Washing the coal at lower liberation sizes and use of very fine cuts in the washing process to produce low ash coking coal in the range of 10–12%.

2. However, the economic viability of such coal beneficiation need to be evaluated and must be acceptable to the consumers.

3. But, in this process coking coal will migrate to middling product which is not desired keeping in view the overall conservation of coking coal.

4. Studies are being made by processing the coking coal after grinding the coal to −100 micron size and then beneficiate it in flotation circuit in the same way as done in the beneficiation of precious minerals like copper, gold, etc.

5. Studies are also being made on technology of grinding the rejects of coking coal washeries to −100 micron size and then beneficiate by flotation to recover precious coking coal which otherwise goes waste. However, pilot plant studies need to be carried for such technologies

and economic viability. Due to high price of coking coal, this technology has a good prospect.

## 7. Conclusion and way forward

i.   Ambitious plans for increasing steel production and thermal power generation throw a big challenge to both coking coal and non-coking coal availability in the country.

ii.  Fluctuating price and availability of internationally graded coal necessitates the blending of imported coking coal with domestic coking coal to the extent possible to improve both coking coal and thermal coal availability in the country.

iii. Irritants for development of the new coal mines and washeries both by the public and private sectors need to be removed so that new mines are opened and new washeries are setup. Implementation of major coal projects along with washeries to be expedited.

iv.  New coal washeries must be equipped with state of art technologies and attention on fine coal beneficiation must be given to increase the yield of washed coal.

v.   Enhancing washing capacity in nationalized coal sector by utilization of land and other infrastructure of old washeries and mines must be pursued. The tendering processes need to made simpler to attract foreign investment and for bringing state of art technologies in new washeries.

vi.  SAIL to expedite its coking coal production from its captive coking coal mines. Efforts to be made to reduce consumption of coking coal in steel manufacture by improving combustion technologies.

vii. Use of high-efficiency, low-emissions (HELE) generation technologies like super critical, ultra super critical, IGCC need to be expedited in power sector.

viii. To cope up with immediate demand, import of both coking and non-coking coal seems to be inevitable to keep up with demand for overall development of country.

ix.  Both SAIL and Nationalized Coal Industry should acquire foreign coal assets to supplement local production.

x.   Data Bank of Coal Characteristics, coalfield wise in the country need to be developed which could form a base for framing new beneficiation technologies. Facilities at CMPDIL could be used for the purpose.

xi.   Government of India must allocate more funds in Research & Development Sector, so that infrastructure available in National Laboratories, CIMFR, CMPDIL and leading technical institutions are put to use effectively. Both coal producers and consumers must come forward to setup new washeries as it is established beyond doubt that both coking and non-coking coal need to be beneficiated keeping in view the environmental concerns and their overall commercial interest.

xii.  Government of India should expedite formulation of long awaited national coal washing policy.

xiii. Expeditious implementation of Jharia Action Plan needs to be expedited.

xiv.  Environment concerns specially $CO_2$ emission need to be vigorously followed to sustain use of coal based power generation.

# 21

# Coking coal's journey from ground to blast furnace

Hardarshan S. Valia

*Coal Science Inc., Highland, Indiana, USA*

**Abstract:** Rigid coke quality requirements are placed on coke producers as ironmakers try to increase productivity and reduce costs by reducing the coke rate and by increasing the pulverized coal injection. The challenge to a coke producer is in using a coal or blend that on carbonization would consistently produce a low cost-high quality coke with safe oven pushing performance. Hence, strict coal quality monitoring procedures are followed by coke makers on the incoming coal from the mine as well as assurances from a coal producer of consistent coal quality from the future mining areas. The article provides a review of coal and coke quality monitoring steps taken at a coal mine, coke plant, and a blast furnace. Few case studies are presented at each stage. Finally, using blast furnace rules of thumb, utilization cost of coke that would be produced from carbonization of the most cost effective coal/coal blend are presented for coal blends and the coal that would produce coke with highest potential cost savings at the blast furnace is recommended for purchasing from a coal producer.

**Keywords:** Coal reserve evaluation, coking coal, coal quality, coke quality

## 1. Introduction

One of the most enduring photographs that down the memory lane of 'Industrial Age' is the molten iron flowing from a blast furnace. The luminescence of flow, the sparks generated by molten iron droplets hitting the trough in the cast-house reminds one of as watching a lava flow from caldera of a volcano. Fast backward into the belly of the blast furnace, and one could visualize coke carbon forms un-leasing its energy to break the powerful atomic bonds between iron and iron oxide, thereby producing metallic iron. But taking more steps backwards into a coke oven, one could imagine the organic entities in coal melt under reducing environment to form 'nematic liquid crystals' that come closer and coalesce into a beautiful strong entity called coke. However, to a naked eye, the same coke at the coke plant and furthermore, the coal at the coal mine may seem like a non-attractive black mass. It is only looking through a polarizing microscope, one marvels at the colorful carbon forms during coal-to-coke carbonization process (Fig. 1). Hence, at the coal mine, it is imperative that one needs to fully understand what desirable properties of coal produce a high strength coke with good carbon forms. Equipped with the knowledge then one could develop a coal mine plan, a wash plan, and a blending strategy that would result in supplying a consistent high quality coal. The coal thus supplied to a coke plant must be able to produce very high quality coke that on introduction into a blast furnace should result in lower coke rate, higher productivity, and lower hot metal cost. In an earlier work, evaluation at coal reserve was described (Valia, 1993). This paper will shed light on managing quality at various stages of coal and coke's journey into its final resting place, the blast furnace (Fig. 2). The following topics will be covered: (1) coal characterization and reserve evaluation with regard to coking properties; (2) using reserve data in mine planning so it meets steel industry's requirement; (3) effect of washing on coking quality; (4) coal preparation at the coke plant; (5) value in use at the blast furnace; and (6) coke properties and coke degradation during coke's passage inside the blast furnace. Basically, the emphasis will be on how to meet the target coking coal quality parameters that if not properly planned and monitored at the reserve/mine/wash plant will affect financially at the end user place, that is, at the coke plant and the blast furnace.

Coal in a seam                  Coal seen under a microscope

Coke in a pile                 Coke seen under a microscope

**Figure 1:** Coal to coke transformation

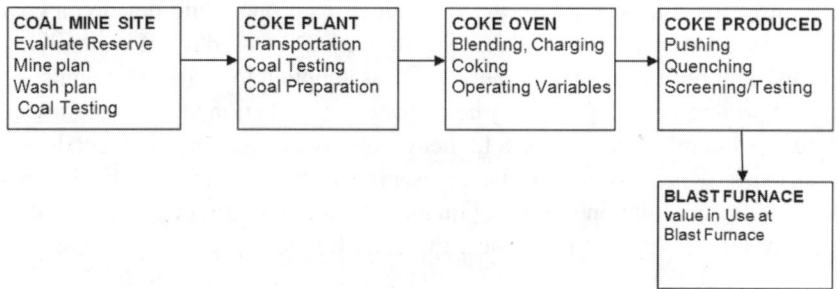

**Figure 2:** Coking coal's journey from ground to blast furnace

## 2. Background and procedure

Before tackling the subject of coal quality control, it is imperative to understand the basics of coal to coke transformation and the role of coke in a blast furnace. Detailed descriptions are presented elsewhere (Wakelin, 1999; Ricketts, 1996; Cheng, 1996; Valia, 1996, 2019). However, brief description of carbonization and role of coke in blast furnace is described herein. Coal when heated under reducing environment (in a slot/by-product coke oven) or under partially reducing environment (HR/NR, heat recovery/ non recovery coke oven) undergoes following changes: first the moisture removal takes place up to 200 °C. Thereafter, at 200–375 °C, dry coal heats up with chemical changes such as initial methane evolution and some CO and $CO_2$ evolution. From 375 to 475°C, the coal decomposes to form plastic layers near each wall as in a by-product but in a heat recovery/non recovery oven, the plastic layer forms on the top of the charge and at the floor of the wall. At 475–600°C, there is a marked evolution of tar, aromatic compounds, followed by re-solidification of the plastic mass into semi-coke. At 600–1100°C, the stabilization phase begins which is characterized by contraction of coke mass, structural development of coke and final hydrogen evolution. In a by-product oven, during the plastic stage, the plastic layers move from each wall towards the center of the oven trapping the liberated gas resulting gas pressure build up which is transferred to the heating wall. However, in HR/NR oven, since carbonization takes place from the top of the coal charge by radiant heat transfer and from bottom by conduction of heat from through the sole floor, coupled with a large unrestricted surface area on the top of coal charge, the carbonization pressure generated does not affect the oven walls. The coke mass after the end of carbonization via both processes is shown in Fig. 3.

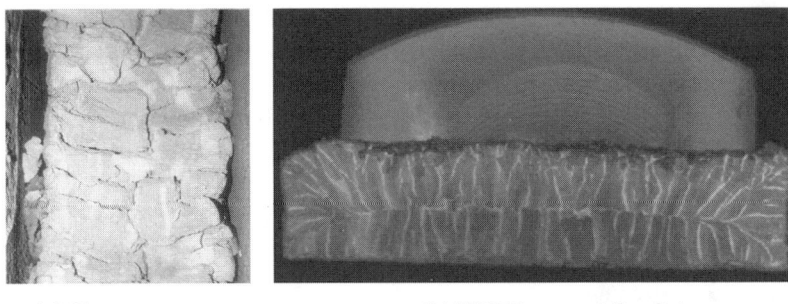

(a) Slot oven                    (b) HR/NR oven at SunCoke

**Figure 3:** Position of re-solidified plastic layer in coke ovens (slot versus HR/NR oven)

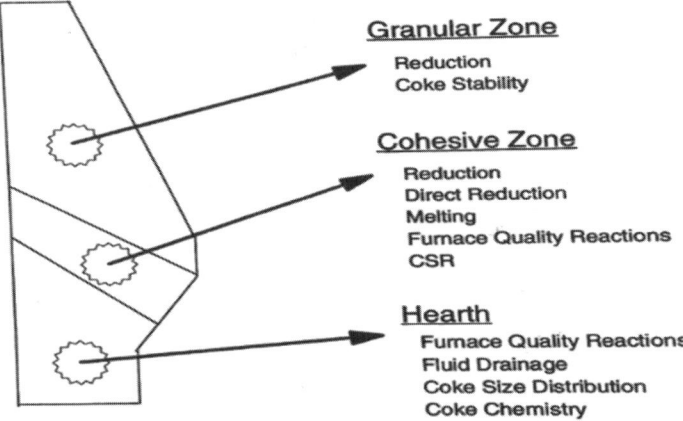

**Figure 4:** Blast furnace operating zones and coke behavior

A world class blast furnace operation demands highest quality of raw material. Coke is the most important raw material fed into the blast furnace in terms of its effect on blast furnace operation and hot metal quality. Coke with iron ore and flux is fed from the top of the blast furnace and molten iron and slag is tapped from the bottom called hearth. Figure 4 is a generalized sketch showing role of coke properties as the burden descends inside the blast furnace. The burden of raw material basically passes through five zones as follows: (a) granular zone where coke and iron ore retain their distinct shape, (b) cohesive zone the top part has softened whereas the lower part has started melting, (c) active coke zone supplying coke to the raceway, (d) raceway, an area where hot air enters through tuyeres and combusts the coke supplying reducing gases and thermal energy, and (e) hearth and deadman, an area of packed coke providing permeability and voids for molten metal and slag to flow into the hearth (Cheng, 1996). A high quality coke should be able to support smooth descend of the burden with as little degradation as possible while providing the lowest amount of impurities, highest thermal energy, highest metal reduction, and optimum permeability for flowage of gaseous and molten products. The coke properties desired for a large blast furnace are shown in Table 1.

**Table 1:** Coke quality specifications for 4000 m³ blast furnace

|  | **Mean** | **Tolerance** |
| --- | --- | --- |
| Average coke size (mm) | 52 | 45 – 60 |
| Plus 4" (% by wt.) | 1 | 4 max |
| Minus 1" (% by wt.) | 8 | 11 max |
| Stability | 60 | 58 min |
| CSR | 65 | 61 min |
| Chemical |  |  |
| Ash (% by wt.) | 8.0 | 9.0 max |
| Moisture (% by wt.) | 2.5 | 5.0 max |
| Sulfur (% by wt.) | 0.65 | 0.82 max |
| Volatile matter (% by wt.) | 0.5 | 1.5 max |
| Alkali (K2O + Na2O) (% by wt.) | 0.25 | 0.40 max |
| Phosphorus (% by wt.) | 0.02 | 0.33 max |

*Note:* Not included hardness (abrasion resistance) mean value 68 and 66 min preferred.

## 3. Discussion

The emphasis here is that for a coal company to sell its metallurgical coal to a steel or coke company, its personnel have to provide a coal with greatest value-in-use at the blast furnace. Equipped with an understanding the role that coke plays in a blast furnace and the impact of coal quality on coke quality and coke plant operation, a coal producer can assure a consistent high quality coal shipment. Understanding the variables will help in optimization of the coal reserve and development of a sustainable mine plan.

## 3.1 Proper identification of macerals at the mine

It is a worldwide practice to do standard white light petrographic evaluation of coal. However, a mine owner should know that fluorescence light (blue light) analysis may reveal certain unusual coking properties thereby increasing the value of the coal. Many of the fluorescent macerals are often difficult to identify in white light because they might appear uniformly dark. Hence,

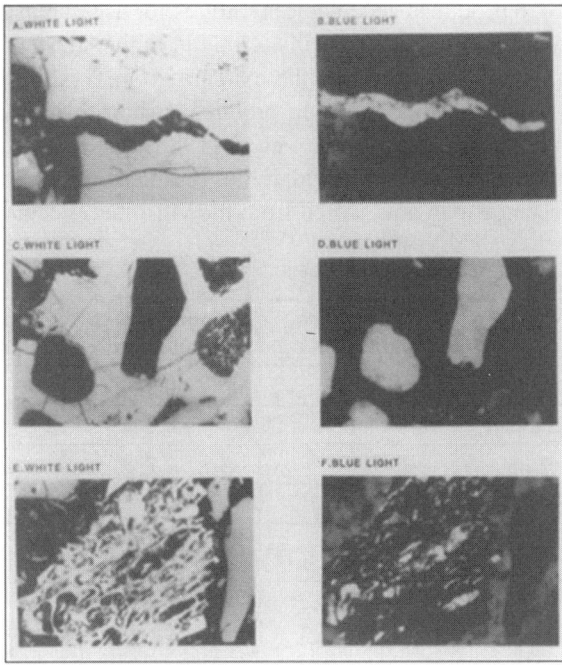

**Figure 5:** Exudatinite in Western USA (2/1 ratio of M/B coal)

normal maceral analysis in white light may result in an underestimation of the liptinite content. One case study is presented below where liptinite macerals were counted under both white light and under blue light. The two coals from Western USA are of high rank and under white light do not show any liptinite. However, subsequent examination in blue light resulted in identification of a significant amount of liptinite in the form of exudatinite (Fig. 5, Table 2). Similarly in very low rank coal, such as Indiana UB coal, the liptinite content in white light is 6% whereas under blue light, the content is very high (21%). Similar observations were reported in literature where liptinite in a coal is not detected in white light but significant amount is detected under blue light (Crelling, 1983; Spackman et al., 1976; Stanton, 1982). It is suggested that for coals containing more than 5% liptinite macerals in white light, a combination of white and fluorescence light analysis should be done (Spackman et al., 1976; Stanton, 1982). It was recognized that the pilot oven carbonization of Western USA coals with low rank Illinois No. 6 produced ASTM coke stabilities that in some cases were unexpected based on standard white light petrographic evaluation (Fig. 6). The most interesting observation was that only 20% addition of Western USA coal (2/1 ratio of M to B Bed) with low rank Illinois No. 6 produced acceptable stabilities for use in blast furnace. The actual stability at coke plant is usually 2 points higher than those obtained via pilot oven. It was observed that the exudatinite in Western USA coal on carbonization reacted with vitrinite to produce carbon forms typical of low volatile coals. It was shown by Kaegi et al. (1988) that micro-carbonization studies of vitrinite containing exudatinite produces carbon with higher rotational reflectance than anticipated from the vitrinite reflectance.

**Table 2:** Liptinite macerals in selected coals

|  | Reflectance (MMVR, %) | Liptinite, % in white light | Liptinite, % in blue light |
|---|---|---|---|
| Western USA M seam | 1.21 | 0.0 | 0.8 |
| Western USA B seam | 1.42 | 0.0 | 10.0 |
| Illinois # 6 | 0.74 | 2.7 | 6.2 |
| Indiana UB | 0.59 | 6.0 | 21.0 |

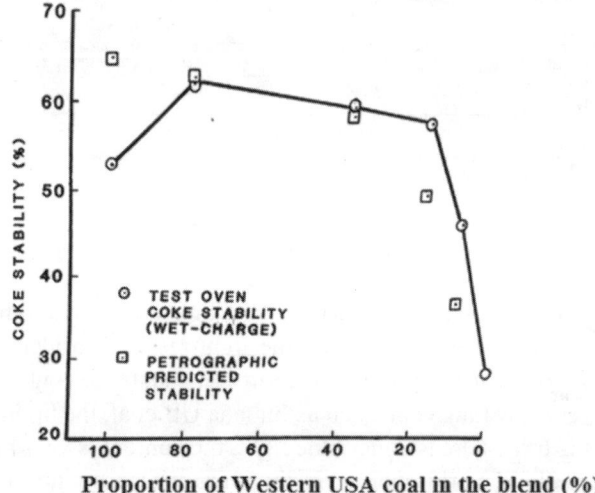

**Figure 6:** Effect of blend proportion on coke stability for blends of Western USA (2/1 ratio of M/B) coal with Illinois # 6

Indiana UB coals are also known to show abnormally higher coking properties for its rank. A UB coal (0.63% mean maximum vitrinite reflectance) when carbonized with 50% of LVM Alabama coal produced excellent quality coke with acceptable maximum oven wall pressure (Table 3). Valia (2005) showed that such low rank Indiana UB coals can produce good quality cokes with ternary blends containing higher rank coals varying in rank from HVM,

MVM, and LVM content. The unusually high coking properties are attributed to high amounts of liptinite resulting in higher fluidity, higher fluid range, and higher contraction (Valia, 2005; Valia and Mastalerz, 2004). Also it should be kept in mind that the bitumen in vitrinite can falsify the reflectance readings but the fluorescence intensities might be interpreted as suggesting higher rank for the respective coals. So these coals may be behaving as of higher ranked coal. It is advised that the coal company should also characterize their coal using fluorescence microscopy.

**Table 3:** Addition of high liptinite content lower rank coal to a LVM coal (Valia, 2005)

|  | Indiana UB (100%) | 50% Indiana UB = 50% low VM Alabama coal |
|---|---|---|
| Grind(%, −3.35 mm) | 80.8 | 91 |
| Dry oven B.D. (kg/m3) | 73.7 | 794 |
| Oven wall pressure (kPa) | 2.20 | 7.25 |
| Coking time (h) | 17.9 | 17.02 |
| CSR | 46 | 66 |
| Stability | 25 | 62 |
| Hardness | 65 | 72 |
| SHO contraction (%) | −10.10 | ND |

## 3.2 Reserve evaluation with regard to coking properties

Besides monitoring normal coal quality variables (such as petrographic; rheological; chemical – VM, ash, sulfur, phosphorus; ash chemistry), high quality coking operations, and steel mill personnel are focusing on deducing predicted coke quality, such as CSR and blend stability, in a coal reserve. This is primarily done to assure consistency of quality now and into the foreseeable future. Below are few of the examples.

CSR: Since CSR plays important role inside the blast furnace, the purchasing personnel at a steel pant would like to know if CSR varies a lot. One example of CSR variation in a coal mine is shown in Fig. 7 (Wozek and Zuke, 1990). Because of higher CSR values and low CSR variability in the reserve, the coal was a favorite choice for use at the coke batteries of a steel plant. Note that the CSR was calculated for the US mine by using Inland Steel CSR Prediction Model (Valia, 1989) that takes into account Gieselar Fluid Temperature range, Alkali Index, and sulfur values of coal from different drill holes and the mine face.

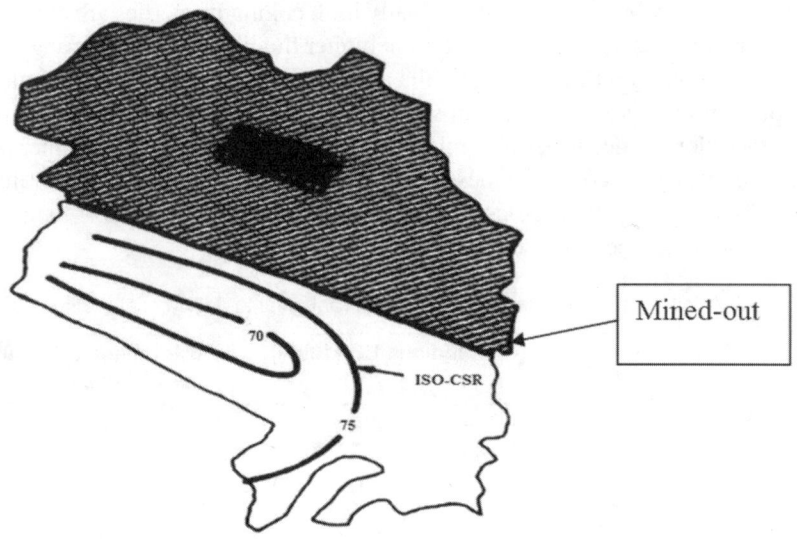

**Figure 7:** Coal reserve evaluation for CSR (Wozek and Zuke, 1990)

Stability: Similar exercises are done for stability predictions. As stability plays an important role both before it reaches blast furnace and then inside the blast furnace, the purchasing personnel at a steel plant would like to know if stability varies a lot. For stability predictions, petrographic data along with ash and sulfur values are used. In this particular example, using US Steel Stability Prediction Model (Schapiro et al., 1961), first individual stability values were calculated for one seam of a medium VM coal at different drill holes and the mine face. Then blend stability values were calculated by incorporating 40% of this seam of the medium VM coal with 60% of the other high VM coals being used at the coke batteries of the steel plant. The reserve evaluation map (Fig. 8) shows that in order to obtain a coke with predicted blend stability of 57–59, coal from this seam should be mined from those areas where the reflectance value lie within the range of 1.26–1.36%. Coal from the central part of the seam is suitable for the steel plant's purpose. The northern and southern parts of the future mining areas would produce poorer coke quality. Hence, a proper mining plan need to be developed that utilizes coal in such a way so that the mine is not left out with poorer quality coal. Note that reflectance and inert values from petrographic analysis also give indication of possible wall pressure exerted during coke making which is predicted by using a relationship developed by Thompson and Benedict (1975). Similar reserve evaluations were done for many other coal properties relevant to coke making.

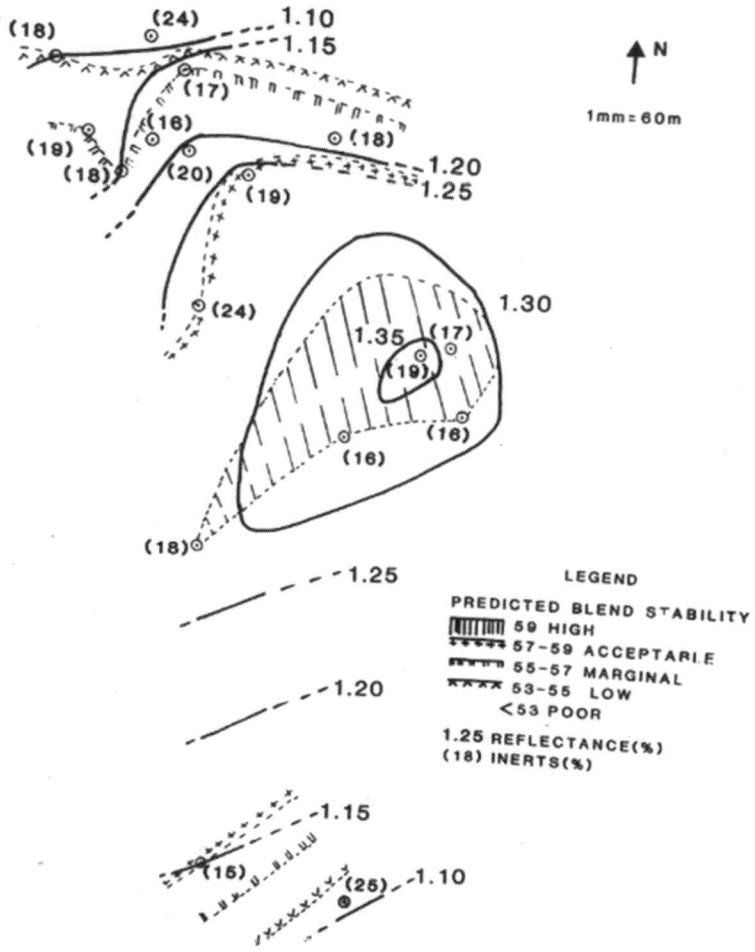

**Figure 8:** Coal reserve evaluation for stability and other parameters

## 3.3 Effect of washing on coking quality with reference to predicted CSR

A coke maker is interested in knowing the effect of washability on coal it is going to receive from the mine. Besides measuring ash and sulfur content, it is imperative to explore what washability would do to rheological, petrographic, and chemical properties that affect CSR, coke stability, coke chemistry and also coking pressure and coke pushing performance. With respect to flotation of macerals, the vitrinite performs the best followed by liptinite, and the inertinite the worst in that order (Arnold and Aplan, 1989). How does flotation affect the coking potential of a coking coal? One example of an Indian coal washed to produce different levels of resultant ash and its effect on coking potential is shown in Table 4. The data was supplied through a joint study by SAIL and CPMDI of India. The data indicates that decreasing coal ash (from 16.70% to 14.20% to 10.20%, db) results in dramatic improvement in CSR (from 37 to 62 to 66). Note the CSR was predicted using Inland CSR Prediction model that utilizes fluid range, alkali index, and sulfur content of coal (Valia, 1989). The predicted CSR improvement is due to an increase in rheological properties (higher maximum fluidity and fluid range) and a decrease in alkali index. The higher fluid range facilitates formation of nematic liquid crystals and their coalescence into larger carbon forms. The higher the carbon form, higher is the resistivity to $CO_2$ gasification. On the other hand, the drop in alkali index lowers the intensity of catalytic gasification of carbon forms. It is imperative that when washability studies are done at a coal preparation plant, besides ash, sulfur, yield, and size fraction, the coking properties should also be taken into account.

**Table 4:** Lowering coal ash in one coal seam and its effect on resultant coal properties and coke CSR

|  | Coal 1-A | Coal 1-B | Coal 1-C |
|---|---|---|---|
| Ash (%, db) | 16.70 | 14.20 | 10.20 |
| Pred. CSR | 37 | 52 | 66 |
| Fluid range (°C) | 61 | 82 | 97 |
| Max fluidity (ddpm) | 12 | 217 | 875 |
| Alkali index | 2.25 | 2.04 | 1.54 |
| Sulfur (%, db) | 0.62 | 0.62 | 0.64 |
| VM (%, db) | 16.60 | 17.40 | 19.20 |
| Reflectance (%, MMVR) | 1.43 | 1.43 | 1.43 |
| Vitrinite (%) | 31.00 | 35.00 | 42.60 |
| Inertinite (%) | 58.90 | 56.40 | 51.30 |
| Phosphorus (%, db) | 0.184 | 0.162 | 0.139 |

Note: Data obtained through a joint study via SAIL R&D and CPMDI, India.

## 3.4 Effect of oxidation on coking quality at coal piles or in coal mine

Because an appreciable portion of coal is stored in large piles either at coal mine or coke plant for various periods of time, it is important to measure properties that deteriorate due to weathering and assess its effect on coke properties. Valia (1990) showed CSR deteriorates with storage time for wide range of coals, high VM to medium VM coals and were associated with a drop in fluid range. Similarly, prolonged exposure of coal to weathering would result in formation of oxyvitrinite that would result in deterioration of coke quality. The coal may also contain pseudovitrinte, formed during exposure to oxygen in coalification. It behaves more inertly than vitrinite. Its presence results in deterioration of coke stability (Benedict et al., 1968).

## 3.5 Coal preparation and blend design at the coke plant

One example of how a coke plant operator selects a coal for use at a slot oven battery blend is shown in Table 5. The base blend uses 52%HVM B-21%HVM A-12%LVM2-15%LVM3. By using the coal quality parameters, coke quality was predicted. Following models were used for stability prediction (Schapiro et al., 1961), CSR prediction (Valia, 1984), and coke yield prediction (Walker, 1996). With the base blend being used at the coke plant, the coke CSR is 59 which is lower than the blast furnace minimum CSR requirement of 60. Note in this Coke Plant Blend Modeling, only emphasis is being paid on coke quality and the coke yield with approximate values for wall pressure (using reflectance and inert content). The wall pressure is not shown in Table 5. Also the coal costs are not shown. It is clear that in order to improve CSR, the new LVM1 coal was evaluated and its inclusion in the blend, with slight fine-tuning, would improve the CSR to 61. The stability would drop by one point but still meets the blast furnace target value of 60. Provided the LVM1 coal is cost competitive, it will be the choice for inclusion in the blend.

**Table 5:** One example of coal blend selection modeling

|      | HB  | HA  | LV 1 | LV 2 | LV 3 | CSR | Stability | Coke ash | Coke sulfur | Coke phos. | Coke alkali | Coke yield | Blend VM | Blend refl. |
|------|-----|-----|------|------|------|-----|-----------|----------|-------------|------------|-------------|------------|----------|-------------|
| Base | 52% | 21% | x    | 12%  | 15%  | 59  | 61        | 9.32     | 0.69        | 0.015      | 0.27        | 72.3       | 30.07    | 1.14        |
| New  | 54% | 18% | 18%  | x    | 10%  | 61  | 60        | 9.22     | 0.69        | 0.012      | 0.24        | 72.14      | 30.04    | 1.16        |

The most cost effective blend is then selected for further evaluation through a BF operation model. BF rules of thumb from AISI survey (Poveromo, 1995) are used to evaluate the utilization cost of cokes that would be produced from carbonization of the most cost effective blends as selected through the Coke Plant Operation model. The AISI survey suggested a consensus of coke rate coefficient for CSR of 3.0 lbs/NTHM (<58 CSR) and 1.5 lbs/NTHM (>58

CSR) and for ash, 20 lbs/NTHM. Using predicted coke quality (from Table 5) and coke rate (not shown here), the quality adjusted price of blends are determined. The blend that would produce coke with highest potential cost savings at the BF is recommended for implementation at the coke plant.

Once the coal with highest potential for saving at BF is selected, the coal starts its journey from mine to the coke plant. It should be noted that the normal practice at many coke plant is that a coal company personnel at the mine need to fax key coal quality parameter to the coke plant personnel. Only after the quality data is approved, the coal train leaves for the coke plant. Once it has arrived at the coke plant, the quality monitoring begins vigorously at various stages from unloading, stacking, blending, crushing, charging, coking, coke quenching, and coke screening. Fish bone diagrams for every coal quality parameter for every coke property is monitored so as to produce consistent high quality coke with lowest variation in quality.

## 3.6 Descent of coke inside a blast furnace

What happens to coke once it is charged into a blast furnace? Coke degradation can be determined either by characterizing coke at the stockhouse before it is charged and then raking coke at the tuyere level and/or by inserting a coke core probe from tuyere to center of the furnace during outage time (Cheng, 1997). The results from such a study by Chaubal and Valia (2003) for a large BF (4000 m$^3$) are reproduced in Table 6. It should be noted that this blast furnace uses two cokes and also uses pulverized coal injection. The size reduction in coke feed sample (average of properties for two cokes) versus tuyere-raked sample is 28%. Samples show an increase in ash, alkali and a drop in hot strength properties, sulfur, silica, and titania.

**Table 6:** Coke quality changes between feed coke and tuyere-raked coke (Chaubal and Valia, 2003)

|  | Feed coke | Tuyere-raked coke |
|---|---|---|
| Coke size (mm) | 55.4 | 39.8 |
| CSR | 68.45 | 58.2 |
| CRI | 21.35 | 33.7 |
| Ash (%) | 9.36 | 13.49 |
| Sulfur (%) | 0.57 | 0.45 |
| Alkalies (%) | 1.98 | 3.46 |
| Silica (%) | 53.68 | 36.5 |
| Titania (%) | 1.37 | 1.04 |

Coke core in the steel pipe inserted at tuyere shows five different zones along the blast furnace radial representing positions of the Raceway, Bird's Nest, and the Deadman (Fig. 9). The boundary between the Raceway and the Bird's Nest is placed at 0.75 m (between Zones II and III). The Bird's Nest extends from 0.75 m to 1.5 m. Deadman represents Zone V. Raceway zone, as compared to Bird's Nest, is characterized by highest amount of char, fines, alkalies, iron and slag content, and small coke cemented to iron. Deadman, as compared to Bird's Nest zone, is characterized by drop in char, iron, and slag content, the presence of angular coke, and coke not cemented by metal.

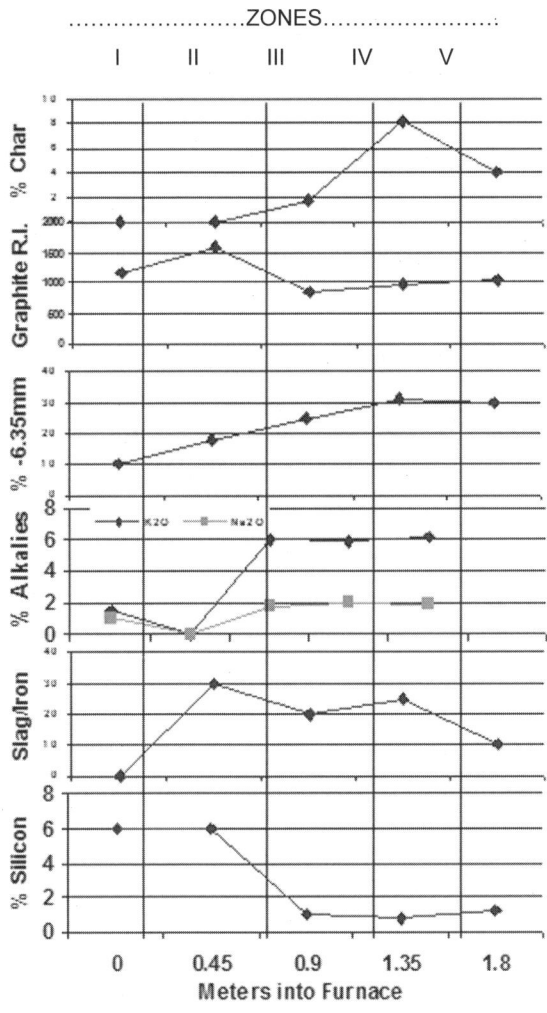

**Figure 9:** Tuyere core sample analysis (Chaubal and Valia, 2003)

They also observed an interesting fact that in the Deadman zone, along with char, coke, and metal droplet (Fig. 10A) there was the presence of coal particles (Fig. 10B). The coal particles in Fig. 10B did not show any sign of pyrolysis or carbonization suggesting possibility of extremely low temperatures in certain areas of Deadman zone.

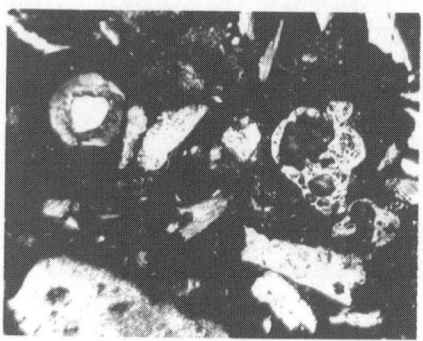

**Figure 10A :** Char, coke, metal droplets from blast furnace Deadman Zone
(Chaubal and Valia, 2003)

**Figure 10B:** Coal and coke from blast furnace Deadman zone
(Chaubal and Valia, 2003)

It is interesting to note that during pulverized coal injection periods, examination of blast furnace flue dust from the dust catcher also indicated traces of unaltered coal particles travelling through the raceway upwards.

## 4. Summary

A coke producer selects coal based on its contribution to the resultant coke quality, with safe oven pushing performance, that would result in highest potential cost savings at the blast furnace. In other words, coal producer must give assurances of consistent high quality now and from the future mining areas. The coal quality monitoring system at each stage is rigidly followed with examples. Starting at the reserve, coal need to be fully characterized, including measurement of abnormal coking properties that a coal may possess; e.g., unusually high amount of liptinite macerals that may enhance coking properties or presence of pseudovitrinite or oxyvitrinite that may adversely affect the coking properties. Example of role of exudatinte in enhancing the coke quality is presented. The next step is, besides studying normal coal quality variations, understanding predicted CSR and predicted coke stability variations in the coal reserve. Case studies of variations of CSR and stabilities in a reserve are presented. Once the coal is mined and blended and washed, need to understand how washability may affect predicted coke quality. Examples of CSR improvements with washability are presented. Finally coal selection and blend selection is described that utilizes the Coke Plant model and the Blast Furnace model. For coal producers to become a preferred vendor for coke/steel industry, they need to familiarize with users selection procedure and the role of coke in the blast furnace as described in the article.

## 5. Acknowledgement

The author would like to thank AercelorMittal USA and Sun Coke management for supporting the research work done during author's professional career days at ArcelorMittal USA R&D laboratory, East Chicago, Indiana. And gratitude is extended to colleagues who helped in carrying the research work, laboratory tests, and plant trials at AercelorMittal and at Sun Coke.

## 6. References

[1] Arnold, B. and Aplan, F., 1989, The hydrophobicity of coal macerals, Fuel, Vol. 68, No. 5, p. 651–618.

[2] Benedict, L.G., Thompson, R., Shigo, J., and Aikman, R., 1968, Pseudovitrinite in Appalachian Coals, Fuel, Vol. XLVII, March 1968, pp. 125–143.

[3] Chaubal, P. and Valia, H., 2003, Studies on Blast Furnace Coke Degradation – A case study with low reactivity stamp charged beehive coke, METEC Congress, Dusseldorf, Germany.

[4] Cheng, A., 1996, Coke quality requirements for blast furnaces, in Selecting Coals for Quality Coke, Iron & Steel Society 1996 Ironmaking Short Course, Pittsburgh, PA, March 24, 1996, pp. 1–44.

[5] Crelling, J.C., 1983, Current uses of fluorescence microscopy in coal petrology, Journal of Microscopy, Vol. 132, Pt. 3, pp. 252–266.

[6] Kaegi, D., Valia, H., and Harrison, C., 1988, Maceral Behavior and Coal Carbonization," ISS-Ironmaking Conference Proc., Vol. 47, 1988, pp. 339–349.

[7] Poveromo, J., 1995, AISI Coke Quality Survey, issued by AISI, October 1995.

[8] Ricketts, J.A., 1996, Historical Development and Principles of the Iron Blast Furnace, Blast Furnace Ironmaking Short Course, Vol. 1 – Principles, Design, and Raw Materials, McMaster university, May 1996, pp. 1-1 to 1-82.

[9] Schapiro, N., Gray, R., and Eusner, G, 1961, Recent developments in coal petrography, blast Furnace, Coke Oven, and Raw Materials Conference, TMS-AIME, Vol. 20, pp. 89–109.

[10] Spackman, W., Davis, A, and Mitchell, G.D., 1976, The fluorescence of liptinite macerals, Brigham Young University Geologic Studies, Vol. 22, PT. 3, pp. 59–75.

[11] Stanton, R.W., 1982, Application of fluorescence microscopy to the study of the components of coal, Microbeam Analysis, Heinrich, K.F.T. (Ed.), San Francisco Press, San Francisco, pp. 330–332.

[12] Thompson, R. and Benedict, L., 1975, Coal composition and its influence on cokemaking, ISS-AIME Proc., Vol. 34, pp. 112–121.

[13] Valia, H.S., 1989, Prediction of Coke Strength after Reaction with $CO_2$ from Coal Analysis at Inland Steel Company, Iron & Steelmaker, pp. 77–87.

[14] Valia, H.S., 1990, "Effects of Coal Oxidation on Cokemaking," ISS-Ironmaking Conference Proc., Vol. 49, pp. 199–209.

[15] Valia, H.S., 1993, Coal Quality Reserve Evaluation for the Iron and Steel Industry, Proc. 10th Annual International Pittsburgh Coal Conference, Pittsburgh, PA., pp. 305–310.

[16] Valia, H.S., 1996, Coke Production for Blast Furnace Ironmaking, Blast Furnace Ironmaking Short Course, Vol. 1 – Principles, Design, and Raw Materials, McMaster university, May 1996, pp. 10-1 to 10-19.

[17] Valia, H.S. and M. Mastalerz, 2004, "Indiana Coals and the Steel Industry," Indiana Geological Survey Special Report Number 64, Bloomington, Indiana, 28 pp.

[18] Valia, H.S., 2005 "Coal Cost Reduction Using Low Rank Coals," AIST 2005 Conf. Proceedings, Vol. 1, pp. 83–89.

[19] Valia, H.S., 2019, Nonrecovery and Heat Recovery coke making technology, p. 203–292, in New Trends in Coal Conversion, Suarez-Ruiz, I, Diez, M.A., and Rubiera, F., (Ed.), Woodhead Publishing an imprint of Elsevier Duxford, U.K., 511 p.

[20] Wakelin, D.H., 1999, The Making, Shaping and Treating of Steel, Ironmaking Volume, 11th edition, AISE – AIST, Warrendale, PA, USA, 811 p.

[21] Walker, D.N., 1996, High CSR coke from Non Recovery cokemaking process, Paper distributed at the 3rd International Cokemaking Congress, Ghent, Belgium.

[22] Wozek, J. and Zuke, A., 1990, An integrated approach to coal quality and its impact on primary operations at Inland Steel Company, ISS-AIME Proc., Vol. 49, pp. 235–242.

# 22

# Washing of coking coal: A sustainable business endeavor

Swarup Kumar Datta[1] and Nikkam Suresh[2]

*[1]BCCL, Dhanbad, India*
*[2]IIT(ISM), Dhanbad, India*

**Abstract:** Coking coal is a scarce commodity in India. Conscious efforts of Government of India to improve the unfavorable balance of payment of the country and increasing quality and environmental concerns are resulting in a paradigm shift towards effective utilization of indigenous coking coal and its washing capabilities to cater the steel sectors. Coal sectors in India today operate in a fast changing business environment where there is continual empowerment of consumers and green lobby, there is socio-political thrust on effective utilization of resources, quality of product is an important issue, and aspirations of the nation centers around a new clean India.

To meet the macro environmental challenges prevailing in coal business today, to accomplish its mission statement and to add value to its coal business, Bharat Coking Coal Limited (BCCL), the prime supplier of coking coal in India and a fully committed coal company to supply quality-coal to its consumers has embarked upon an ambitious journey to produce and market 100% washed coal only. BCCL shall be the first coal PSU in INDIA to produce washed coal only in medium to long term time perspective of 10 years from now. The resent paper summarizes the efforts made to depict the coking coal scenario of India with the need for effective utilization of resources, present coking coal demand and supply scenario, necessity for development of washing culture in India, a road map for coal washing in BCCL, impact of coking coal washing in the in 3P's (Profit, People, Planet), technologies to be adopted in the design of future washeries, favorable economics through coal washing for a coal producing company, challenges to overcome, and opportunities to harness for meeting the future requirement of coking coal.

The paper also highlights the real life issues of India coal washeries. Promoting green business through production of washed coal will not only promote brand BCCL towards making it consumers' dream destination, but also promote ethical consumerism.

**Keywords:** Coal washing, Pradigm shift, Consumer, Mission, BOM concept, Deshaling

## 1. Background

With the expansion of existing steel plants and addition of greenfield projects in India, supply of quality coking coal to steel plants will becomes a matter of concern. Indian coking coals are characterized by their high ash and lower reactive contents. The coking properties of indigenous coking coals are also inferior. Thus, to meet the requirements of coking coal for blast furnace coke, use of imported coking coal as blend with indigenous coking coal is a technological compulsion, as the domestic quality has higher ash content and is not suitable for steel industry with their present iron making technologies. Conscious efforts of GoI (government of India) to improve the unfavorable balance of payment and increasing quality and environmental awareness, have lead paradigm shift towards higher utilization of washed and blended coal in steel sector in India.

## 2. Introduction

Coal sectors in India today operate at a fast changing business environment, where there is continued empowerment of consumers and green lobby, there is a socio-political thrust on effective utilization of resources. Quality of product is an important issue and acquires center position in building new clean India. In such, a challenging business environment, the *Mission Statement* of Bharat Coking Coal Limited, a prime producer and supplier of coking coal, reads as under:

*"To produce and market the planned quantity of coal and coal products efficiently and economically in an eco-friendly manner with due regard to safety, conservation and quality."*

To meet the macro environmental challenges prevailing in coal business today, to accomplish its mission statement and to add value to its coal business, Bharat Coking Coal Limited, a fully committed coal company to supply quality coal to its consumers has embarked upon an ambitious journey to produce and market with 100% washed coal in long term time perspective (0–10 yrs) with the following visions and missions:

- To produce and market 100% washed coal in long term time perspective (0–10 yrs),
- Enhanced supply of washed coking coal to steel industry,
- Minimise import of metallurgical coal,
- Meet environmental obligations,
- Construction of new coal washeries,
- Renovation of existing washeries.

## 3. Need for effective utilization of coking coal reserves

Coking coal reserves of the country constitutes only 12% of the total coal resources out of which 50% possess high ash and difficult to wash LVMC (low volatile medium coking) coal. If not washed, the natural destination for the LVMC coal having coking properties at 17–19% ash level will be power sectors. Thus, it is imperative that coking coal reserves be rationally utilized through washing of inferior quality for long term sustainability.

## 4. Present coking coal supply scenario of India

- Although BCCL is the prime coking coal producer, entrusted to meet the demand of sectors, but around 21% of India's coking coal requirements are met through import.
- Coking coal is an important ingredient in steel making through blast furnace routes accounts for 30–50% of the cost of production for steel.
- The revival of steel demand for India has further increased forecasted requirements of coking coal consumption stimulating coking coal imports.
- Total coking coal requirement by 2030–31 as per NSP 2017 is 155–166 Mt.
- The country is losing a large portion of currency for foreign exchequer by importing coking coal.
- Year-wise supply of coking coal (Mty) in India is given in Table 1.

**Table 1:** Year-wise supply of coal (indigenous and imported)

| Coking coal | Year-wise supply of coking coal (Mty) | | | | | | |
|---|---|---|---|---|---|---|---|
| | 2012–13 | 2013–14 | 2014–15 | 2015–16 | 2016–17 | 2017–18 | 2018–19 (Estimated) |
| Indigenous | 16.90 | 15.49 | 12.02 | 12.37 | 12.51 | 13.60 | 15.0 |
| Import | 35.56 | 36.87 | 43.72 | 43.51 | 44.11 | 46.5 | 51.20 |
| Total | 52.46 | 52.36 | 55.74 | 55.88 | 56.62 | 60.10 | 66.20 |

## 5. Present scenario of coking coal washing in Coal India Limited (CIL)

While steel making technologies with alternative fuels are being developed world wide, coking coal would continue to be important in Indian context for quite some time. CIL is currently producing about 50 Mtpa of coking coal which is likely to be enhanced to the tune of 68 Mtpa by 2019–20. Only 5 Mtpa are being washed in the existing washeries and supplied to Steel sector.

The balance coking coal along with non-coking coal is being supplied to the power sector under FSAs and other miscellaneous consumers.

In BCCL the present status of production of washed coking coal is not encouraging. Approximately 2.0 Mtpa washed coking coal is being produced from BCCL washeries by washing 10.3 Mtpa. This is because of poor yield of washed coal and their capacity utilizations which are as given below:

Coking coal washed in BCCL (Mty): 10.3.

Average yield of washed coal (%): 23.5%.

Capacity utilization (%): 17%.

Main reasons for present dismal condition of washeries in BCCL are as below:

- All the washeries have outlived their rated operable life.
- Most of the superior grade coals of upper seams are exhausted.
- Available raw coals from lower seams are not suitable for washing in existing system.
- Fine coal circuits have not been successful.
- Absence of timely renovation.

## 6 Status of present and future coal washeries in BCCL

Conscious efforts of GoI to improve the unfavorable business of production of the country and increasing quality and environmental concerns are resulting in a paradigm shift towards higher utilization of washed and blended coal in steel sector in India. In consonance with the changing business scenario, BCCL, the prime producer of coking coal in India has BCCL, which has planned for enhancement of washing capacity through construction of new washeries and renovation of its existing washeries to the tune of 31.03 Mtpa by 2024–25. More details on construction of new washeries in two different phases and renovation of existing washeries are given in Table 2.

**Table 2:** Current Status of coking coal washeries in BCCL

| Sl. No. | Type of washery | Phase | Name of washery | Capacity (Mtpa) | Expected year of Commissioning | Washed Coal (Mty) | Washed Coal (Power) (Mty) |
|---------|-----------------|-------|-----------------|-----------------|--------------------------------|-------------------|---------------------------|
| 1 | New washery | | Madhuband | 5.00 | 2020-21 | | |
| 2 | | First | Patherdih - I | 5.00 | 2019-20 (Construction completed, commissioning in progress) | 7.25 | 7.88 |

*Contd...*

*Contd...*

| Sl. No. | Type of washery | Phase | Name of washery | Capacity (Mtpa) | Expected year of Commissioning | Washed Coal (Mty) | Washed Coal (Power) (Mty) |
|---|---|---|---|---|---|---|---|
| 3 | | | Patherdih – II | 2.50 | 2019-20 | | |
| 4 | | | Dahibari | 1.60 | Commissioned (Aug'18) | | |
| 5 | | | Bhojudih | 2.00 | 2020-21 | | |
| 6 | | | Dugda | 2.50 | 2020-21 | | |
| 7 | | | Moonidih | 2.50 | 2020-21 | | |
| 8 | | Second | Kalyaneswari | 3.60 | 2023-24 | 1.08 | 1.44 |
| 9 | Washeries to be renovated | Existing old washeries | Sudamdih | 1.60 | End of 2020 | 3.00 | 1.26 |
| 10 | | | Moonidih | 1.60 | | | |
| 11 | | | Mohuda | 0.63 | | | |
| 12 | | | Madhuband | 2.50 | | | |
| | Total | | | 31.03 | | 11.33 | 10.58 |

The wheel of BOM concept based washeries has started to roll. Out of the seven washeries planned in first phase, six washeries are scheduled to be set up on BOM concept. The uniqueness of 5.0 Mtpa Patherdih coal washery is besides being first BOM washery it is the highest capacity coking coal washery in India. However, the requirement of indigenous washed coking at varying ash levels is given in Table 3.

**Table 3:** Requirement of indigenous washed coking at varying ash levels

| Sl. no. | Total coking coal requirement by 2030–31 as per NSP 2017 (Mt) | Requirement of indigenous washed coking coal for iron and steel sector of the country by 2030–31 (Mt) | | |
|---|---|---|---|---|
| | | At 18% ash level of clean coal | At 15% ash level of clean coal | At 13% ash level of clean coal |
| 1 | 155–166 | 31–33 | 40–43 | 59–63 |

Higher demand of coking coal at low ash level may act as growth opportunity for indigenous coal producers in India.

## 7. Washing technologies in new washeries

BCCL has already awarded five washeries under BOM concept. These washeries shall be equipped with state of the art technologies to wash the impurities from LVMC raw coal having ash content 42±4% (Table 4).

**Table 4:** Washing technologies in new washeries

| Washery | Size range | Washing scheme | Technology | Capital cost | % Yield | | |
|---|---|---|---|---|---|---|---|
| | | | | Rupees in crores | WC | WCP | Rejects |
| Madhuban washery (5.0 Mtpa) | 75–13 mm coal | Deshaling | HM cyclone | 262.99 | 40.30% | 19.40% | 40.30% |
| | 13–1 mm coal | Washing | HM cyclone | | | | |
| | 13–1 mm coal | Rewashing | HM cyclone | | | | |
| | 1–0.25 mm coal | Small coal washing | TBS | | | | |
| | (–)0.25 mm coal | Fines beneficiation | Froth floatation | | | | |
| Patherdih washery (5.0 Mtpa) | 50–13 mm coal | Deshaling | Jig | 131.66 | 22.46% | 51.59% | 25.95% |
| | 13–0.5 mm coal | Washing | HM cyclone | | | | |
| | (–)0.5 mm coal | Fines beneficiation | Froth floatation | | | | |
| Dahibari washery (1.6 Mtpa) | 50–13 mm coal | Deshaling | Deshaling jig | 113.36 | 22.10% | 41.40% | 36.50% |
| | 13–0.5 mm coal | Washing | HM cyclone | | | | |
| | (–)0.5 mm coal | Fines beneficiation | Water only cyclone, spiral concentrator | | | | |
| Patherdih washery (2.5 Mtpa) | 13–2 mm coal | Washing | HM cyclone | 243.00 | 30.60% | 52.15% | 17.25% |
| | 13–2 mm coal | Rewashing | HM cyclone | | | | |
| | 2–0.5 mm coal | Small coal washing | Spiral concentrator | | | | |
| | (–)0.5 mm coal | Fines beneficiation | Froth floatation | | | | |
| Bhojudih washery (2.0 Mtpa) | 20–2 mm coal | Washing | HM cyclone | 242.55 | 19.70% | 55.60% | 24.70% |
| | 20–2 mm coal | Rewashing | HM cyclone | | | | |
| | 2–0.5 mm coal | Small coal washing | Spiral concentrator | | | | |
| | (–) 0.5 mm coal | Fines beneficiation | Froth floatation | | | | |

## 8. Future prospect

1. Huge reserves of LVMC coal in BCCL: Future opportunity lies in mutilation of huge reserves of LVMC coal in BCCL and in meeting the demand supply gap of metallurgical coal in India. By 2030, coking coal requirement in India is expected to treble from current 57 Mt to about 161 Mt (Fig. 1). After 100 % implementation of Jharia action plan, majority of prime coking coal locked up in these areas (approx. 232.764 Mt) shall also be available for mining. The supply of indigenous coking coal to the Indian steel sector shall change drastically.

2. Demand of low ash coking coal by steel sector: Ash is a most significant parameter that has high impact on BF coke rate and productivity. Coking/caking properties improve with reduction in ash content. Ash is the main source of mineral impurities, sulfur and alkali oxides in coal/coke which largely impacts CRI/CSR of coke and BF operation. As such with 1% increase in coal blend ash BF coke rate is increased by about 3%.

3. Competition from Imported Coal: The two main computations met with imported coal are:

   (i)  Cost competition, and

   (ii) Quality competition.

4. Economies of scale: To overcome the challenge of low profitability, the strategy should be to achieve economies of scale in washery business to reduce operating cost and thereby, diminish net cost of production and enhance gross profit margin. Further, opportunities should be explored to go for at least 5.0 Mty capacity washeries.

5. Lack of process engineers (mineral engineers) in coal washery operation; Nona availability of qualified and trained coal preparation engineers has resulted poor washing efficiencies incurring huge losses and underutilization of natural resources.

   • Achievement of low organic efficiency.

   • Loss of coal values in coal slurry.

6. Lack of project management expertise in implementation of washery projects resulting in delays.

## 9. Challenges

1. Disposal of rejects produced from these washeries will be a major challenge for successful operation of the new washeries. Coal India

Limited is formulating a rejects policy in this regard. Academia may play a pivotal role in gainful utilization of rejects.

2. Paucity of indigenous magnetite.

3. Washing culture: In state owned public sectors in India, there is lack of washing culture i.e. reluctance to sell and use of washed coal by the consumers. The cost of coal production is increasing day by day resulting in top line pressure for the coal producers. It is high time that the coal producers should look beyond mining and dispatch of the mined out coal through siding to the consumers. It is the time to look for production and marketing of value added products. Coking coal washing shall not only add value to the economic sustainability for the coal producers, it shall ensure optimum utilization of our scarce coking coal reserves and thereby, reducing FOREX exodus of India (Table 5).

**Table 5:** Value addition capacity of new washeries under BOM concept

| Projected profit/loss for hypothetical 1.6 Mtpa LVMC washery (under BOM concept) | | | | |
|---|---|---|---|---|
| Capacity (lakh te. per annum) | 16.00 | Percent yield of washed coal | 22.4 | |
| Basic raw coal cost (Rs) (G-13 grade; approx. 46% ash) | 1335.00 | | | |
| Selling price of washed coal (Rs) | 6465.00 | Yield% washed coal (power) | 41.60 | |
| Selling price of washed coal (Power) (Rs) (40% ash) | 2775.00 | | | |
| Elements of cost | For 100 % capacity utilization | | For 85% capacity utilization | |
| | Cost per tonne (Rs) | Amount (INR in lakhs) | Cost per tonne (Rs) | Amount (Rs in lakhs) |
| Basic raw coal ( opportunity) cost | 1,335.00 | 21,360.00 | 1,335.00 | 18,156.00 |
| Transportation charges ( 1–2 km lead) | 45.00 | 2,250.00 | 45.00 | 1,912.50 |
| Total raw coal cost | 1,380.00 | 23,610.00 | 1,380.00 | 20,068.50 |
| Cash operating cost (approx) | 220.00 | 3,520.00 | 220.00 | 2,992.00 |
| Depreciation (as per straight line method) | 114.68 | 1,834.88 | 134.92 | 1,834.88 |

*Contd...*

*Contd...*

| Projected profit/loss for hypothetical 1.6 Mtpa LVMC washery (under BOM concept) | | | | |
|---|---|---|---|---|
| Commitment charges by BCCL for less supply of coal | 0.00 | 0.00 | 45.07 | 108.17 |
| Total cost of input | 1,714.68 | 28,964.88 | 1,779.99 | 25,003.55 |
| Less: credit for washed coal (power) | 1,154.40 | 18,470.40 | 1,154.40 | 15,699.84 |
| Net cost of raw coal input | 560.28 | 8,964.48 | 625.59 | 8,507.99 |
| Cost per tonne of washed coal | 2,501.25 | | 2,792.80 | |
| Sales/selling price of washed coal/te | 6,465.00 | 23,170.56 | 6,465.00 | 19,694.98 |
| Profit/loss on sale of washed coal | 3,963.75 | 14,206.08 | 3,672.20 | 11,186.98 |
| Profit/loss per annum | | 14,206.08 | | 11,186.98 |

# 10. Conclusion

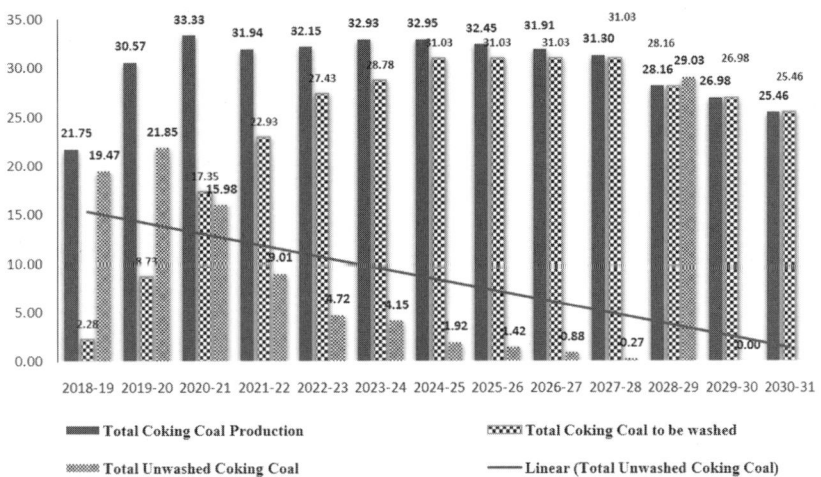

**Figure 1:** Coking coal production and washing in BCCL till 2030 (in Mt)

(a)   Besides being a major contributor to the financial health of BCCL, the upcoming washeries in BCCL will surely be a great leap to minimize the demand supply gap of coking coal of our country.

(b) Further, will help in providing energy security as well as financial security to our country. Above all, the clean coal technologies used in these washeries will act as a safeguard against environment pollution. These washeries are ecologically sustainable growth initiatives of BCCL directed towards effective utilization of scarce coking coal reserves of our country.

(c) Large number of coal washery shall follow the two washeries, i.e., 5.0 Mtpa Patherdih washery and 1.6 Mtpa Dahibari washery making a paradigm change in the indigenous coking coal supply scenario of the country by 2021–22, provided the challenges are converted to opportunities.

(d) In the long run, BCCL coal will no more be black, it will be completely green. Promoting green business through production of washed coal will not only promote brand BCCL towards making it consumers' dream destination, but also promote ethical consumerism.

"Brand BCCL Shall Be Consumers' Dream Destination!!"

# 23

# Assessment of technical feasibility of washing low grade coking coals (W-IV/LVMC) to 13% ash content

Ashim Kr. Chakraborty, Ratan Basu, Abha Prasad

*CMPDI, Ranchi, India*

**Abstract:** Coking coal is a scarce commodity in India. The coking properties of indigenous coking coals are also inferior. Thus, to meet the requirements of coking coal for blast furnace coke, use of imported coking coal as blend with indigenous coking coal is a technological necessity as the domestic coal quality has higher ash content and not suitable for direct feed in steel industry with present technology. The average ash content of coking coal required for coke making is about 10%. Presently, imported coking coal having ash about 8–9% is blended with indigenous coking coal having ash about 18% for coke making and used in blast furnace.

A study was carried out to assess the technical feasibility of washing low grade indigenous coking coal (washery Grade-IV i.e. ash between 28% and 35%/ low volatile medium coking having ash beyond 35%) to 13% ash content so that requirement of imported coking coal could be reduced. Based on the study, it was opined that washing of low grade coking coal to get clean coal of 18% ash level will give viable techno-economics for both coal companies well as consumers. For washing washery Grade-II (i.e. ash between 21% and 24%)/washery Grade-III (i.e. ash between 24% and 28%) coking coal to 13% ash content, new washeries may be set up in those areas where such coals are available with potential of better yield leading to favorable techno-economics and the consumer is willing to accept coal of 13% ash content at higher price, on long-term basis.

**Keywords:** W-IV, LVMC, low grade coking coal

## 1. Background

Coking coal is the vital raw material for steel making in metallurgical industries and is used to produce coke, which is used as reductant in blast furnaces for production of steel. Demand of coking coal is rapidly growing with capacity expansion of existing steel plants and addition of Greenfield projects in India. Indian coking coals are poor in quality; characterized by high ash content, inferior caking properties, low rank, and low reactive contents apart from poor amenability to washing due to presence of near gravity material. Since indigenous metallurgical coking coal is a scarce commodity in India, import of coking coal in India has been showing an increasing trend to compensate for the lack of good quality coal from the country's own mines. Imported coking coal is cost efficient due to its low ash and has superior coking properties when compared to Indian coking coal.

Laboratory studies on LVMC coals from the Jharia and East Bokaro coalfields conducted for CIL's R&D funded projects [1,2] revealed that a considerable quantity of high ash (>35%) low volatile medium coking (LVMC) coal, which does not fall under the classified coking coal category and used as power coal, exhibits improvement in coking propensity on washing and may therefore be blended for steel making through blast furnace route. Moreover, utilization of low volatile high rank (LVHR) coals and high volatile low rank coal after proper beneficiation in blend can reduce import of coking coal to a significant extent.

The average ash content of coking coal for coke making is about 10%. Presently, imported coking coal having about 8–9% ash is blended with indigenous coking coal having about 18% ash for coke making which is used in blast furnace. The price of imported coking coal keeps on fluctuating and sometimes becomes uneconomical for the consumer. To sustain production of blast furnace coke at competitive cost, optimization of the usage of lower ash indigenous coking coals in coal blends was thought of. With this objective, a study was carried out to assess the technical feasibility of washing low grade indigenous coking coal (washery Grade-IV/low volatile medium coking) to 13% ash content so that requirement of imported coking coal could be reduced.

## 2. Testing of coal samples

Testing of a number of coal samples from various mines of Bharat Coking Coal Ltd. (BCCL) and Central Coalfields Ltd. (CCL) and projection of balance of products at 13%, 15%, and 18% ash levels were carried out by Central Institute of Mining and Fuel Research (CIMFR) and Indian Institute of Technology-Indian School of Mines (IIT-ISM), Dhanbad. The various washability tests of raw coal revealed that the ash% varies widely and is as high as 52%, with majority of the coal falling under washery Grade-IV and ungraded.

## 3. Process flowsheet

Based on the test results, a process flowsheet was envisaged (Fig. 1) which is briefly described hereafter:
- Receiving of −100 mm coal in the washery and screening at 50 mm.

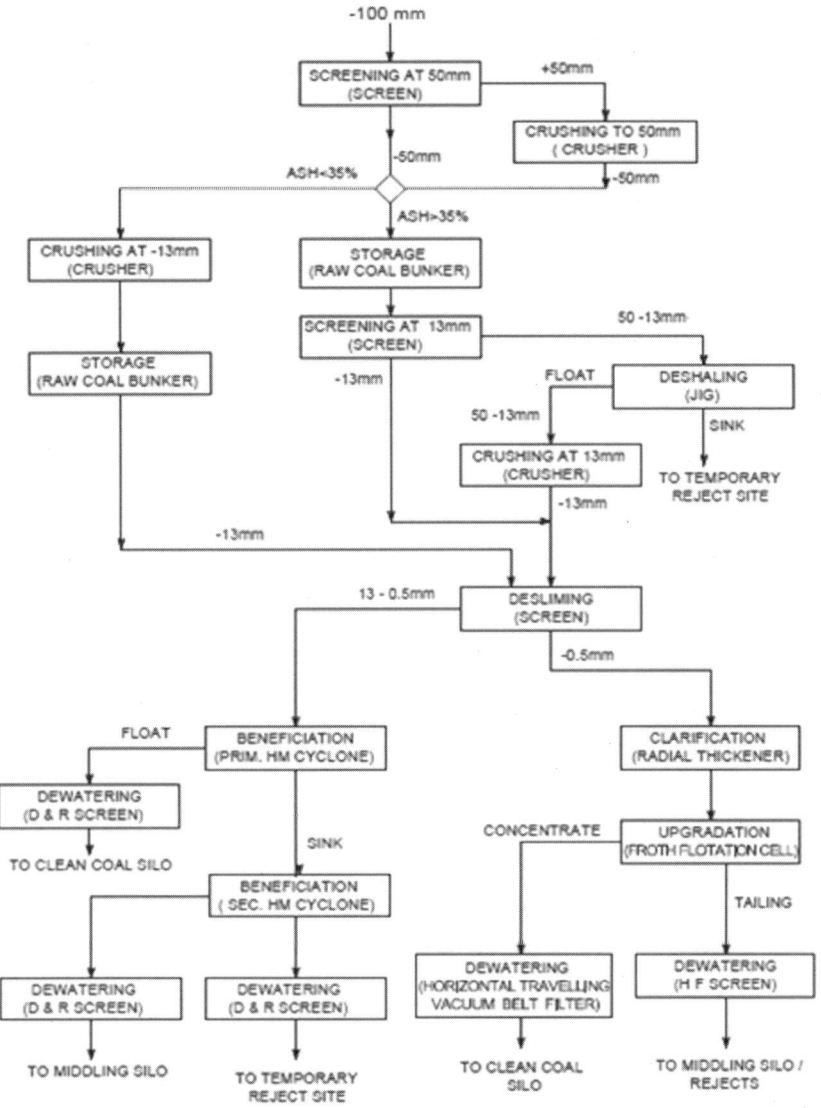

**Figure 1:** Process flowsheet

- Crushing of +50 mm coal down to 50 mm.
- In case raw coal ash is more than 35%, screening of −50 mm coal at 13 mm; deshaling of 50–13 mm coal in jig and mixing the deshaled floats crushed down to 13 mm with natural −13 mm.

- In case raw coal ash is less than 35%, crushing of raw coal down to 13 mm.
- In both cases, desliming of −13 mm coal to obtain two size fractions viz. 13–0.5 mm and −0.5 mm.
- Treatment of 13–0.5 mm size fraction in two stage heavy media cyclones to get clean coal, middlings, and rejects.
- Upgradation of −0.5 mm in froth flotation cells to get clean coal and middlings/rejects.

## 4. Projection of balance of products

The projection of balance of products was done based on test results of coal samples of different mines. washery Grade-IV and ungraded coal are to the tune of approx. 80% of coking coal reserves. There is limited production of washery Grade-I, washery Grade-II, and washery Grade-III coal as upper seams having good quality coal have already exhausted and wherever available is scattered. Moreover, the production of washery Grade-I, washery Grade-II, and washery Grade-III coal from different mines is linked to existing washeries and in future, production will consist mainly of washery Grade-IV and ungraded/LVMC coal. As such, balance of products was considered for washery Grade-IV and ungraded/LVMC coal only.

One number each of washery Grade-IV and ungraded/LVMC coal from BCCL mines and two numbers of washery Grade-IV coals and one number of ungraded/LVMC coal from CCL mines were considered for techno-economics in the study. The balance of products at 13%, 15%, and 18% ash levels for the above samples are given hereafter in Tables 1–5. The same are depicted in Figs. 2–6.

**Table 1:** Balance of products for washery Grade-IV coal (Mine-1, BCCL)

| Products | At 13% ash level | | At 15% ash level | | At 18% ash level | |
|---|---|---|---|---|---|---|
| | Wt% | Ash% | Wt% | Ash% | Wt% | Ash% |
| Clean coal | 27.9 | 13.0 | 39.4 | 15.0 | 51.6 | 18.0 |
| Middlings | 56.5 | 33.5 | 34.9 | 34.0 | 11.5 | 34.0 |
| Rejects | 15.6 | 59.8 | 25.7 | 54.8 | 36.9 | 50.8 |
| Total | 100.0 | 31.9 | 100.0 | 31.9 | 100.0 | 31.9 |

**Table 2:** Balance of products for ungraded/LVMC coal (Mine-2, BCCL)

| Products | At 13% ash level | | At 15% ash level | | At 18% ash level | |
|---|---|---|---|---|---|---|
| | Wt% | Ash% | Wt% | Ash% | Wt% | Ash% |
| Clean coal | 13.0 | 13.0 | 19.4 | 15.0 | 26.6 | 18.1 |
| Middlings | 51.3 | 33.7 | 37.1 | 34.0 | 25.0 | 34.0 |
| Rejects | 35.7 | 65.7 | 43.5 | 61.8 | 48.4 | 59.4 |
| Total | 100.0 | 42.4 | 100.0 | 42.4 | 100.0 | 42.1 |

**Table 3:** Balance of products for washery Grade-IV coal (Mine-3, CCL)

| Products | At 13% ash level | | At 15% ash level | | At 18% ash level | |
|---|---|---|---|---|---|---|
| | Wt% | Ash% | Wt% | Ash% | Wt% | Ash% |
| Clean coal | 20.1 | 13.0 | 29.0 | 15.1 | 40.0 | 18.0 |
| Middlings | 58.3 | 34.0 | 46.3 | 34.0 | 24.3 | 34.0 |
| Rejects | 21.6 | 52.8 | 24.7 | 55.5 | 35.7 | 51.5 |
| Total | 100.0 | 33.8 | 100.0 | 33.8 | 100.0 | 33.8 |

**Table 4:** Balance of products for washery Grade-IV coal (Mine-4, CCL)

| Products | At 13% ash level | | At 15% ash level | | At 18% ash level | |
|---|---|---|---|---|---|---|
| | Wt% | Ash% | Wt% | Ash% | Wt% | Ash% |
| Clean coal | 16.7 | 13.0 | 24.6 | 15.0 | 35.6 | 18.1 |
| Middlings | 62.4 | 34.0 | 51.3 | 34.0 | 32.2 | 34.0 |
| Rejects | 20.9 | 56.3 | 24.1 | 58.1 | 32.2 | 55.1 |
| Total | 100.0 | 35.2 | 100.0 | 35.1 | 100.0 | 35.1 |

**Table 5:** Balance of products for ungraded/LVMC coal (Mine-5, CCL)

| Products | At 13% ash level | | At 15% ash level | | At 18% ash level | |
|---|---|---|---|---|---|---|
| | Wt% | Ash% | Wt% | Ash% | Wt% | Ash% |
| Clean coal | 16.4 | 13.0 | 20.4 | 15.1 | 27.8 | 18.0 |
| Middlings | 53.5 | 33.3 | 47.6 | 33.7 | 34.3 | 33.5 |
| Rejects | 30.1 | 63.7 | 32.0 | 62.4 | 37.9 | 59.5 |
| Total | 100.0 | 39.1 | 100.0 | 39.1 | 100.0 | 39.1 |

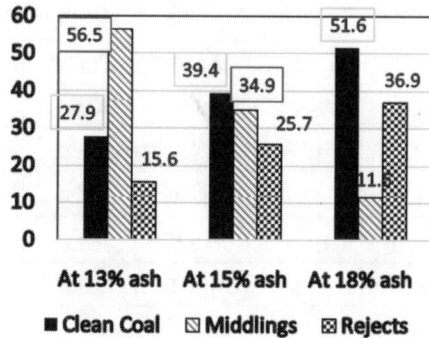

**Figure 2:** Balance of products – washery Grade-IV coal (Mine-1, BCCL) ungraded/ LVMC coal (Mine-2, BCCL)

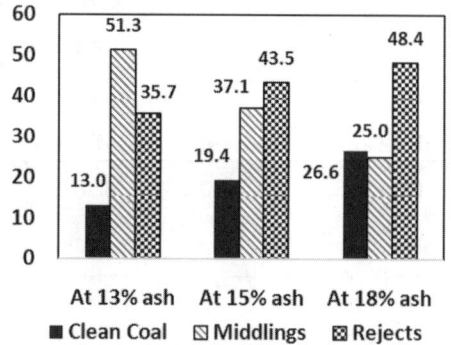

**Figure 3:** Balance of products – washery Grade-IV coal (Mine-1, BCCL) ungraded/ LVMC coal (Mine-2, BCCL)

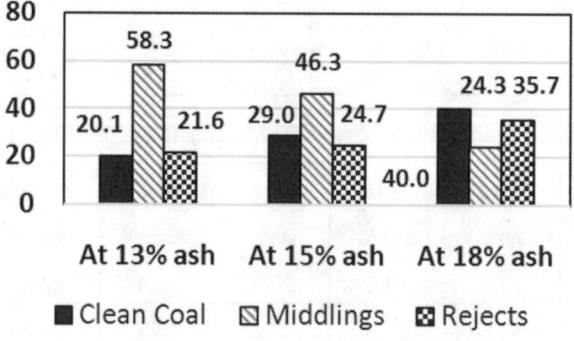

**Figure 4:** Balance of products – washery Grade-IV coal (Mine-3, CCL) washery Grade-IV coal (Mine-4, CCL)

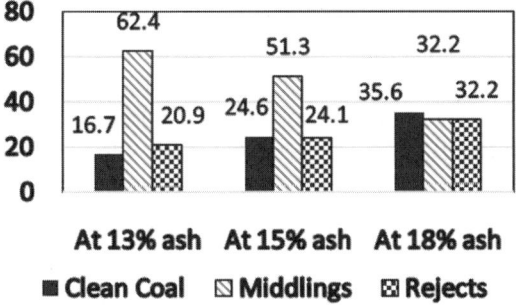

**Figure 5:** Balance of products – washery Grade-IV coal (Mine-3, CCL) washery Grade-IV coal (Mine-4, CCL)

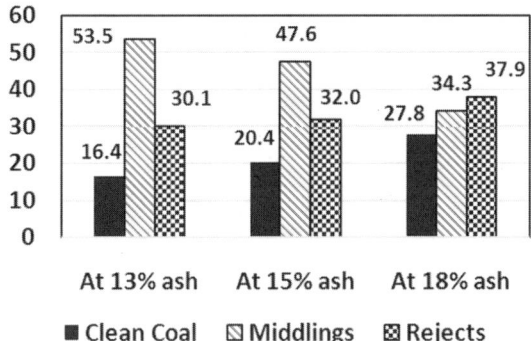

**Figure 6:** Balance of products – ungraded/LVMC coal (Mine-5, CCL)

It was seen that there is considerable decrease in yield of clean coal at 13%, 15%, and 18% ash levels when raw coal ash deteriorates from washery Grade-IV to ungraded. Simultaneously, there is also notable decrease in yield when clean coal ash is kept at 13% in comparison to 18%. The quality of raw coal feed varies widely depending on the seams being mined. Hence, the yield of clean coal will also vary widely from time to time.

Depending on the quality of raw coal feed (washery Grade-IV and ungraded) and its washability characteristics, the theoretical yield of clean coal varies from 13% to 28% at 13% ash. The corresponding theoretical yield of middlings at 34% ash varies from 62% to 51% which is more than double the quantity of middlings generated at 18% ash clean coal and may vary on case to case basis. This indicates that huge quantity of coal having metallurgical properties migrates to middlings which can otherwise be used for metallurgical purpose, if washed to obtain clean coal at 18% ash level.

## 5. Techno-economic feasibility

A techno-economic feasibility study was carried out considering a 2.5 Mty capacity washery, which is a standard module for coking coal now-a-days. Capital investment was broadly estimated considering the envisaged process flowsheet and infrastructure facilities such as land, railway siding, etc. The economics at 80% capacity utilization of washeries for BCCL and CCL was worked out and is given in Table 6.

**Table 6:** Economics at 80% capacity utilization

| Sl. no. | Particulars | BCCL | | CCL | | |
|---|---|---|---|---|---|---|
| | | Mine-1 (W-IV) | Mine-2 (ungraded) | Mine-3 (W-IV) | Mine-4 (W-IV) | Mine-5 (ungraded) |
| 1 | Raw coal cost (Rs/t) | 3734.94 | 1872.44 | 3033.54 | 3033.54 | 1872.44 |
| 2 | Raw coal ash (%) | 31.9 | 42.4 | 33.8 | 35.1 | 39.1 |
| 3 | Desired selling price at 12% IRR (Rs/t) at 80% capacity utilization | | | | | |
| | At 13% ash | 16107.93 | 18220.92 | 18086.72 | 21405.97 | 14242.76 |
| | At 15% ash | 12232.10 | 13374.01 | 13151.75 | 15232.66 | 11899.51 |
| | At 18% ash | 10048.57 | 10458.67 | 10393.66 | 11324.66 | 9504.36 |

It can be seen from Table 6 that the desired selling price of clean coal at 13% ash level is high and varies from about Rs 14,000–21,000 per tonne depending upon washability characteristics and ash value of washery Grade-IV/LVMC coal. Desired selling price at different ash levels is depicted in Fig. 7. In general, the desired selling price of clean coal produced on washing washery Grade-IV coal is less than the desired selling price of clean coal produced on washing ungraded coal.

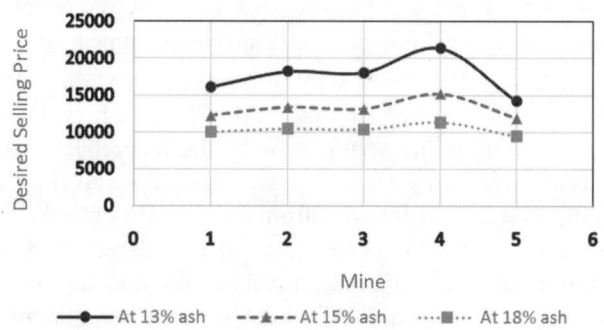

**Figure 7:** Desired selling price of clean coal at varying ash levels (Rs/t)

# 6. Conclusion

Washing of coking coal to produce clean coal at 13%, 15%, and 18% may be considered based on the overall economics of the project which is greatly dependent on yield. To have a balance between cost of production, yield and ash, project specific study needs to be carried out in a holistic manner and project specific selling price of clean coal may be finalized to have a win-win situation both for Coal Company and consumers.

In case of washing low grade coking coal to obtain clean coal at 13% ash, there is huge generation of middlings when compared to that at 18% ash clean coal. Thus, significant quantity of coal having coking properties will report to the middlings which could otherwise be used for metallurgical purpose. Moreover, marketability of such huge quantity of middlings needs to be looked into.

In view of the techno-economics, it was opined that washing of low grade coking coal (washery Grade-IV and low volatile medium coking) at 18% ash level will give viable techno-economics for coal companies as well as consumers. For washing washery Grade-II (i.e. ash varying between 21% and 24%)/washery Grade-III (i.e. ash varying between 24% and 28%) coking coals to get clean coal at 13% ash, new washeries may be set up in the areas where there is potential of better yield leading to favorable techno-economics and the consumer is willing to accept clean coal at higher price on long-term basis.

# 7. Acknowledgement

We appreciate and thank our colleagues from BCCL, CCL, Central Institute of Mining and Fuel Research (CIMFR), Dhanbad; IIT-ISM, Dhanbad; SAIL, etc. who provided insight and expertise that greatly assisted the paper.

# 8. References

[1] Report on assessing the technical feasibility of washing coking coals to 13% ash content prepared by committee constituted by Ministry of Coal, November 2017.

[2] CIL R&D Funded Project "Resource Assessment, Characterization and Blending Studies of Low Volatile Coking Coal for their use in Steel Industry, September 2002.

[3] CIL R&D Funded Project, "Effective Utilization of Low Rank and Low Volatile High Rank Indian Coking Coal for BF Coke Making" with RDCIS-SAIL, Ranchi as its co-implementing agency, February 2013.

# 24

# Improving the quality of blend components for coking at the CPP using special mineral additives and semicoking method

V.A. Kozlov and W. Garber

*Coraline Engineering Company, Russia*
*FTT-Ing.-Büro Feuerungs – und Trocknungstechnologien*
*(Combustion and Drying Technologies), Germany*

**Abstract:** Investigations of qualitative characteristics of caking coal from some new large deposits in the Russian Federation have revealed a limitation of their share in the coal blend for coking. These limitations are related to the chemical composition and degree of metamorphism of coal matter. Estimated reserves of such coking coals amount to billions of tons. In this paper we propose new methods for physicochemical and thermal conversion of clean coal from such deposits in order to improve the quality and increase the sales volume of the products of coal preparation plants. The method of physicochemical changes in the composition of coal is based on the addition of special mineral additives to the low-ash clean coal. The thermal method involves a semicoking process, which reduces the content of volatiles in coal to the required values. A mixture of clean coal with semicoke from the same coal will significantly improve coke reactivity index (CRI) and coke strength after reaction (CSR) indices in coke produced at coke plants. This will increase the share of component in the coal blend for coking and will increase coal preparation plant (CPP) output, which is of important when preparing coal with initially unfavorable physicochemical characteristics.

**Keywords:** Coke quality indicators, coal blend for coking, semicoke, ash content, volatile matter, fluidized bed drying and preparation

## 1. Introduction

The determining condition for production of quality coke is selection of coal charge components before coking. Furnace charge is formed from coals of different ranks, the share of which defines their importance at the coke market.

Investigations of qualitative characteristics of caking coal from some new large deposits in the Russian Federation have revealed a limitation of their share in the coal blend for coking. These limitations are related to the chemical composition and degree of metamorphism of coal matter. Estimated reserves of such coking coals amount to billions of tons. In this paper we propose new methods for physicochemical and thermal conversion of clean coal from such deposits in order to improve the quality and increase the sales volume of the products of coal preparation plants.

## 2. The problem of blend components quality for coking

Stankevich and Zolotukhin (2015) proposes to use values complex indices of coal for preparing coking blend. The indices of technological coal value are calculated on the basis of $M_{25}$, $M_{40}$, and $M_{10}$ coke strength (ISO 556: 1980; GOST RF 5953-93, 1996), coke reactivity index (CRI) and coke strength after reaction (CSR) (ISO 18894: 2006; GOST RF 54250-2010, 2011). Petrographic characteristics of coals are also considered: reflectance of vitrinite index $R_{o.r}$, the content of caking components CC, output of volatile matter $V^{daf}$, plastic layer thickness $y$, ash content of furnace charge $A^d_{ch}$ and chemical ash composition. By chemical ash composition the basicity of ash $I_b$ is calculated as a ratio of contents of main and acid oxides in the ash of furnace charge. The basicity of ash is closely connected with CRI and CSR indices.

The technological value of coals is determined by deviations in their characteristics from the value of the reference furnace charge index. For example, reference furnace charge from coals of the Kuznetsk Basin will have the following composition by coal ranks, as shown by Stankevich (2015): Zh (fat coal) – 30%, GZh (gas fat coal) – 9%, K (coke coal) – 18%, KO (coke semi-lean coal) or SS (weakly caking coal) – 17%, and KS (coke weakly caking coal) – 26%. This furnace charge has values: $R_{o.r} = 1.158$; $S_R = 0.192\%$; Vt = 66%; $V^{daf} = 26.5\%$; $y=18$ mm, etc.

Coke with $M_{25} > 89$; $M_{10} < 7.5$; CSR > 60 and CRI < 30 indices at moist coke quenching is produced from basic reference furnace charge in the slab-coke ovens with the 17-h period of coking. At that, final values of CSR and CRI indices of coke have a great influence on the level of purchase price of coal concentrates being components of this furnace charge.

Many research works show that CRI and CSR indices strongly depend on the value of the "index of basicity", the calculation of which as previously shown (Prasad et al., 1999) proposes to perform by the following formula:

$$I_b = 100 \cdot A^d_{ch} \cdot I_b/(100 - V^{daf}) = A^d_c \cdot I_b, \% \qquad (1)$$

where $A^d_c$ is modified ash content of coke; $I_b$ is the basicity of ash.

To produce quality coke with CSR > 65% it is necessary for furnace charge to have an index of basicity $I_b < 2.5\%$ ($I_b = 0.115$), as shown by Ulanovskiy and Likhenko (2009).

As show by Golovin and Maloletnev (2007) in Table 1 provides for the shares of coal ranks in the composition of furnace charge, which use for production of coke at the coke plants in Russia.

**Table 1:** Share of coal ranks in furnace charge on the coke plants in Russia

| Name coke plant | Share of coal ranks in furnace charge (%) | | | |
|---|---|---|---|---|
| | K, KS | Zh, GZh | KO, SS | Other coal ranks |
| Novolipetsk by-product coke plant | 21 | 34 | 15 | 30 |
| Kemerovo by-product coke plant | 42 | 25 | 16 | 17 |
| Altay by-product coke plant | 28 | 24 | 20 | 28 |
| Magnitogorsk metallurgical plant | 18 | 55 | 10 | 17 |
| Nizhny-Tagil metallurgical plant | 18 | 52 | 22 | 8 |
| Chelyabinsk metallurgical plant | 43 | 23 | 17 | 17 |

In recent years, one can observe deficiency of caking coal ranks GZh, Zh, and KZh in the by-product coke plants. In this regard the studies on their replacement in furnace charge, for example, by oil additives like tar are conducted. However, involvement in development of new coking coal fields in the Republic of Tyva and in other regions of the Russian Federation will enable to replenish a lack of caking components of furnace charge for coking.

It turned out that coals of certain plots in new fields, for example, Kaa-Hemsky and Mezhegeysky fields located in the Republic of Tyva have unfavorable chemical composition of ash and increased output of volatile substances. Coal of the Kaa-Hemsky field of GZh and Zh ranks have the basicity of ash $I_b$ = 0.944, this can considerably reduces the value of this coal. This value is 8 times bigger than the indices for coal of GZh and Zh ranks produced at the Raspadskaya, Osinnikovskaya, Polosukhinskaya, Abashevskaya, and Vorkutinskaya mines as for reference furnace charge $I_b$ = 0.115.

So, for coal in the Kaa-Hemsky field CRI = 44.4% and CSR = 32.3% that is significantly worse than the indices of reference furnace charge for which CRI < 30% and CSR > 60% as it was stated above.

Trial coking furnace charge with the use of caking coal of Kaa-Hemsky field showed that its use in furnace charge should not exceed 10% in order not to significantly worsen coke indices.

As shown by Tarazanov (2018) in his analytical review, coal industry in Russia processed 91.8 million tons of coking coals in 2017. It is predicted that the growth of production and preparation of coking coals in 2020 will make about 110 million tons, in 2025 – about 120 million tons, in 2030 – about 125 million tons. The production of clean coal of coking ranks made 58.4 million tons in 2017. From this quantity only 31.8 million tons were delivered to by-product coke plants. Considering a share of caking coal ranks in furnace charge at a level of 30–50%, we can assume that the need of these coals in 2018–2019 for by-product coke plants in Russia will make 10–15 million tons.

However, the above restrictions in the furnace charge share for coking coal of the Kaa-Hemsky field assume the sales volume of clean coal of this coal at the Russian market at a level of not more than 1.5 million tons per year and with doubtful prospects of sales at the international market. Coal reserves in this field are estimated in tens of billions tons that justifies the search of methods for "transformation" of this coal into a more quality product. It is economically reasonable to install technological lines for coal "transformation" at the new mining and processing enterprises under construction in order to lower transport costs of coal and to increase the added value cost of products of the coal preparation plants.

## 3. Three ways to solve this problem

Considering the above, we have chosen the following three ways of quality improvement of coking coals similar to coals of the Kaa-Hemsky field, at a stage of project of coal preparation plants. They are:

1.  Coal preparation to have ash content in the clean coal as low as possible.
2.  Blending the low-ash clean coal with special mineral additives at arrival to the coal storage, which change the chemical composition of coal, as proposed by Kozlov and Novak (2013).
3.  In order to decrease quantity of volatile matter in coal we offer to perform the semicoking for clean coal in the fluidized bed plants.

*The first way* assumes the use of the most effective process of coal preparation at present – heavy-media separation for big- and small-size coal in the magnetite suspension. In practice it is obviously possible to ensure functioning of dense medium vessels and cyclones at a low separation density of 1.30–1.35 with maintenance of conditions for suspension stabilization and full automation of the process to maintain its density.

*The second way* assumes technical support of addition of special mineral additives to the low-ash clean coal before its stowage at the storage.

*The third way* assumes to carry out semicoking of the low-ash clean coal instead of coal drying for the purpose removal of moisture and volatile matter from it. We suggest carrying out the process of coal semicoking in the fluidized bed plants, as previously shown (Jenkins et al., 1956).

As the first two ways are generally understandable for coal preparation as regards the practical use, we can consider the third way relating to adjacent by-product coke production in details.

*Coal semicoking* is a method for physicochemical change of coal by heating up to 500–550°C without air access. The result of this process is semicoke with output of 50–70%, primary pitch 5–25%, combustible gas, and tar water.

It is possible to receive semicoke directly at the coal preparation plants at the drying and furnace sections when drying coal on the standard drying units. Production of semicoke provides for considerable expansion of the range of products at the coal preparation plants with the improved consumer characteristics and cost.

## 4. "Cooperation of resources" technology

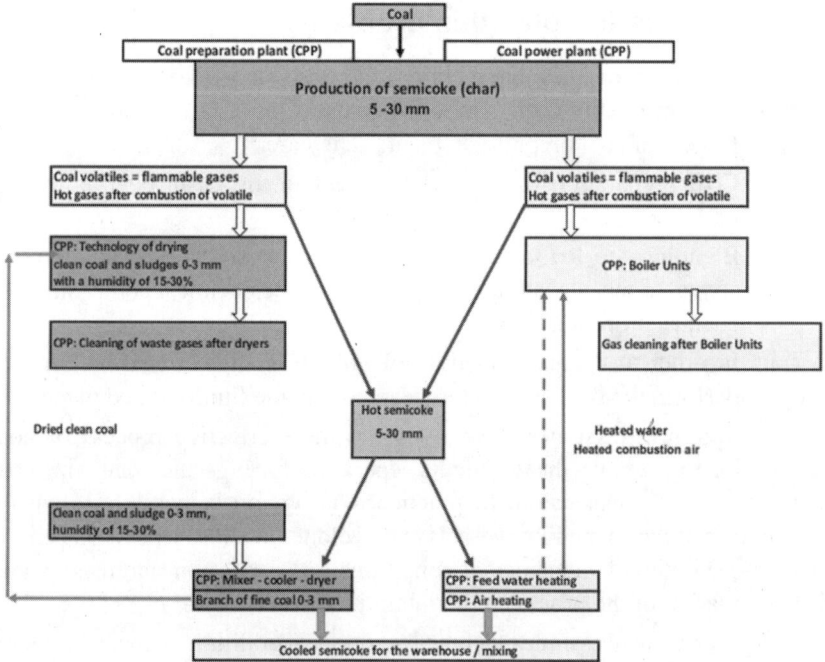

**Figure 1:** Shame of the technology for production of semicoke by the cooperation of resources of the coal preparation plant (CPP) and thermal power plant (TPP)

When producing semicoke *the "cooperation of resources" technology* (Fig. 1) is used. It has all advantages of double use of coal preparation plant

(CPP) and thermal power plant (TPP) products and equipment without additional construction.

*Resource-1*: Volatile components removed from coal are burned as gas fuel in the hot gas generators for drying coal and semicoking plants, and also in the combustion chambers of the boiler plants as main fuel.

*Resource-2*: To purify waste gases at production of semicoke the existing equipment of technological lines of coal drying or the equipment of the boiler plant is used.

*Resource-3*: When producing semicoke in the fluidized bed furnace of the boiler plant hot semicoke transfers its heat to water or air for their heating and dries wet coal, which at the same time cools semicoke when transporting to the storage.

The technologies of resources cooperation are implemented with the minimum capital investments at any coal preparation plant (CPP), thermal power plant (TPP), boiler room, and coal or black oil boilers at the metallurgical enterprises.

# 5. Technologies for production of semicoke

## 5.1 Boiler-based fluidized bed technology

The *fluidized bed combustors* of the power technological boilers can be used for production of semicoke which is removed from the bed through the gateway. The volatile matters emitted from coal are burned as gas fuel of the boiler. In the fluidized bed there is an intensive process of heat exchange between solid particles of 5–30 mm in size blown from below by air through the nozzles. Heat transfer in the fluidized bed is five times more intensive than by means of convection between hot gases and solid coal particles in the moving dense bed. Constant motion of coal particles in the fluidized bed interferes with their baking among themselves; formation of big cakes of semicoke is avoided.

Standard equipment is used to feed lump coal into the fluidized bed: throwers, chutes, screw conveyors, etc. A basic element of fluidized bed mechanisms is a hearth grid with nozzles through which air is supplied from below. The start of the fluidized bed combustor begins with heating up to 450–550°C of the inert aggregate – slag  or sand on the grid. Heating is carried out by supplying gases from below from the starting combustion chamber, on which the starting torch is installed. After heating the inert aggregate the first portions of fuel are supplied to it from above. The start from a cold state takes from 20 to 40 min.

Total time for coal particles being in the fluidized bed and combustion chamber can be regulated in the wide range depending on the required characteristics of semicoke. It gives a possibility of producing semicoke with the different residual content of volatile matter, enables to regulate a degree of burning out of liquid and volatile organics in initial coals. Semicoke is removed down under the grid of the fluidized bed through water-cooled gateways and water-cooled fittings on the cooling conveyor or in the mixers where semicoke is mixed up with wet coal.

The fluidized bed is also used to remove sulfur from coal. Addition of ground limestone in the fluidized bed enables to bind sulfur and to remove its part from the process in the form of gypsum in the composition of dry ash.

The proposed technology of semicoke production meets the highest environmental requirements as two-level combustion of volatile components of coal and pitches provide for the low content of nitrogen oxides in the waste gases.

The positive moments of semicoke production in the fluidized bed combustor of the power technological boiler are as follows:

- losses with mechanical underburning do not exceed 1.5%;
- limit of superheated steam temperature regulation extends;
- regulation range extends – 30–100%;
- nitrogen oxide emissions are twice less than that of stokers and torch furnaces, when due to two-phase burning and low temperatures of the bed in the whole adjusting range of loads, at any air excess in the furnace the $NO_x$ maximum concentration does not exceed 200 mg/m$^3$;
- considerable losses with chemical underburning are excluded, concentration of carbon oxide due to afterburning does not exceed 100 ppm;
- a part of used heat will be removed via the pipe screens located in the fluidized bed where heat exchange is at least five times more intensive than on the heating surfaces contacting with gases.

## 5.2 Production of semicoke in the hot gas generators with the fluidized bed combustor

The coal-operated *hot gas generators* (Fig. 2) are used in the fluidized bed technology.

Hot gas generators with the temperatures of 500–950°C at coal combustion in the fluidized bed combustors and dry removal of 60–80% of ash under the grid are intended for the drying workshop at the coal preparation plants (Fig. 3). Such gas generators enable not to only burn low-quality coals and high-ash waste, as shown by Garber and Kozlov (2017), but they can be also used when producing semicoke.

The use of hot gas generators for semicoke production repeats operation of the boiler furnace with the fluidized bed. Semicoke after removal of volatile matter is removed from the fluidized bed through the gateway. Hot gases from burning of volatile substances and pitches are supplied to the drying plant and waste gases are purified in the gas purification of the drying line.

*Combination of drying with semicoke production* provides for pure hot gases which bring the minimum quantity of ash to the dried product similar to combustion of natural gas. Mineral components of coal remain in semicoke without turning into flue ash. In the mode of sublimation and burning out of volatile matter from coal, the generators, except hot gas generators, produce semicoke required in metallurgy at the minimum ingress of ash in products of drying or granulation. It enables to use coal in clean technologies where the ingress of ash in products is undesirable.

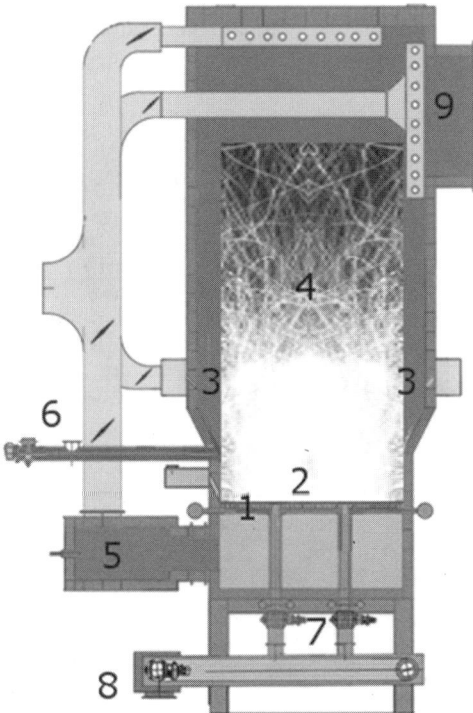

**Figure 2:** 30 MW hot gas generator with a fluidized bed furnace for semicoke production. 1, Air distribution grid; 2, fluidized bed zone, 3, air supply nozzles for afterburning; 4, afterburning zone of volatile substances; 5, starting combustion chamber; 6, supply of raw coal; 7, cooling channels exit the semicoke; 8, conveyor unloading semicoke; 9, hot gas outlet with adjustable temperature

## 5.3 Scope of semicoke using

At present semicoke is used in the following productions:

- at production *ferroalloys*, with improvement of technical and economic indices;
- at preparation *of furnace charge for coking* in the amount of 20–30%, replacing clean coals providing production of coke with the improved quality indices;
- at production *of coke briquettes*, semicoke as a main filler;
- at injection *of pulverized coal fuel* in blast furnaces (PCF).

In the technology of semicoke production in the fluidized bed the separate release of liquid and gaseous products of thermal decomposition is not carried out, they are burnt directly in the fluidized bed combustor. It provides for minimum investments and maximum simplicity of the process in comparison with other ways of semicoking with gas and pitch processing in chemical products with high commodity value representing complex chemical productions.

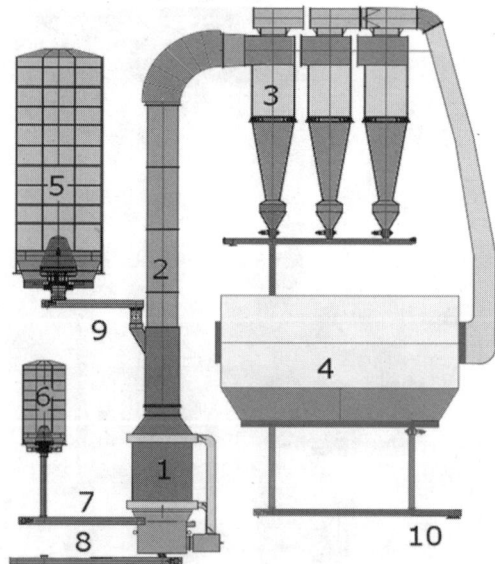

**Figure 3:** Hot gas generator with fluidized bed for vertical flow dryer. 1, Hot gas generator; 2, vertical flow tube dryer; 3, cyclones; 4, bag filter; 5, bunker of wet fine coal (sludge, clean coal); 6, bunker lump coal; 7, the supply of coal in a fluidized bed; 8, Unloading of semicoke; 9, supply of coal slurry for drying; 10, discharge of coal sludge after drying the material

The mixture of clean coal of caking ranks and semicoke will have higher cost at the market, 30–50 USD/t higher than the cost of low-ash clean coal, and for high quality semicoke the price will be higher by 50–150 USD/t. Semicoke can replace nut coke at the price of 230 USD/t, lump semicoke replaces foundry coke at the price of 430–450 USD/t, and pulverized coal fuel (PCF) from semicoke at injection in the blast furnaces replaces more than 15% blast-furnace coke in furnace charge at the price of 350–400 USD/t.

## 6. Conclusion

1. *Semicoke production* is the first step to the technologies of deep thermal coal processing in the high-cost chemical products with steadily growing demand. Thermal division of coal into solid semicoke residue and combustible volatile gases for burning or further complex processing is a process with good economic indices and recoupment of investments in 2–2.5 years.

2. *The proposed technology of resources cooperation* provides a possibility of gradual increase in semicoke production, starting from small volumes without significant investments. The fluidized bed technology is one of the variants of implementation of economical semicoke production.

3. *For coal preparation plants and mining enterprises:*
   - *Increase twice in the total cost* of marketable products of the coal preparation plants.
   - *Reduction of transportation costs* owing to coal compression at coal semicoking and decrease in mass of transported moisture.
   - *Ensuring of the stable product quality* by moisture, ash, sulfur, caloric content, and in case of coking coals – also by main coke quality indices.
   - *Increase in sales volume of products at the coal preparation plants* owing to the increased caloric content and stable quality indices of coking coals.
   - *Up to 50% of furnace charge for coking could be receive from one supplier using mixtures of low-ash coking coals and semicoke.* For example, for coal of the Kaa-Hemsky field the proposed technology will enable to increase the potential sales of coal preparation plant products up to 15 million tons per year.

## 7. References

[1] Garber V., Kozlov V.A. Burning of high ash sludge as a way to waste-free technology of coal preparation. Coal – Russian Coal Journal, 2017, No 8, pp. 140–145.

[2] Golovin G.S., Maloletnev A.S. Complex processing of coals and increasing the efficiency of their use. Manual. Under the general editorship of V.M. Shchadov. M, NTK "Track", 2007, p. 292.

[3] ISO 556: 1980; GOST 5953-93. Coke (greater than 20 mm in size), Determination of mechanical strength. M, IPK Ed. RF Standards, 1996.

[4] ISO 18894: 2006; GOST RF 54250-2010. Coke – Determination of coke reactivity index (CRI) and coke strength after reaction (CSR). M, RF Standartinform, 2011.

[5] Jenkins G.I., Boyer A.F. et al. Fluidized bed coal oxidation. An International Conference, organized by the National Coal Board, Great Britain and held at its Coal Research Establishment at Stoke Orchard, Cheltenham, England. June, 1956.

[6] Kozlov V.A, Novak V.I. Patent of the Russian Federation No 2530109. The method of preparing coal, including high-sulfur, for coking. 2013.

[7] Prasad H.N., Singh B.K., Chattejee A. Cokemaking International, 1999, No 2, pp. 50–59.

[8] Stankevich A.S., Zolotukhin Yu.A. Complex indicator of the technological value of coal and clean coal of coal preparation plants. Russian Bulletin – Ferrous Metallurgy, 2015, No 9, pp. 15–25.

[9] Tarazanov I.G. The results of the work of the Russian coal industry in January–December 2017. Coal – Russian Coal Journal, 2018, No 3, pp. 58–73.

[10] Ulanovskiy M.L., Likhenko A.N. Change in the mineral composition of coals during enrichment and coking. Coke and Chemistry – Russian Coal Journal, 2009, No 6, pp. 13–20.

# 25

# Re-classification of Indian coking coal

Shekhar Saran[1], Sunil Kumar Jayswal[2], Abha Prasad[3], Jayant Prasad[4]

*[1]Chairman-cum-Managing Director, CMPDI, Ranchi, India*
*[2]General Manager, CMPDI, Ranchi, India*
*[3]Chief Manager, CMPDI, Ranchi, India*
*[4]Deputy Manager, CMPDI, Ranchi, India*

**Abstract:** Coking coal is a critical input in metallurgical industry especially in steel sector where it is used primarily as energy source and also as reducing agent in blast furnaces. Presently, coking coal has been classified into six grades up to 35% ash. It has been found that a considerable quantity of high ash (>35%) low volatile medium coking (LVMC) coal, which does not fall under the classified coking coal category and used as power coal, exhibits improvement in coking properties on washing down to about 18% ash and may therefore be blended with good coking coal for utilization in steel making through blast furnace route. This paper outlines the genesis of coking coal gradation and proposes re-classification of coking coal to include high ash LVMC coal.

**Keywords:** LVMC, coking coal, re-classification

## 1. Introduction

Coal, a fossil fuel is the vital raw material for steel making in metallurgical industries apart from fuelling the thermal power stations for generation of electricity and other uses. Coal is characterized by its use either as "metallurgical/coking coal" or "thermal/non-coking coal". Thermal coal is used by electricity generators to produce electricity. Coking coal is a critical input in metallurgical industry especially in steel sector where it is used primarily as energy source and also as reducing agent in blast furnaces.

India is endowed with vast geological reserves of coal – estimated to be 319 billion tonnes (Bt) as on 1st April 2018. The Jharia and Bokaro coalfields constitute the major resources of coking coal in India. The share of coking coal in the coal inventory is 34.5 Bt (10.8%), out of which 27.5 Bt is medium coking coal. Out of this 27.5 Bt medium coking coal, about 6.4 Bt is low volatile medium coking (LVMC) coal. Thus, the reserves of LVMC coal amount to about 19% of the total coking coal reserves, which is quite substantial. This LVMC coal having ash more than 35% does not fall under classified coking coal category.

## 2. Need for re-classification of Indian coking coal

As per present Indian practice, initially the classification and codification of coal into coking and non-coking coal is done as per BIS standard (IS 770:2013) based on basic parameters like mean random reflectance of vitrinite, volatile matter% (dry mineral matter free), gross calorific value (dmf) and supplementary parameters such as moisture% at 96% relative humidity and 40°C (mineral matter free basis), Gray-King low temperature coke type and crucible swelling number. After classification of coal as coking coal on the basis of above parameters, grading of coking coal is done on the basis of ash% as Indian coals being of drift origin, in general has high ash%. Thus, Indian coking coal has been classified into six grades based on ash% as per Gazette of India Extraordinary, Part-II, Section 3, Sub-section (ii) dated 30.12.2011, as given in Table 1.

**Table 1:** Existing gradation of Indian coking coal

| Sl. no. | Grade | Ash% |
|---------|-------|------|
| 1 | Steel-I | Up to 15 |
| 2 | Steel-II | Exceeding 15 up to 18 |
| 3 | Washery-I | Exceeding 18 up to 21 |
| 4 | Washery-II | Exceeding 21 up to 24 |
| 5 | Washery-III | Exceeding 24 up to 28 |
| 6 | Washery-IV | Exceeding 28 up to 35 |

Presently, considerable quantity of high ash (>35%) LVMC coals are being used in thermal power plants, instead of augmenting the supply of metallurgical coal for coke making, thus wasting the scarce indigenous coking coal resources. Laboratory studies on LVMC coals from the Jharia and East Bokaro coalfields conducted by various organisations and CIL's R&D funded projects [1,2] showed that after beneficiation when the ash level is brought down to about 18%, these coals exhibit improvement in coking properties and may therefore be utilized in the blend for preparation of metallurgical coke for steel making through blast furnace route.

Due to stagnant availability of indigenous metallurgical coal, import of coking coal in India has been showing an increasing trend. Encouraged by the studies, 6 nos. of non-linked washery (NLW)/LVMC washeries are coming up in Bharat Coking Coal Ltd. (BCCL) for washing of such coal to promote use of indigenous coking coal in steel making and reduce our dependence on import of coking coal. The average ash of raw coal feed to these washeries varies in the range of 37–46%.

Coal Controller's Organisation, which is a subordinate office of the Ministry of Coal, took initiative for sampling and analysis of coal from eight such LVMC mines of BCCL. The testing and analysis were carried out by Council of Scientific & Industrial Research – National Metallurgical Laboratory (CSIR-NML), Jamshedpur and Indian Institute of Technology – Indian School of Mines (IIT-ISM), Dhanbad and the coals were found to be:

"B5 – Class: Sub-bituminous/bituminous; Type: Medium volatile; Nature – weakly to medium caking; calorific value (dry mineral matter free): 8500–8700 kcal/kg; volatile matter% (Dmf): 22–33; Gray-King coke type: C–F; moisture (60% relative humidity): <2; utilisation: Blending" as per IS:770-1977 [3].

However as per IS:770-2013 [4] the results were falling between Sl. (v) Medium volatile bituminous, medium to strongly caking and Sl. (vi) Medium volatile bituminous, strongly caking."

But as per present grading system, the same LVMC coals are not categorized as coking coal due to their high ash% and are being used as thermal coals which is a severe loss to our country's coking coal reserves. In order to promote washing of such indigenous LVMC coal reserves for use in steel making, the need was felt to revisit the present coking coal grading and to suitably modify it to accommodate the LVMC coal.

## 3. Related work

### 3.1 Genesis of Indian coking coal gradation [5]

In the early 1920s, export of Indian coal dwindled. Though of small volume (around 0.7 million tonnes), it had provided a useful outlet for the surplus coal. Indifferent quality of Indian coal was considered to be one of the contributing factors. The government, therefore, set up a Coal Grading Board as recommended by the Indian Coal Committee, 1924. Initially, the Grading Board certified coal for shipment but the coverage was gradually extended to all coals except coal produced in the Singareni Coalfield and the North Eastern part of India. The classification was based on 'ash/calorific value' and stipulated different specifications for 'low volatile' and 'high volatile' coals as given in Table 2.

**Table 2:** Classification of coal (1924)

| Grade | Low volatile coals (Barakar measures) | | | High volatile coals (Raniganj Measures) | | |
|---|---|---|---|---|---|---|
| | Ash % | Cal. value (cals/g) | Moisture % | Ash % | Cal. value (cals/g) | Moisture % |
| Selected grade | Up to 13 | Over 7000 | The moisture in all cases should not exceed 2% | Up to 11 | Over 6800 | Under 6 |
| Grade I | 13–15 | Over 6500 | | 11–13 | Over 6300 | Under 9 |
| Grade II | 15–18 | Over 6000 | | 13–16 | Over 6000 | Under 10 |
| Grade III | Over 18 | Under 6000 | | Inferior to grade II | Under 6000 | – |

Later, the grading scheme stipulated three different sets of specifications: one for low-moisture (not more than 2%) Bengal and Bihar coals, second for high moisture Bengal and Bihar coals, and a third for coals from outlying fields (almost entirely high moisture coals). For low moisture coals, including coking coals, the specification were based on ash only and provided for six grades (Selected A, Selected B, Grades I, II, IIIA, and IIIB). However, coking coal grades extended only up to Grade II and not below. For high moisture coals, the grading was based on ash plus moisture content (at 60% RH and 40°C). For high moisture coals only four grades were specified (Bengal and Bihar – Selected A, Selected B, Grade I and Grade II; outlying fields Selected, Grade I, Grade II, and Grade III).

In 1962, the specifications for coking coals were spelt out separately by Coal Board and different grades were specified as given in Table 3.

**Table 3:** Classification of coal in 1962

| Grade | Ash% |
|---|---|
| A | Does not exceed 13 |
| B | Exceeds 13 but does not exceed 14 |
| C | Exceeds 14 but does not exceed 15 |
| D | Exceeds 15 but does not exceed 16 |
| E | Exceeds 16 but does not exceed 17 |
| F | Exceeds 17 but does not exceed 18 |
| G | Exceeds 18 but does not exceed 19 |
| H | Exceeds 19 but does not exceed 20 |
| HH | Exceeds 20 but does not exceed 24 |

Subsequently from 1st July, 1975, J and K grades were introduced with ash ranges, 24(+) to 28% and 28(+) to 35%, respectively. This basis i.e. ash% for gradation of coking coals continue even today but arising out of the recommendations of the Chakrabarty Committee on coal prices appointed by the government, the structure of grades were changed in July 1979. The coking coal grades (A–K) were restructured into six grades within the same maximum ash content of 35% as given in Table 4 and is also presently the basis of gradation of coking coal.

**Table 4:** Classification of coal in July, 1979

| Sl. no. | Grade | Ash% |
|---------|-------|------|
| 1 | Steel-I | Upto 15 |
| 2 | Steel-II | Exceeding 15 up to 18 |
| 3 | Washery-I | Exceeding 18 up to 21 |
| 4 | Washery-II | Exceeding 21 up to 24 |
| 5 | Washery-III | Exceeding 24 up to 28 |
| 6 | Washery-IV | Exceeding 28 up to 35 |

## 3.2 Indian standard procedure (ISP), May 1989 [6]

The Committee on Assessment of Resources (CAR) of Coal Council of India (CCI) drew up the Indian standard procedure (ISP) for coal reserve estimation in 1957. The government and the public sector organizations involved in the coal exploration at that time namely Geological Survey of India (GSI), Indian Bureau of Mines (IBM), and National Coal Development Corporation (NCDC) adopted the ISP for classification of the coal resources.

The present practice of assessment of coal resources is primarily based on the revised guideline issued by the Task Force (constituted by Central Geological Programming Board of GSI in 1989 to redefine the ISP criteria).

Major Points of ISP, May 1989 relevant to classification of coking coal are as follows:

(i) Classification of coking coal is done as hereunder:

| Grade | Ash% | Sp. Gravity |
|-------|------|-------------|
| Steel Grade-I | Upto 15 | 1.42 |
| Steel Grade-II | 15–18 | 1.44 |
| Washery Grade-I | 18–21 | 1.46 |
| Washery Grade-II | 21–24 | 1.50 |

*Contd...*

*Contd...*

| Grade | Ash% | Sp. Gravity |
|---|---|---|
| Washery Grade-III | 24–28 | 1.53 |
| Washery Grade-IV | 28–35 | 1.58 |
|  | 35–40 | 1.65 |
|  | More than 40 | 1.72 |

(ii) Wherever the overall ash content of the seam exceeds 35%, the coal is to be placed under non-coking category for the purpose of reserve estimation even though the coal exhibits medium coking properties. The grading of such coals is carried out on the basis of useful heat units as per Govt. of India Notification of July, 1979.

(iii) In the minutes of the meeting of the Task force on assessment of coal resource in India, annexed with ISP, 1989, it is also mentioned that:

"For classification of coking coal, Shri Roy Chaudhuri suggested that the Low Temperature Carbonisation (L.T.C), coke type, unit volatile matter and plastometric test should be kept in view. These parameters would help in identifying the different categories of metallurgical coal viz. prime, medium and semi coking coal. The task force resolves that the coking coals which contain more than 35% ash should be classified separately as high ash coking coal".

## 4. Observations

Following may be observed from above:

1. The genesis of grading of coking coal indicates that the basis of classification of Indian coking coal has always remained as 'ash%'.

2. ISP, 1989 envisaged classification of coking coal beyond 35% ash in two grades

3. Analysis of test results of LVMC coal samples of BCCL by CCO proves that LVMC coals have coking propensities.

4. CIL's R&D funded projects indicate that there are superior coking characteristics of the low volatile high rank coals (LVHR) once their ash level is brought down as near to 18%, and can be used in the blend for preparation of metallurgical coke after proper beneficiation.

## 5. Conclusion

Keeping the above in view, it is suggested to continue with the existing grading of coking coal upto 35% ash and further add two grades as Washery Grade-V (exceeding 35% up to 42% ash) and Washery Grade-VI (exceeding 42% up to 49% ash) to accommodate LVMC coal under classified coking coal category. The proposed re-classification of coking coal is given in Table 5.

**Table 5:** Proposed re-classification of coking coal

| Sl. no. | Grade | Ash% |
|---|---|---|
| 1 | Steel-I | Upto 15 |
| 2 | Steel-II | Exceeding 15 up to 18 |
| 3 | Washery-I | Exceeding 18 up to 21 |
| 4 | Washery-II | Exceeding 21 up to 24 |
| 5 | Washery-III | Exceeding 24 up to 28 |
| 6 | Washery-IV | Exceeding 28 up to 35 |
| 7 | Washery-V | Exceeding 35 up to 42 |
| 8 | Washery-VI | Exceeding 42 up to 49 |

# 6. Acknowledgement

We appreciate and thank our colleagues from Coal Controller's Organisation (CCO), Kolkata; Central Institute of Mining and Fuel Research (CIMFR), Dhanbad; IIT-ISM, Dhanbad; etc. who provided insight and expertise that greatly assisted the paper.

# 7. References

[1] CIL R&D Funded Project "Resource Assessment, Characterization and Blending Studies of Low Volatile Coking Coal for their use in Steel Industry, September 2002.

[2] CIL R&D Funded Project, "Effective Utilization of Low Rank and Low Volatile High Rank Indian Coking Coal for BF Coke Making" with RDCIS-SAIL, Ranchi as its co-implementing agency, February 2013.

[3] Indian Standard: 770-1977 Classification and Codification of Indian coals and lignites.

[4] Indian Standard: 770-2013 Classification and Codification of Indian coals, lignites and anthracites.

[5] Coal Mining in India – a CMPDI publication published on the occasion of 12th World Mining Congress, New Delhi, November 1984.

[6] Indian Standard Procedure (ISP) for coal resource/reserve estimation, CMPDI, May 1989.

# 26

# Low-grade coal as a precursor for preparation of value-added products for high-end use

Tonkeswar Das and Binoy K Saikia

*CSIR-North East Institute of Science and Technology,*
*Jorhat, India*

**Abstract:** The research is aimed to the production of high impact carbon nanomaterials (CNMs) during coal preparation technology. The CNMs have obtained significant attention among researchers since their discovery due to their unique optical, electrical, thermal, and mechanical properties. The purpose of the research is the development of a simple and less expensive chemical coal processing technology to reduce the sulfur and mineral matter (ash) content from high sulfur Indian coals (NER). In addition, another purpose of the research is to examine the formation of carbon nanomaterials from low-grade coals by using the advanced level analytical characterization techniques.

In the course of the study, laboratory-scale processes for the beneficiation of several types of high sulfur Northeast India (NER) coal were conducted including the molten caustic leaching (MCL), oxidation-cum-extraction (OCE) process, wet-chemical ultrasonic simulation, ultrafiltration, and purification processes. All types of produced CNMs were characterized by using advanced level characterizations techniques. The results showed the removal of unwanted sulfur and mineral matter (ash) content significantly from the low-grade coals making it cleaner. The results also showed the formation of technogenic carbon nanomaterials (TCNMs) including carbon nano-balls, carbon nanotubes, chemically converted graphene-like carbon nanosheets, nanodiamnod, carbon dots, etc., which are suitable for various high-end uses. These low-grade coal-derived value-added products were also satisfactorily tested in different field of application.

The developed wet-chemical ultrasonic simulation process for the high sulfur Indian coal for the synthesis of different types of CNMs will lead to a new and futuristic coal processing/preparation technology.

**Keywords:** High sulfur coal, low-grade coal, carbon nanomaterials, value added products from coal, coal preparation

## 1. Introduction

Coal has various important uses round the world. The foremost vital uses of coal are in the generation of electricity, production of steel, manufacturing of cement, and also as a liquid fuel. As per the new policies scenario of International Energy Agency, the global primary energy demand is projected to increase day by day in future due to the rapid economic growth in many countries (Dong, 2011). In this scenario, the fossil fuels will play a vital role in the primary energy demand despite their decrease share, where coal will remain the second most important fossil fuel after oil in global energy resources. To meet increasing demand for coal, coal production is projected to increase, primarily in China and India.

There is a growing importance of cheaper coal feedstock i.e. low-quality coal due to the sharp fall in the reserves-to-production ratios of hard in many important coal mining regions in the world (Dong, 2011). To increase the reserves-to-production ratios of hard coal, mining operations have to be undertaken in deeper coal seams. As a results, the mining cost as well as price of the hard coal increase, which become an adverse impacts on plant economics in many countries.

The low-quality coals could be of vital importance to both energy security and economic development of some countries or regions, where these coals may be the only significant indigenous energy resources. Low-grade or low-quality coals include lignite and sub-bituminous coals as they contain less carbon and low energy content than higher grade coal. All low-quality coals including Sub-bituminous and lignite coals are generally considered into low-quality coals because of high moisture content, low heating value, and high mineral matter content. These coals usually require the application of specific technologies for their successful use in power generation and other industrial processes. Anthracites and semi-anthracites coals are also sometimes classified as low-quality coals based on the ignition and burnout problems.

As per the current status of utilization of low-quality coal worldwide, it is observed that in most of the countries, the largest market for low-quality coals is for power and cogeneration purposes (Mills, 2011). A number of drying, cleaning, and upgrading processes with high efficiency have been developed in Australia, Germany, the United States of America, and Japan. In India, low-quality coal is also used for the power generation, manufacturers of cement, textiles, chemicals, and ceramics.

Although low-quality coals could serve as a low cost fuel and help improve the energy security issue, their use in existing coal facilities often offerings a great challenge. Most of the challenge is related to the undesirable properties of low-quality coals. These include high moisture content, low calorific value, aggressive ash characteristics such as high fouling or slagging propensity, high sulfur/nitrogen/mercury content, low Hardgrove Grindability Index, low volatile matter, and high ash content (of low-quality coals).

Because of the unusual combination of the mineral matter and sulfur component, in different coals throughout the world (depending on the geologic conditions) (Kan et al., 2010; Mukherjee et al., 2001; Baruah et al., 2006,

2007; Saikia et al., 2013; Duran et al., 1986; Thomas, 2012), considerable work has been carried out for coal beneficiation including de-sulfurization and de-ashing for sustainable utilization and to reduce the harsh environmental effects in different industries/countries (Kan et al., 2010; Mukherjee et al., 2001; Baruah et al., 2006, 2007; Saikia et al., 2013; Duran et al., 1986; Thomas, 2012). The limiting factors of the physical and chemical methods of coal beneficiation, which are mostly lethal to humans, the development of a simple and less expensive chemical coal processing technology is very much essential to reduce the sulfur and mineral matter (ash).

Considering all these factors and to give more emphasis to our earlier works, here, we report the removal mineral matter (ash) and sulfur components of few high sulfur northeast Indian coals by using molten caustic leaching (MCL), oxidation-cum-extraction (OCE) process, wet-chemical ultrasonic simulation, ultrafiltration, and purification processes simultaneously. We also report the formation of high impact carbon nanomaterials (CNMs) during the coal preparation technology.

The Northeast part of India has a good reserve of tertiary coals with high sulfur. The northeast Indian coals contain low ash and medium to high sulfur (2–8%) with mainly 75–90% in the organic-form (Saikia et al., 2014, 2016). So, it could not be directly used in the thermal plants without removal of sulfur and minerals (ash) due to stringent environmental regulations. Several beneficiation studies of Northeast Indian coals has been reported (Saikia et al., 2014). However, no any suitable and eco-friendly coal beneficiation process has been developed till now to use in the field. Therefore, it is essential to encourage and emphasize on value addition of Northeast Indian low-quality coal for its gainful and sustainable utilization and to meet the future energy demands.

## 2. Materials and method

The selection of NER coal samples and their beneficiation process such as molten caustic leaching (MCL) (Das et al., 2016a), oxidation-cum-extraction (OCE) process (Das et al., 2016b), wet-chemical ultrasonic simulation, ultrafiltration, and purification processes (India and US patent, file No. IN 201711007354; US 15/704,364) has been described in details elsewhere (Das et al., 2016a,b; India and US patent, file No. IN 201711007354; US 15/704,364), by us. The characterization techniques of the raw coals and their treated product have also been described in details in our reported studies (Das et al., 2016a,b). Here, we only highlighted the results of physico–chemical properties and electron beam studies (TEM/HRTEM) of the treated coal samples.

## 3. Results and discussion

### 3.1 Physico–chemical properties

The Physico–chemical characteristics of the raw and treated coal samples are summarized in Table 1. The properties of the raw coals indicated that they are low ash and high sulfur coals. In MCL process, the mixture of alkali was able to reduce the ash yield and total sulfur of coal around 70% and 37.80% respectively (Das et al., 2016a). Whereas, in OCE process, we observed that the percentage of ash yield and sulfur contents are found to be considerably decreased up to 67–85% and 53–58%, respectively (Das et al., 2016b). However, the extent of de-sulfurization and de-ashing of coal depends upon its type/rank.

**Table 1:** Physico–chemical properties of the raw and treated coal samples

| Coal samples | Proximate analysis (%) | | | | Ultimate analysis (%) | | TS (%) | Processes |
|---|---|---|---|---|---|---|---|---|
| | M | VM | Ash | FC | C | H | | |
| Coal-NK2 (raw coal) | 3.80 | 40.60 | 10.80 | 44.80 | 65.70 | 5.70 | 7.80 | Molten caustic leaching (MCL) (Das et al., 2016a) |
| NP (MCL product) | 5.60 | 33.10 | 3.30 | 58.00 | 71.60 | 5.30 | 4.80 | |
| Coal-NK1 (raw coal) | 4.20 | 30.68 | 18.67 | 46.45 | 61.20 | 4.51 | 3.26 | Oxidation-cum-extraction (OCE) process (Das et al., 2016b) |
| Coal-NK-NP | 17.12 | 43.00 | 2.63 | 37.25 | 51.80 | 5.54 | 1.51 | |
| Coal-NG (raw coal) | 9.29 | 41.69 | 3.80 | 45.23 | 68.80 | 6.12 | 3.59 | |
| Coal-NG-NP | 12.46 | 46.25 | 1.23 | 40.06 | 55.50 | 5.64 | 1.50 | |
| Coal-T60 (raw coal) | 1.75 | 2.95 | 41.19 | 54.11 | 77.75 | 5.16 | 3.62 | Wet chemical ultrasonic simulation (India and US patent, file No. IN 201711007354; US 15/704,364) |
| Coal-T20 (raw coal) | 2.35 | 2.33 | 50.27 | 45.05 | 80.90 | 8.19 | 1.90 | |
| Coal-NK1 (raw coal) | 4.20 | 18.67 | 30.68 | 46.45 | 61.20 | 4.51 | 3.26 | |
| Coal-NG (raw coal) | 9.29 | 3.80 | 41.69 | 45.23 | 68.80 | 6.12 | 3.59 | |

M: moisture; VM: volatile matter; FC: fixed carbon; C: carbon; H: hydrogen, and TS: total Sulfur.

## 3.2 Electron microscopic study (SEM and TEM/HRTEM)

During our detailed electron microscopic analysis (TEM/HRTEM), we have observed the presence of carbon nanomaterials in the products obtained from MCL, OCE, and wet-chemical ultrasonic simulation process. Figure 1a and b shows the HRTEM images of the carbon nanoballs in the range of 5–10 nm formed in the MCL product. However, these carbon nanoballs are found to be made of random and radial carbon layers after analysis at 5–10 nm

(Figure 1e and f) (Das et al., 2016a). Figure 2a and b shows the typical TEM images of CNMs formed along with clusters of carbon (Figure 2a). The diameters of these CNTs-like materials are also estimated to be in the range of 18–24 nm (Fig. 2c). The two touching carbon fibers in some typical TEM images also indicated the presence of branch carbon nanotubes (BCNTs) as shown in Fig. 2d.

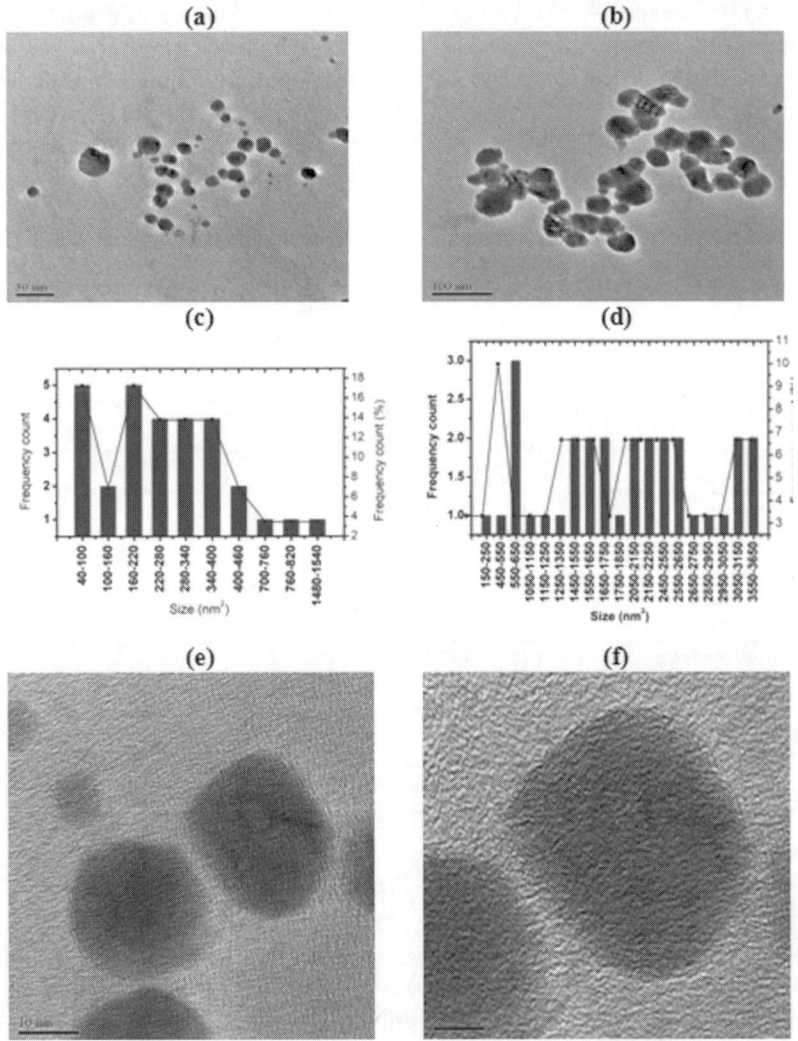

**Figure 1:** (a, b) HRTEM images of carbon nanoballs (NP) and (c, d) their corresponding size distributions, and (e, f) HRTEM analysis of the carbon nanoballs at 5–10 nm (random and radial carbon layers) (Das et al., 2016a)

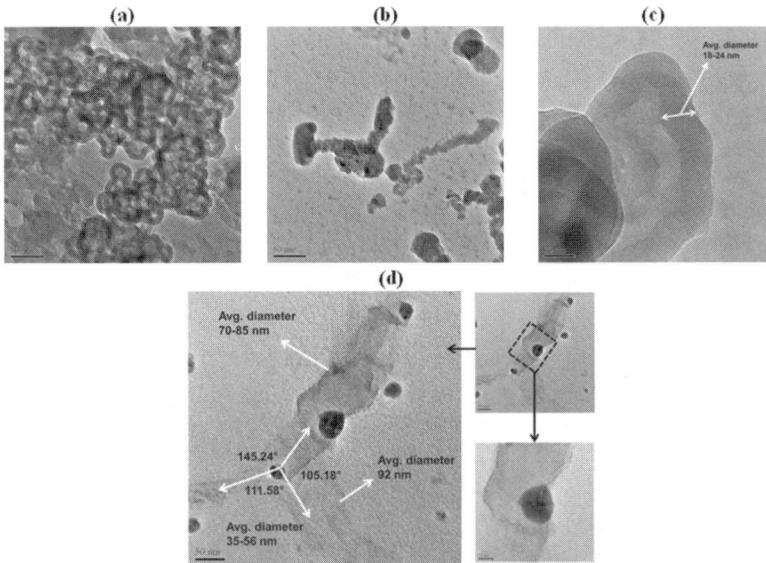

**Figure 2:** HRTEM images of CNTs (a–c) and BCNTs (d) in CNMs NP sample
(Das et al., 2016a)

In OCE process (Das et al., 2016b), the morphologies of the coal-NG-NP sample, as observed in the HRTEM images clearly show the cluster of onion-like fullerene (OLF) structures (Fig. 3a). The onion-like fullerene structure displays quasi-spherical, polyhedral type morphologies with a hollow center (Fig. 3a). The size of the onion-like fullerene is estimated to be in the range of about 5–20 nm. The outer diameter of these onion-like structures is estimated to be in the range of 3–15 nm. Figure 3b shows that the coal-NK-NP sample contains clusters of chemically converted graphene-like nanosheets.

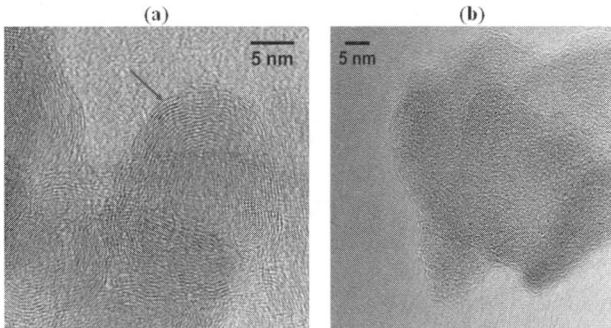

**Figure 3:** (a) HRTEM image of onion-like fullerene (OLF) and (b) chemically
converted graphene-like nanosheets (Das et al., 2016b)

In wet-chemical ultrasonic simulation process, the electron beam analysis revealed the formation of carbon quantum dots (CQDs) as well as graphene quantum dots (GQDs) from the used coal feedstock (Fig. 4 a and b) (India and US patent, file No. IN 201711007354; US 15/704,364). The diameters of these fabricated carbon nanostructure were estimated to be in the range of 1–30 nm (Fig. 4a and b). The produced carbon quantum dots and graphene quantum dots have crystalline as well as amorphous carbon addends on their edge. It is to be mentioned that the obtained CQDs and GQDs are similar to the carbon dots structure as reported elsewhere (Wang et al., 2017; Dong et al., 2014; Xu et al., 2004).

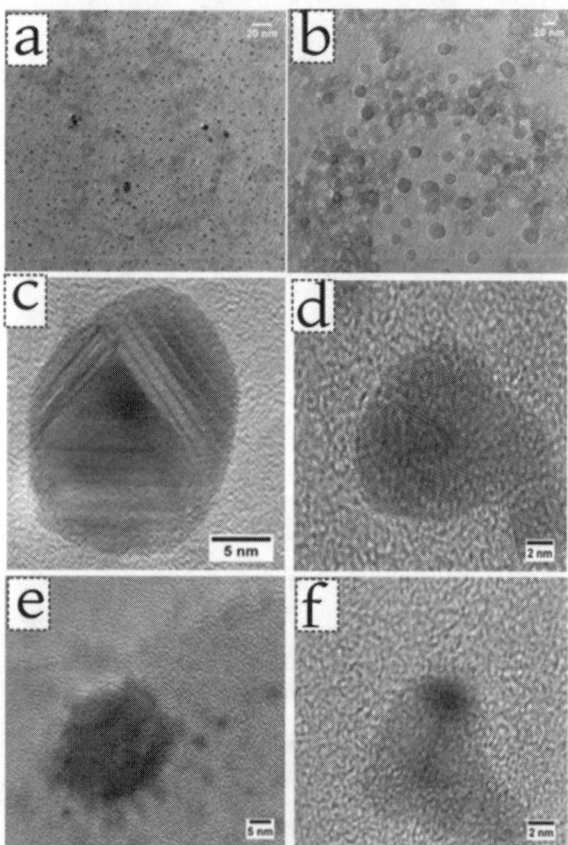

**Figure 4:** (a and b) HRTEM image of CQDs and GQDs, (c–f) HR-TEM images of some large-size carbon nanocrystal having the size of 4–15 nm (Das and Saikia, 2017)

Figure 4 c–f shows the typical HRTEM images of some large-size carbon nanocrystal formed along with the CQDs and GQDs (Das and Saikia, 2017).

The nanocrystal particles are found to be monocrystalline and polycrystalline in nature with multiple twins. The nominal sizes of the carbon nanocrystals were found to be 4–15 nm. The planner spacing of the crystal lattice fingers was measured to be in the range of 2.0–2.3 Å, which is good agreement with the lattice planes of various diamond phases including cubic diamond (111) (2.06 Å), lonsdaleite (002) (2.06 Å), and lonsdaleite (100) (2.18 Å) as reported elsewhere (Das and Saikia, 2017; Kumar et al., 2013), by us and others. These measurements indicated that the nanocrystals are considered to be nanodiamonds, rather than common graphene quantum dots (Das and Saikia, 2017).

From the electron beam studies, it was observed that the present processes may facilitate the formations of technogenic CNMs from the low-quality coals used in this experiment. The novelty of these products is their simple preparation technique from the low-quality subbituminous coals. To further confirm the formation of CNMs from the coals, these CNMs were further characterized by using XRD, Raman, FTIR, XPS, solid-state $^{13}$C NMR, UV–visible spectroscopy, and FL spectroscopy as reported elsewhere by us.

## 4. Conclusion

In summary, the observations suggest that the high impact CNMs could be prepared from low-quality coals by using molten caustic leaching (MCL), oxidation-cum-extraction (OCE) process, wet-chemical ultrasonic simulation, ultrafiltration, and purification processes. New carbon materials such as carbon quantum dots offer the opportunity to develop new markets for traditional feedstock. However, the detail analysis of the structural features and relevant characteristic that opens diverse applications will be a subject of future studies.

## 5. Acknowledgements

The authors are grateful to the Director, CSIR-NEIST for his constant encouragement during the research work. The financial assistance from CSIR-New Delhi is duly acknowledged (MLP-6000-WP-III). Authors express their thankfulness to SAIF-NEHU for their instrumental support.

## 6. References

[1] Bimala P. Baruah, Binoy K. Saikia, Prabhat Kotoky, and P. Gangadhar Rao, Aqueous leaching on high sulfur subbituminous coals, in Assam, India. Energy Fuels, Vol. 5, pp 1550–1555, 2006.

[2] Bimala P. Baruah, Puja Khare, Desulfurization of oxidized Indian coals with solvent extraction and alkali treatment. Energy Fuels, Vol. 21, pp 2156–2164, 2007.

[3] Tonkeswar Das, Binoy K. Saikia, Bimala P. Baruah, Formation of carbon nano-balls and carbon nano-tubes from northeast Indian Tertiary coal: value added products from low grade coal. Gondwana Research, Vol 31, pp 295–304, 2016a.

[4] Tonkeswar Das, Purna K. Boruah, Manash R. Das, Binoy K. Saikia, Formation of onion-like fullerene and chemically converted graphene-like nanosheets from low-quality coals: Application in photocatalytic degradation of 2-nitrophenol. RSC Advances, Vol. 6(42), pp 35177–35190, 2016b.

[5] Tonkeswar Das and Binoy K. Saikia, Nanodiamonds produced from low-grade Indian coals. ACS Sustainable Chemistry and Engineering, Vol 5 (11), pp 9619–9624, 2017.

[6] Nigel S Dong, Utilization of low rank coals (Clean Coal Centre, 2011).

[7] Joseph E Duran, Sumit Ray Mahasay, Leon M.Stock, The occurrence of elemental sulfur in coals, Fuel, vol. 65, pp 11678, 1986.

[8] H. Kan, C.M. Wong, N. Vichit-Vadakan, Z. Qian, The PAPA project teams shortterm association between sulfur dioxide and daily mortality: the public health and air pollution in Asia (PAPA) study. Environmental Research, Vol. 110, pp 258–264, 2010.

[9] Stephen J Mills, Global perspective on the use of low quality coals (Clean Coal Centre, 2011).

[10] Samit Mukherjee, S. Mahiuddin, P. C. Borthakur, Demineralization and desulphurization of sub bituminous coal using hydrogen peroxide. Energy & Fuels, Vol. 15, pp 1418–1424, 2001.

[11] Binoy K. Saikia, Tonkeswar Das, Sonali Roy, Bardwi Narzary, Hari Prasana Deka Boruah, Manabjyoti Bordoloi, Junmoni Lahkar, Dipanker Neog, Danaboynia Ramaiah, A process for the preparation of Blue-Flourescence emitting Carbon Dots (CDs) from sub-bituminous Tertiary high Sulfur Coals. Patent filed in India and US (IN 201711007354; US 15/704,364).

[12] Binoy K. Saikia, Nilakshi Kakati, Kakoli Khound, Bimala P. Baruah, Chemical kinetics of oxidative desulfurization of Indian coals. International Journal of Oil, Gas and Coal Technology, vol. 6, pp 720–727, 2013.

[13] Binoy K Saikia, Arju M. Dutta, Lakshi Saikia, Shahid Ahmed, Bimala P Baruah, Ultrasonic assisted cleaning of high sulphur Indian coals in water and mixed alkali. Fuel Processing Technology, vol. 123, pp 107–113, 2014.

[14] Binoy K. Saikia, Adilson C. Dalmora, Rahul Choudhury, Tonkeswar Das, Silvio R. Taffarel, Luis F. O. Silva, Effective removal of sulfur components from Brazilian power-coals by ultrasonication (40 kHz) in presence of $H_2O_2$. Ultrasonics Sonochemistry, vol. 32, pp 147–157, 2016.

[15] Larry Thomas, Coal geology. 2nd ed. Wiley-Blackwell, 2012.

# 27

# Computation of coefficients of separation products size grading in the process of preparation of run-of-mine coal machine grades

D.O. Polulyakh[1], A.S. Buchatskiy[2], O.D. Polulyakh[3]

*[1]National TU «Dnipro Polytechnic, Dnipro, Ukraine*
*[2]Satellite LLC, Moscow, Russia*
*[3]GP "Ukrniiugleobogashchenie", Dnepropetrovsk, Ukraine*

**Abstract:** Based upon averaging 544 balances of ranging the products size composition, the size grading coefficients are computed:

(a) bottom-screen products in the process of 50-, 25-, 13- and 6-mm size preliminary dry screening;

(b) bottom-screen products of vibrating screens, aqua screens, aqua screen unit, with vibrating screens, aqua mechanical screens in the process of 13-mm size run-of-mine coal preliminary wet screening;

(c) bottom-screen products of screens in the process of 0.5-; 1.0- and 2.0-mm fine machine size grade de-slurring;

(d) overflow products of aqua screens of different diameters in the process of coal slurry ranging.

These dependences may be used for computation of qualitative and quantitative parameters of operations related to preparation of run-of-mine coal machine grades.

**Keywords:** Coal, screening, balances of ranging, size grading coefficients

## 1. Introduction

Process parameters of coal preparation plant operation depend to a large extent upon efficiency of preparation of run-of-mine coal machine grades prescribing its distribution through the preparation processes. This distribution shall be taken into account, when the actual balance of coal preparation products and qualitative and quantitative, as well as water-sludge flow charts of the designed or modified coal preparation plants are computed.

Actual data of plant operation with the equivalent raw material and equipment, coal preparation plant design codes [1], as well as different regulatory documents [2,3] and scientific literature [4–6] are used for their computation. However, variation of coal quality towards increase of rock content, moisture content and fine grade content has caused obsolescence of previously used rates. Computation of consistency of size grade distribution through products separation in process of preparation of run-of-mine coal machine grades under the existing conditions shall be the actual research and production task and its addressing shall facilitate approximation of design and actual coal preparation parameters.

Machine grade preparation at the coal preparation plants shall generally include four process operations: dry screen splitting, large-size grain machine grading, fine grain machine grading and coal slurry ranging.

The first process operation is performed with dry screening, the second is performed with wet screening, the third is with de-slurring and the fourth is with aqua screening.

To compute parameters of the above process operations, it shall be required to evaluate size grades of the feedstock separated into bottom-screen or underflow products, as well as top-screen product moisture content. These values are computed with due account for averaged product separation grain-size composition balances shown in [7–13].

**Table 1:** Computation of parameters of feedstock size grading into the bottom-screen product $\varepsilon_1$ in the process of preliminary dry coal screening

| Separation size (mm) | Products | Parameters | Grade (mm) | | | | | | | |
|---|---|---|---|---|---|---|---|---|---|---|
| | | | +50 | 25–50 | 13–25 | 6–13 | 3–6 | 1–3 | 0–1 | Total |
| 50 | Feedstock | Feedstock-relative output (%) | 13.8 | 12.5 | 19.5 | 229 | 8.2 | 12.8 | 10.3 | 100.0 |
| | Top-screen | Product-relative output (%) | 32.0 | 23.2 | 22.2 | 16.5 | 2.8 | 2.5 | 0.8 | 100.0 |
| | Bottom-screen | Product-relative output (%) | 2.3 | 5.7 | 17.8 | 27.0 | 11.6 | 19.3 | 16.3 | 100.0 |
| | | Feedstock-relative output (%) | 1.4 | 3.5 | 10.9 | 16.5 | 7.1 | 11.8 | 10.0 | 61.2 |
| | | $\varepsilon_1$ design (unit fraction) | 0.10 | 0.28 | 0.56 | 0.72 | 0.87 | 0.92 | 0.97 | |
| | | $\varepsilon_1$ (unit fraction) | 0.10 | 0.30 | 0.55 | 0.70 | 0.85 | 0.90 | 0.95 | |
| 25 | Feedstock | Feedstock-relative output (%) | 13.6 | 16.1 | 20.8 | 23.3 | 7.6 | 11.5 | 7.1 | 100.0 |
| | Top-screen | Product-relative output (%) | 21.9 | 25.1 | 24.5 | 20.6 | 3.7 | 3.2 | 1.0 | 100.0 |
| | Bottom-screen | Product-relative output (%) | – | 1.3 | 14.8 | 27.7 | 14.0 | 25.1 | 17.1 | 100.0 |
| | | Feedstock-relative output (%) | – | 0.5 | 5.6 | 10.5 | 5.3 | 9.5 | 6.5 | 37.9 |
| | | $\varepsilon_1$ design (unit fraction) | 0 | 0.03 | 0.27 | 0.45 | 0.70 | 0.83 | 0.91 | |
| | | $\varepsilon_1$ (unit fraction) | 0 | 0.05 | 0.25 | 0.45 | 0.70 | 0.85 | 0.90 | |

*Contd...*

*Contd...*

| Separation size (mm) | Products | Parameters | Grade (mm) | | | | | | | |
|---|---|---|---|---|---|---|---|---|---|---|
| | | | +50 | 25–50 | 13–25 | 6–13 | 3–6 | 1–3 | 0–1 | Total |
| 13 | Feedstock | Feedstock-relative output (%) | 7.0 | 10.1 | 20.0 | 18.7 | 14.7 | 15.1 | 14.4 | 100.0 |
| | Top-screen | Product-relative output (%) | 9.9 | 14.3 | 27.5 | 21.2 | 12.5 | 9.5 | 5.1 | 100.0 |
| | Bottom-screen | Product-relative output (%) | – | – | 2.0 | 12.6 | 20.1 | 28.6 | 36.7 | 100.0 |
| | | Feedstock-relative output (%) | – | – | 0.6 | 3.7 | 5.9 | 8.4 | 10.8 | 29.4 |
| | | $\varepsilon_1$ design (unit fraction) | 0 | 0 | 0.03 | 0.20 | 0.40 | 0.56 | 0.75 | |
| | | $\varepsilon_1$ (unit fraction) | 0 | 0 | 0.05 | 0.20 | 0.40 | 0.55 | 0.75 | |
| 6 | Feedstock | Feedstock-relative output (%) | 10.8 | 7.9 | 15.5 | 14.9 | 19.1 | 18.8 | 13.0 | 100.0 |
| | Top-screen | Product-relative output (%) | 12.9 | 9.5 | 18.6 | 16.9 | 18.3 | 14.6 | 9.2 | 100.0 |
| | Bottom-screen | Product-relative output (%) | – | – | – | 4.8 | 23.0 | 40.0 | 32.2 | 100.0 |
| | | Feedstock-relative output (%) | – | – | – | 0.8 | 3.8 | 6.6 | 5.3 | 16.5 |
| | | $\varepsilon_1$ design (unit fraction) | 0 | 0 | 0 | 0.05 | 0.20 | 0.35 | 0.41 | |
| | | $\varepsilon_1$ (unit fraction) | 0 | 0 | 0 | 0.05 | 0.20 | 0.35 | 0.40 | |

Equipment loads and operation parameters met the certificate data. In the process of dry preliminary screening, the near-mesh separation grain size made 50, 25, 13 and 6 mm, for wet screening – 13 mm, for de-slurring – 0.5; 1.0 and 2.0 mm, for aqua screening of the coal slurry – 1000, 710, 630, 500, 350, 250, 150 and 100-mm diameter aqua screens.

Total 544 balances of the grain-size composition of products split with screens and aqua screens operated at 102 coal preparation plants were considered.

Size grades separated into bottom-screen or underflow products shall be computed by the formula:

$$\varepsilon_{1.i} = \frac{\gamma_{1.u.i}}{\gamma_{u.i}} \tag{1}$$

where $\gamma_{1.u.i}$ is the output (relative to the feedstock) $i$ of the size grade, transferred to the bottom-screen or overflow products, % and $\gamma_{u.i}$ is the output $i$ of feedstock size grade, %.

The ash-content of size grades transferred to the bottom-screen or overflow products $A_{1.i}^d$, is equal to the ash-content of appropriate feedstock size grades $A_{feed.i}^d$, i.e.

$$A^{d}_{1.i} = A^{d}_{feed.i} \ \%. \tag{2}$$

Moisture content reduction factors for the top-screen product $\varepsilon_{i\,.w}$ in the process of dry screening

$$\varepsilon_{i\,.w} = \frac{W_i}{W_{feed}} \tag{3}$$

where $W_{H}$, $W_{feed}$ is the moisture of appropriate top-screen and feedstock products, %.

Table 1 shows parameters of size grading into bottom-screen products in the process of preliminary dry screening.

Figure 1 shows relations of coefficients of near-mesh 50-, 25-, 13- and 6-mm size grading into bottom-screen products, thus, as far as the screen operating surface hole size increases, these coefficients get higher. In addition, as far as the near-mesh size is reduced, the bottom-screen product output is drastically decreased. For such size grading the difference between outputs of bottom-screen adjacent screens makes in average 10% (absolute).

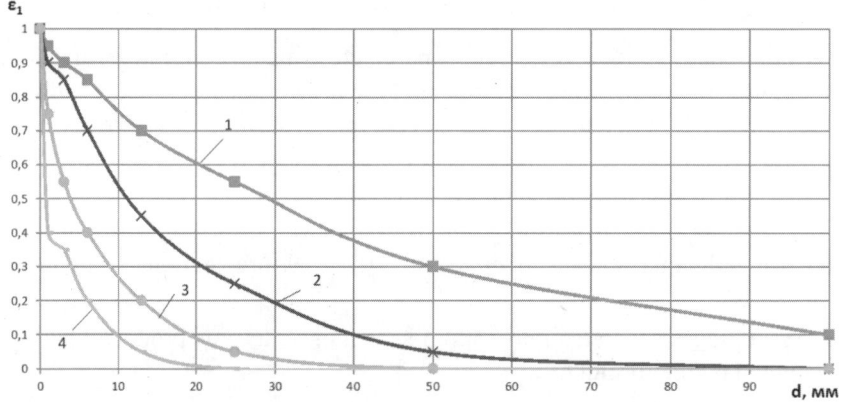

**Figure:** 1 $\varepsilon_1 = f(d)$ Dependence at different size grading
(1 – 50 mm, 2 – 25 mm, 3 – 13 mm, 4 – 6 mm)

Table 2 shows computation of moisture content reduction factor for top-screen product $\varepsilon_{i\,.w}$ by the Eq. (3). The provided data show that as far as near-mesh size grading is decreased, top-screen product moisture content reduces as compared with the feedstock moisture content due to separation of the wettest size grades.

Performed studies allow for the conclusion that in the process of preliminary dry screening, 2-time increase or decrease of size grading results in absolute 10% reduction or rise of bottom-screen product output, as appropriate.

**Table 2:** Computation of moisture content reduction factors for top-screen product ($K$w) in the process of preliminary dry coal screening

| Products | Parameters | Size grading (mm) | | | |
|---|---|---|---|---|---|
| | | **50** | **25** | **13** | **6** |
| Feedstock | $W$feed (%) | 6.1 | 6.2 | 7.2 | 6.4 |
| Top-screen | $W$H (%) | 5.7 | 5.7 | 6.5 | 5.4 |
| | $\varepsilon$H,W (fraction unit) | 0.4 | 0.92 | 0.90 | 0.85 |

Table 3 shows parameters of size grading into the bottom-screen product in the process of preliminary wet screening.

**Table 3:** Computation of parameters of size grading into bottom-screen product in the process of preliminary wet run-of-mine coal screening

| Products | Parameters | Size grade (mm) | | | | | | | | Top-screen product moisture content (%) |
|---|---|---|---|---|---|---|---|---|---|---|
| | | +50 | 25–50 | 13–25 | 6–13 | 3–6 | 1–3 | 0–1 | Total | |
| **Vibrating screens** | | | | | | | | | | |
| Feedstock | Feedstock-relative output (%) | 6.4 | 14.4 | 13.4 | 14.9 | 11.7 | 13.5 | 25.7 | 100.0 | |
| Top-screen | Product-relative output (%) | 18.4 | 38.6 | 25.1 | 8.6 | 3.8 | 2.3 | 3.2 | 100.0 | |
| Bottom-screen | Product-relative output (%) | 0 | 1.5 | 7.2 | 18.2 | 15.9 | 19.5 | 37.7 | 100.0 | 12.4 |
| | Feedstock-relative output (%) | 0 | 1.0 | 4.7 | 11.9 | 10.4 | 12.7 | 24.6 | 65.3 | |
| | $\varepsilon_1$ design (unit fraction) | 0 | 0.07 | 0.35 | 0.80 | 0.89 | 0.94 | 0.96 | | |
| | $\varepsilon_1$ (unit fraction) | 0 | 0.05 | 0.35 | 0.80 | 0.90 | 0.94 | 0.96 | | |
| **Vibrating screen machines** | | | | | | | | | | |
| Feedstock | Feedstock-relative output (%) | 10.1 | 14.0 | 14.8 | 20.1 | 12.1 | 12.9 | 16.0 | 100.0 | |
| Top-screen | Product-relative output (%) | 23.3 | 32.3 | 31.4 | 8.6 | 2.3 | 1.4 | 0.7 | 100.0 | |
| Bottom-screen | Product-relative output (%) | 0 | 0 | 2.1 | 28.9 | 19.6 | 21.7 | 27.7 | 100.0 | 9.8 |
| | Feedstock-relative output (%) | 0 | 0 | 1.2 | 16.4 | 11.1 | 12.3 | 15.7 | 56.7 | |
| | $\varepsilon_1$ design (unit fraction) | 0 | 0 | 0.08 | 0.81 | 0.92 | 0.95 | 0.98 | | |
| | $\varepsilon_1$ (unit fraction) | 0 | 0 | 0.10 | 0.80 | 0.90 | 0.95 | 0.98 | | |

*Contd..*

| Products | Parameters | +50 | 25–50 | 13–25 | 6–13 | 3–6 | 1–3 | 0–1 | Total | Top-screen product moisture content (%) |
|---|---|---|---|---|---|---|---|---|---|---|
| **Aqua screens with rectangular screening surface** | | | | | | | | | | |
| Feedstock | Feedstock-relative output (%) | 6.5 | 13.8 | 16.4 | 16.1 | 14.2 | 15.5 | 17.5 | 100.0 | |
| Top-screen | Product-relative output (%) | 15.2 | 32.2 | 31.1 | 12.8 | 4.0 | 2.6 | 2.1 | 100.0 | |
| Bottom-screen | Product-relative output (%) | 0 | 0 | 5.4 | 18.5 | 21.9 | 25.2 | 29.0 | 100.0 | 24.3 |
| | Feedstock-relative output (%) | 0 | 0 | 3.1 | 10.6 | 12.5 | 14.4 | 16.6 | 57.2 | |
| | $\varepsilon_1$ design (unit fraction) | 0 | 0 | 0.19 | 0.66 | 0.88 | 0.93 | 0.95 | | |
| | $\varepsilon_1$ (unit fraction) | 0 | 0 | 0.20 | 0.65 | 0.90 | 0.93 | 0.95 | | |
| **Aqua screens with cone-shaped screening surface** | | | | | | | | | | |
| Feedstock | Feedstock-relative output (%) | 0.6 | 10.2 | 13.2 | 13.7 | 12.9 | 21.1 | 28.3 | 100.0 | |
| Top-screen | Product-relative output (%) | 2.4 | 41.0 | 45.4 | 6.4 | 2.0 | 1.6 | 1.2 | 100.0 | |
| Bottom-screen | Product-relative output (%) | 0 | 0 | 2.5 | 16.1 | 16.5 | 27.6 | 37.3 | 100.0 | 10.2 |
| | Feedstock-relative output (%) | 0 | 0 | 1.9 | 12.1 | 12.4 | 20.7 | 28.0 | 75.1 | |
| | $\varepsilon_1$ design (unit fraction) | 0 | 0 | 0.14 | 0.88 | 0.96 | 0.98 | 0.99 | | |
| | $\varepsilon_1$ (unit fraction) | 0 | 0 | 0.15 | 0.90 | 0.95 | 0.98 | 0.99 | | |
| **Aqua screen machine with vibrating screens** | | | | | | | | | | |
| Feedstock | Feedstock-relative output (%) | 4.2 | 14.6 | 14.0 | 16.8 | 13.6 | 16.8 | 20.0 | 100.0 | |
| Top-screen | Product-relative output (%) | 13.1 | 45.3 | 31.4 | 6.5 | 1.6 | 0.9 | 1.2 | 100.0 | |
| Bottom-screen | Product-relative output (%) | 0 | 0 | 5.7 | 21.7 | 19.3 | 24.3 | 29.1 | 100.0 | 10.8 |
| | Feedstock-relative output (%) | 0 | 0 | 3.9 | 14.7 | 13.1 | 16.5 | 19.6 | 67.8 | |
| | $\varepsilon_1$ design (unit fraction) | 0 | 0 | 0.28 | 0.86 | 0.96 | 0.98 | 0.98 | | |
| | $\varepsilon_1$ (unit fraction) | 0 | 0 | 0.30 | 0.85 | 0.95 | 0.98 | 0.98 | | |
| **Aqua mechanical screens** | | | | | | | | | | |
| Feedstock | Feedstock-relative output (%) | 2.2 | 8.7 | 14.5 | 15.0 | 12.7 | 19.3 | 27.6 | 100.0 | |
| Top-screen | Product-relative output (%) | 8.0 | 31.6 | 49.1 | 8.0 | 1.5 | 0.7 | 1.1 | 100.0 | |
| Bottom-screen | Product-relative output (%) | 0 | 0 | 1.4 | 17.6 | 17.0 | 26.4 | 37.6 | 100.0 | 7.8 |
| | Feedstock-relative output (%) | 0 | 0 | 1.0 | 12.8 | 12.3 | 19.1 | 27.3 | 72.5 | |
| | $\varepsilon_1$ design (unit fraction) | 0 | 0 | 0.07 | 0.85 | 0.97 | 0.99 | 0.99 | | |
| | $\varepsilon_1$ (unit fraction) | 0 | 0 | 0.05 | 0.85 | 0.95 | 0.99 | 0.99 | | |

Figure 2 shows dependences of coefficients of near-mesh 13-mm size grading into the bottom-screen product with different equipment used for preliminary wet screening, resulting in the least bottom-screen product wasting with the aqua mechanical screens. Table 3 shows the lowest top-screen product moisture content on aqua mechanical screens.

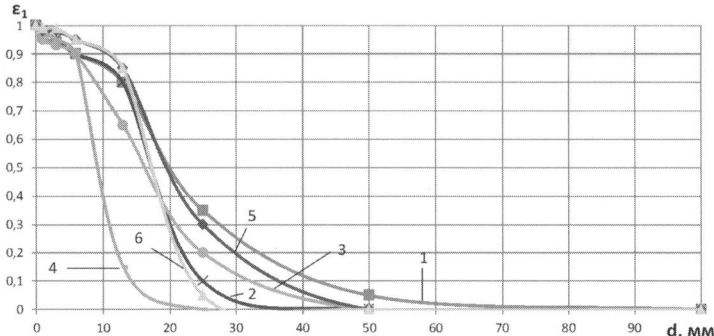

**Figure 2:** $\varepsilon_1 = f(d)$ Dependence for near-mesh 13-mm size grading with different preliminary wet screening equipment (1 – vibrating screens, 2 – vibrating screen machines, 3 – aqua screens with rectangular screening surface, 4 – aqua screens with cone-shaped screening surface, 5 – aqua screen machine with vibrating screens, 6 – aqua mechanical screens)

Table 4 shows computation of coefficients of size grading into the bottom-screen product in the process fine machine grade de-slurring.

**Table 4:** Computation of parameters of size grading into the bottom-screen product in the process of fine machine grade de-slurring

| Grading size (mm) | Products | Parameters | Grade (mm) | | | | | | | | Top-screen product moisture content (%) |
|---|---|---|---|---|---|---|---|---|---|---|---|
| | | | 13–25 | 6–13 | 3–6 | 1–3 | 0.5–1 | 0.25–0.5 | .0–0.25 | Total | |
| 0.5 | Feedstock | Feedstock-relative output (%) | 3.3 | 20.3 | 17.9 | 17.5 | 13.1 | 11.4 | 16.5 | 100.0 | 40.0 |
| | Top-screen | Product-relative output (%) | 4.2 | 26.0 | 22.9 | 22.3 | 13.5 | 6.8 | 4.3 | 100.0 | |
| | Bottom-screen | Product-relative output (%) | 0 | 0 | 0 | 0.4 | 11.8 | 27.8 | 60.0 | 100.0 | |
| | | Feedstock-relative output (%) | 0 | 0 | 0 | 0.09 | 2.57 | 6.06 | 13.08 | 21.8 | |
| | | $\varepsilon_1$ design (unit fraction) | 0 | 0 | 0 | 0.005 | 0.196 | 0.532 | 0.794 | | |
| | | $\varepsilon_1$ (unit fraction) | 0 | 0 | 0 | 0.01 | 0.20 | 0.55 | 0.80 | | |

*Contd...*

| Grading size (mm) | Products | Parameters | Grade (mm) | | | | | | | | Top-screen product moisture content (%) |
|---|---|---|---|---|---|---|---|---|---|---|---|
| | | | 13–25 | 6–13 | 3–6 | 1–3 | 0.5–1 | 0.25–0.5 | .0–0.25 | Total | |
| 1.0 | Feedstock | Feedstock-relative output (%) | 7.8 | 17.4 | 18.1 | 22.0 | 11.1 | 11.1 | 12.5 | 100.0 | 35.0 |
| | Top-screen | Product-relative output (%) | 10.7 | 23.9 | 24.6 | 25.4 | 8.3 | 4.6 | 2.5 | 100.0 | |
| | Bottom-screen | Product-relative output (%) | 0 | 0 | 0.6 | 12.9 | 18.7 | 28.6 | 39.2 | 100.0 | |
| | | Feedstock-relative output (%) | 0 | 0 | 0.16 | 3.52 | 5.09 | 7.77 | 10.66 | 27.2 | |
| | | $\varepsilon_1$ design (unit fraction) | 0 | 0 | 0.009 | 0.160 | 0.458 | 0.700 | 0.853 | | |
| | | $\varepsilon_1$ (unit fraction) | 0 | 0 | 0.01 | 0.15 | 0.45 | 0.70 | 0.85 | | |
| 2.0 | Feedstock | Feedstock-relative output (%) | 6.7 | 18.6 | 13.4 | 23.4 | 12.1 | 10.8 | 15.0 | 100.0 | 25.0 |
| | Top-screen | Product-relative output (%) | 10.7 | 29.6 | 19.1 | 28.0 | 6.8 | 3.5 | 2.3 | 100.0 | |
| | Bottom-screen | Product-relative output (%) | 0 | 0.4 | 3.9 | 15.6 | 21.0 | 23.0 | 36.3 | 100.0 | |
| | | Feedstock-relative output (%) | 0 | 0.15 | 1.46 | 5.85 | 7.87 | 8.63 | 13.54 | 37.5 | |
| | | $\varepsilon_1$ design (unit fraction) | 0 | 0.008 | 0.109 | 0.250 | 0.651 | 0.799 | 0.903 | | |
| | | $\varepsilon_1$ (unit fraction) | 0 | 0.01 | 0.10 | 0.25 | 0.65 | 0.80 | 0.90 | | |

Based upon Table 4, one may conclude that as far as de-slurring size is increased, the top-screen product moisture content is decreased.

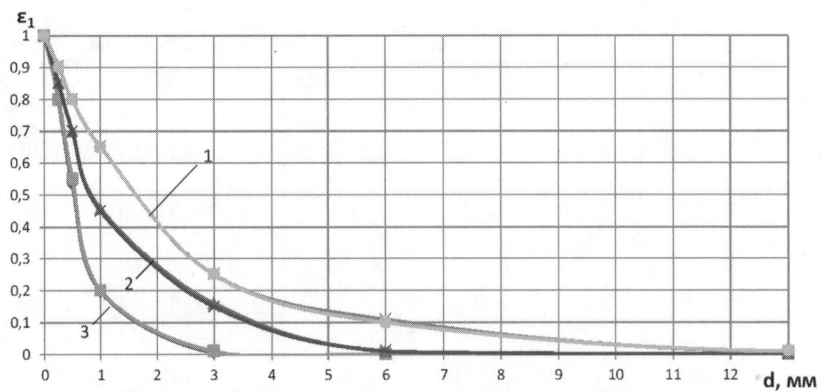

**Figure 3:** $\varepsilon_1 = f(d)$ Dependence in the process of fine machine grade de-slurring (1 – 2 mm, 2 – 1.0 mm, 3 – 0.5 mm)

Figure 3 shows dependences of coefficients of 0.5-; 1.0- and 2.0-mm de-slurring grain size grading into the bottom-screen product, consequently, it follows that as far as the de-slurring grain size is increased, the grading coefficient is increased as well.

Table 5 shows design and recommended coefficients of size grading into aqua screen overflow product. Recommended values are prescribed with due account for the design values inclusive of up to 0.00 and 0.05 round off in cases when rounding off does not affect the trend of grading coefficient variation.

Figure 4 shows dependences of coefficients of size grading into the overflow product ($\varepsilon_1^d$) of ranging aqua screens starting with the medium feedstock coal slurry grain diameter ($d_g$), it follows from these dependences that as far as $d_g$ increases, $\varepsilon_1^d$ grading coefficient increases as well. This dependence remains for all diameters of ranging aqua screens.

Based upon the data of Tables 1–5, the procedure for computation of qualitative and quantitative parameters of run-of-mine coal screening with screens and coal slurry ranging with aqua screens includes the following.

**Table 5:** Coefficients of coal slurry grading into aqua screen overflow product

| Aqua screen type | Data | Grading coefficients ($\varepsilon_1^d$) (unit fraction) | | | | | | |
|---|---|---|---|---|---|---|---|---|
| | | Size grade (mm) | | | | | | |
| | | +3.0 | 1.0–3.0 | 0.5–1.0 | 0.25–0.5 | 0.125–0.25 | 0.063–0.125 | −0.063 |
| GTS-1000 | Design | 0.91 | 0.77 | 0.65 | 0.63 | 0.42 | 0.29 | 0.12 |
| | Recommended | 0.90 | 0.75 | 0.65 | 0.55 | 0.40 | 0.20 | 0.12 |
| GTS-710 | Design | 0.93 | 0.77 | 0.67 | 0.48 | 0.40 | 0.25 | 0.14 |
| | Recommended | 0.93 | 0.77 | 0.67 | 0.57 | 0.45 | 0.25 | 0.14 |
| GTS-630 | Design | 0.95 | 0.82 | 0.71 | 0.71 | 0.50 | 0.35 | 0.16 |
| | Recommended | 0.95 | 0.80 | 0.70 | 0.60 | 0.50 | 0.28 | 0.16 |
| GTS-500 | Design | 1.0 | 0.85 | 0.74 | 0.63 | 0.49 | 0.29 | 0.18 |
| | Recommended | 1.0 | 0.85 | 0.75 | 0.65 | 0.52 | 0.30 | 0.18 |
| GTS-350 | Design | 0.97 | 0.92 | 0.80 | 0.70 | 0.54 | 0.39 | 0.20 |
| | Recommended | 1.0 | 0.90 | 0.80 | 0.70 | 0.55 | 0.35 | 0.20 |
| GTS-250 | Design | 1.0 | 0.95 | 0.86 | 0.73 | 0.61 | 0.39 | 0.21 |
| | Recommended | 1.0 | 0.95 | 0.85 | 0.75 | 0.60 | 0.40 | 0.21 |
| GTS-150 | Design | 1.0 | 1.0 | 0.89 | 0.78 | 0.63 | 0.44 | 0.13 |
| | Recommended | 1.0 | 1.0 | 0.90 | 0.80 | 0.65 | 0.45 | 0.22 |
| GTS-100 | Design | 1.0 | 1.0 | 1.0 | 0.83 | 0.69 | 0.50 | 0.20 |
| | Recommended | 1.0 | 1.0 | 1.0 | 0.85 | 0.70 | 0.50 | 0.23 |

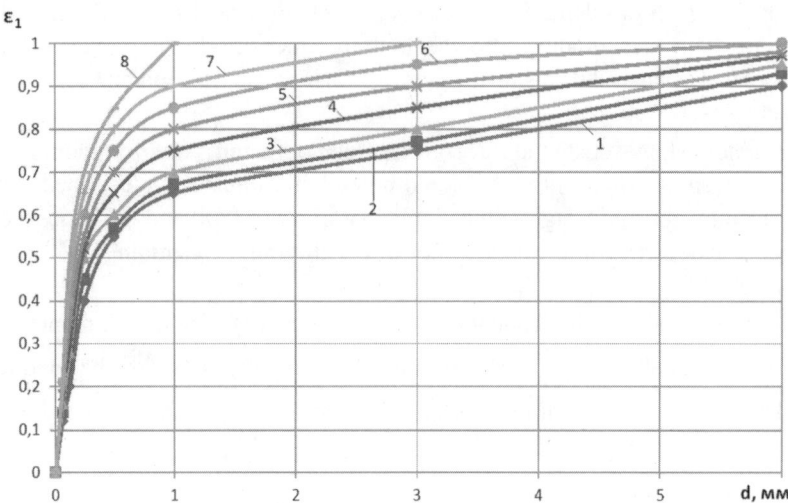

**Figure 4:** $\varepsilon_1 = f(d)$ Dependence for aqua screens of different diameters (1 – GTS-1000, 2 – GTS-710, 3 – GTS-630, 4 – GTS-500, 5 – GTS-350, 6 – GTS-250, 7 – GTS-150, 8 – GTS-100)

$i$-Size grade output into $\gamma_{\text{output.}i}$ overflow product

$$\gamma_{1i} = \varepsilon_{1i} \cdot \gamma_{feed.i} \text{ %,} \tag{4}$$

where $\varepsilon_{1i}$ is the coefficient of $i$-size grading into bottom-screen or overflow product, unit friction, shall be assumed in line with Tables 1–5; $\gamma_{feed.i}$ is the $i$-size grade output in the feedstock product.

$i$-Size grade output into top-screen or over-split product $\gamma_{2i}$

$$\gamma_{2i} = \left(1 - \varepsilon_{1i}\right) \cdot \gamma_{feed.i} \text{ %.} \tag{5}$$

The ash-content of $i$-size grades in the bottom-screen or overflow products $A_{1i}^d$ and top-screen or over-split products $A_{2i}^d$ shall be equal to the ash-content of $i$-size grade in the feedstock product $A_{feed.i}^d$; i.e.

$$A_{1i}^d = A_{2i}^d = A_{feed.i}^d \text{ %.} \tag{6}$$

Bottom-screen or overflow product output $\gamma_1$

$$\gamma_1 = \sum_{i=1}^{i=n} \gamma_{1i} \text{ %.} \tag{7}$$

The ash content of the bottom-screen or overflow product $A_1^d$

$$A_1^d = \left( \sum_{i=1}^{i=n} \gamma_{1i} \cdot A_{1i}^d \right) : \gamma_1 \ \%. \tag{8}$$

Top-screen or over-split product output $\gamma_2$

$$\gamma_2 = \sum_{i=1}^{i=n} \gamma_{2i} \ \%. \tag{9}$$

The ash content of the top-screen or over-split product $A_2^d$

$$A_2^d = \left( \sum_{i=1}^{i=n} \gamma_{2i} \cdot A_{2i}^d \right) : \gamma_2 \cdot \ \%. \tag{10}$$

Upon definition of the grain size compositions of the split products and using fraction compositions of run-of-mine coal size grades, fraction compositions of machine grades and their preparability classes may be computed, hence, the rates of product separation wasting of subsequent preparation operations.

## 2. Conclusion

1. Parameters of run-of-mine coal size grading into the bottom-screen product in the process of preliminary dry and wet run-of-mine coal screening and fine machine grade de-slurring, as well as into the aqua screen overflow product in the process of coal slurry ranging have been computed.

2. The averaged moisture content of the top-screen products of the screens under study has been evaluated.

3. These dependences may be used for computation of qualitative and quantitative parameters of run-of-mine coal machine grade preparation operations and utilizing fraction compositions of run-of-mine coal, fraction compositions of machine grades and their preparability classes may be computed, hence, the rates of product separation wasting of subsequent preparation operations.

## 3. References

[1] VNTPZ-94 (Departmental Plant Process Design Codes), Coal Preparation Plant Process Design Codes. Kharkov: Yuzhgiproshakht Design Institute. 1993. 156 p.

[2] SOU 10.1.00185755:002-2004 Вугільні продукти збагачення. Методика розрахунку показників якості. К.: Мінпаливенерго України. 2004. 46 р.

[3] RD 03-306-99 (Guidance), Instruction on Coal (Slurry) Processing Loss Computation and Rating. M.: Gosgortekhnadzor of Russia. 1999. 34 p.

[4] Polulyakh A.D.,Workshop for Computation of Qualitative, Quantitative and Water-Slurry Process Flow Charts of Coal Preparation Plants: Training Aid/Polulyakh A.D., Pilov P.I., Egurnov A.I. D.: National TU, 2007. 504 p.

[5] Polulyakh A.D.,Workshop for Environmental Engineering in the Process of Natural Resources Preparation; Training Aid/Polulyakh A.D., Pilov P.I., Egurnov A.I., Polulyakh D.A. D.: National TU, 2011. 89 p.

[6] Polulyakh A.D., Workshop for Computation of Mined Coal Quality Parameters: Training Aid/Polulyakh A.D, Polulyakh D.A. D.: National TU, 2016. 144 p.

[7] Polulyakh A.D., Distribution of Size Grades in the Process of Coal Preliminary Dry Screening/Polulyakh A.D., Berlin A.M., Polulyakh O.V./ Збагаченнякориснихкопалин: Research Collection Book, 2017. No. 66(107). P. 64–73.

[8] Polulyakh A.D., Distribution of Size Grades in the Process of Coal Preliminary Wet Screening with Vibrating Screens/Polulyakh A.D., Polulyakh D.A. Machine and Technology Vibrations, 2017. No. 3(86). P. 102–109.

[9] Polulyakh A.D., Distribution of Size Grades in the Process of Run-of-Mine Coal Aqua Screening/Polulyakh A.D., Polulyakh D.A. Збагаченнякориснихкопалин: Research Collection Book, 2017. No. 66(107). P. 48–56.

[10] Polulyakh A.D., Distribution of Size Grades in the Process of Coal Preliminary Wet Screening with Aqua Screen Machine with Vibrating Screens/A Polulyakh A.D., Polulyakh D.A., Geomechanics: IGTM Research Collection Book, 2018. No. 138. P. 212–217.

[11] Polulyakh D.A., Distribution of Size Grades in the Process of Run-of-Mine Coal Aqua Mechanical Screening/Polulyakh D.A., Machine and Technology Vibrations, 2017. No. 4(87). P. 76–81.

[12] Polulyakh A.D., Distribution of Size Grades in the Process of Fine Machine Grade De-Slurring/Polulyakh A.D., Polulyakh O.V./Збагаченнякориснихкопалин: Research Collection Book, 2017. No. 66(107). P. 56–64.

[13] Polulyakh A.D., Computation of Qualitative and Quantitative Parameters of Coal Slurry Ranging in Aqua Screens/Polulyakh A.D., Moiseenko O.V., Polulyakh D.A. Ukraine Coal. 2018. No. 3. P. 6–8.

# 28

# Ash in coal: A source for REE?

Claus Bachmann

*J&C Bachmann GmbH, Pforzheim, Germany*

**Abstract:** Efficiency of brushless electric drives, lanthanium in battery electrodes, computer screens are examples for driving forces on the increasing demand on rare earth elements. Chinese mines are the major suppliers with a market share of approximately 80%. However, even though the output in China can be increased there will be a gap between production and worldwide demand in the near future. Therefore an intense search for alternative resources is going on. Research done by Luttrell, Bethell, Laudal, Holuszko et al. [1,2] in North America have shown that some coal seams contain REE concentrations which are worth to be recovered and have also investigated recovery methods. Due to the difficult situation of coal in Europe activities on rare earth element recovery do not concentrate on local coal production but on the composition of imported coal.

Research and development activities on coal based REE abundance take place also in Europe.

Keywords:

**Keywords:** REE, fly ash, CPT-XRF, bioleaching

## 1. Rare earth elements

The so called rare earth elements are the chemical elements of the third subgroup of the periodic table and the lanthanoid group. These elements are not really rare but their concentrations are normally small. As an example, the amount of cerium, ytrium or neodynium in the earth shell is higher than the amount of lead, copper or arsenic. However, deposits where these normally as mineralic oxides existing elements can be mined economically are quite rare and these deposits are not equally distributed around the world. Therefore the production of rare earth metals is often linked to the chemical preparation of ore containing other metals with higher concentration.

Today more than 80% of the rare earth elements are produced in China. Although Australia shows a growing share in the world market the total production is more or less stable over the past years (Fig. 1).

Technical development, especially electric cars, cell phones and communication is going to increase the demand strongly. There will be an increase in the demand of about 50% for certain elements from 2020 to 2030 just for clean technologies (Fig. 2).

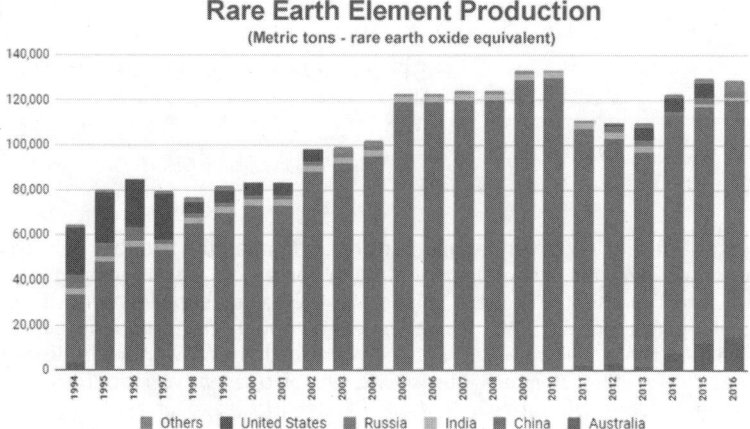

**Figure 1:** Rare earth element production [7]

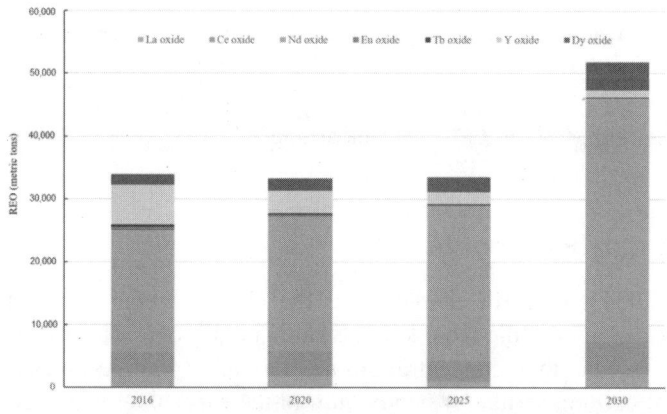

**Figure 2:** Global demand for REO from clean technologies in 2016, 2020, 2025 and 2030 [8]

Given the fact that the development of preparation technologies and the construction of preparation sites under environmental and political aspects is quite time consuming the mining industry is challenged now to make new sources for REE accessible.

## 2. REE in coal

Investigations conducted by Honacker et al. [9] using coal samples from the eastern coal fields of the US have shown an average REE content of 346 ppm. Two concentration methods were evaluated to concentrate the REE's. Detailed analysis of the samples have shown that the concentration of REE in the ash correlates strongly with the ash content of the coals; below 20% ash a significant increase of REE concentration was observed [10]. Furthermore, it could be proven that the majority of samples showed a REE concentration which is significantly increased compared to the earth crust's average.

Similar results were derived from coal fields in British Columbia by Kumar et al. [12].

Using hydrophobic hydrophilic separation (HHS) the concentration could be increased from 346 to 9.539 ppm. The concentration could be increased to 35.443 ppm using acid leaching.

## 3. REE in fly ash

Although part of the ash produced in power station as residual of the coal combustion process is used in cement and road construction materials the majority of the ash is put to deposits. For example, the RWE power plants in the Cologne region (Germany) produce 5.3 million tons of ash per year to be deposited. Corresponding figures can be found also for all other coal fired power stations.

This fine material is expected to get denser and form a compact mineral due to pussolanic effects over time. This ash deposits could be used as source for REE provided that the REE concentration is high enough to make extraction economically feasible.

## 3.1 Evaluation of ash deposits

Evaluation of ash deposits given the fact that REE are concentrated in mineral particles like monazite, xenotime and bastnasite with particle sizes in the range of nanometers to micrometers economic extraction of the elements will be strongly dependent on the size distribution of the raw material. Therefore fly ash deposits may be a preferred source of REE extraction.

Not only in the US but also in Europe several R&D projects were conducted to evaluate the content of REE in fly ash.

Franus et al. [3] have investigated 12 samples of fly ashes from lignite and bituminous coal combustion in Poland. Variation in REE content was quite high; the sample with the lowest content of REE contained only 101 ppm while the highest content reached 543 ppm which might be of economic interest. These results are in accordance to the results presented by Luttrell et al. [10].

It was pointed out that the samples were not representative for the whole stockpile and therefore systematic investigations have to take place in order to evaluate the homogenity of the deposit. Furthermore, there is no general recipe for recovery of the REE from the stockpile; tailor made methods will be required according to the special characteristics of the stockpile. It should be mentioned that also other strategic as well as hazardous elements are contained in those fly ashes which needs to be considered.

## 3.2. Deposit characterization

It is obvious that recovery of REE from ash deposits as well as from other deposits can only be done economically if the deposit is carefully evaluated and areas with high concentrations of rare earth elements are identified. Then especially the enriched material can be processed while the ashes with low concentrates remain untreated.

Sampling and chemical analysis of such a deposit is time consuming and expensive. Therefore the examination and characterization of the deposit using a combined CPT-XRF probe (Fig. 3) have been proven to be an economic alternative. The CPT (cone penetration test) consists of pushing a cone into the ground with a constant speed. Beside other sensors the probe is equipped with an XRF spectrometer. Equipped with an X-ray tube and a silicon drift detector the spectrometer monitors the elements in the deposit passing by the window.

**Figure 3:** CPT XRF probe with control unit and computer [6]

Present CPT XRF probes are limited to X-ray energies up to 40 keV which is not ideal for the measurement of REE's fluorescence. Therefore, J&C Bachmann is currently developing a high energy version which will strongly improve the response of the REE and which will be used in the German/ Brasilian MoCa project [11] which is supported by the German Federal Ministry of Research. The energy dispersive spectrum allows to identify and quantify the different elements. Figure 4 shows a typical spectrum gained with a CPT XRF probe.

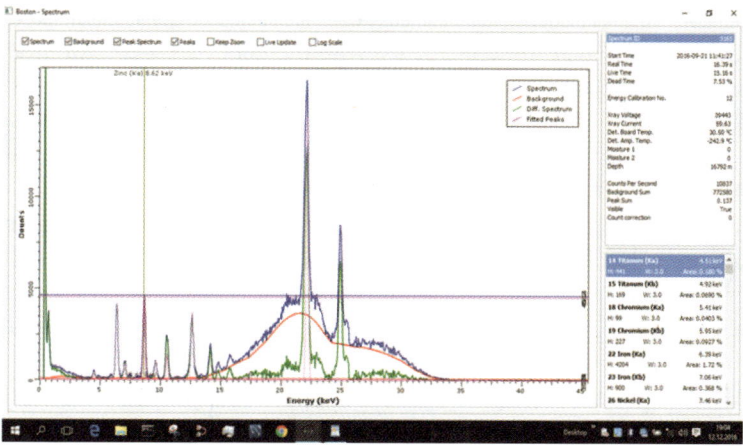

**Figure 4:** CPT XRF spectrum

The spectra are obtained in short time intervals and therefore the variation of concentrations along the depths can be monitored. Figure 5 shows plots of metal concentrations gained using the logging software BOSTON which was developed by J&C Bachmann.

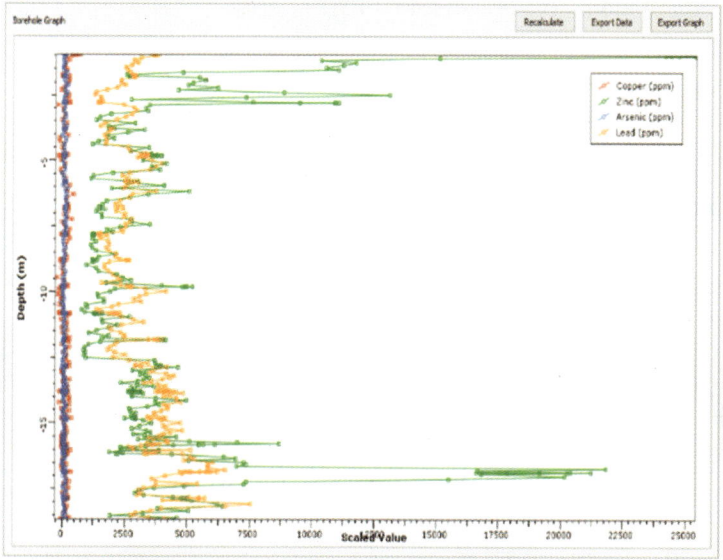

**Figure 5:** Borehole log derived with BOSTON software

## 3.3 REE recovery using biomining

Based on the fact that even deposits which are rich in REE show still comparably small concentrations highly efficient and economical methods will be required to recover these elements from the deposit. Microbial processes can help attain a higher yield of these elements from soil, sediments, slag heaps and mineral waste. One of the increasingly used technologies in mining to dissolve out such substances is the biological raw material extraction from the underground, so called biomining.

In biomining one distinguishes two different types of application. The first one is bioleaching, which means the microbial caused transformation of insoluble, valuable metal into its soluble form. The second one is biooxidation, whereby valuable metals like gold are dissolved out by microbial dissolution of valueless metals.

As an important prerequisite for the application of in-situ-technology the heterogeneous tailing body must be prepared for in-situ-leaching processes, specifically a well water penetrable tailing body and a dense bottom lining or water-impermeable layer. Both prerequisites can be met by applying modern geotechnology/geobiotechnology. The adaptation of existing geotechnologies such as hydrofracking, geoblocking and geobiocementation shall be adapted and upscaled for its use in tailing bodies.

Beside others SENSATEC, Kiel, is specialized on the design, test and realization of biomining processes for REE.

## 3. Conclusion

Analyses show in different investigations that rare earth elements are more concentrated in coal ash than in the earth crust. Furthermore, preparation respectively combustion leads to products of fine particle size where REE containing minerals are already liberated. This is a prerequisite for economic concentration of these elements. Further developed probes help to characterize a deposit quickly and cost efficient.

However, although interesting research is done in the major coal producing and coal consuming industrial regions and successful pilot projects could be launched there is still a huge demand for additional research and development if rare earth elements will be recovered from coal and its by-products.

This work has to be started now in order to ensure that the increasing demand on REE's will be available in future.

# 4. References

[1]  Session 16: Recovery of Rare Earth Elements I, Clearwater Clean Energy Conference, Clearwater 2018.

[2]  Session 22: Recovery of Rare Earth Elements II, Clearwater Clean Energy Conference, Clearwater 2018.

[3]  Franus, Wojciech, M.M. Wiatros-Motyka and M. Wdowin: Coal fly ash as a resource for rare earth elements, Environmental Science and Pollution Research, June 2015, 22, 9464–9474.

[4]  Roth, Elliot, M. Macala, R. Lin, T. Bank, R. Thompson, B. Howard, Y. Song and E. Granite: Distributions and Extraction of Rare Earth Elements from Coal and Coal By-Products, 2017 World of Coal Ash (WOCA) Conference in Lexington, KY – May 9–11, 2017.

[5]  Tolhurst, Linda: Commercial Recovery of Metals from Ash, 2015 World of Coal Ash (WOCA) Conference in Nashville, TN – May 5–7, 2015.

[6]  Alisch, Uta: Innovation for the Mining Market, German Day at Mining Indaba, 6 February 2018.

[7]  https://geology.com/articles/rare-earth-elements/.

[8]  Zhou, Baolu, Z. Li and C Chen: Global Potential of Rare Earth Resources and Rare Earth Demand from Clean Technologies, Minerals 2017, 7, 203.

[9]  Honacker, Rick, J. Grappo, R-H. Yoon, G. Luttrell, A. Nobel and J. Herbst: Rare Earth Element Production from Coal, Professional Engineers in Mining (PEM), Lexington, Kentucky, August 26th, 2016.

[10] Gerald H. Luttrell, M. J. Kiser, R-H. Yoon, A. Bhagavatula, M. Rezaee and R. Q. Honaker: Concentrations of Rare Earth Elements generated by U.S. Coal Preparation Plants, Coal Prep Conference, Louisville, KY 2016.

[11] https://www.tu-clausthal.de/en/press/tu-news/details/2315/.

[12] V. Kumar, A. Kumar, and M.E. Holuszko (2018): Occurrence of rare-earth elements in selected British Columbian coal deposits and their derivative products; in Geoscience BC Summary of Activities 2017. Minerals and Mining, Geoscience BC, Report 2018-1, p. 87–100.

# 29

# On the desirability of abandoning the use of imperfection as an indicator of the effectiveness of gravity coal separation in water pulsation jigs

Stanisław Głowiak[1] and Barbara Tora[2]

[1]*Automation Company "BGG", Katowice, Poland*
[2]*AGH University of Science and Technology, Kraków, Poland*

**Abstract:** The 'Ecart Probable' $(E_p)$ is a generally used measure of the gravity separation process efficiency. The value of that index depends on the selection of the operating point in the process, characterised by the parameter called the 'separation density'. Next another index 'Imperfection' $(I)$ was implemented. 'Imperfection' is strongly associated with the evaluation of water jigs. Imperfection is not better than $E_p$ and it considerably falsifies the evaluation of jigs separation, especially for the densities lower than 2.0 Mg/m³. For that reason, it is recommended to consider the imperfection's properties. The basic advantages of the imperfection rate were supposed to include the independence from separation density and to be unambiguous assessment of enriching device. None of these advantages really exists. That results from the fact that the essence of imperfection occurrence is only a logarithmic conversion of the partition curve, introduced to obtain its symmetric form. The partition curve asymmetry is a natural feature of the coal. Besides, it has not been justified yet why the symmetric form of the partition curve should be more useful for the process evaluation. The procedures of empirical determination of imperfection do not ensure a symmetric form of the partition curve, which is indispensable for obtaining correct values of imperfection. Negative side of imperfection is the existence of various definitions of that index in various enrichment processes.

This paper presents mathematical transformations that show the relationships existing between the Ecart probable and imperfection. Further part is an attempt of critical evaluation of imperfection.

**Keywords:** Ecart Probale, Tromp, Imperfection, Evaluation, Assumptions, Enrichment

## 1. Introduction

The publication of papers by Tromp (1937) and Terra (1938) in 1937–1938 became the origin of the development of new methods adopted for the evaluation of the gravity separation process. The importance of these methods is consisted in the evaluation of the most essential part of the process, which is the separation of the material being enriched, based on grain density, without using any quantitative–qualitative evaluations of final products. The $E_p$ as a gravity enrichment index, introduced by Terra, called the 'Ecart probable', based on Tromp's partition curves, became the most commonly used index in mineral processing. It was and has been still applied mainly to compare various gravity separation processes and various enrichment devices used in such processes. However, we need to mention that this index was previously known as 'quartile deviation' in the probability theory and it has been used as a measure of the dispersion of the random variable distribution, presently most often characterised by variance or standard deviation. In the military application of the probability theory, that index is the radius of the circle or of the sphere into which a half number of fired missiles is landed. What is an important, although unintended effect of not completely correct application of the probability theory in the description of that portion of the coal enrichment process is an excessively common understanding of this process as a random phenomenon. Meanwhile, in reality, the enrichment process in a jig is well determined, although it still contains a certain considerable random component, resulting mainly from the random properties of the material being enriched. These properties are the causes of the existence of relationships between the parameters that specify the deterministic part of the enrichment process and the phenomenon which is completely random and involves only the behaviour of single particles in the enrichment process.

## 2. Reasons of introducing the concept of imperfection to the theory of separation curves

In Tromp's and Terra's approaches, the partition curves were approximated by the functions associated with normal distribution. These functions were introduced because of the similarity of separation curves drawn for a heavy product to the distribution function of normal distribution. That similarity was especially visible in separation curves of the enrichment devices using heavy liquids, operating usually with much better enrichment effectiveness that all other types of gravity devices. It was found, however, that experimental separation curves, obtained especially in studying the enrichment effectiveness of water pulsation jigs, were much different from the shapes of the distribution function of normal distribution.

A considerable asymmetry of the experimental approximations of partition curves was an important reason of the difficulties in applying the $E_p$ index. That asymmetry quite unnecessarily turned the attention of the researchers to the possibility of partition curve symmetrisation. For the definition of Ecart probable $E_p = (\delta_{75}-\delta_{25})/2$, the condition $\delta_{75}-\delta_{50}=\delta_{50}-\delta_{25}$, marking the separation curve symmetry, is completely insignificant.

For unknown reasons, the curve asymmetry, approximating the empirical partition curve, became the most serious issue in the practical application of the $E_p$ index. The simplest solution of that matter was to adopt the well-known asymmetric distributions that used to be applied in mineral processing, as various forms of the gamma distribution (Tarjan, 1974; Gottfried, 1978). That actually did not happen probably because such distributions required fairly complex calculations. At the same time, it had been noticed the possibility of obtaining relatively symmetric forms of experimental distributions, described by the distribution function of normal distribution, in which modified is the $x$-axis of that distribution. Such a modification was found in the determination of partition numbers for another variable, $\log(\rho-1)$ (Belugou and Ulmo, 1950).

It is also worth pointing at another important aspect of the issue under discussion, which is the relationship of the probable error with separation density, essentially associated with the partition curve asymmetry. That problem was not studied with necessary care in the 1950s. It was rather omitted, by the implementation of the concept of imperfection, which was supposed to correct the dependence of $E_p$ index on separation density, in the authors' intention, so that the dependence became insignificant. Such actions led to omission of the physical phenomena causing the dependence of particle dispersion on their density until today, although such dependence is actually the reason of the changes in probable errors occurring in the partition density function. All actions intended to obtain a single, correct and universal partition curve model, based on the distribution function of normal distribution, visible in the bibliography on that issue, were charged with the basic fault, i.e. the lack of evidence that the existence of such a model is possible at all.

The partition curve asymmetry is especially visible at higher particles densities and it can be caused by a larger dispersion of particles at such densities, actual asymmetry of particle distribution in the jig bed layers, or larger measurement errors associated with the necessity to separate these particles in high-density liquids. It is also important to notice that nothing has changed in the separation curve theory for a number of years, and no works have been published to explain fundamental issues of the influence of physical phenomena occurring in jigs on the shape of separation curves, or on the significance of partition curve asymmetry in the evaluation of the enrichment efficiency. The introduction of partition curves on a logarithmic scale and the consequential enrichment evaluation index, called 'imperfection', can be accepted as a result of trying to omit essential and difficult theoretical issues, such as the influence of the changes in the dispersion of particular density fractions on the partition curve shape. We should also notice that no solid study has been published as yet on the evaluation of the properties of imperfection and partition curves in the logarithmic scale, although 70 years have passed since the implementation of those conceptions in practice. That situation may result from the lack of detailed advantages of imperfection, in comparison to the properties of the probable error, in the studies of the authors introducing the concept of that new index (Belugou and Ulmo, 1950). However, when implementing such an essential change in the evaluation of the enrichment process, it was necessary to present all specific details with justification.

## 3. Mathematical justifications of the faults of imperfection

With no doubt really existing is $E_p$ index dependens on separation density. There are a lot of results of experimental investigations known from references which confirm existence of that dependence. Really unknown widely is only the shape of funktion describing this dependens, but from same now existing results of current unpublished up to date studies follows that experimentally confirmed is the approximately linear shape of this funktion. This shape is also the result of simple theoretical investigation presented in final part of this paper. Now, the mathematical explanation of the form of existing two different notations of imperfection understanding will be presented. Below are presented Figs. 1 and 2 with some mathematical transformations. Both figures present an asymmetric separation curve and a symmetrised curve, obtained as a result of the replacement of density by a logarithm of that variable, decreased by the constant c. Assumed is a definition of symmetry for that issue is equality $a = b$. At this point, we will introduce a certain generalisation to our further considerations. It will be useful for the consideration of generally understood separation curve symmetry.

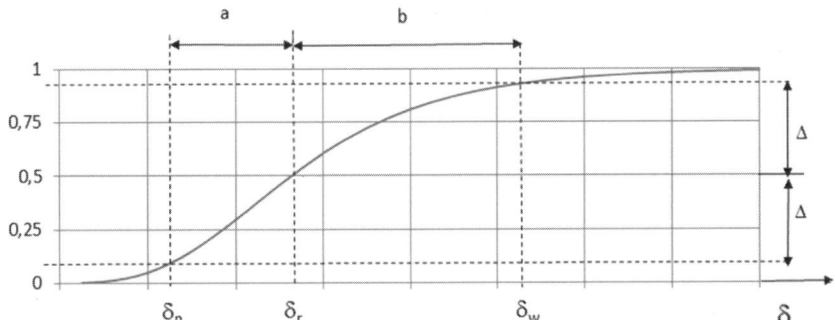

**Figure 1:** Asymmetric separation curve of a jig

The Ecart probable $E_p$ of symmetric distribution corresponds to the difference between separation density and such density for which the increase or decrease of the separation number amounts is equal to 0.25. It means that $\Delta=0.25$ in both Figs. 1 and 2. If the curve is symmetric in the whole density range in the logarithmic scale, then, for each $0<\Delta<0.5$, there exists such $E$, different from $E_p$, that corresponds to the partition numbers of $0.5+\Delta$ and $0.5-\Delta$. Such general enlargement of $E_p$ will be used in further considerations. It is worth noticing that dispersion understood in that manner is also applied in practice, as is confirmed by Swansons study (Swanson et al., 2008).

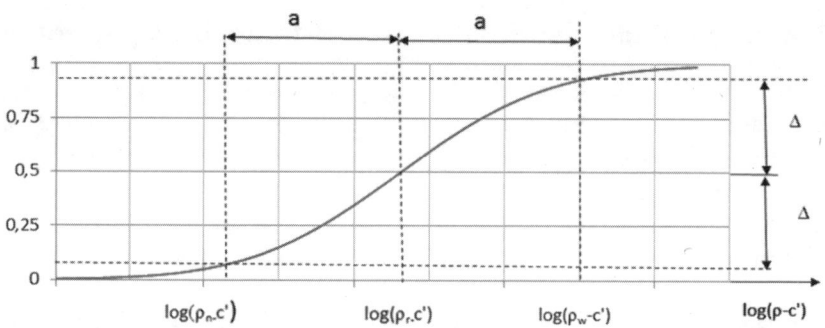

**Figure 2:** Symmetric separation curve of a jig, for *x*-axis, in the logarithmic scale

Further considerations will be purposefully conducted in the scale of $\log(\rho-c)$, where $c$ is a constant, fulfilling the condition of $0<c<\delta_{min}$. This will allow to explain the existence of the two imperfection definitions in literature using only one proof.

Both curves shown above are the results of the same experiment. The differences occurring in them are only the consequences of a different presentation of results. In that situation, it is possible to determine the relationships between the dispersion which characterise the curves in question.

Based on the definition of index $E$, analogous to $E_p$ and Fig. 1, we obtain the following:

$$a+b = 2E = \delta_w - \delta_n \qquad (1)$$

Based on Fig. 2 and with the assumption that in the partition curve drawn for the logarithmic variable, the following relationships occur, describing the symmetric partition curve:

$$2a = 2E \qquad (2)$$
$$\log(\rho_w - c) = \log(\rho_r - c) + E \qquad (3)$$
$$\log(\rho_n - c) = \log(\rho_r - c) - E \qquad (4)$$

In this place, it is necessary to explain different markings of density on the $x$-axis in both graphs. Both variables, $\delta$ and $\rho$, have the same numerical value, although the density $\delta$ has a dimension, while the density $\rho$ should be dimensionless (it is possible to calculate its logarithm only for the dimensionless variable) and it is the quotient of density $\delta$ and the unit density. Such a conversion must be applied to preserve the dimensional correctness of the equations presented here. For the same reason, the index $E$ and the constant $c$ have the dimension of density (or of specific weight), while $E$ and the constant $c$ are dimensionless. These remarks result from the conventions applied in almost the whole references, where $E_p$ has the dimension of density (or specific weight).

It is worth pointing out an essential assumption resulting from the above equations, that is partition curve symmetrisation in the logarithmic scale, within the separation numbers of $0.5 \pm \Delta$, following directly from Eq. (2). A serious consequence of accepted assumptions is the necessity to check whether the empirical partition curve, drawn in the logarithmic scale, fulfils that assumptions and is symmetric. If the assumptions have not been fulfilled, imperfection should not be determined on the basis of these empirical data. In that case, we have, in fact, two separation curves, both being asymmetric. Therefore, the question is raised what is the sense of converting the asymmetric curve into another one, also being asymmetric. However, all the theoretical and normalised procedures of empirical determination of imperfection, known to the authors of this study, completely omit that condition.

Verification of whether the curve in the logarithmic scale is symmetric is simple, because the following relationship occurs, based on Fig. 2:

$$\log (\rho_r - c) - \log (\rho_n - c) = \log (\rho_w - c) - \log (\rho_r - c) = E, \text{ from which}$$

we can conclude that $(\rho_r - c)^2 = (\rho_w - c)(\rho_n - c)$, which means that $\rho_r - c$ is a geometric mean of $\rho_w - c$ and $\rho_n - c$.

The following relationships are obtained from the system of Eqs. (3 and 4), and a simple conversion has been performed for the assumed form $E$:

$$\log \frac{\rho_n - c'}{\rho_r - c'} = -E' = k \ln \frac{\rho_n - c'}{\rho_r - c'} = kE'' \tag{5}$$

$$\log \frac{\rho_w - c'}{\rho_r - c'} = E' = k \ln \frac{\rho_w - c'}{\rho_r - c'} = kE'' \tag{6}$$

where $k = \log e = 0.43429448$, $e$ is the base of natural logarithm, equals to $2.71828182$, and $E = E/k$.

We can conclude from Eqs. (5 and 6) and the comparison of logarithmic expressions that:

$$\rho_n - c = (\rho_r - c')e^{-E''} \tag{7}$$

$$\rho_w - c = (\rho_r - c')e^{E''} \tag{8}$$

After subtraction on both sides of Eqs. (7 and 8), we obtain:

$$\rho_w - \rho_n = (\rho_r - c')\left( e^{E''} - e^{-E''} \right) \tag{9}$$

Since the same relationship occurs with the dimensional values of $\delta_w$, $\delta_n$, $\delta_r$, and c, then:

$$\delta_w - \delta_n = (\delta_r - c)\left( e^{E''} - e^{-E''} \right) = 2E = 2(\delta_r - c)\sinh E'' \tag{10}$$

Hence, $\dfrac{E}{(\delta_r - c)} = I = \sinh E'' = \sinh\dfrac{E'}{k}$ \hfill (11)

$E' = \log(I + \sqrt{I^2 + 1}) = k\ln(I + \sqrt{I^2 + 1}) = kar\sinh I$ ,

where $I = \dfrac{E}{\delta_r - c}$ \hfill (12).

For $\Delta = 0.25$, $c = 1$, and $E = E_p$, where $E_p$ is the Ecart probable, assumed in the whole presently known literature, while, for $\Delta \neq 0.25$, we can also consider other definitions of $E_p$ mentioned already (Swanson et al., 2008).

When analysing Eq. (11), we can notice that, for $c = 1$ and $E = E_p$, we obtain a generally used definition of imperfection applicable to water pulsation jigs, while, assuming $c = 0$, we can obtain another equation known from literature (Jowett, 1968):

$$I = \frac{E_p}{\delta_r} \tag{13}$$

That is assumed in the calculations that are applied to the enrichment processes conducted in rarely used now air jigs. Hence, it is clear the general conclusion that the presently used imperfection definitions result only from the logarithmic form of the x-axis of the partition curve.

## 4. The properties of the partition curve which are contradictory to the concept of imperfection

The other very important reason making us critically evaluate imperfection as an index of coal enrichment effectiveness in a jig is the real partition curve and its considerable natural deformation, in respect of the distribution function of normal distribution. We can suppose from the results of unpublished studies that the experimentally confirmed probabilistic model of particle distribution in the jig bed, produced with taking into account all essential limitations occurring in the grains dispersion process, allows for identification of the following phenomena that were not known earlier:

- The separation curve of a heavy product very rarely represents the form being similar to the whole distribution function of normal distribution, either in linear or logarithmic scale. Most often, it is only a part of the curve that is similar in shape to the distribution function of normal distribution, although cut off either at the bottom or at the top.

- A detailed shape of the distribution function of normal distribution may be obtained only for the following two conditions that are impossible to be fulfilled in practice: particle dispersion in the jig bed has an ideally normal distribution that is also identical for all particle densities in the jig bed.

- The distribution of the density layers that occur in the bed when the particle dispersion is normally distributed in layers allows for a determination of separation density precisely only for some several separation density values, while all other separation density values can be determined only with a certain approximation.

- In prediction of enrichment result, it is not possible to establish in advance any separation density, within the range of all density particles existing in the jig bed.

- The number of the partition curves of the jig operating at a usually meeting dispersion which are distinguishable by measurement does not exceed 5–7 curves that are essentially different from each other.

Consequently, it is clear that, even upon the most exact experiment, one may not always obtain such a partition curve which is symmetric when is presented in the logarithmic scale. Such a situation is not predicted in normalised procedures for the determination of imperfection, and imperfection calculated according to such procedures usually is not correct. For that reason, the Ecart probable $E_p$ is a much better index because, once it is determined, we always obtain a correct result affected only by a measurement error.

One can maintain that the approximation of the empirical partition curve of the distribution function of normal distribution has any sense only because of the ease of application of the mathematical apparatus associated with that distribution. If the symmetrisation of the partition curve occurred always after the application of the logarithmic coordinate transformation, we would obtain a simple and good tool for a broad use in the separation curve evaluation.

Since that is not the case, using of imperfection index can be treated as a certain stage of research on the enrichment process finished without success. Such a statement can be justified by the existence of another theoretical model of symmetric separation curves (Zapała, 1994) and a different definition of the concept of imperfection. That model, based on the random physical phenomena of the separation process is not associated in any way with the distribution function of normal distribution.

Taking into account all above presented details, it is possible to conclude that imperfection $I$ is an index established arbitrarily and associated with the experimental data, similarly to the $E_p$ index. Its single theoretical foundation consists in the above specified mathematical transformation resulting from symmetrisation of partition curve. The same cannot be, however, said about the $E_p$ index whose value results completely from the experiment.

## 5. Other reasons of the lack of justification for the existence of imperfection

Literature does not quote any experimental evidence for better match of the partition curve points established experimentally with the curve in the logarithmic scale. In the majority of quoted curves, one can notice such a large dispersion of the experimental points in respect of the model curve, that the curve can be matched with many diverse models with the same accuracy. That is the most essential problem for the determination of the partition curve, and the conformation of how essential it is consists in the fact that the calculation of separation numbers does not concern the feed composition, being determined experimentally, but rather the so-called replacement feed, reducing the error of the determination of yield of the products that are indispensable for finding experimental partition numbers. An interesting thing confirming the importance of the problem is the fact that the separation curve stated in Tromp study (Tromp, 1937) matches better the distribution function of normal distribution in the linear scale than in the logarithmic one (in the sense of the minimisation of mean square errors).

Consequently, the necessity of abandoning the use of imperfection as an index of the enrichment effectiveness seems to be obvious, at least until both experimental and theoretical evidence of its justification has been presented and that seems to be impossible in the light of the considerations presented above. Removal of imperfection from the enrichment practice in water pulsation jigs will not be easy taking into account the existence of numerous normalised procedures for the determination of imperfection, as well as the force of habitual use of that measure.

Finally, one can also make another remark on the issue under consideration. Such a remark is contained in the question why so many actions were conducted to make reality fit the invented poor model of the phenomenon with symmetric form of the partition curve and the anamorphous conversion of asymmetric curve (Terra, 1954). It would be better to develop research leading to a different model of the phenomena, more compliant with reality, or to accept other evaluation index for use, known in the enrichment theory, such as the Driessen index or its different geometric interpretation known as the $A_{er}$ index (Swanson et al., 2008). When both of those indexes are used, the asymmetric form of the partition curve is not significant. That is even more surprising as the authors of the concept of imperfection also presented the Driessen index in the study that introduced imperfection, calling the former the 'error surface' or the 'Tromp surface' (Belugou and Ulmo, 1950).

The search for a separation model in which partition curve asymmetry does not occur was and has been completely contradictory to the physical nature of the process in question. That contradiction results from failure to remember about the essential influence of the forces affecting the particles of the material being enriched on the result of separation. Among those forces, we may not omit the resultant of the force of buoyancy and of weight visibly linear increasing with the increase of particle density (the force of buoyancy in the water separation process is always lower than weight). As a consequence, vertical relocation of the particle is decreasing in the phase of rising water movement. That has to cause a reduction of material loosening and an increase of the average density of jig bed which always leads to the increase of particle dispersion in particular jig bed layers.

The research on the jig bed model mentioned above indicates that the change of grains dispersion is the reason of the separation curve asymmetry and the change is not essentially associated in any way with separation density. Separation density results from the selection of the cut point of the delaminated material into two products in the enrichment process. Moving that point in the direction of higher density, in the practice of separation in jigs, is effected by increasing the float density through the increase of the float weight by adding weights. Besides, a change of separation density does not essentially influence in any way the average particle dispersion in all jig bed layers.

We need to mention finally that some authors often incorrectly understand dispersion in the enrichment process conducted in jigs. That misunderstanding results from the fact that the partition function, whose graphic presentation has the form of the partition curve, is a function of the density of the particles of the material being enriched, while the dispersion function, whose graph is called the 'dispersion curve', is the function of the density of the layers that develop in the jig bed. Consequently, do not exist the integral–differential relationships between those functions. These relationships are a result of incorrect conversion of the relationships existing in the probability theory into the gravity separation theory.

## 6. Conclusion

1. The index called the 'imperfection $I$' does not introduce any new quality into coal separation effectiveness evaluation, in comparison to the 'Ecart probable $E_p$'.

2. One of the reasons why imperfection was introduced as an index is the alleged symmetric form of the partition curve in the logarithmic coordinates that has no significance at all for the evaluation of enrichment effectiveness.

3. An essential reason of introducing imperfection index was the alleged but non-existing property of independence from separation density.

4. We can infer from the considerations presented above, that it is possible to use imperfection only in the form of $I=E_p/\delta_{50}$ for both water and air jigs; however, imperfection defined as the $I=E_p/(\delta_{50}-1)$ should not be used for water pulsated jigs.

## 7. References

[1] Belugou M., Ulmo M. (1950), Représentation des résultats d'une épuration. Conférence Internationale sur la Préparation des Charbons Paris, Juin 1950, A3, p. 16–20.

[2] Gottfried B.S. (1978), A generalization of distribution data for characterizing the performance of coal cleaning equipment. Mineral Processing, 5 (1978) 1–20.

[3] Hughes F.S. (1967), The partition curve. Coal Preparation, March/April 1967, p. 63–67.

[4] Jowett A. (1968) An appraisal of partition curves for coal-cleaning processes. International Journal of Mineral Processing, 16 (1986) pp. 75–95.

[5] Swanson A., Atknson B., Weale W. (2008), Design and Operational Data for the Optimum Utilisation of Large Diameter Dense Medium Cyclone, XV International Coal Preparation Congress and Exhibition, China University of Mining and Technology Press, Pekin, p. 297.

[6] Tarjan G. (1974) Application of distribution functions to partition curves. International Journal of Mineral Processing, 1 (1974) p. 261–265.

[7] Terra A. (1938) Essai d'une thèorie du lavage. Revue de l'industrie minèrale, Nr 425, p. 391–401.

[8] Terra A. (1954) Significance of the Anamorphosed "Partition Curve" and the "Ecart Probable" in Washery Control. Second International Coal Preparation Congress Essen, C3 (1954) p.1–8.

[9] Tromp K.F.(1937) Neue Wege für die Beurteilung der Aufbereitung von Steinkohlen. Glückauf, nr 6 (1937) p. 125–131.

[10] Zapała W.T. (1994) Theoretical Model of the Curve Concerning Separation of the jig. Preprints of the 12th International Coal Preparation Congress. 1994 Cracov vol. 1 C-4 p. 243–251.

# 30

# Technological characteristics of coal concentration under the conditions of ever-changing raw material resources base

O. Yehurnov[1] and O. Polulyakh[2]

*[1]ANA-TEMS LLC, Dnipro, Ukraine*
*[2]Society of the Ukrainian Mineral Engineers, Dnipro, Ukraine*

**Abstract:** It is known from the solid mineral processing theory and practice that the best qualitative and quantitative indicators are provided at processing the raw material of constant qualitative and quantitative composition.

However, in actual practice, we encounter the continuous variation of plant raw material supply base and ever-changing requirements for the commercial coal products quality, and ecology predetermines elimination of above-standard losses of combustible matter with coal preparation rejects. Let us consider the technological characteristics of coal concentration caused by these changes.

**Keywords:** Concentration, Resource, Underutilisation, Washer, Hydrocyclones, Grain-size distribution

## 1. Plant underutilization and overutilization

In Ukraine, maximum number of coal preparation plants falls within the period of 1980–78 (Table 1). These processed 161.5 million tons of ROM (run-of-mine) coal [1]. In 2012, there were 55 coal preparation plants remaining, which processed 73 million tons of ROM coal at the production capacity of 142.3 million tons, i.e., utilization of plants made 51.3% [2]. Herewith, it should be noted that in 2012, total coal output in Ukraine made about 84.6 million tons, and even if total mined coal was washed, utilization of plants would only made 59.5%.

Utilization of operating plants ranged from 1.1% (Trudovskaya) and 2.1% (Novopavlovskaya) to 125.7% (Dobropolskaya) and 127.5% (Kurakhovskaya). Such utilization range is abnormal, and it results in the above-standard losses of combustible matter with coal preparation rejects in case of overutilization and unreasonable operating costs in case of underutilization.

Hence, it is premature to speak of construction of additional new coal preparation plants in Ukraine. The primary question must be their reconstruction and modernization.

**Table 1**

| Depth of cleaning[a] (mm) | Year of operation[b] | | | | | |
|---|---|---|---|---|---|---|
| | 1960 | 1970 | 1980 | 1990 | 2000 | 2012 |
| 0 | 28 (45.9%) | 32 (43.4%) | 39 (50%) | 41 (55.4%) | 32 (51.6%) | 30 (54.5%) |
| 0.5 | 1 (1.65%) | 8 (10.5%) | 5 (6.4%) | 5 (6.8%) | 4 (6.5%) | 6 (10.9%) |
| 1 | 1 (1.65%) | – | – | – | – | – |
| 3(4) | 1 (1.65%) | 1 (1.3%) | 1 (1.3%) | 1 (1.4%) | – | – |
| 5 | 1 (1.65%) | – | – | – | – | – |
| 6 | 21 (34.4%) | 24 (31.6%) | 24 (30.8%) | 18 (24.3%) | 17 (27.4%) | 14 (25.5%) |
| 10 | – | 2 (2.7%) | – | – | – | – |
| 13 | 8 (13.1%) | 8 (10.5%) | 8 (10.2%) | 9 (12.1%) | 9 (14.5%) | 5 (9.1%) |
| 25 | – | – | 1 (1.3%) | – | – | – |
| Total plants | 61 (100%) | 76 (100%) | 78 (100%) | 74 (100%) | 62 (100%) | 55 (100%) |
| Including without coking plant CPP | 49 (80.3%) | 66 (86.8%) | 70 (89.7%) | 68 (87.2%) | 60 (96.8%) | 54 (98.2%) |
| Processed coal quantity (thous. t) | 80,051 | 133,201 | 161,564.5 | 159,263 | 47,882.6 | 72,962.4 |

[a]Presented data relates to the coal preparation plants, but not to the plants at minoc, rock-disposal dumps, and external slurry ponds.
[b]Indicated in brackets is percent of plants with this depth of cleaning (%).

## 2. Number of supplier mines

Amid the disordered stochastic system of the coal preparation plant raw material resources base formation applied in Ukraine, which does not account for the washed coal compatibility, transportation cost effectiveness, and optimal capacity utilization requirements, unfavorable conditions adversely affecting the coal preparation performance were established in recent years. For example, in 2012, 14 coking plants of total 20 ones took coking coals of three and over grades for processing. The number of supplier mines made up to 20–26 ones at some plants in power-generating area.

With a lack of sufficient volume of storage capacities and homogenizing-and-metering devices at the plants, such deliveries adversely affected the enterprise smooth work and process performance metrics, and resulted in violation of applicable quality standards and consumer's market requirements.

## 3. Engineering level of coal preparation plants in Ukraine

The engineering level of coal preparation plants in Ukraine is characterized by the dynamics of cleaning depth and development of main concentrating processes (Tables 1 and 2).

Table 2

| Processa | Number of processes by years | | | | | |
|---|---|---|---|---|---|---|
| | 1960 | 1970 | 1980 | 1990 | 2000 | 2012 |
| Washer troughs | 31 | 10 | 2 | – | – | – |
| Coarse coal jigging | 23 | 30 | 30 | 30 | 29 | 26 |
| Fine coal jigging | 24 | 36 | 52 | 53 | 42 | 43 |
| Unclassified jigging | 6 | 24 | 18 | 10 | 8 | 4 |
| Coarse coal dense-medium separation | 3 | 17 | 27 | 30 | 29 | 25 |
| Fine coal dense-medium separation | – | – | 3 | 1 | 1 | 3 |
| Wet screw separation | – | – | – | 2 | 6 | 22 |
| Flotation | 24 | 37 | 42 | 41 | 19 | 20 |
| Drying | 19 | 33 | 37 | 36 | 19 | 12 |
| Pneumatic concentration | 1 | – | – | – | – | – |
| Concentrating hydrocyclones | 1 | 3 | 3 | 4 | 2 | 4 |
| Concentration tables | – | 6 | – | – | – | – |
| Hydrosizers | – | – | – | – | – | 5 |
| Cone separators | – | – | – | – | – | 1 |

[a]It is referred to the coal preparation plants, but not to the plants at mines, rock-disposal dumps, and external slurry ponds.

Table 1 implies that number of plants with the cleaning depth of 0 mm increases constantly, while number of plants with the cleaning depth of 6 mm and 13 mm decreases. Cleaning depths of 0.5, 1, 3, 4, 5, 10, and 25 mm are virtually not used. This trend of cleaning depth indicates that concentration of all ROM coal grades is required. At the same time, more advanced concentration methods are used (see Table 2). In so doing, by 1980, most of washer troughs were replaced with dense-medium separation units or hydraulic jig washers. By 1970, the use of pneumatic concentration was abandoned in Ukraine; by 2012, number of plants using the unclassified jigging was reduced drastically. Consistent trend is observed in the use of the dense-medium concentration in separation units for coarse machine grade and in hydrocyclones for fine machine grade.

Proportions of various methods in total concentration balance vary mainly as a result of current redistribution of concentration volumes between plants with different combinations of concentrating processes. Principal methods: foam separation plant in the anthracite concentration process chart (OP CCM "Sverdlovskaya", CPP of the "Russia" mine); introduction of other slurry concentration methods: hydrocyclones (MPM "Luhanskaya", "Slavyanoserbskaya"), hydrosizers (CCM "Samsonovskaya", MPM "Krasnoluchskaya", CCM "Duvanskaya", coal preparation shop No. 2 of the Avdeevskiy By-product Coke Plant), screw separators (CCM "Dobropolskaya", MPM "Belorechenskaya", CCM "Chumakovskaya"), cone separators (CCM "Kievskaya"); use of dense-medium cyclones for the middlings rewashing (CCM "Chumakovskaya"), recessed-plate filtering presses (coal preparation shop No. 2 of the Avdeevskiy By-product Coke Plant) for the liquid rejects dewatering, pneumatic separators (at mines and rock-disposal dumps) for concentration of coal and carboniferous materials, concentration tables (at concentrating plants) for concentration of settlings from external slurry ponds, and putting into operation of the CPP "Svyato-Varvarinskaya" equipped with dense-medium cyclones, Crossflow hydrocyclone-classifiers and column flotation machines. Extension of application scope of these methods means certain progress in the concentration technology development, thus creating opportunities for the subsequent improvement in qualitative and quantitative indices of coal preparation plants.

## 4. ROM coal grain-size distribution

ROM coal grain-size distribution determines the amount of this size machine grade, for which concentrating and auxiliary equipment of suitable capacity is selected, and width of conveying equipment belts. In case that ROM coal size is changed, it results in overutilization of some concentration lines and underutilization of other ones; this effect comes into particular prominence when very coarse coal is replaced with ultrafine one, and the plant water–

slurry system fails to handle the circulating water regeneration. The system is burden with slurry, which is distributed with the concentration products thus increasing the concentrate ash content and reducing the reject ash content.

## 5. ROM coal gravimetric composition

When the gravimetric composition is changed at the same washability category, special priority must be given to the ash content of float and sink fractions determining the concentrate quality and combustible matter loss with rejects, respectively. The sharp increase in ash content of float fractions can result in production of faulty products.

Quantitative ratio of float, intermediate and sink fractions in ROM coal predetermines the variation of concentrate, middlings, and rejects outputs, respectively, and poses the risk to transport capabilities of some conveying equipment lines.

## 6. ROM coal washability

It is known that under otherwise equal conditions, the degree of cross-contamination of concentration products depends on the ROM coal washability category: The more difficult washability category is (higher T index), the higher is contamination of concentrate and rejects, particularly in case of separation of two concentration products.

For example, in accordance with [3], the contamination of float fractions at concentration of the 1–13 mm fine machine grade makes 1.5% for the 'easy' coal washability category ($T < 5\%$) and 3% for 'very difficult' category ($T \geq 15\%$), i.e., it shows twofold increase. It can be assumed that with index $T = 45\%$, content of float fractions in rejects would also be doubled and then would make 6%.

ROM coals supplied to the coal preparation plants in Ukraine can have the washability index $T \geq 45\%$.

For example, Table 3 presents data [4] on washability category determination for ROM coals from the GP "Lvovugol" mines, which contain sapropelite with density equivalent to the middlings density, i.e., 1500–1800 kg/m$^3$. Table 3 shows that the $T$ index increases for a coarse machine grade: for the "Velikomostovskaya" mine – from 53.1% to 83.6%; for the "Mezhirichanskaya" mine – from 81.2% to 89.2%; for the "Vidrodzhennya" mine – from 2.6% to 55.8%; for the "Stepnaya" mine – from 10.5% to 40.3%.

**Table 3:** Dynamics of washability category of ROM coal from the GP "Lvovugol" mines

| Indicators | 2003 | 2004 | 2005 | 2007 |
|---|---|---|---|---|
| **"Velikomostovskaya" mine** | | | | |
| ROM coal ash (%) | 46.4 | 50.8 | 51.6 | 50.6 |
| 25 mm grade yield (%) | 28.56 | 29.86 | 34.9 | 30.67 |
| Including sapropelite (%) | 11.86 | 18.23 | 23.92 | 17.83 |
| 25 mm grade ash (%) | 50.6 | 59.6 | 60.8 | 61.6 |
| Including coal (%) | 8.0 | 18.7 | 21.2 | 5.0 |
| Including sapropelite (%) | 47.8 | 47.7 | 51.9 | 49.2 |
| Including rock (%) | 79.6 | 80.9 | 85.0 | 84.0 |
| 13 mm grade washability category | Very difficult | Very difficult | Very difficult | Very difficult |
| *T* index | 53.1 | 83.6 | 79.6 | 83.4 |
| 1–13 mm grade washability category | Very difficult | Very difficult | Very difficult | Very difficult |
| *T* index | 28.6 | 32.9 | 30.7 | 22.5 |
| **"Mezhirichanskaya" mine** | | | | |
| ROM coal ash (%) | 54.6 | 45.0 | 47.1 | 55.2 |
| 25 mm grade yield (%) | 34.93 | 21.33 | 20.41 | 27.15 |
| Including sapropelite (%) | 18.09 | 11.48 | 7.41 | 17.51 |
| 25 mm grade ash (%) | 64.2 | 62.1 | 61.0 | 64.8 |
| Including coal (%) | 5.7 | 13.5 | 13.2 | 5.5 |
| Including sapropelite (%) | 49.7 | 60.9 | 52.0 | 54.6 |
| Including rock (%) | 81.2 | 86.6 | 85.7 | 89.2 |
| 13 mm grade washability category | Very difficult | Difficult | Difficult | Very difficult |
| *T* index | 79.0 | 12.4 | 14.1 | 59.5 |
| 1–13 mm grade washability category | Very difficult | Moderate | Moderate | Very difficult |
| *T* index | 23.2 | 6.8 | 9.3 | 16.2 |
| **"Vidrodzhennya" mine** | | | | |
| ROM coal ash (%) | 44.5 | 55.0 | 60.2 | 58.0 |
| 25 mm grade yield (%) | 20.66 | 27.10 | 22.32 | 21.28 |
| Including sapropelite (%) | 9.76 | 12.22 | 7.12 | 10.28 |
| 25 mm grade ash (%) | 55.6 | 71.7 | 70.3 | 68.8 |
| Including coal (%) | 6.9 | 17.3 | 6.1 | 7.4 |
| Including sapropelite (%) | 24.5 | 61.0 | 48.8 | 58.9 |

*Contd...*

*Contd...*

| Indicators | 2003 | 2004 | 2005 | 2007 |
|---|---|---|---|---|
| Including rock (%) | 85.7 | 83.0 | 86.5 | 83.1 |
| 13 mm grade washability category | Easy | Very difficult | Very difficult | Very difficult |
| *T* index | 2.6 | 23.0 | 23.7 | 55.8 |
| 1–13 mm grade washability category | Easy | Very difficult | Difficult | Very difficult |
| *T* index | 3.3 | 16.2 | 11.3 | 21.8 |
| **"Stepnaya" mine** | | | | |
| ROM coal ash (%) | 44.0 | 50.5 | 58.9 | 43.9 |
| 25 mm grade yield (%) | 20.45 | 10.51 | 15.44 | 19.32 |
| Including sapropelite (%) | 7.04 | 2.23 | 2.25 | 1.51 |
| 25 mm grade ash (%) | 61.8 | 71.5 | 82.4 | 71.9 |
| Including coal (%) | 17.3 | 13.6 | 23.1 | 19.8 |
| Including sapropelite (%) | 42.3 | 52.9 | 58.8 | 40.0 |
| Including rock (%) | 85.9 | 86.3 | 89.5 | 85.0 |
| 13 mm grade washability category | Very difficult | Very difficult | Very difficult | Difficult |
| *T* index | 17.2 | 18.4 | 40.3 | 10.5 |
| 1–13 mm grade washability category | Very difficult | Moderate | Very difficult | Easy |
| *T* index | 23.3 | 9.7 | 19.6 | 1.8 |

In case of the fine machine grade, the $T$ index range for the listed mines makes 22.5–32.9%; 6.8–23.2%; 3.3–21.8%; 1.8–23.3%. The obtained data confirm that the washability index $T$ values can far exceed those determined in accordance with [3].

In 2007, the GP "Ukrniiugleobogashchenie" developed the "Procedure of calculation of the expected qualitative and quantitative indicators of products of the ГП "Lvovugol" ROM coal concentration at the CCM "Chervonogradskaya" of the ZAO "Lvovsistemenergo" based on the [4] data and taking into account sapropelite content", which recommends increase of concentration products cross-contamination indices depending on sapropelite content in the machine grade (Table 4).

Table 4 shows that the cross-contamination of concentration products increases even beyond $T \geq 15\%$, and that the existent coal washability categories do not cover total range of their values, while the cross-contamination indices established based thereon fall short of actual ones.

**Table 4:** Contamination of products at the separation density of 1800 kg/m3

| Sapropelite content in coarse machine grade $\beta^c_{кл.к.}$ (%) | Contamination (%) | | Sapropelite content in fine machine grade $\beta^c_{кл.м.}$ (%) | Contamination (%) | |
|---|---|---|---|---|---|
| | coarse concentrate $\Delta\gamma_{m\,(к.к)}$ | Coarse rejects $\Delta\gamma_{л\,(отх.к.)}$ | | Fine concentrate $\Delta\gamma_{m\,(к.м)}$ | Fine rejects $\Delta\gamma_{л\,(отх.м.)}$ |
| 5 | 3.0 | 2.0 | 5 | 6.0 | 3.0 |
| 10 | 5.0 | 2.7 | 10 | 8.0 | 4.0 |
| 15 | 7.0 | 3.4 | 15 | 10.0 | 4.5 |
| 20 | 9.0 | 4.0 | 20 | 12.0 | 5.0 |
| 25 | 11.0 | 4.5 | 25 | 14.0 | 5.5 |
| 30 | 13.0 | 5.0 | 30 | 16.0 | 6.0 |
| 35 | 15.0 | 5.5 | 35 | 18.0 | 6.5 |
| 40 | 17.0 | 6.0 | 40 and over | 21.0 | 7.0 |
| 45 | 19.0 | 6.5 | | | |
| 50 | 21.0 | 7.0 | | | |
| 55 and over | 25.0 | 9.0 | | | |

With reference to the foregoing, the "Ukrniiugleobogashchenie" Institute proposes establishment of eight washability categories, in particular: T = 0–5% – easy; T > 5–10% – moderate; T > 10–15% – difficult; T > 15–25% – very difficult; T > 25–40% – ultra-difficult #1; T > 40–60% – ultra-difficult #2; T > 60–85% – ultra-difficult #3; T > 85–100% – ultra-difficult #4.

In addition, respective standard concentration product contamination levels are defined for each new coal washability category.

# 7. Quality of machine grades preparation from ROM coal

Modern theoretical conceptions of concentrating processes physics allow for the conclusion that the highest qualitative and quantitative indicators of concentrating processes are achieved at concentration of narrow size grades prepared in accordance with the these processes' requirements as to the initial product. From this perspective, preparation of machine grades becomes one of critical preparatory operations at the state-of-the-art coal preparation plants. Quality of machine grades preparation is determined by both classification efficiency and dewatering efficiency, and researches […] found that the improvement of both coarse (Fig. 1) and fine (Fig. 2) machine grades preparation performance results in the concentration process $E_{pm}$ decrease.

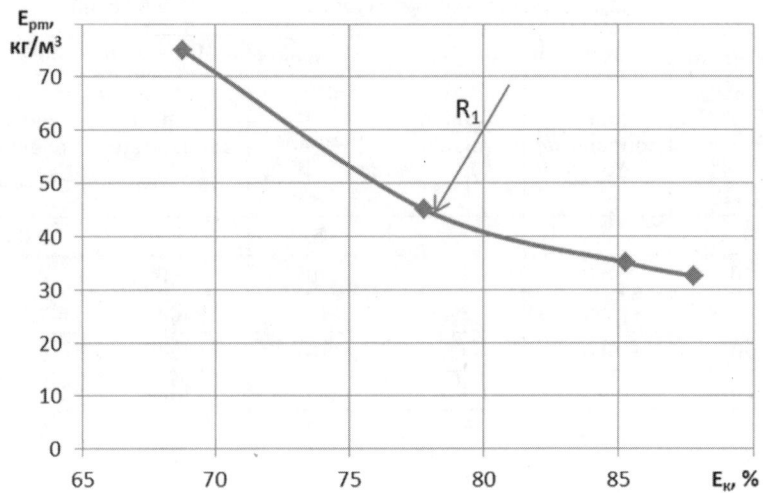

**Figure 1:** $E_{pm.\kappa}$ index vs. coarse machine grade preparation performance index $E_{\Pi MK.\kappa}$

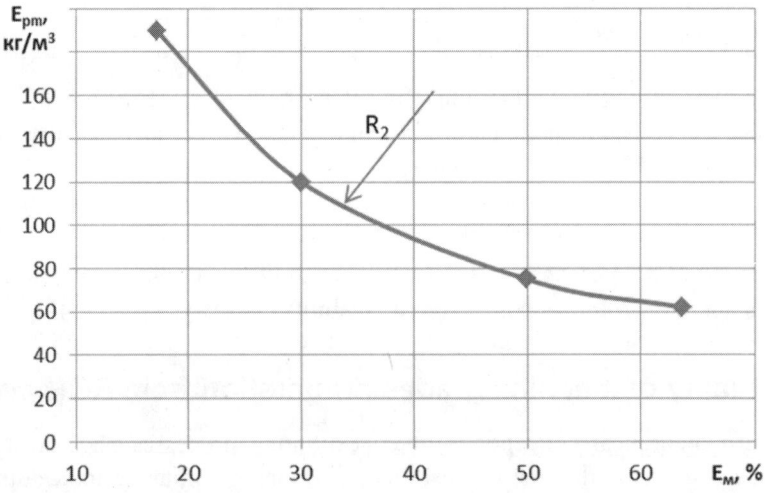

**Figure 2:** $E_{pm.\text{м}}$ index vs. fine machine grade preparation performance index $E_{\Pi MK.\text{м}}$

Thus, the gravity concentrating process $E_{pm}$ root-mean-square deviation determined by its separation characteristic is inversely exponential to performance index of corresponding coal machine grade preparation by size and moisture content, where the inversely exponential curve radius value depends on the number of machine grade size constraints: at unilateral constraint, it is greater than that for bilateral constraint.

## 8. Rock quantity

Rock quantity in the coal machine grades makes up to 40% and over (up to 85%), which preclude the operation of hydraulic jig washers and extension of rock conveyor belt width.

## 9. Float fraction ash content

The increase in ash content of float fractions necessitates the separation density reduction in concentrating machines in order to produce commercial concentrate with standard ash content and predetermines the increased combustible matter losses with coal preparation rejects or increased standard values of ash content for commercial coal products.

## 10. Presence of readily-soaking rocks

For example [6], the Western Donbas coal beds, particularly, those in western Pavlograd District are characterized by high water cut. Readily-soaking wall rocks, increased coal matter hardness and low density (1050–1150 kg/m$^3$) determine considerable ROM coal moisture content (up to 16–17%) and high content of fine grades (below 80 μm): up to 70% in primary slurry and up to 80% in secondary slurry (mainly clayey mud with ash content of over 60–65%).

Increased ROM coal moisture content and coal matter hardness facilitate the rock constituent swelling and abrasion during the transportation to the CCM "Pavlogradskaya". Rocks entering the production process are already prepared for the intensive soaking and abrasion in water environment, which results in concentrate ash content increase, magnetite suspension contamination, and degradation of magnetite washing-off from the concentration products and their dewatering, and suppression of flotation and coagulation processes. This situation is compounded by the continuous deterioration in raw material quality, i.e., by the coal ash content increase due to coal-cutting with a wall rocks, and hence, increased content of readily-soaking rock components.

Efficient concentration of such coals and slurries requires the plant to have such process chart that would provide the maximum practicable clayey mud removal from the system at the process starting, separation and removal of fine and coarse rocks as soon as is practicable, and would provide the clean circulating water production. The latter requirement is mandatory, since circulation of finished concentration products is not allowed, minimum intermediate storage tanks for slime water are required, and deslimed slurry flotation and recovery of coarse coal particles discharged to classification unit drains due to low density are required.

Moreover [7], for these reasons, prior to wetting with water at the hydro-preparation area, coal fed to the preparatory screen separation is mass prone to caking. During the transportation to the plant, accumulation in storage bins and prescreening, conglomerates of up to 200 mm and over are formed of sticky clay with disseminated coal and hard rock particles. These can survive after passing the dense-medium separation units. Sometimes, percent of conglomerates with diameter exceeding the near-mesh grain size (10–13 mm) makes up to 20% of over-sized material. Floating conglomerates result in the increased concentrate ash content, sinking conglomerates result in the reduced coarse reject ash content, i.e., increased losses of combustible matter with coal preparation rejects.

## 11. Presence of sapropelite

For example [8], ROM coals from the GP "Lvovugol" mines contain sapropelite, which can have high ash content (40–60%) at density of 1500–1600 kg/m³ and low ash content (20–40%) at density of 1600–1800 kg/m³. The sapropelite density range coincides with the density of middling fractions (1500–1800 kg/m³), of which content determines the coal washability categories. Sapropelite content and ash for some mines of the GP "Lvovugol" are shown in Fig. 3, while Figs. 4 and 5 show dependencies of reject and concentrate contamination, respectively, on sapropelite content obtained under the conditions of the CCM "Chervonogradskaya".

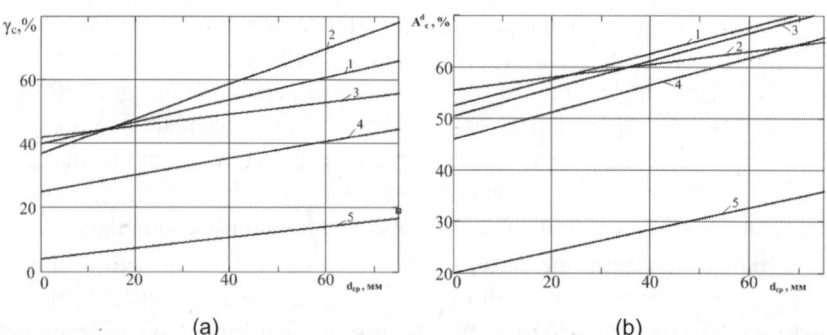

(a)                                (b)

**Figure 3:** Sapropelite content $\gamma_c$ and ash-content $A^d_c$ vs. average size grade diameter $d_{cp}$: a – $\gamma_c = f(d_{cp})$; 6 – $A^d_c = f(d_{cp})$; 1 – "Velikomostovskaya" mine; 2 – "Mezhirichanskaya" and "Chervonogradskaya" mines; 2, 4 and 5 – "Vidrodzhennya", "Zarechnaya", and "Stepnaya" mines

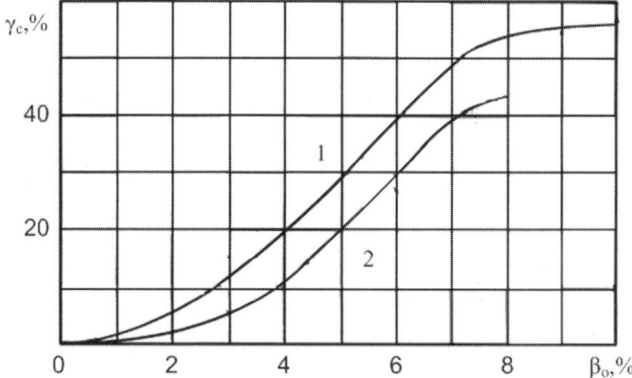

**Figure 4:** Rejects contamination $\beta_o$ with fractions below 1800 kg/m³ vs. sapropelite content $\gamma_c$: 1 and 2 – fine and coarse rejects

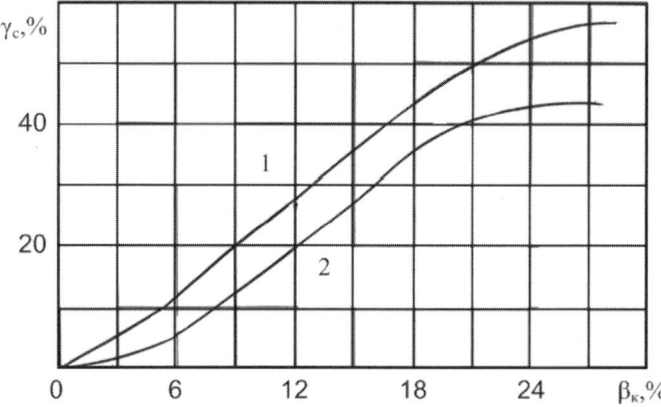

**Figure 5:** Rejects contamination $\beta_\kappa$ with fractions over 1800 kg/m³ vs. sapropelite content $\gamma_c$: 1 and 2 – coarse and fine rejects

Figure 4 shows that at the same sapropelite content within a machine grade, contamination of fine rejects with −1800 kg/m³ fraction is higher than contamination of coarse rejects. The obtained dependencies are S-shaped curves with the maximum reject contamination increase observed at the initial machine grade sapropelite content at the level of 10–45%.

Figure 5 shows that at the same sapropelite content within a machine grade, contamination of coarse concentrate is higher than contamination of fine concentrate.

Thus, the increase in size grade results in increase in its sapropelite content and ash. It is found that the higher sapropelite content in coarse and fine machine grades, the worse are performance of dense-medium separators and hydraulic jig washers, respectively.

## 12. Plant unsuitability

For example [9], supply of Western Donbas ROM coals containing readily-soaking rocks to the coal preparation plants of Central Donbas resulted in their performance deterioration. It is particularly typical for plants with process chart designed for concentration of fine slurries and circulating water regeneration by flotation method. Table 5 shows performance indicators of foam separation areas at those plants before and after introduction of the Pavlograd coals into their raw material resources. Increased percent of Pavlograd coals in the plant charge results in the reduced ash content of flotation rejects; for example, during the period under consideration, flotation was stopped in 1997 at four plants of total 12 ones due to the high ash content of flotation rejects (over 70%) obtained at the CCM "Bryankovskaya", "Kondratievskaya", and "Mikhailovskaya" with percent of Pavlograd coal in their charges ranging from 12.8% to 58.7% at the low plant utilization rates making 115, 179, and 664 thousand ton per year, respectively.

**Table 5**

| CCM | 1975 | 1980 | 1985 | 1990 | 1995 | 1997 |
|---|---|---|---|---|---|---|
| "Kurakhovskaya" | 78.8/0 | 72.4/0 | 69.8/31.0 | 69.7/8.3 | Shut-down | |
| "Selidovskaya" | 70.0/0 | 68.5/0 | 69.1/39.8 | 68.3/40.4 | 71.0/0 | Shut-down |
| "Pavlogradskaya" | 59.0/100 | 62.3/100 | Shut-down | | | |
| "Krasnolimanskaya" | 69.2/0 | 68.8/0 | 63.3/11.1 | 70.0/0 | 68.6/0 | 68.9/0 |
| "Komsomolskaya" | 75.2/0 | 76.8/0 | 81.1/0 | 79.0/0.3 | 71.6/0 | 71.3/0.2 |
| "Mikhailovskaya" | 70.6/0 | 72.5/0 | 74.8/20.8 | 71.8/8.1 | 70.0/32.3 | 70.4/58.7 |
| "Belorechenskaya" | 71.1/0 | 71.0/0 | 72.0/1.6 | 68.7/27.4 | 70.8/57.5 | 69.5/41.7 |
| "Luhanskaya" | – | 72.0/0 | 68.2/36.7 | 61.1/36.9 | Shut-down | |
| "Bryankovskaya" | 75.4/0 | 75.5/0 | 76.8/0 | 74.2/0 | 67.4/0 | 70.1/42.6 |
| "Krivorozhskaya" | 74.0/0 | 74.8/0 | 73.2/0 | 72.5/0 | 72.7/0 | 71.5/1.2 |
| "Dobropolskaya" | 68.4/0 | 68.2/0 | 70.8/0 | 67.1/0 | 59.9/0 | 63.9/13.8 |
| "Kondratievskaya" | 77.9/0 | 76.1/0 | 76.1/0 | 75.0 | 72.3/0 | 72.0/12.8 |

*Note:* Numerator – reject ash content, %; denominator – percent of Pavlograd coals in the raw material resource base, %

Degradation of performance indicators of foam separation areas at those CPPs in Donetsk and Luhansk regions where charge includes Western Donbas coals is primarily caused by unsuitability of process charts of these plants for foam separation of such coals. When the Pavlograd coal content in the raw material resource base exceeds 10–15%, deterioration of qualitative indicators of foam separation is observed.

In order to improve the Central Donbas coal preparation plants performance, it is reasonable to exclude the GKhK "Pavlogradugol" coals from the raw material resource base. It is recommended to centralize their processing at three or four plants, of which process charts should be adapted for the respective technology.

The use of deslimed flotation is the solution of circulating water regeneration problem at the plants concentrating the Pavlograd coals. Feed is deslimed by means of two-stage classification process in hydrocyclones, for example, of 630 and 350 mm in diameter.

## 13. Thermal drying of coal products

Limited application or complete exclusion of drying units causes problems with moister content in finished commercial coal products, since at high (30% and over) content of 0–1 mm grade in a concentrate, the mechanical dewatering methods not always meet the imposed requirements.

In 2012, dried coal moister content made 9.9% as compared with 8.7% in 2010. In addition, considerably higher moister content of dryer cake is observed at some plants: MPM "Luhanskaya" – 11%, CCM "Pavlogradskaya" – 12.7%, CCM "Samsonovskaya" – 11.3%, ChAO "Krivorozhskoye" – 10.9%, UP CCM "Chumakovskaya" LLC – 14.5%. This worsen the adverse trend of ballast moisture content increase in commercial concentrates, particularly, without other measures for coal products moisture content reduction (introduction of efficient centrifuges, filtering presses, etc.), and at limited amounts of financing of works on modernization of drying units being operated on the shutting-down threshold.

Directions of minimization of coal concentration technological characteristics impact must be realized by implementation of technological and engineering solutions and organizational decisions.

The technological solutions shall include as follows:
- conversion to concentration with 5 machine grades;
- definition of mine groups with the compatible grade, grain-size, and fraction composition;
- establishment of concentration sequence for coals of defined mine groups;
- application of deslimed flotation process;

- coal preparation plant water–slime circuit closing by means of recessed-plate filtering presses.

The engineering solutions shall include as follows:

- application of state-of-the-art technologies and equipment for machine grades preparation and concentration, and dewatering of concentration products;
- timely maintenance servicing of equipment;
- observance of the plant process operating procedure;
- application of automation and computerization of production processes and all technologies in general;
- application of state-of-the-art analyzers providing direct real time in-flow measurements of required qualitative and quantitative parameters of process equipment products;
- annual integrated testing of plant process chart with the calculation (based on testing results) of its qualitative and quantitative flowsheet and water–slime circuit with location of process bottlenecks and development of measures on elimination thereof.

The organizational decisions shall include as follows:

- not delivery for concentration of those coals, for which this plant is not designed;
- extension of washability categories number from four to eight ones;
- development of standard procedure and determination of slurry formation coefficient for process charts of coal preparation plants;
- development of instruction on calculation of standard and above-standard losses of combustible matter during the coal concentration at coal preparation plants;
- development of new corporate standard (SOU) in replacement of SOU 10.1.00185755.002-2004 taking into account the new coal washability categories, slurry formation coefficient of process charts of coal preparation plants and new cross-contamination standards at gravity processes, in particular: heavy-media, jigging, pneumatic separation, steeply inclined separators (SIS), concentration tables, cone and screw separators, concentrating hydrocyclones, and hydrosizers;
- development of process operating procedures for coal preparation plants and machines in accordance with СОУ 10.1.00185755-004:2006 (currently, there are process operating procedures at 33 plants);

- publication of the specialized scientific-and-technical and educational literature on coal preparation;
- establishment of regular refresher training courses for engineering and technical personnel;
- keeping the scientific and technical archives of design, development and research institutes and organizations;
- branching of process charts of coal preparation plants in order to provide flexibility in response on the changes in ROM coal and commercial coal products quality.

Presently, the requirement for branching of the process charts of coal preparation plants is determined by following objectives:

- solution of certain process task on the improvement of qualitative indicators of some preparatory, main or auxiliary process product;
- need for commercial product output increase by means of reduction of combustible matter losses with coal preparation rejects;
- commercial product line expansion determined by the market conditions or consumer's requirements.

When the first problem of process charts branching is solved by introduction of one or two process operations into the plant process chart, then the solution of second and third problems requires the introduction of production processes or technological complexes, and most significantly selection of additional machine grades from ROM coal or separation products and bringing their grain sizes to correspondence with the duty parameters of the used or additionally installed equipment in order to provide its optimum separation characteristics.

The process chart branching requires the additional investment of financial, material, power and human resources, and increase in operational costs. Experience shows that the introduction of additional production process could cost about 10 million UAH on the average, and introduction of technological complex could cost 15–30 million UAH depending on its purpose and preparedness of site or building and structures for its installation. In Ukraine, there are all necessary scientific-and-engineering and production resources for the solution of these problems.

## 14. Conclusion

In Ukraine, there is sufficient scientific, design, machine-building, construction, and production capabilities for solution of any scientific and production problems on concentration of ROM coals and carboniferous materials on a level with the highest world standards.

## 15. References

[1] Polulyakh A.D. Analysis of coal preparation plants operation in Ukraine for 50 years, Polulyakh A.D. Transactions. GP UkrNIIugleobogashchenie. Donetsk: Skhidnyi Vydavnychyi Dim, 2013. P. 2. Coal concentration and processing. P. 472–486.

[2] Technical and economic analysis of coal preparation plants operation in Ukraine for 2012. Luhansk: UkrNIIugleobogashchenie, 2013. 114 p.

[3] COУ 10.1.00185755.002-2004 "Coal concentration products. Quality indicators calculation procedure". Kiev: Mintopenergo, 2004. 47 p.

[4] Polulyakh A.D. On the ROM coal washability categories, Polulyakh A.D., Polulyakh D.A. Mineral processing: Collection of Science and Technology Articles. 2011. No. 44(85). P. 13–16.

[5] Polulyakh D.A. On the effect of ROM coal machine grades preparation performance on their concentration performance, Polulyakh D.A. Vibrations in the engineering and technologies. 2019.

[6] Zhuravel V.A. Particularities of the concentration of Western Donbas coals with high content of readily-soaking rocks, Zhuravel V.A., Kleshnin A.A., Polulyakh A.D. Coal of Ukraine. 1990. No. 11. P. 43–46.

[7] Polulyakh A.D. Particularities of the coal prescreening at the CCM "Pavlogradskaya", Polulyakh A.D., Kolesnik D.N. Coal of Ukraine. 1987. No. 10. P. 18–19.

[8] Polulyakh A.D. Particularities of the concentration of sapropelite coals from the Lviv-Volyn Basin, Polulyakh A.D., Avramenko O.J., Pererva A.Yu. Coal of Ukraine. 2008. No. 3. P. 13–15.

[9] Kurchenko I.P. On the concentration of the Western Donbas coals, Kurchenko I.P., Polulyakh A.D., Vasko I.P. Coal of Ukraine. 2000. No. 9. P. 50–52.

# 31

# Screens as devices for grain classification and dewatering

Piotr Pasiowiec[1], Barbara Tora[2], Jerzy Wajs[1], Klaudia Bańczyk[1]

*[1]Progress Eco Sp. z o.o. Sp. K, Tuczępy, Poland*
*[2]AGH University of Science and Technology, Kraków, Poland*

**Abstract:** The aim of the article is to present a diverse spectrum of application of welded wedge wire slotted screens with an emphasis on new solutions in terms of their design and the modern ways of their installation. In the article we present welded wedge wire screens as key elements of machines for classification and filtration processes of coal enrichment, the extraction and refining of oil and uranium. We want to show screens, as an element that is subject to continuous development process, in order to ensure maximum effectiveness of processing devices, simple and easy maintenance, trouble-free operation and economic efficiency.

Wedge wire screens are the universal product that has found wide scope of application in many industries: from coal and uranium enrichment, through the mining processes and processing of petroleum and natural gas (shale gas), food industry, sugar industry, beverages, paper, environmental protection, chemical to the elements used in construction industry and architecture.

**Keywords:** Screens, separation, classification, dewatering

## 1. The use of industrial wedge wire screens: Introduction

Industrial screens and products base in them have very wide scope of application in many industrial processes and many industries. The changing needs of processing plants always have shaped and continue to shape the development of this range of products. Industrial screens can work in very different conditions, as well as systems, both dynamic and static. Very often, are the key elements of the treatment processes such as: preliminary classification, final classification, sieving, separation, filtration of liquids and gases (Pasiowiec et al., 2017a,b, 2018).

Experience and knowledge of these processes allows to understand how important it is to maintain their continuity and what loss carries the installing of low quality screens or incorrect installation. The quality understood in such way is today the basic criterion of supplier choice. This applies to any industry in which screens are used. In addition to the mining industry it is worth mentioning the extensive use in screening aggregates i.e.: limestone, sandstone, dolomite, porphyry, quartzite, granites, melaphyre, granodiorite and basalts. In addition, versatile use in the steel industry where m.in are used in the recycling process of metallurgical waste. Among the screens buyers you can list the companies of such industries as: coke engineering, mechanical engineering, power engineering, construction and insulation, petrochemical and refining, food industry, including sugar, beverages, milk, fruit and vegetables, the chemical industry and wood and paper (Bańczyk et al., 2017). What it's worth to emphasise practical knowledge of advanced stainless steels processing techniques, high aesthetics and development of products have effected their use in architecture and construction.

## 2. Characteristics of industrial screens

First type of screen used in the industry was perforated sieves and simple woven screens. The growing requirements of the clients of processing companies in the field of technical parameters of final products and the development of technology in industrial plants led to the dynamic development of the industrial screens.

The main types of industrial sieves include:
- Steel screens: woven screens, fine wire screens, harp screens, piano screens, flat top screens (TL), welded screens, wedge wire screens, non-mesh screens and finger screens
- Plastic screens: polyurethane tensioned screens, polyurethane modular screens, rubber tensioned screens and rubber modular screens

## 3. Characteristics of wedge wire slotted screens by the case of Pro-SLOT® produced by Progress Eco

Slotted screen is the area with regularly spaced gaps (referred to as mesh or slot), of equal dimension and shape, used in classification and dewatering machinery. The sieve is the most important working part of the screening device, on which the separation process takes place. Therefore, correct design of parameters of the sieve, for specific technological conditions, depends on the effectiveness of the ongoing process (classification and filtration) (Tora et al., 2003).

Effectiveness (accuracy) of screening depends on the shape and layout of the holes of the sieves, their dimensions and their number on the specified area. The transition of screened medium largely depends on the shape of the holes and geometric forms of the lateral sides of these holes. The type of material, of a sieve, affects the durability of the geometric forms of the holes, the smoothness of the screen surface, the edges of the holes or slots. In the industry of mineral tailings industrial screens are present in each process, in which the classification is needed of product or waste of geometric characteristics, specific size of hole, from feed on the sorting device.

The precision and reproducibility of the slots with a minimum value of 30 microns (profile Sb6, Sb8, Sb10), along with the use of the profile wires of 2 microns tolerance, cause the development of many technological processes without the use of slotted screens would be impossible.

## 4. Technical characteristics of wedge wire slotted screens

Welded wedge wire screens are produced on the basis of the method of electrofusion welding, which involves welding profiled working wire to the support wires. The result of this is strong screens able to carry heavy loads.

Welded wedge wire screens are characterised by the ability to carry heavy loads, a large open area ratio, low susceptibility to clogging, perfectly flat and smooth surface, high precision, increased efficiency and accuracy of separation and dewatering (Fig. 1).

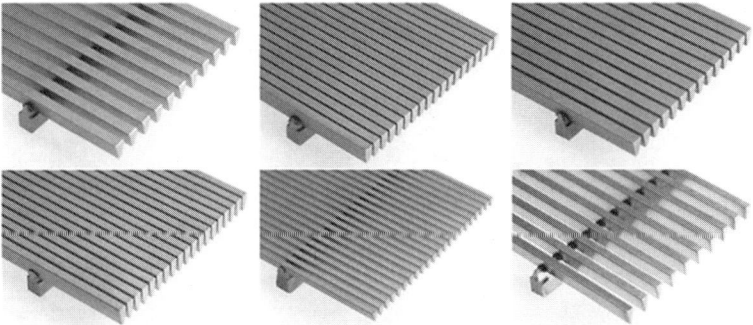

**Figure 1:** Progress Eco Pro-SLOT® wedge wire screens with different profiles and slots (www.progresseco.pl)

Profile wires are divided into two main groups. First is the working profile that creates the filtration surface. Those are made of wedge profile wires type Sb, Sbb and special profiles. Second is the support wires, made of profile wire type Q, Sb and special profiles, as shown in Fig. 2. The range of working profiles and support profiles are presented in Tables 1–3. The most commonly used material for manufacturing welded wedge wire screens is stainless steel and acid-resistant steel (chrome or chrome-nickel still with molybdenum, titanium and manganese) range of slots is in the range of 0.02–20.0 mm.

## 5. Types of working wires and support wires

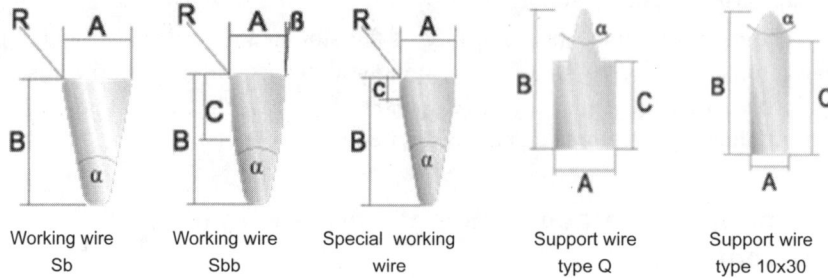

Figure 2: Types of working and support wires

Wedge wire slotted screens can be identified as: flat wedge wire screens and wedge wire tubes/cylinders. Flat wedge wire screens are produces by welding profiled working wire to support wire at 90°. Details of the structure of the screens are shown in the following diagram (Fig. 3).

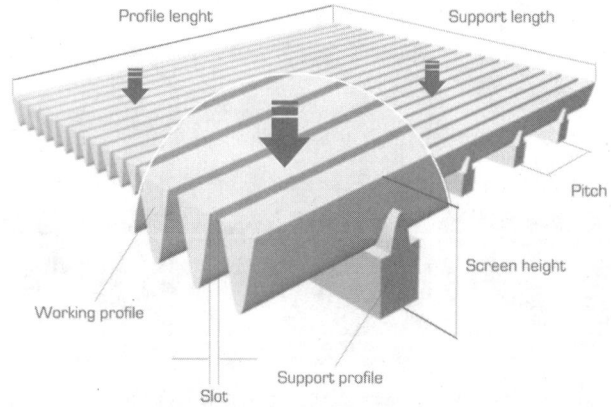

Figure 3: The structure of the screens

Wedge wire screens welding technology allows to use various types of working wires, achieving various slots sizes in one screen, the use of various supporting wires and various distance between the support wires in one screen, high flatness of the screen (no waviness of working wires between support wires) and accurate quality control of welds.

## 6. Wedge wire tubes/cylinders

Wedge wire tubes are formed by simultaneous winding spiral and welding to the support wires placed along a pipe. This technology allows to produce the wedge wire screens for applications where very high accuracy of the slot and high durability of the screen is required. Thanks to the modern technology of welding, you can get any required distance between the working wires and very accurate and repeatable slot. They are applied mainly in the processes of filtration, filter elements in the food industry, environmental protection and others. In an environmental protection wedge wire screens are used in sewage treatment plants for municipal and industrial waste water treatment. Wedge wire tubes are manufactured in four types, shown in Fig. 4: circumferential slot, flow from inside, circumferential slot, flow from outside, slot parallel to the $z$-axis, flow from the outside, slot parallel to the $z$-axis, flow from the inside.

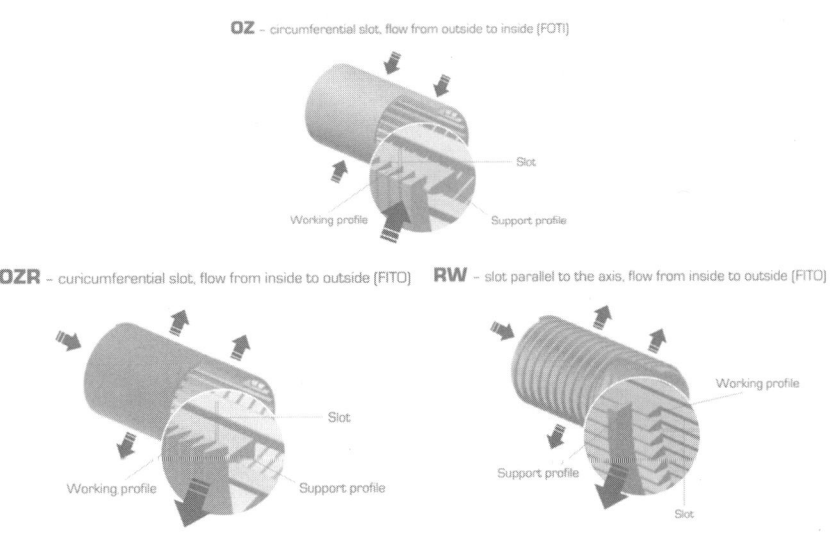

**Figure 4:** The use of wedge wire screens in equipment and products for coal enrichment

## 7. Screen frames for vibrating screening machines

Flat wedge wire screens designed for the vibrating screening machines are produced in segments (modules), allowing for a quick replacement. Depending on the design they can be reinforced in the flat, polyurethane frame or frame with specially contoured shape. Examples of flat wedge wire screen frames are shown in Fig. 5.

**Figure 5:** Examples of Progress Eco Pro-SLOT® wedge wire frames
(www.progresseco.pl)

A modern and very functional design of wedge wire screens is fitting it in a polyurethane frame with the wedge fastening system. Advantages of the wedge: lightweight, quick and easy assembly and disassembly, no joints. Figure 6 shows the idea of building the deck of welded wedge wire screens in a polyurethane frame with the wedge fastening system.

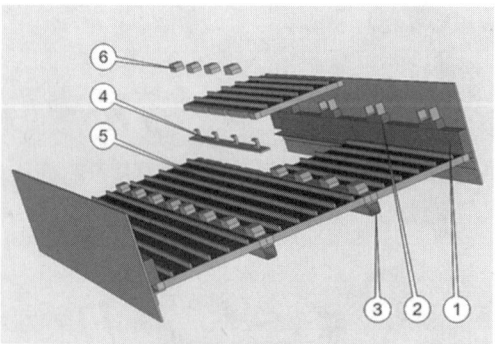

1. Side scuff plate
2. The key
3. Steel profile
4. The adapter wedge system
5. Segment sieves slot
6. Wedge fitting

**Figure 6:** The idea of building the deck of welded wedge wire screens in a polyurethane frame with the wedge fastening system

## 8. Screen decks for jigs

Pulsator water jigs are used mainly for raw coal enrichment and abstraction of gravel and sand, with simultaneous release of organic pollutants and minerals on the basis of gravity separation of coal on the fractions of a specific density. Pulsing water movement is caused by compressed air, repeatedly metered by marshalling yards to the water–air chamber, located under the screening deck. The water for the enrichment process and transportation of coal layer, is brought to each section of the jig in quantities suited to the size and composition of the feed material. Depending on the quality of raw coal and its densymetric composition and desired parameters of enrichment products is used double product enrichment or triple product enrichment.

Flat wedge wire screens with profile wires welded in angle β to the support wire, work as a barrier at jig decks. The flow rate depends on the size of the slot. Adjustable angle β is in terms of β = 0–12° (standard β = 10°). The scheme of such screens is illustrated in Fig. 7.

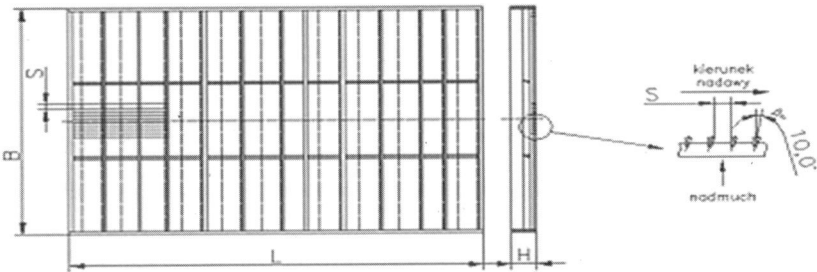

**Figure 7:** Wedge wire screens with profile wires welded in angle β

## 9. Wedge wire inserts for static/parabolic screens

Wedge wire inserts for static screens are mainly used for dewatering particulate matter from liquids. The principle of operation of static screen shown in Fig. 8 comes down to separation of the solids from the liquid on the wedge wire screen (Hycnar et al., 2015).

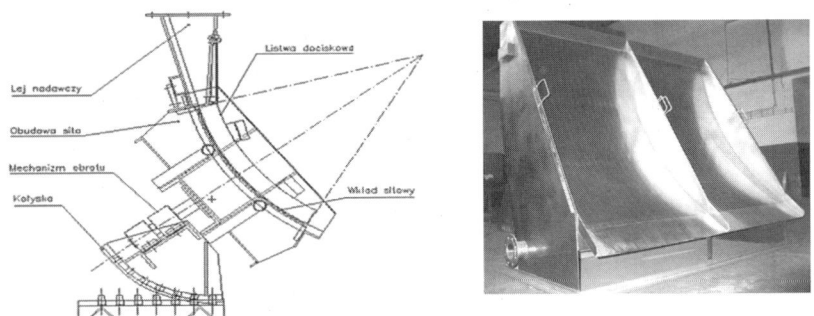

**Figure 8:** Progress Eco wedge wire screen (www.progresseco.pl)

In the process of mechanical processing of coal wedge wire static screens are used for fine coal sedimentation and silt control before flotation. The quality of separation depends on the parameters of the suspended matter and the type of the sieve. The use of wedge-shape profile wire and wedge-shaped slot, allows to avoid clogging of slots and self-cleaning of the screen. Wedge wire screen allows to isolate the elements that are smaller than the size of the slot i.e. their dimension is ~60% of the size of the slot. The required performance can be achieved by appropriate size of the screen or number of screening machines. Static wedge wire screen is a gravity type of device not using electricity. For intensive processes with "difficult" suspensions, in which clogging of the slots can appear you can apply a spray system for washing and cleaning the working surface of the screen. Type of the selected profile wire and the slot determines the open area of the screen and its performance. The main advantages of static wedge wire screens are: high efficiency of dewatering, low noise and vibration, high reliability and low operating costs.

## 10. Gutter screens and chutes

Gutter inserts are made of wedge wire screens with the slot parallel to the axis of the flow of the filtered medium from the inside (RW) and outside (RZ). They can also have the reinforcing structure. The screen is produced in the technology of welding wedge profiled wire type Sb to the support wire. Gutter screens are mostly placed at the bottom of tanks, chutes and conveyors, where in addition to transport, it is a need for further dewatering or separation. Figure 9 shows the schema and examples of wedge wire screens of type RW reinforced.

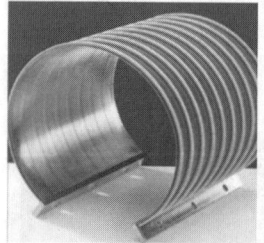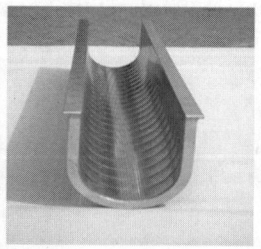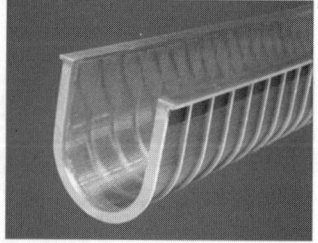

**Figure 9:** Progress Eco Pro-SLOT® gutter screens type RW reinforced (www.progresseco.pl)

## 11. Inserts and baskets for centrifuges

In dewatering centrifuges inserts in the shape of cones or cylinders are used. They work in the dynamic systems and are used for the final dewatering of the feed material. They can be made with a supporting structure, integral and reinforced by fins, rings, flanges, forming an integral element of the screen, or without supporting structure, as screening inserts. In this situation the only part that may be qualified for replacement is the wedge wire screening basket (Tora et al., 2003). Examples of the types of wedge wire baskets for centrifuges are illustrated in Fig. 10.

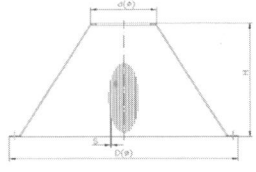  

**Figure 10:** Examples of Progress Eco Pro-SLOT® centrifuges wedge wire baskets (www.progresseco.pl)

Due to the strength in the centrifuges with a larger value of rotation, the wedge wire baskets with the working surface covered with an additional layer of abrasion-resistant up to a thickness of 200 µm (chromium, tungsten carbide) are used.

## 12. Centrifugal dewatering screen OSO

Centrifugal dewatering screens shown in Fig. 11 are static processing devices used for sedimentation and dehydration of fine coal. The devices work without the use of electricity, do not require constant maintenance and do not produce noise (Pasiowiec, 2008). The basic elements of the OSO wedge wire screen are: power nozzle, the steering wheel cover, wedge wire screen insert, steering wheel, basket cumulative container with solids removal solution.

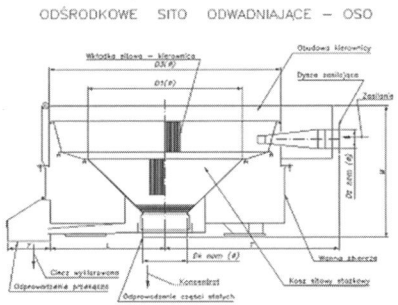 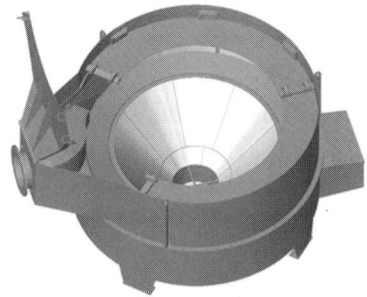

**Figure 11:** Progress Eco Pro-SLOT® centrifugal dewatering screen (OSO) (www.progresseco.pl)

In recent years, wedge wire slotted screens OSO are subject to significant constructional changes and are built as sectional elements screwed or fastened by studded items, wedge wire–polyurethane elements. The solution shown in Fig. 12 allows you to quickly replace the used segment and not the entire screen.

**Figure 12:** Progress Eco modular inserts for centrifugal dewatering screens (www.progresseco.pl)

## 13. Conclusion

The need to constantly raise the technological effectiveness of wedge wire slotted screens, the quality of execution and possibility of reducing operating costs, forces the continuous development of their design solutions. The work constructional and technological works on the selection of materials for the production of industrial screens and their design are also on the search for optimal solutions to ensure their functionality, easy and efficient method of assembly and failure-free operation. The leader on the Polish market, and in recent years also the European, in the development of welded wedge wire slotted screens is Progress Eco.

The demand for electric energy and energetic resources will grow. It is worth once again emphasise that it would be impossible to process and achieve the quality parameters of energy coal, oil, natural gas and uranium without the use of welded wedge wire screens. Number and variety of industrial applications of welded slotted screens make it a versatile product that meets the highest requirements of the most modern industrial materials subject to the ongoing development and improving their design. Slotted wedge wire screens play a key role in ensuring the highest quality of enriched raw materials.

# 14. References

[1] Bańczyk K., Wajs J., Pasiowiec P., Orlik R., Tora B. 2017, Charakterystyka przesiewaczy produkcji Progress Eco zastosowanych do klasyfikacji i odwadniania odpadów górniczych w Zakładzie Przeróbki Hermanicka Hałda w Ostrawie. Konferencja Energetyczna, Zakopane 2017 r.

[2] Hycnar J.J., Pasiowiec P., Bańczyk K., Wajs J., Tora B. 2015, Zwiększenie skuteczności odwadniania i klasyfikacji zawiesiny wody odciekowej w instalacjach odwadniania żużla przy zastosowaniu sit OSO, XXIX konferencja z cyklu: Zagadnienia surowców energetycznych i energii w gospodarce krajowej pt. Paliwa dla energetyki – mix energetyczny. Zakopane.

[3] Jonczak P., Pasiowiec P., Śmiejek Z. 2004, Technologiczne i ekonomiczne racje istnienia nowych rozwiązań w obszarze stosowania sit produkcji Progress Eco S.A.; Nowoczesne systemy przeróbcze surowców mineralnych z uwzględnieniem problemów ochrony środowiska; KOMEKO 2004 Ustroń.

[4] Materiały reklamowe, prace badawcze i dokumentacje firmy Progress Eco S.A.; http://progresseco.pl/ (available: January 19, 2019).

[5] Pasiowiec P. 2008, Analysis of Work and Optimization of Centrifugal Dewatering Sieve, Doctoral dissertation, Ostrava 2008.

[6] Pasiowiec P., Jerzy Wajs J., Bańczyk K., Borkowski W., Bogusław A., Tora B. 2015a, Rozbudowa układu klasyfikacji i odwadniania w Zakładzie Przeróbczym PG Silesia na bazie przesiewaczy wibracyjnych produkcji Progress Eco. KOMEKO.

[7] Pasiowiec P., Wajs J., Bańczyk K., Borkowski W., Bogusław A., Tora B. 2015b Rozbudowa układu klasyfikacji i odwadniania w Zakładzie Przeróbczym PG Silesia na bazie przesiewaczy wibracyjnych produkcji Progress Eco; Innowacyjne i przyjazne dla środowiska techniki i technologie przeróbki surowców mineralnych: bezpieczeństwo – jakość – efektywność. Monografia, KOMAG 2015.

[8] Pasiowiec P., Bańczyk K., Wajs J., Gawlista S., Tora B., Burek A. 2015c, Comparative analysis of dewatering efficiency and distribution of materials in centrifugal dewatering sieve with steel and polyurethane insert; 19th Conference on Environment and Mineral Processing, VŠB, TU Ostrava.

[9] Pasiowiec P., Bańczyk K., Tora B., Brożyna J., Wajs J. 2017a Uniwersalne zastosowanie sit szczelinowych zgrzewanych w procesach wydobycia i przeróbki węgla kamiennego, ropy naftowej, gazu ziemnego oraz uranu, Konferencja Energetyczna, Zakopane.

[10] Pasiowiec P., Tora B., Wajs J., Bańczyk K., Strączyński L., 2017b. Screening as a Key Element of Construction of Machinery Used in Processing Processes for Obtaining Optimum Product Parameters, InżynieriaMineralna – Journal of the Polish Mineral Engineering Society, No 2(40), pp. 61–72. DOI: 10.29227/IM-2017-02-07.

[11] Pasiowiec P., Tora B., Wajs J., Bańczyk K., Strączyński L. 2018, Achievements and Participation in the Development of Mineral Processing and Waste Utilization of the Institute of Non-Ferrous Metals in Gliwice, InżynieriaMineralna – Journal of the Polish Mineral Engineering Society, No 1(41) p. 13–34, DOI: 10.29227/IM-2018-01-02.

[12] Tora B., Pasiowiec P., Śmiejek Z. 2003, The possibilities of using the centrifugal dewatering sieve in the system of classification; 7th Conference on Environment and Mineral Processing, VŠB, TU Ostrava, 2003.

[13] Tora B., Borkowski W., Bogusław A., Pasiowiec P., Wajs J., Bańczyk K. 2014, Procesy klasyfikacji i odwadniania węgla na przesiewaczach wibracyjnych produkcji PROGRESS ECO w zmodernizowanym Zakładzie Przeróbki Mechanicznej Węgla PG Silesia. Materiały XXVIII Konferencji z cyklu: Zagadnienia surowców energetycznych i energii w gospodarce krajowej. Zakopane, 2014 r. Wyd. Instytutu GSMiE PAN, Kraków, s. 81–90.

# 32

# Dry vibration screening of coal and semicoke

Leonid Vaisberg, Victoria Lazareva, Andrei Gerasimov

*REC "Mechanobr-Tekhnika", St. Petersburg, Russia*

**Abstract:** Comparative studies of the physical and mechanical, and surface properties of Kuzbass coal before and after heat treatment have been carried out. Thermal modification of coal without access to the air increases its porosity and at the same time changes the ratio of amorphous and crystalline phases composing coal, which creates favourable conditions for further dry processing.

The influence of thermal modification of coal on the process of dry vibrating screening is considered. The efficiency of the semicoke screening, regardless of the type and frequency of vibrations, is 10–16% higher than the screening efficiency of unmodified coal.

**Keywords:** Dry vibratory screening, coal, semicoke, relief structure, rheological properties

## 1. Introduction

At present, all worldwide industrial technologies for the beneficiation of high-ash coal raw materials are based on water processes, such as hydraulic size classification, gravity separation and flotation. The use of "wet" processes generates a number of severe environmental and economic consequences during their processing, transportation, as well as during the waste storage (Arsentyev et al., 2016a). Creation of low-water technologies for beneficiation of solid minerals is the priority development direction of the mining and processing industry, ensuring the raw materials and energy resources saving. Heat treatment of coal without air access allows to significantly change its physical and mechanical properties, in particular, to reduce by 20–30% the specific electricity consumption for their crushing, as well as to increase calorific capacity. A rational approach to the design of dry coal preparation schemes is the inversion of the sequence of basic technological operations – thermal preparation of coal is carried out in the head of the process, before classification and after dry disintegration of coal, which converts the clay component to a state suitable for electromagnetic and electrostatic separation (Arsentyev et al., 2016b, 2017; Syroezhko et al., 2015; Gerasimov et al., 2016). Implementation of these transformations opens up new technological possibilities in the further processing of the heat-treated product: reduction of energy consumption for crushing and grinding due to a decrease in mechanical stability; improving the efficiency of screening due to changes in the rheological properties, the possibility of deep removal of the ash fraction using only dry beneficiation methods; the possibility of production of processed semicoke from the processed carbon fraction; the possibility of using burnt mineral fraction for the production of special binders, additives to concrete, building products.

Thermal modification of coal increases its porosity and simultaneously changes the ratio of amorphous and crystalline phases composing coal, which creates favourable conditions for further dry beneficiation. In this case, the technologically preferred range of heat treatment are temperatures of 500–550 °C (Gerasimov et al., 2016).

Vibration classification plays a significant role and at the same time is a problematic operation with a relatively low screening efficiency in a chain of technological operations as part of dry beneficiation technologies for unmodified coal. This is due to the need to classify coal in a thick layer due to the low bulk density of the raw materials, in contrast to the screening of ore material, which proceeds on the majority of the sieve surface in a monolayer.

The efficiency of the vibratory screening of bulk materials is directly associated with both bulk physical and mechanical properties of raw materials (bulk and specific density, coefficient of internal friction), and with the surface properties of mineral particles and sieve, implemented through friction coefficient of a sliding on the sieve (Blehman et al., 2003; Vaisberg et al., 2004).

## 2. Materials and methods

Within the framework of this paper, the influences of the physical and mechanical properties of coal from the Kuznetsk deposit before thermal treatment, and thermally modified coal on the process of dry vibration screening were studied. The initial sample of coal in petrographic terms was represented mainly by vitrine and fusain. The mineral part is represented by aluminium silicates with a small amount of pyrite, and makes 15%.

To understand the regularities of the sieve vibration classification of coal, its physical properties essential for the screening process were studied (Table 1). The angle of internal friction of the bulk material was calculated through the angle of natural slip, measured in a transparent swivel cuvette. The conditional coefficient of sliding friction was determined for the movement of isometric particles of coal on a steel sieve with square cells (Baldaeva et al., 2017).

**Table 1:** The effect of coal processing temperature on its physical properties

| Technical characteristics | Source coal | Semi-coke – 500 | Semi-coke – 550 |
|---|---|---|---|
| Apparent density, 103 kg/m$^3$ | 1.18 | 0.95 | 0.99 |
| Conditional coefficient of sliding friction | 0.38 | 0.34 | 0.32 |
| Angle of internal friction, degrees | 44 | 42 | 42 |
| Internal friction coefficient | 0.97 | 0.90 | 0.90 |
| Bulk density, 103 kg/m3 | 0.76 | 0.74 | 0.73 |

During the heat treatment of coal, corresponding to the mode of semicoking, physical changes in the coal mass occur, significantly affecting the possibilities of its further processing. In this regard, it represented interest to study the relief structure (roughness) of the coal surface and its change during heat treatment (Figs. 1 and 2). We believe that in relation to the processes of vibratory displacement and the classification of raw material having size from fractions of a millimetre to 10–20 mm, the most promising is the study of surface relief at micron and submicron levels.

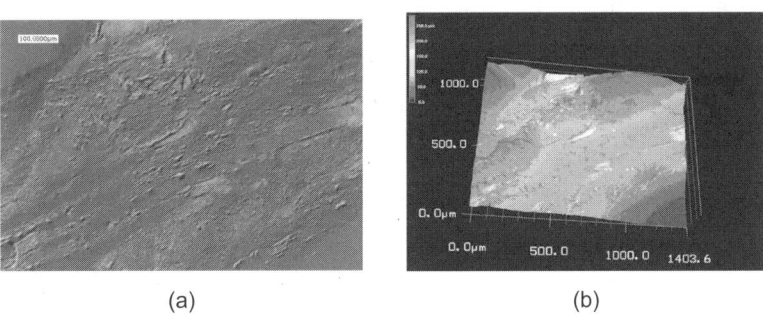

(a)                                    (b)

**Figure 1:** Laser imaging (a) and 3D model (b) of coal surface

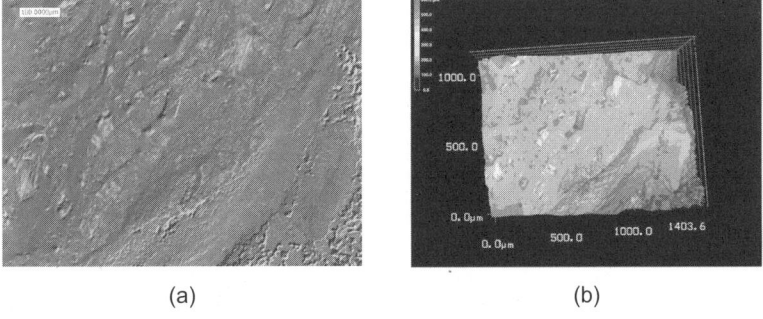

(a)                                    (b)

**Figure 2:** Laser imaging (a) and 3D model (b) of semicoke surface

Surface roughness was quantified using a laser scanning 3D microscope Keyence VK-x200. This installation allows you to obtain a protocol for measuring the surface roughness in many ways, which are used in international practice for assessing the properties of surfaces of various substances, mainly machined (Gladkova et al., 2018).

Visual assessment of the surface relief of laser imaging samples and their 3D models demonstrates an interesting fact – the surface roughness of the original coal is mainly associated with the peaked ledges of micron size, and the roughness of coal thermally modified at 500°C is largely related to the dimples. In addition, semicoke has a pronounced surface bituminization. Thermal effect on the sample of coal affects the structure of the pore space; semicoke porosity ratio increases by 7 times.

## 3. Results and discussions

Studies of the classification of coal crushed to −20 mm on a sieve with 1.6 mm cells were carried out using vibrating screens of semi-industrial type GIL-051 with orbital duct vibrations and GIL-051 with straight-line vibrations. The angle of inclination of the sieve GSL-051 made 17°, and of the self-synchronizing screen GSL-051 made 2° in the direction of unloading with the amplitude of vibrations of 4 mm and different frequencies of vibrations. The specific feed rate made 1 t/m²/h. The granulometric composition of the source and heat-treated coal was similar (Table 2).

**Table 2:** Sieve characteristic of samples of the original coal and semicoke

| Particle size (mm) | Class output (%) | |
|:---:|:---:|:---:|
| | Coal | Semicoke |
| −20+10 | 4.0 | 4.5 |
| −10+5.0 | 31.0 | 31.8 |
| −5.0+2.5 | 23.9 | 21.4 |
| −2.5+1.6 | 8.5 | 9.9 |
| −1.6+0.63 | 14.1 | 14.2 |
| −0.63+0.315 | 7.5 | 8.0 |
| −0.315 | 11.0 | 10.2 |
| Source | 100.0 | 100.0 |

The efficiency of screening of heat-treated coal, regardless of the type and frequency of vibration, is significantly higher than the screening efficiency of unmodified coal (Table 3).

**Table 3:** The comparative efficiency of the vibration classification of the source coal and semicoke in the −1.6 mm class

| Duct vibration frequency (s−1) | Efficiency of screening (%) | | | |
|---|---|---|---|---|
| | Source coal | | Semicoke | |
| | GIL-051 | GSL-051 | GIL-051 | GSL-051 |
| 13.3 | 77.5 | 74.1 | 86.3 | 85.5 |
| 16.7 | 71.3 | 69.0 | 82.9 | 82.0 |
| 20.0 | 71.2 | 71.4 | 81.9 | 81.6 |

At frequencies of vibrations lower than 13.3 $s^{-1}$, the transportation of coal across the sieve surface worsens, and at frequencies above 20 $s^{-1}$, the partial release of product through the edges of the screen occurs.

## 4. Conclusion

During heat treatment (low-temperature pyrolysis), high-ash coal changes its physical and mechanical properties. In coal under the influence of temperature in the process of semicoking, the initial fracture structure is transformed into cracked-porous structure with an increase in the number and size of pores. A sample of coal subjected to low-temperature pyrolysis has different rheological properties compared with the original coal. The efficiency of screening heat-treated coal, regardless of the type and frequency of vibration, is by 10–16% higher than the screening efficiency of unmodified coal.

Thus, heat treatment of coal makes it possible to implement energy-efficient low-water technologies for the beneficiationof high-ash coal, and, in particular, it provides an increase in the efficiency of dry vibration screening.

## 5. Acknowledgement

The study was supported by the grant of the Russian Science Foundation (project No. 17-79-30056).

## 6. References

[1] Arsentyev V. A., Vaisberg L. A., Ustinov I. D., Gerasimov A. M. Prospects of water use reduction in coal preparation. GornyiZhurnal. 2016a. No. 5. pp. 97–101.

[2] Arsentyev V. A., Vaisberg L. A., Ustinov I. D., Gerasimov A. M. Perspectives of Reduced Water Consumption in Coal. XVIII International coal preparation congress. 2016b, vol. 2, pp. 1075–1081.

[3] Arsentyev V. A., Gerasimov A. M., Dmitriev S. V., Mezenin A. O., Strakhov V. M. New approach to coal preparation of different metamorphism stage. Coke and chemistry. 2017. No. 12. pp. 466–469.

[4] Baldaeva T. M., Gladkova V. V., Otroshchenko A. A., Ustinov I. D. Mineral coal thermal modification effect upon its vibratory screening efficiency. ObogashchenieRud. 2017. No. 1. pp. 3–7.

[5] Blehman E. E., Vaisberg L. A, Lavrov B. P., Vasilkov V. B., Yakimova K. S. Universal Vibrating Stand: Experience of Use in Researches, Some Results. Nauchno-technicheskievedomosti SPBGTU. 2003. No. 3 (33). pp. 224–227.

[6] Gerasimov A. M., Abrosimov A. A., Pimenov Yu. G., Strakhov V. M. Changes material composition and porous structure coal on thermal preparation. Coke and Chemistry. 2016. No. 6. pp. 207–212.

[7] Gladkova V. V., Kazakov S. V., Karapetyan K. G., Otroshchenko A. A. Vibratory treatment of a particularly brittle mineral material. ObogashchenieRud, 2018, No. 2, pp. 8–12.

[8] Syroezhko A. M., Gerasimov A. M., Abrosimov A. A. Thermochemical coal treatment prior to dry beneficiation. ObogashchenieRud. 2015. No. 6. pp. 9–13.

[9] Vaisberg L.A., Korovnikov A. N. Fine screening as an alternative to hydraulic size classification. ObogashchenieRud. 2004. No. 3. pp. 23–34.

# 33

# The kinetic analysis of the key components of the centralized-driven flip-flow screen

Chusheng Liu, Jida Wu, Xiaodong Yang, Hongxi Li, Xiaozhou Liu

*China University of Mining and Technology, Xuzhou, China*

**Abstract:** Hole-plugging is a serious problem for the most traditional screening equipments sieving moist and fine coal materials, which results in the low preparation efficiency. Flip-flow screen is kind of screening equipment by using flip-flow motion of the elastic screen surface to implement the separation of materials, with remarkable advantages of the extraordinary vibration intensity of the screen surface, the holes difficult to be plugged, and high screening efficiency for moist and fine materials. A kind of centralized forced flip-flow screen (CFFS) was proposed based on the crank-rocker mechanism. The advantages of CFFS include considerable deformation of the screen surface, stable flip-flow quantity, low working noise, low vibration influence to environment, etc. The principle and construction of CFFS was introduced, and the modal analysis and harmonic response analysis of the key component (the crank and the linkage) was implemented based on finite element method (FEM), respectively. The results illustrate that the resonance frequency is far more than the working frequency, and the stress and strain are all within the safe limit of the material. The prototype was manufactured, the sieving experiment demonstrate that the CFFS perform steadily with screening efficiency of over 80%.

**Keywords:** Flip-flow screen, centralized forced, modal analysis, sieving experiment, water content, fine particles

## 1. Introduction

Coal is the main energy source in China, and has accounted for about 70% in primary energy consumption. The mechanized coal mining technique is commonly used in current. Which results in a high degree of fragmentation of coal seam. And the influence of bunker underground and dropping belt also lead to the high degree of coal fragmentation and the low lump coal rate of coal out of wells. The use of technical measures of hydraulic mining and underground sprinkler dust, lead to an increase in raw coal fine grade coal rock composition and moisture content of the waste rock mud, coal bonded into a group. During dry sieving of the moist fine coal, the problems of sieve paste hole and plugging holes occur seriously with the reduction of screening efficiency. Which also results in that large mount of coal cannot be sorted, the commercial coal moisture increase, and the heat reduces. Consequently the quality of coal is difficult to guarantee. The burden on the system of slime water is increasing, and numerous slimes can be generated. To solve all these problems, as the new type of screen, flip-flow screen is applied to some coal preparation plants in recent years.

The flip-flow screen was originated in the 1960s. It was a screen of throwing material by flipping and flowing of screen surfaces. The flip-flow screens applied in industry currently mainly includes, the second generation Torwell flip-flow screen, the third generation Liwell screen of Germany, ГЭДП screen of Soviet Union, and BiviTEC-CRL screen of Austria. The main feature in common is that the mesh of flip-flow screen is made of scalable urethane rubber. The screen meshes flip and flow alternately in screening process. So that the material on the screen surface could entrain forward motion. The acceleration that screen surface generated can be greater than the cohesive force. That material adhered to the screen could be avoided. The large deflection of screen surface resulted in the formation of "breathing effect". That material clogged the screen could be prevented. The flip-flow screen hold advantages with large capacity of the unit area, high efficiency, and plug prevention. But the disadvantages also remain, such as high energy consumption, high noise, large dynamic load, and screen surface vulnerability.

## 2. The main structure of the CFFS

In the purpose solving the problems mentioned above, China University of Mining and Technology developed centralized forced flip-flow screen (CFFS) with simple structure, smooth movement, low energy cost, and long working life. The main structure of which is shown in Fig. 1.

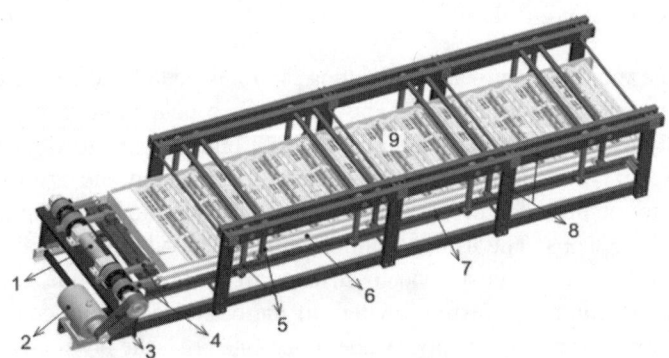

**Figure 1:** The main structure of CFFS. 1, Concentrated drive device; 2, motor; 3, V belt; 4, connecting rod; 5, hanging rod; 6, the in screen box; 7, the out screen box; 8, frame; 9, screen frame

The CFFS consists of motor, crankshaft, connecting rod, suspension mechanism, the screen frame, screen support structure, screen, and base. The screen frame is connected by suspension mechanism with the screen support structure. The energy generated by the motor is transmitted through the belt to the crankshaft. The crankshaft through screen frame drove by connecting rod was to do a reciprocating linear motion, and the screen mounted on the support beam of screen frame was drove to do flipping and flowing moving.

Since the shaft of driveline is designed as the combination of the crankshaft and belt, the symmetry of centering drive and inertial force can be achieved. Thus, the overall force is uniform without twisting moment, and the errors on installation can be reduced. Through which the high reliability, nice integrative performance, smooth motion, and convenience of assemblage can be obtained. Besides the life of the screen mesh can be improved as well.

Results from that, the structure of CFFS is simplified comparing to the custom flip-flow screens, the weight of the CFFS is lighter. A screen with capacity of about 100 t/h weigh less than 2.5 t. As the weight is lighten and the structure is simplified, the energy consumption can be decreased. A screen with surface area of 6 m$^2$ requires 5.5 kW motor power instead of 15 kW that the CZS flip-flow screen required with the same area. Thus, the cost of production and application can be reduced.

The design of the mesh support beam is optimized as well, as shown in Fig. 1. The semi-cylindrical beam is used as the support frame of the screen, thus screen plate had no bending dynamic stress. Through which the reliability and life of screen plate can be improved.

## 3. Kinetic and modal analysis of the key component of CFFS

### 3.1 Kinetic analysis of the CFFS

The driving mechanism, screen frame, and suspension rod of CFFS were simplified. According to their working principle, kinematics model was established, as shown in Fig. 2. Taking the center of the crankshaft as the origin of coordinates O, the coordinate system was constructed. The position of the screen frame relative to the origin can be decomposed into the distance in the $x$ and $y$ directions, which were represented by Sx and Sy, respectively. The values of parameters are shown in Table 1.

**Table 1:** Kinetic model parameters of CFFS

| Parameter | Crank length e (m) | Link length L (m) | Crank angle ω (rad/s) | Rocker length l (m) |
|-----------|-------------------|-------------------|----------------------|---------------------|
| Value | 0.00125 | 0.45 | 57.5 | 0.675 |

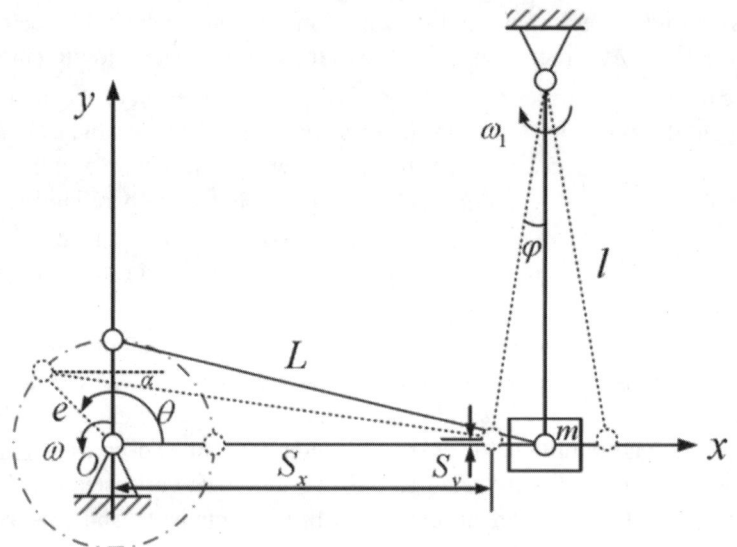

**Figure 2:** The kinetic model of CFFS

$$\begin{cases} S_X = e\cos\omega t + L\sqrt{1 - \dfrac{e^2}{L^2}\sin^2\omega t} \\ S_Y = l(1 - \cos\varphi) \end{cases} \tag{1}$$

According to the geometric relationship in Fig. 2, when the screen box is at the limit position, there is a relationship:

$$e \approx l\sin\varphi_{max} \tag{2}$$

According to the data in Table 1, the maximum value of $S_y$ can be calculated as:

$$S_{Y\,max} \approx l(1 - \sqrt{1 - \left(\frac{e}{l}\right)^2}) = 1.16 \times 10^{-4} \text{ m} \tag{3}$$

Compared with the crank length, $S_y$ has little influence on the motion of the screen box and can be ignored.

The velocity and acceleration in the horizontal direction of the screen box can be obtained by calculating the primary and secondary derivatives of $S_x$, respectively. The expression is shown below:

$$v_x = S'_X = -e\omega\left(\sin\omega t + \frac{e}{2L}\sin 2\omega t\right)$$

$$a_x = S''_X = -e\omega^2\left(\cos\omega t + \frac{e}{L}\cos 2\omega t\right)$$

(4)

## 3.2 Modal analysis of the crank

During working, the inner and outer screen boxes are driven by the eccentric crank and the impact load with accompanying cyclic was imposed to it. The design of the cranks haft is significant for flip-flow screen. For estimating the vibration characteristics of crank to avoid the resonance at the motor operating frequency, the vibration mode and frequency of the crank was obtained by modal analysis.

The three-dimensional model of the crank was established by using Pro/E, which was imported into ANSYS Workbench with the format of STEP. The material of which was set as the default material of structural steel. According to the geometrical characteristics of the model, sweep tetrahedron grid and automatic grid was chosen, respectively.

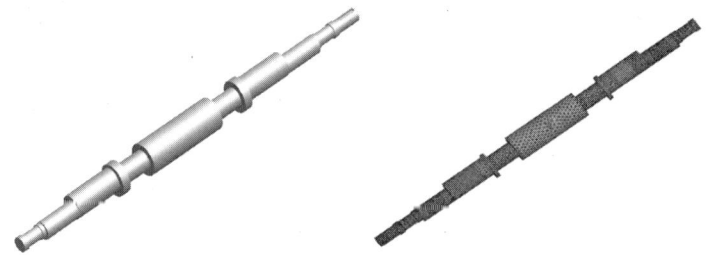

**Figure 3:** Section of the crank model and meshing

As shown in Fig. 3 cylindrical support was load on the place sliding bearing installed, which set the tangential freedom, the axial and radial fixed and frictionless support was load on the place antifriction bearing installed.

The first six order modal vibration mode (Fig. 4) and natural frequencies (Table 2) were obtained after calculation. The first order natural frequency is 219.56 Hz, besides, natural frequencies after the second order are all above 800 Hz, which stay far away from the general vibration frequency of the motor. So no resonance phenomenon can be aroused under normal conditions.

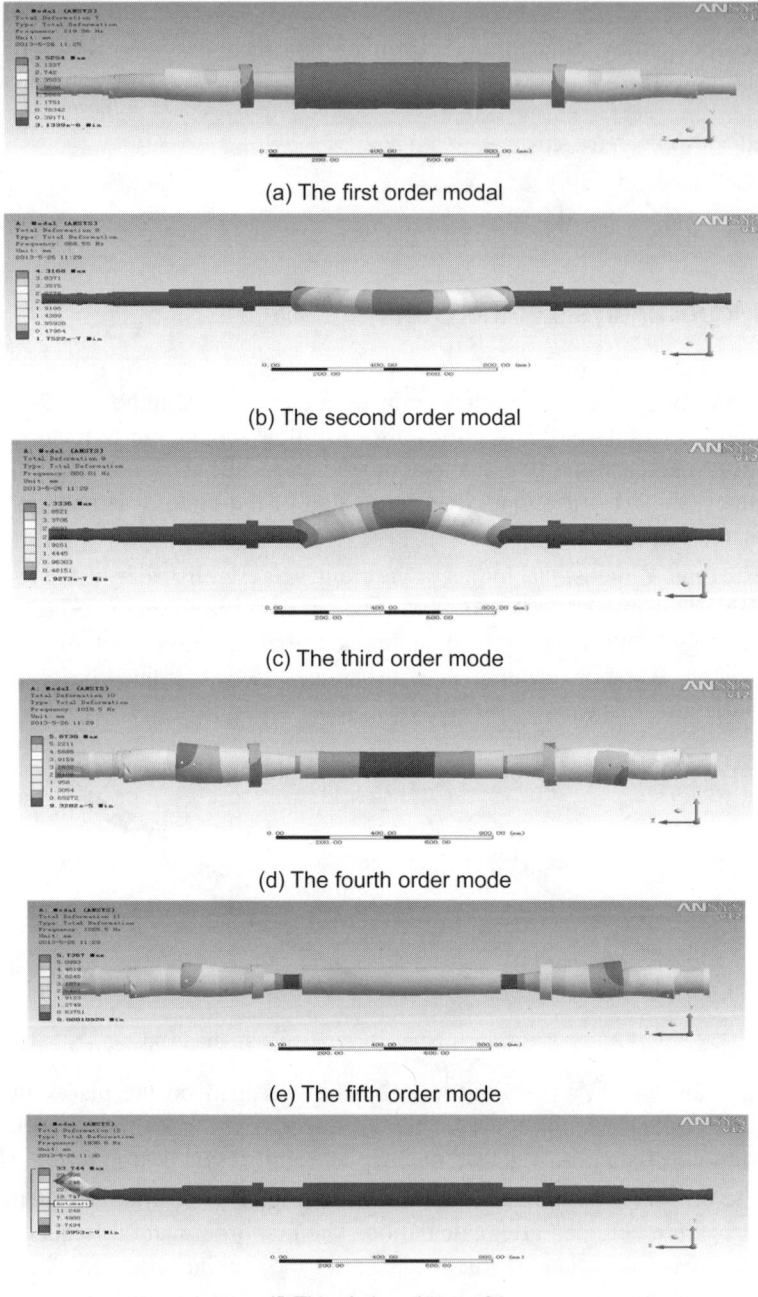

(a) The first order modal

(b) The second order modal

(c) The third order mode

(d) The fourth order mode

(e) The fifth order mode

(f) The sixth order mode

**Figure 4:** The workbench modal analysis results of crank

**Table 2:** Modal analysis results of the crank

| Order | Natural frequency (Hz) | The modal results | Maximum amplitude (mm) |
|---|---|---|---|
| 1 | 219.56 | Swelling–shrinking along the radial as a whole | 3.5254 |
| 2 | 868.55 | Swings around along the Z axis | 4.3168 |
| 3 | 880.01 | Bobbing along the Z axis | 4.3336 |
| 4 | 1018.5 | Two eccentric sections swell–shrink along the Z axis | 5.8738 |
| 5 | 1285.5 | Two eccentric sections swelling–shrinking along the Z axis and swinging | 5.7367 |
| 6 | 1938.6 | Pulley installation swing around the Z axis | 33.744 |

## 3.3 Harmonic response analysis of the crank

In order to further analysis the stress distribution of the crank under sinusoidal variation load, the harmonic response analysis was carried out on the basis of modal analysis. Connected with the previous modal analysis, harmonic response module in the ANSYS Workbench was used to load. Since the inertia force caused by the inner and outer screen boxes was equal and borne by the two eccentric sections. Bearing load was applied on the four eccentric sections of crank, and the value was half of the amplitude along the $Y$ direction. As shown in Fig. 5, the torque was exerted on pulley by moment.

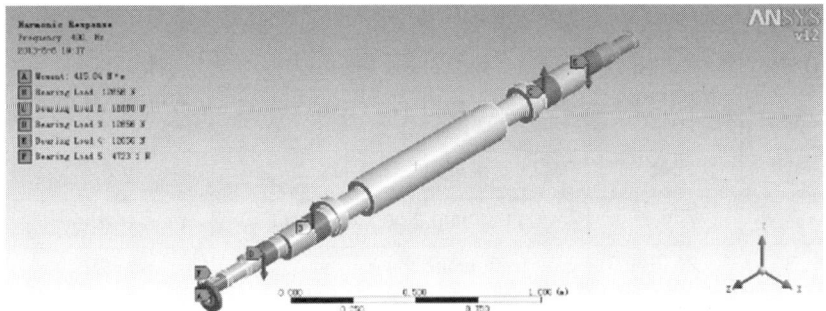

**Figure 5:** Applying inertia force and torque

The frequency response curve was chosen from 400 to 1000 Hz on 20 interval. According to the frequency response curve (Fig. 6), the crank would resonance at 450 Hz. When the crank resonated, danger would occur in the sliding bearing and eccentric junction whose maximum stress value was 753.32 Mpa and maximum strain value was 0.0037. On the basis of the working frequency of the crank points (9.1 Hz) for harmonic response analysis, motor can be found on the excitation frequency of the crank that is much less than the resonance frequency of the crank, which the stress and strain of crank were within the safety of the material.

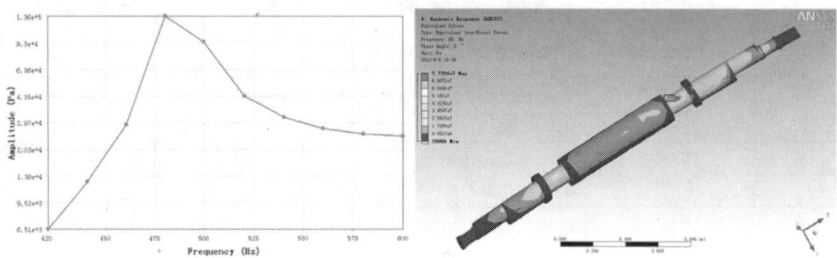

**Figure 6:** Equivalent stress cloud atlas at working frequency

## 3.4 Modal analysis of the linker

Modal analysis was carried out on the linker, the operation such as the import of the model, material and mesh setting are the same as the crank analysis above. Cylindrical that only keep rotational degree of freedom constraint was applied at the end circle, and the frictionless constraint limit the axial displacement was applied on the small circle. The results are shown in Table 3. The workbench modal analysis results of linker are shown in Fig. 7.

**Table 3:** Modal analysis results of the linker

| Order | Natural frequency (Hz) | The modal results | Maximum amplitude (mm) |
|---|---|---|---|
| 1 | 645 | Bobbing along the Y axis | 7.699 |
| 2 | 706.11 | Swings around along the X axis | 14.144 |
| 3 | 2316.5 | Sine twist along the X axis | 11.229 |
| 4 | 2993.7 | Small round side swings around along the X axis | 37.581 |
| 5 | 3087.4 | Small round side separate from each other | 49.985 |
| 6 | 3293.2 | Twisting around the Z axis | 21.318 |

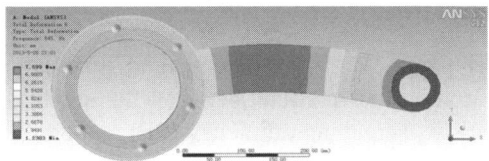

(a) First order mode shape

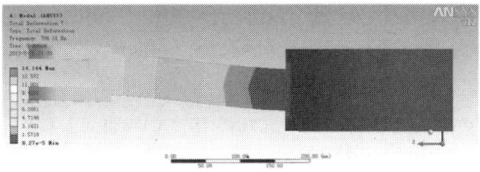

(b) The second order modal

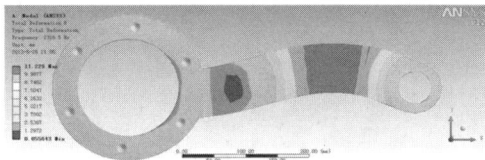

(c) The third order mode

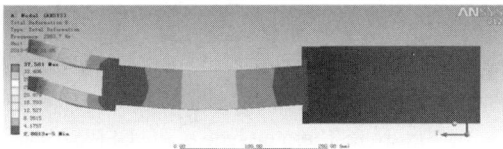

(d) The fourth order mode

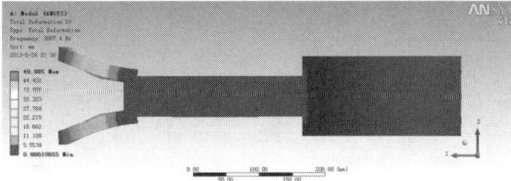

(e) The fifth order mode

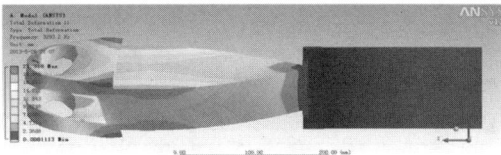

(f) The sixth order mode

**Figure 7:** The workbench modal analysis results of crank

## 4. Sieving experiment of CFFS

The technological parameters of screening and kinematics parameters of flip-flow screen include the relations between the sieving indexes such as dynamic characteristic, screening efficiency (SE), undersize efficiency (UE), oversize efficiency (OE), and size composition of the sieved material, external moisture content, handling capacity (HC), and rational selection of screen slope, vibration angle, vibration frequency, amplitude. The research of these parameters can provides theoretical basis for selection of structural parameters and kinematics parameters of CFFS.

In order to test the practical screening effect of CFFS accurately the prototype was manufactured, as shown in Fig. 8. Sieving experiment was conducted based on GB/T477-2008 (Sieving experiment method for the coal). The sieving experiment results of the raw coal are shown in Figs. 9 and 10.

**Figure 8:** The prototype of the CFFS

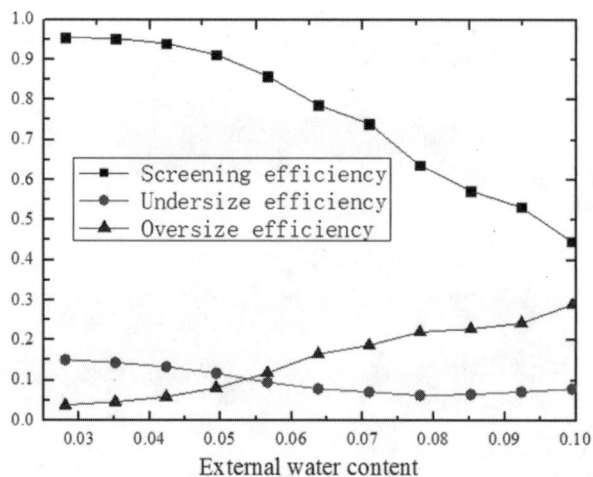

**Figure 9:** Relationship of the external water content to SE, UE, and OE

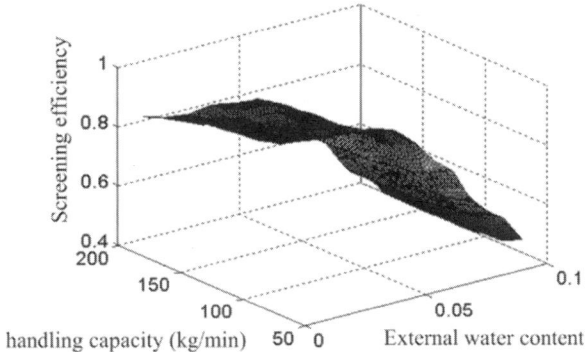

**Figure 10:** Effect of external water content on SE and HC

External water content may has a greater impact on the screening process of flip-flow screen. When the external water content is about 7%, the screening efficiency of the flip-flow screen almost could remain the screening efficiency of 80% with undersize efficiency of less than 10% and oversize efficiency of about 15%. It candidates that CFFS can perform depth screening for dry-cleaning method of moist fine coal excellently.

The screening efficiency of CFFS could maintain a stable level when external water content is constant and handling capacity is increased. When handling capacity is 200 kg/min and external water content was 7%, the screening efficiency of CFFS could reach more than 75%. It shows that CFFS has a higher handling capacity for the moist fine coal. Therefore, CFFS has superior screening efficiency and can support bigger handling capacity at the same time when it is used to perform sieving of −6 mm and −3 mm levels on moisture fine material.

Size composition of input is significant to the screening process. On the one hand, near-sized particle is probable to drive the screen blinding and it show influence on stratification of coarse and fine materials as well. On the other hand, the moist fine material is tend to cause screening blinding because of bonding when the content of it is increased. And it also reduces screening efficiency and increase mismatch ratio because the bonding fine particles may be expelled from output mixed with coarse material. During the screening process of CFFS, the screen holes conduct periodic open–close movement with great tossed acceleration result from that the screen surface has certain tensional amount and does elastic flexural movement. Thus, the screening blinding of various particles can be prevented, and the loosening and stratification of material can be promoted. Through which the screening efficiency can be improved.

Figure 11 shows the influence rule of fine material content under different external water contents to screening efficiency of CFFS. It can be seen that screening efficiency increases with the fine coal content when the external moisture content is constant. Because the elastic screen surface of CFFS has a very high tossed acceleration and it can prevent moisture fine material from blocking screen holes. The screening efficiency reaches the highest level of about 90% when the fine particle content is about 50%. But the screening sufficiency is impacted by material layer because it is too thick to stratification when the fine material content is further increased.

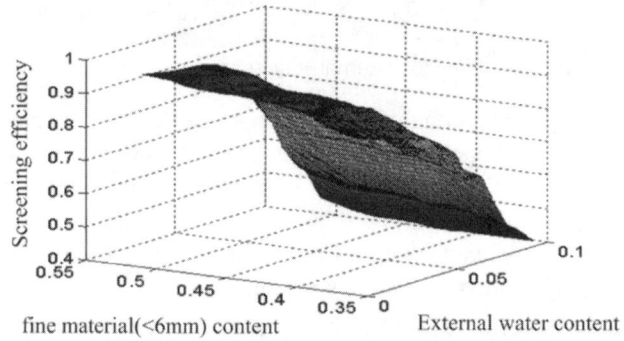

**Figure 11:** Effect of fine material content on screening efficiency under different external water contents

## 5. Conclusion

(1) A finite element model of key parts of centralized forced flip-flow screen based on finite element technology was established, and the first six order natural frequencies and corresponding modes were obtained through numerical simulation. Through which the reliability was proved.

(2) The experimental study was conducted based on the prototype manufactured. It can be demonstrated that when the feed water and processing quantity changes, the CFFS can remain a strong applicability with a stable sieving performance.

(3) A new feasible method of dry screening was proposed by the CFFS, the corresponding numerical simulation and experiment provided a reliable basis for the future promotion of similar products design and research.

# 6. References

[1] Bo Zhang, Guangqing Zhu, Bo Lv, et al. A novel and effective method for coal slime reduction of thermal coal processing. Journal of Cleaner Production, 2018, 198:19–23.

[2] Chaoqun Fan, Hongyu Zhao, Long Ji, et al. Development trend and the factors influencing working effect of flip-flow screen. Coal Preparation Technology, 2013(01):88–90.

[3] Chusheng Liu, Yuemin Zhao, Yunfei Xia. New advances in research and application of flip-flow screen, National Screening Technology Conference, 2010.

[4] Hailin Dong, Yunfei Xia, Chushegn Liu. Study on impact of structural reconstruction of support beam in flip-flow screen on reliability of screen surface. Mining and Processing Equipment, 2012, 40(02):71–74.

[5] Hailin Dong. Study on the Dynamics of the Large Centralized Forced Flip-flow Screen Surface. Doctoral Dissertation of China University of Mining and Technology, 2012.

[6] Junxia Yan, Chusheng Liu, Shimin Zhang, et al. Dynamic analysis of sieving plate of centralized forced flip-flow screen. Mining and Processing Equipment, 2011, 39(04):95–97.

[7] Ministry of Land and Resources of the People's Republic of China. Potential evaluation of national coal resources. 2014.01.07.

[8] Qingru Chen. Consideration of clean coal energy strategy in China. Journal of Heilongjiang Institute of Science and Technology, 2012(04):331–336.

[9] Xiangdong Liu, Zhanrong Liang, Kaifeng Gao. Dry screening of wet raw coal. Coal Preparation Technology, 2013(03):43–45.

[10] Yuemin Zhao, Chusheng Liu, Maoming Fan, et al. Research on acceleration of elastic flip-flow screen surface. International Journal of Mineral Processing, 2000, 59(4):267–274.

# 34

# Two mass screens: A productivity game-changer

Vince Sunter

*iPUT Technologies, Newcastle West, Australia*

**Abstract:** Two mass screening technology have been around for decades, but not seen much application in coal because they cost more than traditional brute force banana screens. They have mostly been used in hard rock and iron ore applications, where they survive when no other does.

Coal mines have traditionally been bottle-necked by various things but typically mining, handling, washing and distribution. Not so much screening, until large DMC's and better bulk mining/handling processes have pushed the bottleneck more onto screening.

Modern coal washing plant designers have pushed the demand for more throughput onto the screens, using traditional methods and made them wider, longer and heavier. Each increase in width has brought substantial teething problems and, as the industry is contemplating going up another step to 4.8m wide screens, two mass screens have come under scrutiny again.

In 2014 iPUT, a design-construct project company, investigated the status of two mass screens and found the leading OEM (original equipment manufacturer) had a mature 4th generation modular design product. The OEM claimed their design would handle at least a 30% throughput increase with the same or better efficiency when installed in the same footprint as an existing brute force banana screen. It used much less energy to do this bigger task, and was inherently more reliable.

iPUT retrofitted one of these screens in 2016 in a coal rejects application, and found that the claims were true. This paper will give an introduction to the two mass screening technology, an outline of the Qld field case study and results, plus discussion of the potential in coal.

**Keywords:** Banana screens, Stratification, Efficiency, Mismatches, Upgrades, Measurements

## 1. Two Mass screening technology: Basic overview

The principal difference between a two mass and brute force banana screen is that the latter has a one-piece trough deck and support structure, whereas the two mass separates the trough deck section and upper 'drive' structure, connecting them with banks of coil springs. A 'brute force' screen activates the whole mass, whereas a two mass screen only needs to activate a (relatively) lighter trough mass to get the particles excited and passing. And the flat deck and lower travel speed means they get a lot more opportunities to pass.

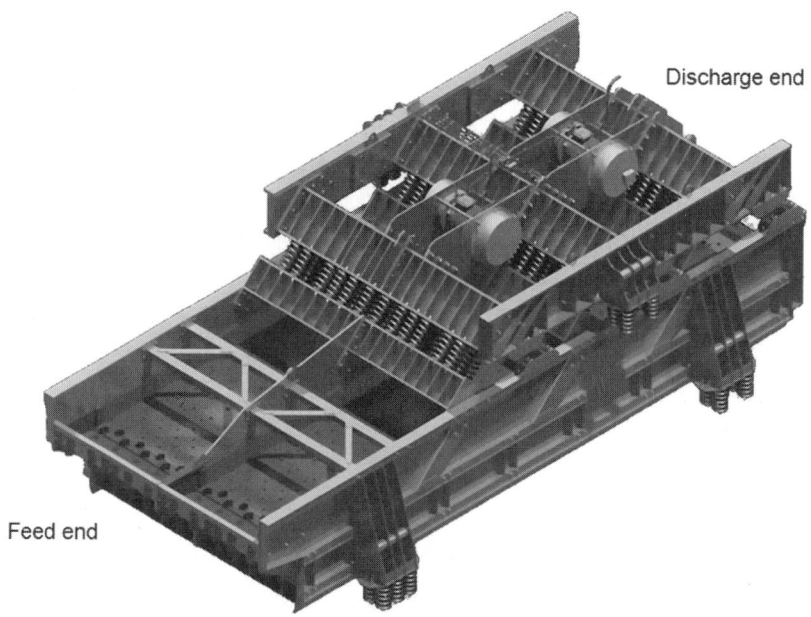

**Figure 1:** Auto CAD illustration of leading OEM two mass screens. Figure 1 explains the basic concepts and Figs. 2 and 3 show examples.

The most important thing going on here is that these screens have a constant bed depth down the entire length, and material stays on the deck for more than double the time of a typical banana screen, making for good segregation and very high screening efficiency.

The physical motion of the exciter and trough mass 'opening and shutting' in sync, means one is largely cancelling out the inertial forces of the other. Hence, only a small amount of vibration gets to the support structure, so they are a lot quieter. The energy to activate a smaller exciter mass typically only needs 1/3 the kW power while doing a bigger job. Much lower panel wear rates occur from the low wear mechanism of a bouncing action on two mass screens vs. high wear from sliding on banana screens, tripling deck life despite the heavier loading.

The top 'driven' mass exciter is relatively low, needing less power. Spring separation of the top drive to bottom trough allows a centre stiffening rib and uniformly applied force. In brute force screens all force is taken in the side-plates, leading to increased premature failure risk.

To enhance throughput further, a two mass screen in coal has a short cascade deck added to the feed end, so the fines pass straight through and the coarse falls on top of the bed. This quickly establishes the desired stratification for higher efficiencies.

Travel speed being slower means a longer rinse zone, providing further screening efficiency.

**Figure 2:** Dahongshan Iron Mine two mass screen install, old style behind (replaced 2010)

## 2. Retrofitting two mass screens: Main design issues

A simple overlay of the proposed two mass screen over the existing banana screen outline reveals the need for relocation of the old drive access walkway and possible chute mods. The OEM can typically accommodate the existing feed chute, being the most expensive item to modify. Modified discharge chutes are preferable to changing the new screen's discharge.

The underflow of the two mass screens is straight and a banana screen curved. An infill piece blends the two mass with the underflow launder. This gives some clearance to divert the rinse zone to start earlier. Replacing the underflow launder would give potential to optimise the two mass screen performances further; exactly how much needs more work to quantify.

Two mass screens are a larger static mass, but impose much lower dynamic loads. Any modern CPP design mismatches screen and structural resonant frequencies, with large crossbeams well able to support the additional mass. Building columns are usually adequate, but may require plating in where they support multiple heavier adjacent screens.

Rinse water circuits also need examining to ensure adequate supply and that the magnetic separators do not become overloaded. Whilst this is always plant specific, there are many techniques to ensure sufficient rinse water without incurring significant cost.

Upstream and downstream equipment capacity also needs review, e.g. discharge conveyors.

**Figure 3:** WA Iron Ore: 8 @ 3.6m × 8.5m two mass screens installed 2016, more ordered; Spec 3100t/h, 2.24sg, −22/−12mm cut, 90%/85% passing; operating at 4000t/h

## 3. A case study, Queensland coal mine site

Adoption of this technology started with the first Australian coal installation at a major miner's Qld coking coalmine in mid-2016. Driven by the mine plan approaching low-yielding lesser grade seam/s, they took the opportunity to both process the material at increased throughput and produce a rewashed thermal product. Previously a more expensive 4.2m wide banana screen conversion was the only possible solution, but now a 800t/h, 3.6m × 6.1m two mass drain and rinse screen is operating very successfully in the rejects stream, down to 1.1 density, with <0.5kg/t magnetite loss, and structurally rated at 1000t/h at maximum bed depth.

Design aspects of the application are illustrated in Figs.4 and 5, with the installation shown in Fig. 6 .

## 3.1 Design process and issues to be accommodated

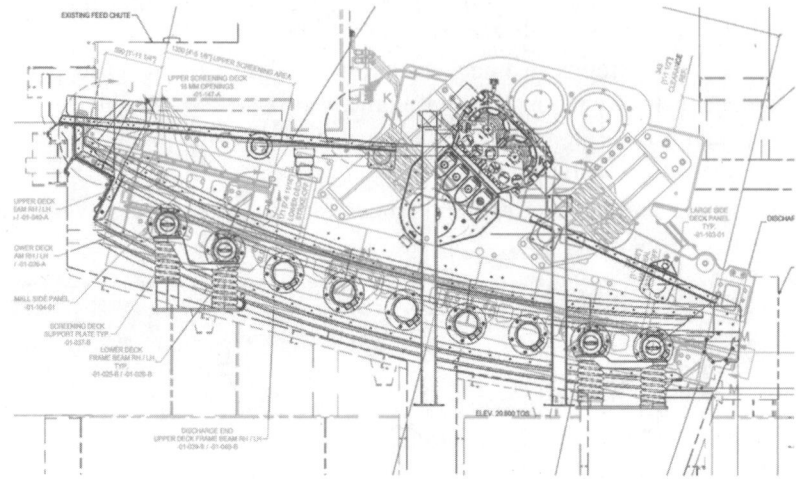

**Figure 4:** Overlay of two mass screen (green) on existing banana screen

## 3.2 Full length spray system development for maximum deep bed rinsing

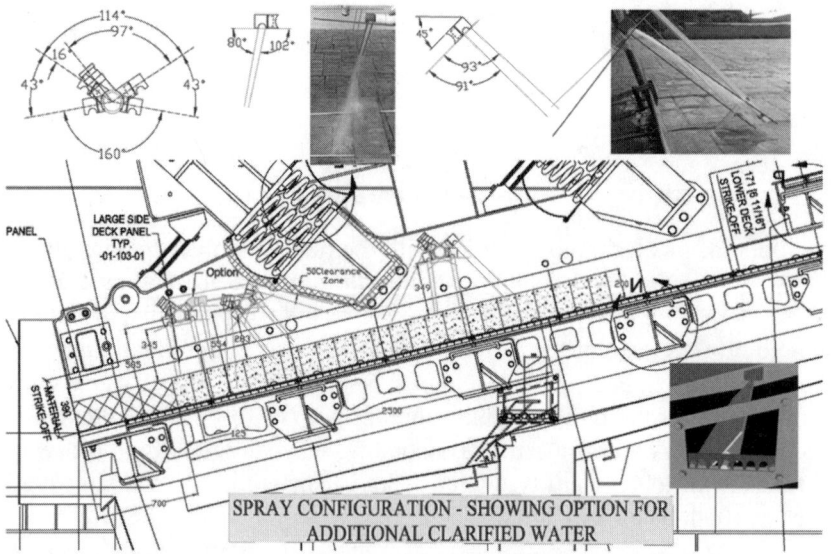

**Figure 5:** The science behind developing a new 'continuous curtains' spray system

**Figure 6:** As installed two mass 3.6m × 6.1m 800t/h rejects drain and rinse screen

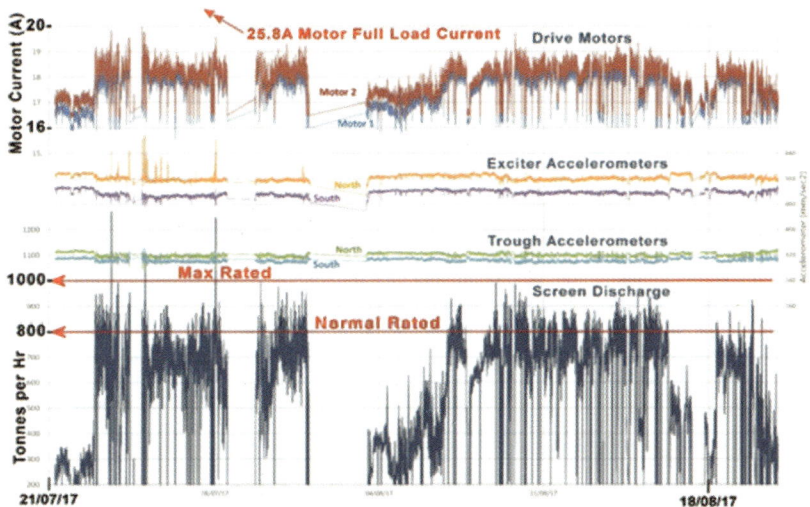

**Figure 7:** Two mass 3.6m × 6.1m D & R rejects screen operation, installed 2016

In operation, the two mass screens performed as expected; see Fig.7. The cascade section let all the −16mm material through and formed a bed of fines that the +16mm then landed on, from a short fall, to cushion the deck panels from impact and stratify the bed in the most desirable way for maximum rinse zone effectiveness. This strategy worked well as spares consumption records show screen deck panel life is triple what it used to be, with 192 days average overall, and 171/213 days left/right. Two DMC's feed the screen, causing some variation; a simple flow diverter was later added in the feed box to even out the feed.

Drain water had visibly fully departed from the bottom deck before it had reached the end of the cascade section. Bed depth was constant down the length of the main deck, at around 250mm depending on density. It took approximately 30s to travel the length of the screen; about triple that observed for the original brute force screen. Measurement of the two mass screen magnetite recovery verified performance specification compliance.

This was the OEM's first experience with a high throughput low density application and, whilst successful, an improved design was developed, which the mine went on to install in 2018. In Fig.8 can be seen even more throughput for less energy needed with the refined design.

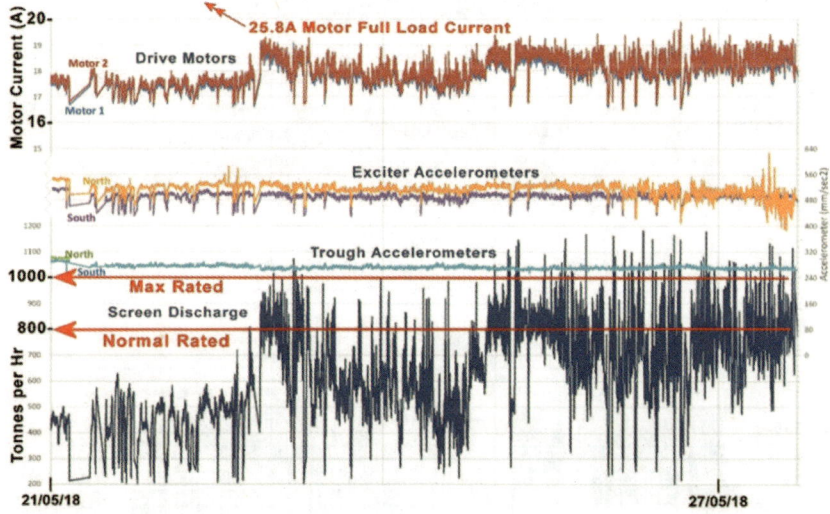

**Figure 8:** Enhanced design two mass 3.6 m × 6.1 m D&R rejects screen, installed 2018

## 3.3 Future potential

Regardless of the exact actual magnetite recovery situation, optimising the rinse zone to suit the screen characteristics could improve magnetite recovery by a significant percentage. Whilst a detailed technical explanation is beyond the scope of this paper, in short, drain water is fully gone by the time the material has left the cascade deck, so a longer rinse zone is possible, and the characteristics of that longer rinse zone deserve careful scrutiny.

## 4. Full CPP upgrades, an existing mine case study

A desktop study showed the opportunity exists to increase output of one major Hunter Valley NSW mine by 60% from ROM hopper to train loader for a cost of $5/t of annual ROM capacity, with a payback period of months. This would be typical for any older style plant.

## 5. New installations

The two mass screens can reliably handle 1900t/h per CPP module with 4.8 m × 7.3 m units. Existing screen technology stretches to get to 1200t/h per CPP module, and it is widely regarded as a consumable item at those throughputs.

The other key opportunity is the process improvements this enabling technology can unlock, e.g. reducing cut size to optimise fines circuit performance or adding a separate cut point (extra deck) and additional DMC circuit, in the same physical size screen location, to extract maximum value from fines and course circuits. A new low-ash product also becomes feasible.

Industry leaders have looked closely at high throughput green-field site proposals. Now that the screening throughput issues are resolved, they are very enthusiastic at the potential. Whilst DMC's will need more power, key suppliers advise they will have no problem accommodating the needs. There are hence no apparent obstacles to move to a new paradigm of operational efficiency with industry dominating productivity.

## 6. How far could we go?

5.4m × 8.5m two mass screens already exist, being installed at a progressive Chinese mine owner's site. The OEM says the modular construction allows them to construct two mass screens up to 6.6m × 10m as this is simply 3 @ 2.2m proven modules. For typical mine parameters, that would likely translate to 2600t/h per module.

Reducing cut sizes to 0.9mm or 0.5mm is a way to take throughput to new levels without fines circuit upgrades for existing plants (Fig.9).

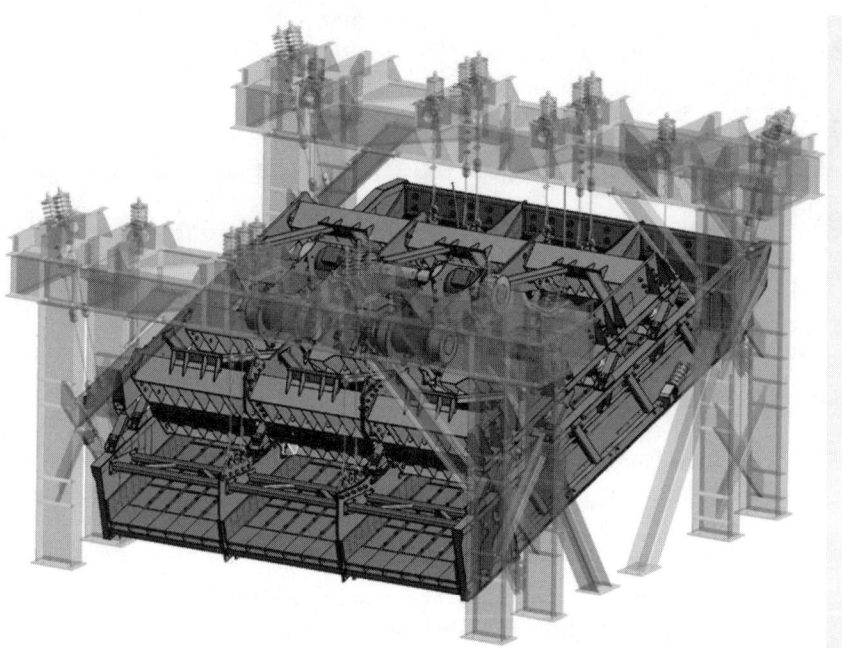

**Figure 9:** 5.4m × 8.5m two mass screen with optional top-hanging system
(to simplify below floor level structure)

*Further resources:* www.iput.com.au/tech.

# 35

# Meeting the challenge of coal & limestone preparation at coal based power plants through Liwell® screens

Partha Banik

*Hein Lehmann India Pvt Ltd., and Shahadat Khan, Ultratech Cement Ltd*

**Abstract:** Coal based independent power producers and captive power plants having circulatory fluidised bed combustion technology worldwide; especially in India and South East Asia region have their prime responsibility to produce their plant specific power generation throughout the year. Accurate sizing of coal for this combustion technology is very important to achieve rated combustion efficiency and avoid clinkering in the combustion bed. During rains, segregation of high moisture coal is a big challenge.

Liwell® Flip Flow screening technology is the most successful technology to meet this challenge at the coal handling plant for coal preparation. Blinding and pegging of conventional screens lead to very low screening efficiencies and thereby insufficient fuel feed to the boiler. The Liwell® flip flow technology through trampoline movement of Flip Flow PU Screen mats provide the necessary acceleration to coal particles to avoid blinding and pegging.

Limestone is fired in the combustion chamber of a boiler to capture SOX & NOX. The low size for limestone necessary can be achieved through Liwell Flip flow Screens which can effectively and efficiently take care of this challenge.

**Keywords:** Circulatory fluidised bed, Liwell® Flip Flow, SOx and NOx, trampoline movement

## 1. Introduction

Thermal Power plants across the world would generate power through steam turbines, which are based on steam generators or boilers. They would require coal and limestone to generate steam for power. Coal provides the necessary energy for the generation of steam and the limestone is added for the capture of SOx and NOx. Environmental statutory norms for the latter are getting stringent every day and have become a critical issue.

Considering the above, finding the correct technology for the separation of both coal and limestone becomes very critical for a thermal power plant. Below we have talked about the combustion technologies, the effect of coal/limestone sizing on power generation and Flip Flow Liwell® Screens designed by M/s. Hein Lehmann, Germany.

## 2. Combustion technologies

Designers, especially for Captive power plants (CPP-10 to 125 MW) would select two or three types of combustion system for the steam generator or Boiler. For the larger sized plants, pulverised fuel (PF) boilers have been very popular over the last 50 years. Atmospheric and Circulating Fluidised bed Combustion System boilers (AFBC/CFBC) have been leaders in 1 to 80 MW segment except for some exceptions like biomass and multi-fuel boilers.

PF fired boilers have been always popular for its ease of operation, as the coal is crushed to fine powders in ball mills and directly injected into the combustion chamber of the boiler. Air blowers take care of the air required for combustion. In the early '90s, prices of coal which were based on their calorific values and ash content started to increase. As a result, users and engineers became more conscious of the word combustion efficiency ushering in new combustion technology.

The revolution started with atmospheric fluidised bed boiler (AFBC) followed by circulating fluidised bed boilers (CFBC) technology. As its name suggests, the complete fluidised bed consisting of sized coal circulates within the boiler for maximisation of carbon combustion in the furnace through a cyclone or U-beam design. In this technology, poor quality of coal including washery rejects could be fired for the generation of power in the boiler. The operation while being economical as it can burn low calorific value coal is comparatively tricky and needs a trained operator. One can easily choke the combustion chamber with higher than required size pieces of coal (plus 8 mm). This phenomenon is called clinkering and is well known to AFBC/CFBC users and operators. This may happen when in the combustion chamber; locally coal particle reaches ash-melting temperatures due to a formation of the nucleus from the higher sized coal/impure particle at any position in the combustion chamber circuit, ash cooler, etc. This results in downtime of power generating a plant.

## 3. Separation of sized coal/limestone

The preparation and sizing of the coal (from low to high calorific value) is very critical for AFBC/CFBC Boilers. Typically, the maximum sizes allowed in these boiler feeds depend on technology provider, but usually varies from minus 6to 8 mm maximum and not more. In addition, during the preparation of the coal through the crushing cycle, care has to be taken to select the right crusher to avoid the generation of fine particles. Once the larger chunks of coal are crushed through a typically reversible hammer mill, it would generate different sizes of coal particles. Also, the bed is kept in a fluid state through air pressures from primary forced draft (FD) fans. These fans make the very fine particles, particularly in AFBC fly out with the flue gas dropping the efficiency of the steam generator. This happens due to unburnt carbon in the Flue gas.

A typical example of Indian coal is given below.

| Indian coal | | |
|---|---|---|
| CPP Feed Belt - Sample data | | |
| Screen Size | After Crushing Circuit | |
| | Sample 1 | Sample 2 |
| (mm) | Wt% | Wt% |
| +200 | | |
| -200+100 | 0.00 | 0.00 |
| -100+80 | 0.00 | 1.08 |
| -80+65 | 4.65 | 4.04 |
| -65+50 | 6.71 | 5.97 |
| -50+25 | 23.52 | 22.59 |
| -25+12.5 | 16.34 | 16.72 |
| -12.5+10 | 3.92 | 3.58 |
| -10+5 | 12.15 | 11.58 |
| -5+4 | 2.54 | 2.91 |
| -4+3 | 3.42 | 3.69 |
| -3+2 | 6.31 | 5.62 |
| -2+0.5 | 11.04 | 11.36 |
| -0.5 | 9.41 | 10.88 |
| | 100 | 100 |

Suppose, a particular coal sample has a minus 8 mm of 70% post crushing and the moisture is more than 10-12% screening efficiency in a standard conventional screen having circulatory motion would be low. During monsoons, choking and clogging of the wire mesh or perforated plate apertures bring the efficiency as low as 35% drastically effecting the power factor and power generation. The following picture shall detail reasons for low efficiencies of conventional screens.

## 3.1 Principles of screening

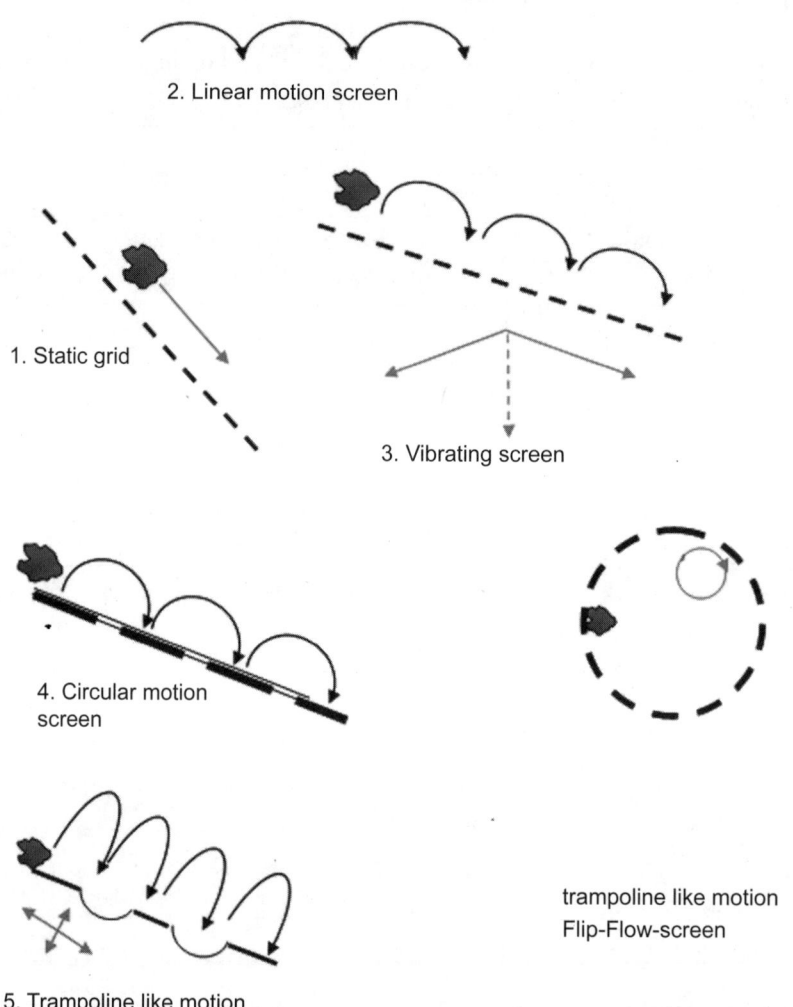

2. Linear motion screen

1. Static grid

3. Vibrating screen

4. Circular motion screen

5. Trampoline like motion Flip-Flow-screen

trampoline like motion Flip-Flow-screen

The above picture shows how screening technology evolved from the static grid in the 1940s and '50s to the current trampoline like flip flow motion design created purely by hi-tech mechanical movements.

## 3.2 Resultant acceleration ratios

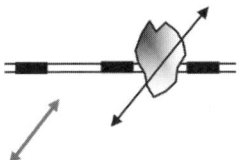 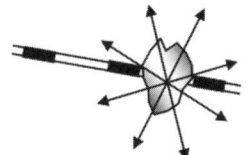 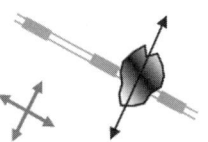

Linear motion screen
$F_A$ = max. 5 g
Machine load = 5 g

Circular motion screen
$F_A$ = max. 5 g
Machine load = 5 g

**Flip-Flow-screen**
$F_A$ = max. 50 g
Machine load = 2.4 g

## 3.3 Your screening problems

**During screening operations, problems will arise due to blocking of apertures in the screen panels, the main causes being pegging and blinding.**

 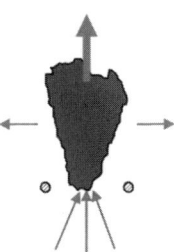

**Pegging** – Clogging of apertures due to conical shaped particles, which get stuck in the screen panel. Releasing it during the acceleration phase

**Prevention** – By means of high acceleration forces which eject the wedged particles.

– Use of flexible "breathing" screen panels allowing the particle to pass through or

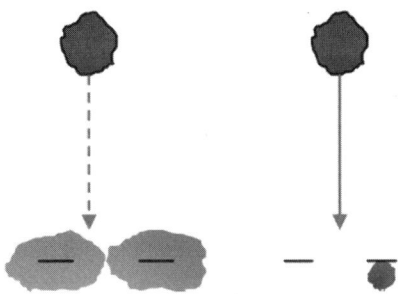

**Blinding** – Blocking of apertures due to a build up of fine, moist material

**Prevention** – By using screen panels made of Polyurethane, rubber etc. which have Anti-adhesive properties

– High acceleration forces (acceleration power > adhesive strength)

– Flexing screen panels causing caked material to break-up

## 4. Flip flow is your solution – guaranted high efficiency screeing for lifetime

The above picture depicts the forces involved and screening technology used to overcome problems of wedging in a conventional screen.

Other than the preparation of coal in a thermal power station, capture of sulphur and nitrogen emanating from boilers through the chimneys creating unwanted air pollution has been under focus for quite some time. Statutory rules in India currently have become more and more stringent requiring more tonnage of limestone being injected in the combustion chamber of the boiler for plants where limestone is crushed and powder produced. It is equally important to maintain a size of 600 microns to minus 1 mm. Higher sizes would result in low SOx-NOx capture. Furthermore, any leakage of higher sizes into the combustion chamber would lead to the formation of a nucleus for clinkering. All these leads to the fact that preparation of limestone is very critical also. The Flip flow technology currently can effectively separate up to 200 microns, up to 2% moisture levels very effectively with efficiencies in plus 90% range.

To summarise, steam generators need coal sized at minus 6 mm or minus 8 mm with very low fines (100%) and limestone required is between 200 microns to 1 mm max.

## 5. Flip flow technology for separation

Flip-Flow Machines type LIWELL® was introduced into the market about 50 years ago by M/s. Hein Lehmann GmbH. The most identifiable feature of this screen compared to conventional screening machines known before is a dynamic moving polyurethane screen panel. Conventional machines generally use screen panels fixed to structural elements of the machine. Therefore, the screen panels typically provide the same amplitude and frequency as the machine itself. Amplitudes and frequencies achieved are limited to what the structure of the machine(s) can accept. Standard machines run with acceleration around 4-5 g. Consequently, the same acceleration is provided by the screen media itself to screen the bulk material.

Especially in case of small cut sizes and/or higher moisture levels of feed material (surface moisture), the mentioned acceleration is not sufficient to avoid clogging of meshes and apertures (Meinel, 1998). The clogged aperture requires downtime of the power plant to clean and remove clogged particles if no standby screen is available.

Flip Flow Machines type LIWELL® are operated with independent dynamically moving polyurethane screen panels. The movement is different from the movement of the machine structure. Consequently, much higher amplitudes and accelerations can be achieved [Schmidt, 1977]. Clogging and plugging of perforations can be avoided to a maximum possible extent without the use of spray water.

In the meanwhile, Flip Flow Machines became the state of art technology for most screening applications with small cut sizes and moist feed materials such as coal, limestone, aggregate, coke, iron ore and others.

## 6. Screen construction

To achieve a significantly higher acceleration Flip Flow machine always consist of flexible perforated screen mats made of Polyurethane. Each side of the mat is fixed on an individual crossbeam. The complete number of cross bars is divided into 2 groups – system 1 and system 2, which are moving during operation in the opposite directions. The result is a permanently changing distance, which relaxes and stretches the screen mats. To generate the contrarious motion the cross beams of system 1 and system 2 are alternately attached to an inner and outer screen frame. An eccentric drive unit at the feed end of the machine actuates both screen frames.

While the screen frames and cross bars are subject to limited accelerations only (about 2 g [Hirsch, 1992]) the screen mats instead create accelerations of much more than 50 g. This extremely high acceleration is a result of the unique movement of screen mat; after relaxation, the mats are tensed while the amplitude between relaxed and tensioned position is significantly high. Just before coming into the tensioned position, the mat reaches its fastest speed [Schmidt, 1977]. The fast vertical speed of the screen mat is abruptly reduced to zero when the mats are tensioned resulting in highest negative accelerations releasing clogged and plugged particles.

For a sufficient flow of material, the machine is installed in an inclined position. The inclination is variable and depends on the volume of feed material, separation size, etc.

## 7. Conclusion

Liwell® Flip Flow Screen has been found to be immensely successful in the separation of minus 6 mm and minus 8 mm coal for atmospheric and circulatory Fluidised Bed Boiler. They have also been found to be very successful in 200 microns to 1 mm or 3 mm of limestone in AFBC/CFBC and PF fired boilers. The machine can be used in coal handling and limestone crushing and segregation plants globally. In India alone, there are approximately 500 Liwell® screens working in Captive and independent power plants.

## 8. References

[1] Hirsch.W.: Flip Flow Screens of the Third Generation. Aufbereitungstechnik 1992 NO. 12.P 686-690.

[2] Meinel.A.Classification of fine medium sized and coarse particles on Shaking screens Aufbereitungstechnik 1998 No.7, P 317-327.

[3] Schmidt.Hch Theoretical aspects of using screens providing for alternate stretching and sagging of the decks. Aufbereitungstechnik 1977 No.7, P 327-332.

# 36

# Ultra low specific gravity dense medium cyclone separation for production of very low ash "supercoal" products

David Woodruff[1], Andrew Dixon[1], Christopher Syran[2], Hemal Madugalla[2]

*FLSmidth UK Ltd., Rugby, UK*
*CleanCarbon AS, Oslo, Norway*

**Abstract:** CleanCarbon is a family-owned Norwegian company that works with speciality carbon materials and has been in existence for over 100 years. CleanCarbon operates a dense medium cyclone (DMC), plant in Sines Portugal, the main port on the Portuguese Atlantic coast. At this site they produce very low ash coal products from imported steam coal of Colombian origin, which are used in specialist applications. The products have an ash content of between 0.9% and 1.24% w/w, achieved by using an FLSmidth Krebs DMC, operating at a separating specific gravity (SG), of 1.26–1.31, using circulating dense medium of 1.20–1.26 SG.

This very low separation SG is needed to get to the required product quality, and requires a highly efficient separation in an environment of approximately 90% "near gravity material" (NGM) ±0.1 SG of the cut-point.

It is also interesting to compare the difficulty of separation at Sines with other coals with difficult washability characteristics and high percentages of NGM, surrounding the desired cut-point, particularly India, where very high percentages of NGM are often seen at the design cut-point when producing low ash coking coal.

**Keywords:** Dense medium cyclone, low ash, high near gravity material, (NGM)

## 1. Plant description

### 1.1 History/overview

The plant has had an interesting history. It was originally designed as a modular dense medium drum plant in 1998, treating larger coal but was converted to a dense medium cyclone (DMC) plant in 2012, when product size and quality requirements changed. This has led to several challenges, mainly as the height of the original plant is very low, with the product drain and rinse screen located at a floor level of approximately 3 m above ground. This means that the circulating medium and dilute medium tank capacities are very limited because of the available height (Fig. 1).

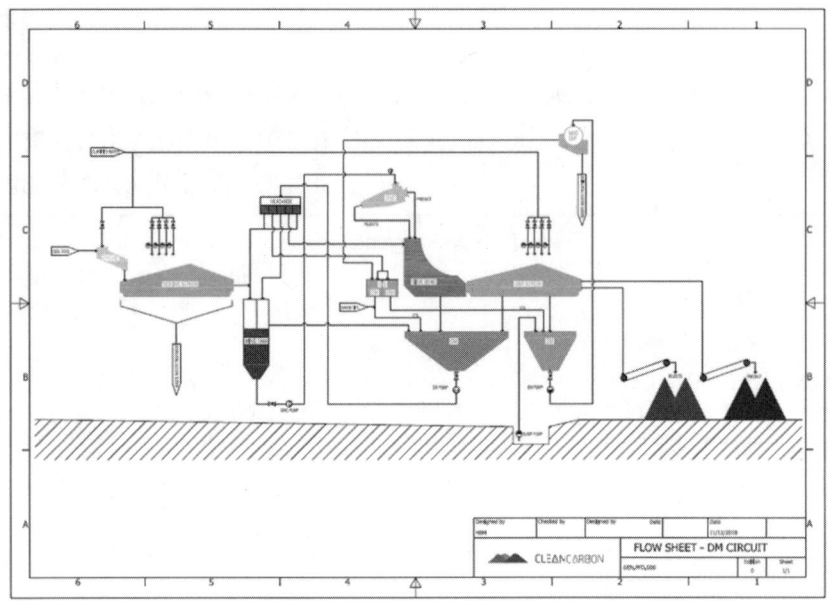

**Figure 1:** DMC section flowsheet

FLSmidth became involved at Sines in 2015 and were asked to provide a solution to increase the DMC capacity to 100 TPH from 60 TPH. The plant upgrade has involved an increase in the main circulating medium and cyclone feed pump capacities. This was complicated because of the limited capacity of the initial circulating medium tank. The tank capacity has been increased in a novel way to accommodate the increased medium volume requirements of a new Krebs 26 inch (660 mm) DMC, to replace the original smaller diameter unit.

The upgrade included the installation of a new cyclone feed pump "Wing" tank to feed the DMC feed pump, which requires an increased volumetric capacity to cater for the increase in DMC size. The DMC feed pump has a variable speed drive to allow the cyclone feed pressure and volumetric throughput to be varied, if necessary, depending on feed quality. Also, the upgrade allowed for an increase in raw coal to medium volumetric ratio for the DMC from the original 1:3 to 1:4, to improve separation efficiency.

In addition, the plant water circuit has been re-built using a new FLSmidth thickener and belt filter press to accommodate the higher and finer feed to the plant, and the subsequent increase in liquid circulation. The water circuit is fully closed with zero slurry effluent, due to the sensitive location of the plant.

## 1.2 Process flow description

The raw feed input to the plant is drawn from the bulk terminal close to the CleanCarbon operation. The raw coal received into the plant has an ash content of 12%, and the rejects, (middlings) from the DMC plant, are re-sold locally. After desliming at 1 mm the raw coal has an ash content of 8–9%.

Raw coal, at a variable top size of 12 mm to 16 mm × 0, is first dry screened to remove as much minus 2 mm material as possible. The sized coal is then fed to the desliming screen, mixed with wash water and the residual minus 2 mm particles are removed. These are pumped to a thickening hydrocyclone and a dewatering screen, and then collected as rejects from the DMC. The combination of the preliminary dry screening and efficient desliming at 2 mm, minimises the build-up of slimes in the circuit.

Desliming screen oversize is pumped to the Krebs 660 mm DMC. The clean coal and discard products are drained and rinsed free of adhering medium solids on a single split drain and rinse screen. The clean coal product is collected by conveyor and transported to the product stockpile.

Rejects, (middlings), from the screen are collected on a separate conveyor and dewatered fine raw coal and belt press cake are then added to the rejects belt.

The very low height of the plant has required that the correct medium sump has had to be duplicated to provide sufficient storage capacity to hold the required volume of circulating dense medium. Each of the two correct medium tanks is fitted with a separate medium circulation pump, which both feed to a common medium distribution headbox. The two tandem medium circulation tanks are connected by a balance pipe.

Medium specific gravity control is achieved by bleeding a balanced amount of correct medium to the dilute medium circuit. The magnetite is recovered as a concentrate of up to 2.4 SG in an Eriez wet drum magnetic separator.

## 2. DMC, geometry and specification

The DMC, (Fig. 2), has been configured with an internal geometry to account for the very high percentage of near gravity material (NGM). This is essential to get the desired underflow to overflow volumetric flow ratio. This is necessary to minimise the offset between the specific gravity (SG) of the circulating medium and the SG of separation in the DMC. To achieve this, the ratio of underflow orifice to vortex finder orifice is higher than in a DMC operating at a higher SG cut point.

**Figure 2:** Photo of DMC at Sines

With the configuration installed, the medium split at 1.24 SG of circulation is 1.162 in the overflow medium and 1.385 in the underflow. This indicates a flow split of 65% to the overflow, which is consistent with the geometry chosen. As can be seen in Fig. 6, the SG of separation which is the 50% partition factor from the partition curve is 1.30, an "offset" of 0.06.

The DMC lining is a full monolithic ceramic system (Fig. 3). This is designed for maximum wear resistance to maintain the necessary split of flows and maintain the separation efficiency, which is essential for such a sensitive duty. These liners are of a modular sectionalised construction to allow for single components, such as the apex and vortex finder, to be replaced without needing to replace the complete cyclone. This promotes optimum wear life and reduces the cost of ownership of the DMC.

**Figure 3:** Photo of Krebs monolithic lining

## 3. Stability of the dense medium suspension

The stability of the dense medium suspension at such a low circulating SG is a particular concern, as the circulating medium at the lowest SG point of 1.2 contains very little magnetite in suspension. The magnetite, which is sourced from Northern Europe, is of the highest quality used anywhere in the world. The raw magnetite SG is 5.5 and has over 95% magnetic content. Because of the dilute nature of the medium solids in suspension, the size distribution has to be a superfine grade, to achieve the stability at 1.20 SG, where the suspension is on the borderline of what is a stable dense medium suspension.

## 4. NGM and its effect on the difficulty of separation

The raw coal washability curves (Fig. 4) constructed from a typical float/ sink analysis of the feed, illustrate that approximately 90% of the raw coal is within ±0.1 SG of the cut-point, to achieve the desired product qualities of 0.93–1.24% ash. The theoretical cut points to achieve this quality range from 1.26 to 1.31 SG.

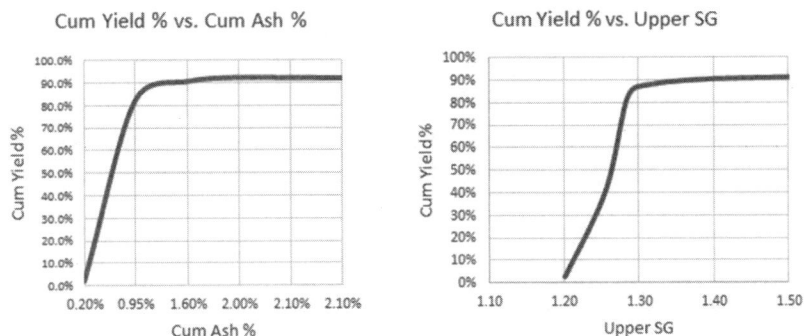

**Figure 4:** Deslimed feed washability curves

Throughout the world the difficulty of separation is defined by the % of NGM and coals are categorised by this feature. Coals which exceed 50% NGM at the desired separation cut-point are characterised as "difficult". As can be seen the 90% NGM at Sines characterises this separation as "extremely difficult".

## 4.1 Plant performance

Figure 5 illustrates the actual partition, (tromp), curve achieved at a typical operating regime in terms of circulating medium specific gravity at Sines.

Figure 5 shows the curve with the circulating medium set at 1.24 SG. The SG of separation in this typical example is 1.30. The curve is relatively symmetrical with slightly more misplaced material evident in the sinks product, than in the floats.

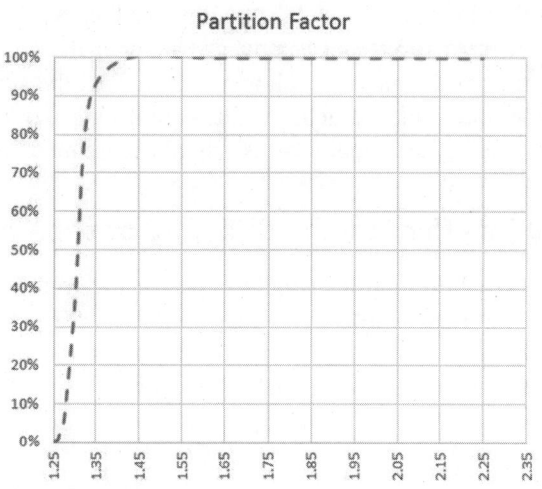

**Figure 5:** Partition curve

The Ecart Probable Moyen (EPM) figure, in this example, at the separating density, is 0.012, which demonstrates the very high efficiency achieved by the plant. This level of efficiency is critical in achieving the required product qualities.

## 5. Other global examples of low ash coal production and high NGM

The products from the CleanCarbon plant are almost unique in the world of coal production. In the large-scale coal producing countries, the major markets are for power station fuel and coking coal. Over the years there have been many plants built using dense medium technology to produce low ash coking coals. However, there have been only a few installations globally which require separation SG's of 1.3. These are usually required only for the very highest grade coking coals.

In a survey from global contacts, it is apparent that there are currently very few plants in the world which separate in DMC's at this low SG. In USA, South Africa, and Australia most low ash plants are operating at SG's around 1.4 (Cresswell, 2018), to produce coking coals at around 5% ash.

One of the regions with the most difficult coals to process in terms of NGM at the required point of separation is India. Indian Coal has been described many times as the most difficult in the world from which to produce low ash products for coking coal.

In order to process coals with this degree of difficulty it is necessary to use DMC's of smaller diameter than the very large units (up to 1.4 m diameter), now in use in many parts of the world. This is due to the higher centrifugal acceleration required to optimise separation performance in the higher NGM environment (Menglers, 1980).

In addition the ratio of raw coal feed to medium needs to be increased from the original Dutch State Mines (DSM), standard, figure of 1:3 to 1:4 minimum. Down-rating the solids throughput of the DMC is also advised to optimise efficiency. In some instances ratios of up to 1:5 are now used. The increase in this ratio does not, however, significantly increase the underflow volumetric solids capacity of the DMC.

Modern DMC's are now mostly designed with involute feed entry, to reduce turbulence, and larger Vortex Finder diameters, to allow for the increased coal to dense medium volumetric ratios commonly installed. The original DSM design had a tangential feed entry and a fixed geometrical configuration of $0.2 \times$ cyclone diameter feed entry, $0.3 \times D$ underflow opening and $0.4 \times D$ vortex finder diameter.

These features of the modern DMC allow higher volumetric throughput, than the original DSM design and as already mentioned, the key orifice dimensions can be adjusted within limits, to optimise the DMC performance.

In 1979 one of the first low ash coking coal plants, in India, was built at West Bokaro, in Bihar state. The plant was designed in the UK with assistance under license from DSM, the inventors of the DMC. In this plant over 80% NGM was evident at a separation SG of 1.42, which was the desired DMC cut-point to achieve a 17% ash coking coal product. This ash content is very high by world standards but not uncommon in India. The coking batteries were specially designed to cater for this higher ash product.

When the plant was commissioned in early 1982, it was evident that the normal design parameters used by DSM were inappropriate for Indian coals of this difficulty. Until that date DSM insisted their very rigid rules had to be followed exactly by their licensees as regards DMC, capacity, geometry, raw coal to medium volumetric ratio, and magnetite preparation. In order to get the West Bokaro plant to the desired level of efficiency of separation, the geometry of the 600 mm diameter DMC's had to be modified to increase the underflow orifice diameter, thereby increasing the ratio of underflow opening to vortex finder, and consequently the offset between SG of circulating medium and SG of separation, and DMC solids capacity.

Similar issues had previously been brought to the attention of DSM from South Africa in the 1950s, after the first application of a DMC in that country at the fuel research institute pilot plant in Pretoria (Van der Walt, 1982), but these were not acknowledged by DSM. Again it was discovered that the original DSM standard geometry was less than ideal with coals exhibiting higher levels of NGM, and inferior magnetite medium solids.

Many of DMC plants had been built by that time under the DSM license, but the vast majority were in the Northern Hemisphere, which by comparison were far easier separations than those in the emerging coalfields of the ancient Gondwanaland continent (India, South Africa and Australia).

Today our knowledge of the more difficult coals is widespread and modern DMC plants can be designed to allow for these more difficult separations. Volumetric throughput, and cyclone geometry, in terms of inlet, vortex finder and underflow orifice diameter, can be tailored to suit the separation and NGM.

## 6. Conclusion

The CleanCarbon plant at Sines in Portugal is a unique coal preparation plant. The SG of separation, and circulating medium SG are probably the lowest in any current operation, in a DMC plant anywhere in the world. The plant successfully produces niche products for small, but highly specialised markets. The efficiency of separation in the DMC is exceptional, and this is remarkable given that this is the first and only CPP ever built in Portugal.

Work continues at Sines, with a proposal to fully automate the DMC by addition of extra medium storage and the introduction of automated SG and tank level control. Experimentation is also ongoing to look at different processes to treat the finer sized coal minus 1–2 mm to see if very low ash products can be produced from this size fraction.

## 7. Acknowledgement

David Woodruff would like to acknowledge the help and personal guidance received from Mr Douglas Jenkinson, who for many decades was the UK representative on the Organising Committee of the International Coal Preparation Conference. Doug passed away in November 2018; one of his final actions was in supporting this paper for admission to the conference. Doug was a renowned figure in the world of coal preparation, from his position of Chief Coal Preparation Engineer of the UK National Coal Board, and his many decades of service to the Coal Preparation industry.

The authors would also like to thank Jinxiang Chen, Research and Development Engineer – CleanCarbon AS Portugal, for his work on sampling and analysis of results.

## 8. References

[1] Cresswell M. (DRA Global Pty). *Personal Communication.* 2018.

[2] Menglers J. *The Influence of Cyclone Diameter on Separation Performance.* ICPC India. 1980.

[3] Van der Walt P. *Advances in Coal Washery Practice in South Africa.* PhD Thesis. 1982.

# 37

# Development of a low specific gravity cutpoint spiral

Barbara J Arnold[1], Wynand Erasmus[2], Faan Bornman[2], Peter J Bethell[3]

*[1]PrepTech, Inc., Apollo, PA, USA*
*[2]Multotec Process Equipment Pty, Ltd., Kempton Park, South Africa*
*[3]Marshall Miller & Associates, Bluefield, VA, USA*

**Abstract:** For coal cleaning plant optimization, achieving equal incremental ash in all plant circuits is vital. It is important to have a technology that can efficiently clean the 1.0 mm × 0.15 mm (2.0 mm × 0.15 mm for South Africa) size fraction at separating gravities lower than ~1.50–1.55, especially for metallurgical coals. As middlings rich coals become the norm, it is essential that lower cutpoints are achieved in this fraction to enable reasonable cutpoints in the dense media cyclone circuit to be achieved. Spirals are used to process this fraction, but typical separating gravities are well above 1.55. Multotec has developed a new spiral for this application, adding three additional turns to a seven-turn, two-stage spiral, and creating a 10-turn spiral. Initial test work indicates that cutpoints of about 1.55 sg or lower can be achieved with good efficiencies and yields at feed rates approaching 2.5 tph and 30 gpm (or higher). Successful operation at these low cutpoints will allow easy-to-operate spirals to be used more effectively in the 1 mm × 0.15 mm. In fact, with middlings rich coals, it is often impossible to produce coking coal quality from a plant if the fines circuit cuts higher than 1.6 sg.

**Keywords:** Spirals, low specific gravity cutpoint, fine coal cleaning, equal incremental ash

## 1. Introduction

Spiral concentrators are widely used for the beneficiation of fine coal in the nominally 1.0 mm × 0.15 mm (2.0 mm × 0.15 mm for South Africa) size fraction and offer a number of advantages for fine coal beneficiation including low operating and capital costs, operational simplicity, excellent tolerance to variation in feed conditions, low maintenance and high reliability. Spirals typically separate effectively at cutpoints between 1.70 and 1.80 specific gravity (SG). Under more carefully controlled conditions of feed rate and percent solids, cutpoints around 1.60 are achievable. The most noted limitation of current spiral models is the inability to achieve cutpoints below these levels.

Traditionally, cutpoints at around 1.70 meant that the spiral's duty was to separate essentially "clean" coal particles from those with a substantial mineral component. On current spiral models, coal particles in the SG range

1.20–1.60 are predominantly water-born. Reducing the cutpoint means that the spiral now has to discriminate between particles of clean coal and particles of "less-clean" coal. For this to happen, the coal particles need to have more contact with the trough surface to enable those particles denser than the new cutpoint to report to either the middling or reject stream.

## 2. Background

Spirals are characterized as a film concentrator, a relatively simple device used to separate particles according to size, density and shape (Glass et al., 1999; Li, 1994). Spirals exhibit one of the most complex flow regimes amongst gravity separators. Spiral concentrator flows display laminar to increasingly turbulent behavior. The primary flow is that of the slurry descending the inclined portion of the trough. Secondary flow occurs radially across the trough. During their passage through a spiral, particles are subjected to hydrodynamic drag and lift, friction, gravity, and centrifugal forces.

In general, the motion of particles depends on the particle properties (size, density, and shape), the operating conditions (slurry density and flow rate), and the spiral design (trough shape, inclination, and number of turns) (Holland-Batt et al., 1984; Richards et al., 1985; Das et al., 2007). With coal spirals, accumulation of relatively lights particles in the middle flow is likely to occur within the range of feasible operating conditions. This makes the coal spiral very suitable for separating organic contaminant from soil when these are sufficiently liberated. The concentration of contaminants in the middle flow can be optimized by adopting splitter positions. A deviation from optimum operating conditions influences separation performance such as an increase in density of separation and a lower sharpness of separation.

## 3. Low cutpoint SX10 spiral

The SX7 spiral model has traditionally been used for coal cleaning in North America (Bethell and Arnold, 2003; Bethell and DeHart, 2004). The SX7 is a seven turn, wide diameter spiral with a "drop-in" repulper and off-take to the centre column. The SX7 was used as a benchmark at 35% solids by mass. Figure 1 shows the SX7 spiral model.

The SX10 spiral model is a longer version of the SX7. This is to allow for an increase in residence time and two stages of high density particle removal. The spiral has two "drop-in" repulpers and off-takes to the centre column. Figure 1 also shows the SX10 spiral model.

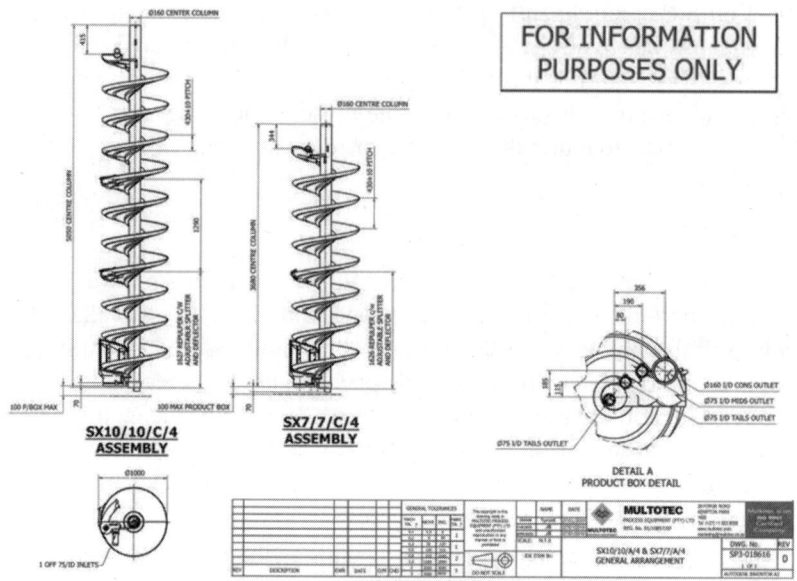

**Figure 1:** The Multotec SX7 and SX10 spiral models

The theory behind the low cutpoint is that the gangue or mineral-containing particles are removed from the spiral trough. This opens up available separation surface and the bed "slides down" towards the centre column. Instead of having a trough filled with mineral-rich particles and coal, particles of "less clean" and clean coal are now present in the trough. Figure 2 shows the principle.

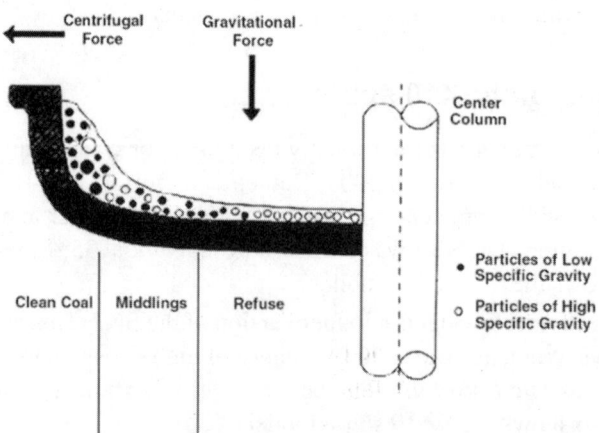

**Figure 2:** Operation principle of the SX10 spiral

## 4. Experimental procedure

## 4.1 Test units

The spiral test units were equipped with a "mouth organ" across the trough in order to collect samples covering the entire profile as shown in Fig. 3. In addition to the samples collected across the trough, a refuse sample was collected from the center column outlet.

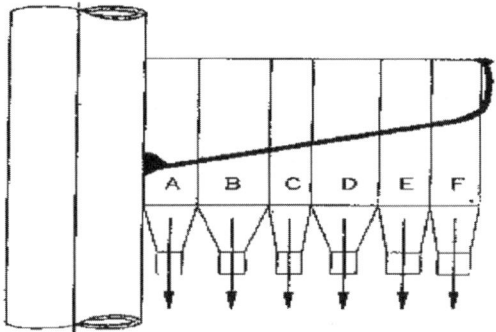

**Figure 3:** Mouth organ product box for test units

## 4.2 South African coal

The SX7 at 35% solids (by mass) was used as a benchmark for the test work. Approximately 1 ton of coal was obtained from a colliery in South Africa. The particle sizes were in the range typically being used for spiral feed. The SX10 was compared to the SX7 at solids concentrations as per Table 1 below. Each of the runs described above was repeated three times in order to ensure repeatability. An additional test at 9.1 m³/h (40 gpm) was also conducted with the SX10 spiral.

**Table 1:** Test work matrix

| Spiral | Solids by mass (w/w%) | Dry solids throughput (tph) |
|--------|----------------------|----------------------------|
| SX7 | 35 | 3.0 |
| SX10 | 35 | 3.0 |

The above mentioned test work was repeated in conjunction with the University of Pretoria. A coal sample from a different colliery was used in the second test. The MX7, MX10, and SX10 spirals were compared against one another. The only difference between the SX7 and MX7 is the SX7 has the auto reject into the center column, while the MX7 has a gulley around the centre column. The SX7 has, therefore, an additional outlet that comes from the centre column. Table 2 shows the operating test parameters.

**Table 2:** Repeat test work matrix

| Spiral | Solids by mass (w/w%) | Dry solids throughput (tph) |
|--------|------------------------|------------------------------|
| MX7    | 29                     | 2.3                          |
| MX10   | 28                     | 2.2                          |
| SX10   | 33                     | 2.6                          |

## 4.3 US coal

Test work was conducted on a Northern Appalachian coking coal. Two feed conditions were tested: Test 1 at 2.2 tph (2.4 stph) and 6.8 m³/h (30 gpm) and Test 2 at 1.4 tph (1.5 stph) and 4.2 m³/h (18.5 gpm) to provide a lower feed rate for comparison. Timed samples were collected across the spiral trough (A and B were combined as the flow was very low) and from the combined first and second stage refuse.

## 5. Results

### 5.1 South African coals

Ramsaywok and Mathumbu (2017) indicated that over the last few years a "low-cut" spiral technology has entered the market. The introduction of the low-cut spiral produced is a recent opportunity for fine coal beneficiation (Palmer and Weldon, 2014). Figures 4 and 5 show the partition curve for the SX7 and SX10 spirals.

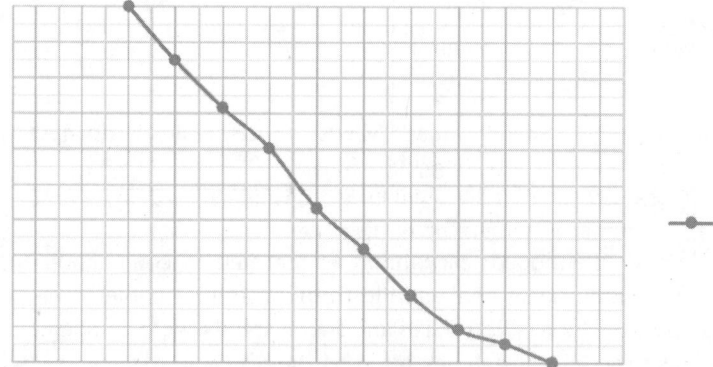

**Figure 4:** Partition curve for the SX7 spiral

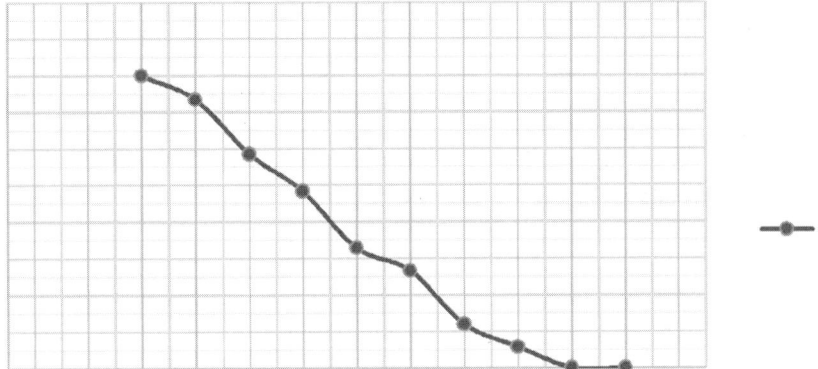

**Figure 5:** Partition curve for the SX10 spiral

Figure 5 shows a slightly lower cutpoint for the SX10 spiral, but with more misplacement of product to discard. The cutpoints and mass yields achieved were in line with the values obtained by Ramsaywok and Mathumbu (2017). A float sink analysis shows a product yield of ±40% and discard yield of ±60% on both spirals. Table 3 shows the results of the South African test work.

**Table 3:** Test work results

| | Feed | | Product (fraction F) | | |
|---|---|---|---|---|---|
| | CV (MJ/kg) | Ash (%) | CV (MJ/kg) | Ash (%) | Mass yield (%) |
| SX7 | 22.45 | 25.18 | 24.41 | 17.87 | 53.73 |
| SX10 | 21.88 | 26.66 | 26.18 | 15.50 | 34.60 |
| MX7 | 22.36 | 26.97 | 27.07 | 15.00 | 61.80 |
| MX10 | 21.74 | 28.80 | 26.90 | 14.10 | 41.80 |
| SX10 | 19.76 | 34.04 | 27.32 | 14.30 | 40.20 |

The test work showed that there was a drop in mass yield in order to obtain lower ash content in the final product. The double off-take on the SX10 spiral removed more gangue or mineral-containing particles in comparison to the SX7. The SX7 had a cutpoint of 1.61 versus 1.53 for the SX10. More misplacement of product to discard followed on the SX10. The imperfection or EPM value for the SX10 was 0.22 versus 0.19 for the SX7 spiral. EPM was defined as EPM=abs($d_{75}-d_{25}$)/2. At 40% solids concentration, the SX10 achieved a cutpoint of 1.49.

## 5.2 US coal

Partition curves were developed for the 1 mm × 0.25 mm (1 mm × 60 mesh) and 0.25 mm × 0.15 mm (60 mesh × 100 mesh) fractions from each test, combining the samples across the spiral trough profile as shown in Figs. 6–8. Low specific gravity cutpoints ($d_{50}$) were found in both tests. There is a loss of 1.3 float material that is most noticeable in Test 2.

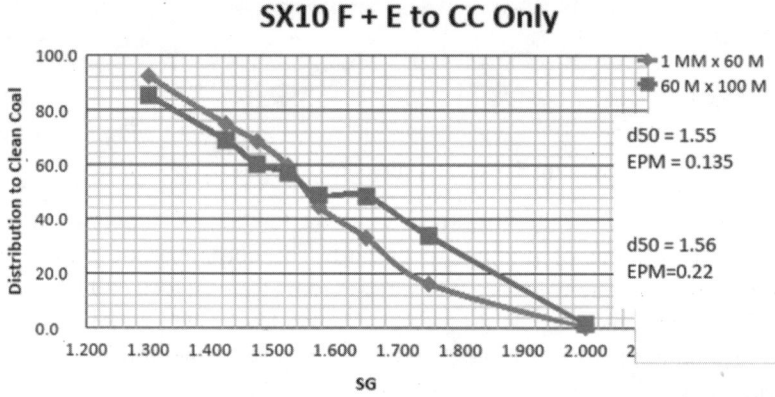

**Figure 7 Test 1:** partition curve (F+E only) 2.2 tph (2.4 stph) and 6.8 m³/h (30 gpm)

It is encouraging that both size fractions have similar cutpoints and somewhat similarly shaped partition curves—performance does not degrade much for the finer fraction. Opportunities to improve the recovery of the 1.3 float material and to improve the separation further exist by recycling middlings as is the practice for the two-stage SX7 spiral in the US. Simulations of this recycle are in progress.

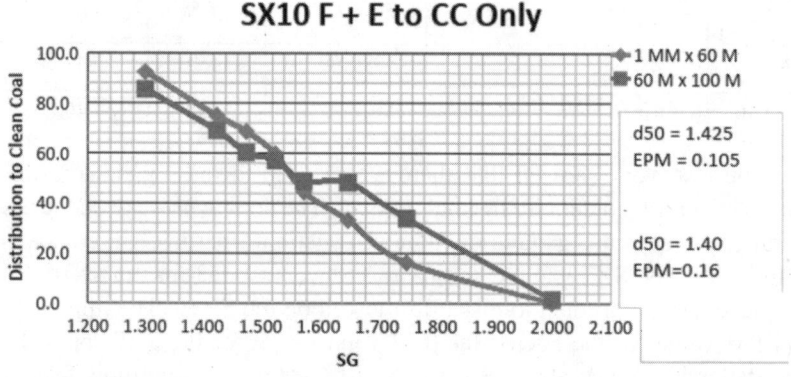

**Figure 8:** Test 2 partition curve (F+E only) 1.4 tph (1.5 stph) and 4.2 m³/h (18.5 gpm)

# 6. Summary and conclusion

Note that no curve smoothing was conducted on this data. We are encouraged that low specific gravity separations (<1.60) are possible at typical spiral feed rates as shown in the South African test work and in the US Test 1. With the usual US practice of recirculation of part of the "middlings" stream (C and D), extremely high recoveries of the low density fractions are expected with the low EP while still maintaining a cutpoint in the desired range. This is possible because the feed to the last three turns is already diluted with the removal of the refuse in the first two stages. The probable error values compare well with values for typical four turn spirals and for typical seven turn spirals (assuming recycling of the middlings, 30–35 gpm, and feed rates of 2.5 stph per start of new feed) for the 1 mm × 100M based on in-house data.

# 7. References

[1] Bethell, P.J. and Arnold, B.J. 2003. Comparing a Two-stage Spiral to Two Stages of Spirals for Fine Coal Preparation. *Advances in Gravity Concentration*, R. Q. Honaker and W. R. Forrest, Eds., SME, Littleton, CO, pp. 107–114.

[2] Bethell, P.J. and DeHart, G. 2004. Fine Circuit Upgrades at Hobet (Beth Station) Plant. Proceedings, CoalPrep 2004, Lexington, KY, pp. 37–49.

[3] Das, S.K., Godiwalla, K.M., Panda, L., Bhattacharya, K.K., Singh, R. and Mehrotra, S.P. 2007. Mathematical modelling of separation characteristics of a coal-washing spiral. *International Journal of Mineral Processing*, 84, pp. 118–132.

[4] Glass, H.J., Minekus, N.J. and Dalmijn, W.L. 1999. Mechanics of coal spirals. *Minerals Engineering,* March 1999, 12(3): 271–280.

[5] Holland-Batt, A.B., Hunter, J.L. and Turner, J.H. 1984. The separation of coal fines using flowing-film gravity concentration. *Powder Technology*, 40(1–3): 129–145.

[6] Honaker, R.Q., Jain, M., Parekh, B.K. and Saracoglu, M. 2007. Ultrafine coal cleaning using spiral concentrators. *Minerals Engineering*, Vol. 20, pp. 1315–1319.

[7] Li, M. 1994. A study of coal washing spirals. PhD thesis, Julius Kruttschnitt Mineral Research Centre, Department of Mining and Metallurgical Engineering, University of Queensland.

[8] Palmer, M.K. and Weldon, W.S. 2014. A new low cut point spiral for fine coal separation. *CPSA Journal*, Vol. 12, No. 4, pp. 27–34.

[9] Ramsaywok, P. and Mathumbu, P. 2017. A straight up single pass comparison of a low-cut spiral and a conventional spiral in the South African coal industry. *Southern African Coal Processing Society*, Coal Processing – the key to profitability, 22–24 August.

[10] Richards, R.G., Hunter, J.L. and Hollant-Batt, A.B. 1985. Spiral concentrators for fine coal treatment. *Coal Preparation*, Vol. 1, No. 2, pp. 207–229.

# 38

# Use of washed thermal coals for power generation in India - status and issues

D.N. Prasad

*Former Adviser, MoC & Presently Adviser, SCCL*

## 1. Introduction

Internationally the coal produced from the mines (Run of Mine Coal) is prepared after crushing, sizing and washing before dispatching the same to the consumers. The pricing of coal is on the basis of saleable product and consumers are at liberty to choose the quality of coal as per their requirements. Washing of coal provides for maintaining the consistency in the quality of coal supplied. A number of studies taken up earlier on the subject of the use of washed coal for power generation have proved it to be beneficial from both environmental and economical points of view.

Indian coals are of high ash content and low heat value in nature compared to American or Australian coals. Added to this, more than 93 percent of the coal mined in India is produced from opencast mines, which also contributes to the dilution of there *in situ* quality. ROM coal extracted from opencast mines in India typically contains ash contents in the range of 35 - 45 percent and low calorific value (2500-5000 kcal/kg).

Generally, high ash content leads to erosion in the power plant components, difficulty in pulverisation, poor emissivity, and flame temperature, low radiative transfer, generation of excessive amounts of fly ash containing large amounts of unburnt carbon, etc. at user end. In addition, the transport of such high-ash ROM coal across long distances is wasteful as it carries large quantities of ash-forming minerals that cause further congestion in already-constrained rail transportation and port-handling systems. The transport of high ash coal across long distances also contributes to the emission of carbon-dioxide ($CO_2$) and other greenhouse gases (GHG) from the mode of transport (rail and road).

Past case studies that have looked at specific power plants in India have shown that the use of beneficiated non-coking coal (a) facilitates use of higher quality fuel with consistent heat value, (b) reduces fuel quantity requirements (handled and transported) for the same heating value, (c) enhances utilisation of installed generation capacity, (d) reduces capital funding requirements in the power plant, (e) reduces cost of transportation (per kWh), (f) decreases fly ash volume in both pre-combustion and post-combustion stages, and (g) reduces land requirement for ash disposal. Even FBCs that are designed to burn low-grade high ash coal can operate more efficiently with higher grade low ash coal.

Government policies encourage beneficiation of coal. The Ministry of Environment, Forest and Climate Change (MOEF&CC) of the Government of India notified the Environmental (Protection) Amendment Rules (EPAR) vide Gazette Notification G.S.R 02 (E) dated 02 January 2014 (CPCB, 2014). As per EPAR (2104), power plants with a capacity of 100 MW or above located between 500-749 kilometres (km) from the pit head "shall be supplied with and shall use *raw or blended or beneficiated coal with ash content not exceeding 34 percent on the quarterly average basis from 05 June 2016.*"On the other hand, a stand-alone thermal power plant (of any capacity), or a captive thermal power plant of installed capacity of 100 MW or above, located beyond 1000 km from the pithead or, in an urban area or in an ecologically sensitive area or in a critically polluted industrial area, irrespective of its distance from the pit-head, except a pit-head power plant, *"shall be supplied with and shall use raw or blended or beneficiated coal with ash content not exceeding 34 percent on quarterly average basis with immediate effect."*

In addition, all new coal plants from the 13th Five Year Plan onwards (commencing April 1, 2017) have been mandated to use supercritical technology, and 144 existing plants have been assigned mandatory efficiency targets which will require the use of higher-quality coal. Despite the benefits and supportive policy interventions that have been in place for over two decades, coal washing has not been adopted on a large scale by coal producers and users (particularly power generators).

## 2. Benefits of using washed coal for power generation

As per the study "Implementation of Clean Coal Technology through Coal Beneficiation" by Montan Consulting GMBH, Germany in October 1998, instituted by Asian Development Bank (ADB), the envisaged benefits of using washed coal for power generation are listed below.

| Parameter | Benefits |
|---|---|
| **1. Transport** | |
| (i)  Reduction in transport cost | Depends on distance and ash reduction. E.g. 1000 km distance and ash reduction from 41% to 34% results in savings of 15%. |
| (ii)  Reduction in CO2 Emissions due to reduced fuel consumption in transportation. | Depends on distance and ash reduction. E.g. 1000 km distance, ash reduction from 41% to 34% results in savings of 15%. |

*Contd...*

*Contd...*

| Parameter | Benefits |
|---|---|
| **2. Power Plant site** | |
| (i)    Decrease in Auxiliary Power | 10% decrease for every 10% reduction in feed coal ash. |
| (ii)   Decrease in Auxiliary Fuel | 50% reduction when using washed coal (present average 4 ml/kwh). |
| (iii)  Improvement in Plant Load Factor | 1.5% improvement for every 10% reduction in feed coal ash. |
| (iv)   Reduction in Operating and Maintenance costs | 20% cost reduction for every 10% reduction in feed coal ash. |
| (v)    Reduction in Capital Investment for New Plants | 5% reduction in Capital Investment when using coal with 34% ash instead of 41%. |
| (vi)   Reduced Land Requirement for Ash Disposal | Using coal of 34% ash instead of coal with 41% reduces land requirement by approximately 30%. |
| (vii)  Reduced Water Requirement for Ash Disposal | Using coal with 34% ash instead of 41% reduces water consumption by approximately 30%. |
| (viii) Reduction in $CO_2$ Emissions | Reduction in the range of 2-3% when using washed coal. |
| (xi)   Improvement in Electro Static Precipitator (ESP) Efficiency | Using washed coal improves ESP Efficiency from 98% to 99%. |

## The major conclusions were that:

(i) Non-coking coal used in power plants should be washed up to 32 ± 2 percent but this should not be a blanket approach for all coal fields, because of varying ROM and washability characteristics.

The policy that power plants located more than 1,000 km from coal mines should use coal with ash content less than 34 percent applied to every power plant will be very expensive and unlikely to provide 'good environmental value' for money spent.

The policy should focus on environmental standards to be achieved allowing consumers and suppliers to find the most economical means to achieve those standards.

(ii) Rejects should have an ash content >60-65 percent to allow safe disposal without the danger of self-combustion. If this is not possible, utilising the rejects in a FBC at the washery site for power generation should be considered.

(iii) Widespread implementation of coal washeries will not occur spontaneously and so the government must (a) encourage private investors to enter the coal washing business (b) internalise environmental costs by enforcing a system of fees and fines reflecting the economic costs of pollution (emissions, land and water use).

(iv) In the context of attracting private operators as partners for Coal India Ltd. (CIL), the build, own and operate (BOO) model will be relatively more difficult on account of the contractual complexity.

The build, own, operate, trade (BOOT) model will be more attractive to investors as it offers a better balance of risks and rewards than the BOO model. Both these models may be based on a partnership of CIL and private capital.

(v) To promote the setting up of washeries, a Clean Coal Fund could be created. The sources of capital for such a fund could include international financing institutions, domestic banks and the state budget.

(vi) In some countries, environmental levies, or fines on plant operators for violating standards are used to fund environmentally desirable projects. This may be an appropriate system for India.

(vii) Among the many organisations and ministries involved in questions over using washed coal, the MOC should have a clear responsibility to co-ordinate and implement policy regarding coal washing.

The study carried out by Central Mine Planning & Design Institute (CMPDI) and National Productivity Council (NPC) in 1988-89 at Satpura TPS of MPEB using 34 percent washed coal in one 210 MW unit brought out the following findings.

| Parameter | From (RoM Coal) | To (with washed coal) | % Improvement |
|---|---|---|---|
| (i) Improvement in Plant Utilisation Factor (PUF) | 73% | 96% | 31.51% |
| (ii) Improvement in Generation in MU/day | 3.71 | 4.83 | 30.19% |
| (iii) Reduction in Specific Coal Consumption in Kg/Kwh | 0.77 | 0.553 | 28.18% |
| (iv) Elimination of Specific support fuel in ml/unit generated | 5 ml | NIL | No need for support fuel |
| (v) Reduction in Rejects | 0.3-0.4% | 0.03% | 91.43% |

*Contd...*

*Contd...*

| Parameter | From (RoM Coal) | To (with washed coal) | % Improvement |
|---|---|---|---|
| (vi) Increase in Boiler Efficiency | 86.57% | 89.51% | 3.40% |
| (vii) Reduction in Smoke & Dust Emission in g/m³ | ESP inlet: 29.78 | 17.23 | 42.14% |
| | ESP Outlet: 1.57 | 0.299 | 80.96% |
| (viii) Reduction in Alpha Quartz | 14.5% | 11% | 24.14% |

The analysis of the National Thermal Power Corporation's (NTPC) Dadri Power Plant which used washed coal with around 34-35 percent ash from Central Coalfield Ltd.'s Piparwar washery revealed the following results.

(a)  Increase in operating hours up to 10 percent.

(b)  Increase in PLF up to 4 percent.

(c)  Increase in PUF up to 12 percent.

(d)  Reduction in breakdown period up to 60 percent.

(e)  Increase in overall efficiency up to 1.2 percent.

(f)  Increase in generation per day 2.4 Million units (MU or million kWh).

(g)  Reduction in support fuel oil 0.35 ml/kWh.

(h)  Reduction in specific coal consumption of 0.05 kg/kWh.

(i)  Increase in total units sent out per day about 2.3 Mus.

(j)  Saving in land area for ash dumping 1 acre/year.

(k)  Reduction in CO emissions (reduced transportations/coal combustion > 600,000 tonne/year).

(l)  Overall benefit resulting from using washed coal of Rs.119 million/year excluding the anticipated reduction in maintenance cost.

(m)  For the 4x210 plant, this represented a savings of Rs 0.02/kWh.

(n)  Savings in demurrage to railways about Rs 7/tonne of coal received.

A case study carried out by CEA on the Dadri plant concluded that if beneficiated coal were used instead of ROM coal, the anticipated improvement in the PLF of the power plant would be of the order of 5 to 10 percent (even more in some cases). The same study also indicated that a mere3 percent improvement in PLF would balance the additional cost of beneficiation of coal at the Piparwar project, which was located about 1,300 km from the plant site. The marked decrease in the cost of generation, with the improvement in the PLF for a given distance, was established through this study.

The Dahanu Thermal Power Station (2X250 MW) reported the following results for use of 30 percent ash washed coal produced at the USAID/DOE sponsored Korba washery. The results included:

(a)  Ash generation reduced by 8.5 percent

(b)  PLF increased by 15.8 percent

(c)  Cost per unit reduced by approximately 10 percent

(d)  Plant availability increased by 6.5 percent

(e)  Specific oil consumption decreased by 65 percent

(f)  Aux power consumption decreased by 5.4 percent

(g)  Power generation increased by 16 percent

(h)  Savings Rs 0.28/kWh

It is evident from the foregoing that use of washed coal in thermal power plants has significant environmental benefits - would reduce carbon emissions per unit energy generation through improved thermal efficiency; reduction in other pollutants SOx, NOx and SPM. Washing also reduces sulphur content in coals. As such sulphur is not an issue with Indian coals. With the use of washed coal which can be combusted efficiently with less air, the formation of NOx could be reduced. Use of low ash coal results in a reduced concentration of SPM in the flue gas leading to a reduced load of particulate in Electrostatic Precipitator (ESP) and thus improving the quality of air. Reducing the ash content in coal through washing results in reduced fly ash generation post-combustion and thus extending the life of ash disposal yards and a lesser amount of areas for ash ponds.

The economic benefits of the use of washed coal for power generation include improved plant operations and thus reduced downtime and improved power generation, reduction in operating and maintenance cost, improved PLF, improved thermal efficiency, decrease in auxiliary power, reduced water requirement, reduced transportation costs due to lesser ash in washed coal. The reduced requirement of land for ash disposal etc.

As per the stipulations of MoEF&CC, coal of less than 34% ash only needs to be supplied to power plants if they are located 500 km. and beyond from the coal source.

The committed linkage by Coal India Ltd. (CIL) to the thermal plants located beyond 500 km. distance away from pit-heads was 236 million tonnes in 2015-16 which was likely to increase to 271 million tonnes by 2019-20. All this coal will have reduced ash content below 34% as a consequence of setting up of additional washeries.

Currently, the availability of coal of less than 34% ash from CIL sources was about 208 million tonnes and by 2019-20 around 63 million tonnes of additional coal needs to be washed. This excludes about 129 million tonnes of coal being supplied to the pit head plants.

The current washed coal output from CIL for thermal coals is around 11 million tonnes and CIL envisages setting up 9 new washeries for thermal coal washing with a raw coal throughput capacity of about 94 million tonnes per annum with an estimated investment of around Rs.1500 crore.

Since setting up of new washeries by CIL for complying with the conditions of MoEF&CC entails huge capital investment, the matter was discussed a number of times with power producers for firm Back to Back commitment for Washed Coal and rejects. The firm commitment needed for off-take from the said washeries was not forthcoming from the power generators. As a result, CIL was skeptical whether the investments being made in washeries would be productive.

NTPC mentioned that they are incurring an additional cost of around Rs.0.35 per unit by using washed coal at their Dadri TPS and thus the cost of generation would be higher if they resort to using of washed coal in other plants.

CIL argued that washing of coal involves additional cost compared to RoM coal on account of additional CAPEX and OPEX for processing the coal. However, the cost cannot be offset by means of additional revenue corresponding to higher heat value after washing of coal. Therefore, the power sector needs to consider this aspect and give their commitment.

It also emerged from the discussions that stipulation of 500 kms distance from the coal sources by MoEF&CC is not providing a level playing field to all the generators since stations located within 500 kms need not use coal of less than 34% ash and have cost advantage compared to those located beyond 500 kms. It was, therefore, opined that there should be a uniform mandate that every tonne of coal being used for power generation should be washed to less than 34% level of ash.

It was further pointed out that like the cost of wheeling of power at about Rs.0.35/Kwh, the cost of washed coal is also adding to about Rs.0.3 to Rs.0.35/ Kwh. This additional cost is making the generators unable to compete for dispatch of power.

Another aspect concerning coal washing is the disposal of washery rejects. The global practice is to use the rejects for filling the mine voids. However, MoEF&CC while granting EC for washeries insists for use of washery rejects for power generation which is not viable in a number of cases. If rejects are of 65-70% ash content, they can be used for filling of mine voids.

Accordingly, a holistic view needs to be taken on the usage of washed coal considering the aspects of the environmental impact of using raw coal vis-à-vis washed coal, washing cost, the impact of the increased cost of power generation, etc.

**Issues for consideration**

    (i) **Environment and Economics** - The use of washed thermal coal for power generation have been proved to be environmentally and economically viable. However, certain issues as listed below are getting attention for consideration of industry.

(ii) **Concerns of consumers** - The directives of MoEF&CC on the basis of distance from the pitheads is working as a disincentive to the power sector since plants located at a distance of less than 500 km from pitheads have better price advantage as they are not required to use the washed coal compared to those located beyond 500 kms from the pitheads who are required to use the washed coal. This is adversely affecting the economics of power generation. Therefore, this needs to be corrected in the overall interest of the economics of power generation.

(iii) **Mindset of the Power Sector** - When they are able to keep the emissions within the prescribed limits why to go for washed coal.

(iv) **International comparability** - Internationally, the concept of trading coal is based on "Saleable coal" i.e. the RoM coal is prepared after mining including washing and the price is fixed accordingly. Consumers are at liberty to choose the required quality of coal. In the present situation, CIL offers both RoM coal as well as prepared/washed coal. However, in spite of switching over to fully variable GCV based grading of thermal coals from the earlier UHV system of grading, the enrichment of heat value/reduction of ashin coals after washing (of below G10 grade i.e. G11 to G17) is not off-setting the cost of washing and the economics is not working out in favour of coal companies. It has become an issue to enhance the price of coal for which power utilities are reluctant to bear the additional cost. The situation is further aggravated due to a slump in the international markets.

(v) **Tax/levy Issues** - Some coal-producing States are proposing to charge the royalty on washed coal instead of raw coal. This step would be adding to the cost of coal and thus affecting the economics of power generation adversely.

(vi) **Disposal of washery rejects** - Some States are of the view that washing of coal at pitheads would be detrimental to the local environment particularly from the point of disposal of washery rejects. They are of the opinion that consumers should carry the coal to the destination and set up washeries at the plant location. This approach defeats the objective of coal washing. Further, MoEF&CC insists for utilisation of washery rejects for power generation. This may not be economically viable. It would be preferable to dispose of the rejects in landfills and backfilling of mined-out areas.

(vii) **Conditions of NCDP** - Coal companies are required to undertake verification of consumers whom coal is being supplied to for their bona fides. In case of consumers who are taking coal and getting it washed at a facility of their choice, it becomes difficult to verify the operations particularly the disposal of rejects by the consumer.

(viii) **Alternative technologies** - Coal suppliers should also examine blending/dry beneficiation or any other better technology for washing coals to make them economically viable.

(ix) **Pricing policy** - Most importantly coal companies should indicate the pricing of raw and washed coals in different projects/areas in order to make the utilities to take a call for use of washed coal as per the notified price and in accordance with their statutory obligations.

(x) **Contractual issues raised by washery construction companies regarding the conditions laid by CIL in their NIT like 100% bank guarantee instead of 10% BG earlier;** finalisation of technology/ washery circuits beforehand; to float tenders only when land, railway siding, statutory clearance are in place; and a review of penalty and bonus clauses. Since these issues are hampering the installation of new coal washeries on BOO basis, CIL/SCCL may put up washeries to suit the raw coal characteristics with their capital, while operation & maintenance of such washeries may be handed over to private players. This will also resolve issues arising out of the disposal of washery rejects since CIL/SCCL are best placed to utilise the rejects in CFBC plants located adjacent to the washery or backfill the unusable rejects in closed mines wherever available.
**The issue of disposal of washery rejects by coal linkage holders has also been a point of contention. To address the same following is suggested.**

(xi) The Washeries may be set up either by CIL/its subsidiary coal companies under BOM/BOO mode under a pre-specified arrangement or by the consumers on their own or by a third party on behalf of the consumers.

(xii) The linkage-holder shall declare the washing arrangements he has entered into with a washery operator on their website and also submit returns to the respective coal companies, District Mining Officer (DMO), State Pollution Control Board (SPCB), CCO and the concerned state electricity regulator or central regulator indicating the quantity of coal supplied to the washery, clean coal obtained and middlings/rejects generated.

(xiii) The linkage holder shall declare the manner in which the middlings/ rejects are being disposed of. If they are being used for power generation, all the details of supply of such middlings/ rejects to the concerned power station, rate at which these middlings/ rejects have been supplied to the power station, purchase of power from such middlings/ reject based stations, the gains, if any, by means of trading the middlings/ rejects.

(xiv)   The linkage holder has to declare the benefit derived from the disposal of middlings/rejects and the methodology for passing on the benefit derived from such transaction to the public through submission of details to the concerned Electricity Regulator.

(xv)    The CCO shall scrutinize, based on the return submitted by the Linkage holder, the total coal fed into the washery, the quantity of clean coal and middlings/rejects produced. The CCO shall recommend appropriate action, to be taken by the concerned coal company, in case of any deviation noted.

(xvi)   **The Way forward** - The commitment of the Government of India to CoP 21 for reduction of emissions and GDP intensity of emissions implies the adoption of all possible measures for low carbon growth. Use of washed coal for power generation is envisaged to reduce the carbon emissions up to 3%. It should be our endeavour to make it possible utilising washed coal for power generation.

(xvii)  Since coal washing is best done by the coal supplier (as per international practice), National Coal Companies with a production share of over 90%, should discharge their responsibilities and install their coal washeries at the pithead (whose operations may be outsourced to ensure higher efficiency and lower cost) and supply washed coal to power plants located at distances greater than 500 km from the pithead (or to power plants located in urban areas) with back to back offtake arrangements with such power plants.

(xviii) Finally, it's the pricing of coal (washed coal and other washery products)by National Coal Companies which needs a critical review.

## 3. References

[1]  ADB Report 1998 – Implementation of Clean technology through Coal Beneficiation

[2]  Different journals of CPSI

[3]  Reports of Working Group on Coal

[4]  Reports of Coal Controller of India

[5]  Annual Reports of MoEF&CC

[6]  Coal beneficiation in India –Status & Way Forward by Observer Research Foundation

# 39

# Comparison of heavy medium consumptions between raw coal desliming and non-desliming processes

Shuyan Zhao

*Beijing Guohua Technology Group (Part of GT Global Group)*

**Abstract:** Based on author and his team's detailed investigation, study and analyses of magnetite consumption data collected from large numbers of coal preparation plants including raw coal desliming and non-desliming process plants, the data presented in this paper clearly show the magnetite consumption primarily depends on the selected processing technologies, flowsheets, equipment selection and effective measures taken to manage and reduce magnetite loss but is not related to selection of raw coal desliming or non-desliming process. In other words, the average magnetite consumptions in whatever processes with raw coal desliming or non-desliming are the same.

**Keywords:** Desliming, Non-desliming, Heavy medium consumption, HM cyclone

## 1. Introduction

In 2008, the raw coal throughput to heavy medium process was about 43% of total raw coal processed and first time surpassed that to jig process in China, which primarily resulted from development & extensive application of high-efficient and simplified gravity-fed 3-product HM cyclones without desliming of raw coal feed during that period of time. This technology already becomes the first choice for Chinese coal plants to select to process raw coal feed specially for difficult-to-wash or very difficult-to-wash coals.

It has been argued for a long time about the magnetite consumption of the gravity-fed 3-product HM cyclone process without desliming of raw coal compared to the conventional HM cyclone process with desliming of raw coal. This argument has never stopped since and during development and application of the 3-product HM cyclone process without desliming of raw coal feed. People, promoting conventional HM cyclone process with desliming of raw coal feed, believe they must select raw coal desliming HM process in order to realize a low magnetite consumption.

Up to today, the gravity-fed 3-product HM cyclone process without desliming of raw coal feed has been extensively used in approximately 600 coal process plants in China, Mongolia and India. We have enough data to compare the magnetite consumptions between the HM cyclone processes with and without desliming of raw coal feeds. The data collected confirm that magnetite consumption primarily depends on the selected processing technologies, flowsheets, equipment selection and effective operation measures taken to manage and reduce magnetite loss but is not related to selection of raw coal desliming or non-desliming process. In other words, magnetite consumptions in whatever processes with raw coal desliming or non-desliming are the same.

## 2. Examples of extremely low magnetite consumption

## 2.1 Examples of Extremely Low Magnetite Consumption without Desliming of Raw Coal

### 2.1.1 Coal Preparation Plant A from Shandong Energy Ling Mining Group

The plant was designed with a capacity of 1.5 MTPA using a gravity-fed 3-product HM cyclone to process non-deslimed raw coal feed, fine coal HM cyclone to process coarse fine fraction and flotation to process fine coal. The raw coal is classified as very-difficult-to-wash coal. The magnetite consumption was slightly less than 0.95 kg/t at the beginning when the plant was started up in 2013. Average magnetite consumption was significantly reduced to 0.45 kg/t during July-December of 2014 after improvement of magnetic separator process, enhancement of water spray on the Drain & Rinse dewatering screens, clarity of the spray water, adjustment of throughput of circuits, control of magnetite quality and improvement of operation management. The average magnetite consumption was reduced to 0.44 kg/t in 2015 with an annual throughput of 1.36 MTPA. The magnetite consumption was further reduced to 0.35 kg/t during Jan-Sept 2017. The lowest magnetite consumption happened in Feb 2017 at 0.22 kg/t. The magnetite consumption of the plant during 2014-2015 is given in Table 1.

**Table 1:** Monthly Magnetite Consumption (kg/t) during 2014 and 2015

| Month \ Year | 1 | 2 | 3 | 4 | 5 | 6 | 7 | 8 | 9 | 10 | 11 | 12 | Average |
|---|---|---|---|---|---|---|---|---|---|---|---|---|---|
| 2014 | | | | | | | 0.27 | 0.54 | 0.51 | 0.48 | 0.45 | 0.43 | 0.45 |
| 2015 | 0.45 | 0.46 | 0.45 | 0.40 | 0.47 | 0.42 | 0.43 | 0.40 | 0.51 | 0.40 | 0.37 | 0.51 | 0.44 |

### 2.1.2 Coal Preparation Plant B from Jizhong Engergy Fengfeng Group

The plant was rebuilt in 2007 to expand its plant throughput to 10 MTPA, which consists of 3 branch coal plants. The three plants were all designed with gravity-fed 3-product HM cyclones without desliming of raw coal feed and fine coal flotation circuits. The magnetite consumption was averaged at about 2 kg/t in 2009. However, the average magnetite consumption of 0.50 kg/t was reached in 2012 after the measures taken to improve process operation and management mentioned above. The average magnetite consumption was further reduced to 0.41 kg/t during April 2013, in which technic loss is 0.31 kg/t and management loss 0.10 kg/t.

## 2.2 Examples of Low Magnetite Consumption with Desliming of Raw Coal Feed

### 2.2.1 Coal Preparation Plant C from Zhaozhuang Mining Group

It was retrofitted in 2014 to increase plant raw coal feed capacity to 2.4 MTPA. The plant was designed with a gravity-fed 3-product HM cyclone to process 50-1mm fraction. The raw coal was deslimed at 1 mm. TBS separators were selected to process coarse fine coal particles while the flotation process circuit was designed to process fine coal.

Average magnetite consumption against plant raw coal feed was 0.43 kg/t during Jan-May 2015, in which magnetite losses from clean coal D&R screen, middling D&R screen and reject D&R screen were 0.15, 0.10 and 0.10 kg/t respectively, which was 0.53 kg/t of magnetite consumption on average if against the cyclone raw coal feed.

### 2.2.2 Coal Preparation Plant D from Jizhong Energy

The plant was retrofitted and designed with gravity-fed 3-product HM cyclone process with desliming of raw coal feed of 0.9 MTPA in 2004. The raw coal feed is deslimed at 0.5 mm with CSS (similar to TBS) to process coarse fine fraction and flotation circuit to process fine coal. The magnetite consumption was as high as 3.66 kg/t when the plant started up and maintained for a long period. To reduce the magnetite consumption, processes were improved, some equipment was remodeled and changed, and management was also improved as well. The magnetite consumption was finally reduced to about 0.5 kg per ton of plant raw coal feed or 0.63 kg per tone of the HM cyclone raw coal feed.

## 3. Example of high magnetite consumption

## 3.1 Examples of High Magnetite Consumption without Desliming of Raw Coal Feed

### 3.1.1 Coal Preparation Plant E from Kailuan Mining Group

The plant was retrofitted in 2010 with gravity-fed 3-product HM cyclone process without desliming of raw coal feed. Fine coal is processed with flotation circuit. High magnetite consumption at 2.6 kg/t was resulted from severe argillization of coal and very high content of high ash clay, in which 0.3 kg/t was lost due to improper management. As high as 2.2 kg/t of magnetite consumption was reached and could not be reduced any further even after great efforts to improve processes and management such as magnetic separator circuit, D&R screen, magnetite quality control and etc.

### 3.1.2 Coal Preparation Plant F of Guizhou Panjiang Clean Coal Ltd.

It was designed at raw coal throughput of 3 MTPA using gravity-fed 3-product cyclone HM process without desliming of raw coal feed. Fine coal fraction was processed in the flotation circuit. The plant was started to operate on March 2010. Magnetite consumption from January to June in 2013 was 2.09 、2.38、2.17、1.97、1.88 and 2.32 kg/t with an average of 2.14 kg/t。

## 3.2 Examples of High Magnetite Consumption with Desliming of Raw Coal Feed

### 3.2.1 Coal Preparation Plant G of Huajing Coking Coal Ltd. Company

The plant was built in 2009 with raw coal feed capacity of 3 MTPA using conventional pressure-fed primary and secondary 2-product cyclones and primary and secondary flotation circuits. Magnetite consumption was extremely high at approximate 6 kg/t at the beginning after it was put into operation due to a large amount of fine coal fraction and easy argillization of reject. The magnetite consumption was finally reduced to 3.2 kg/t with great efforts to improve coarse fine coal circuit and dewatering of D&R screen, keep effective water spray and the spray water quality, use quality magnetite, and improve operation management to minimize magnetite technical loss and supervision loss.

### 3.2.2 Coal Preparation Plant H of Xishan Coal Power Ltd. Company

The plant was built in 1987 with raw coal throughput of 3 MTPA. Gravity-fed 3-product HM cyclone with desliming of raw coal feed was used to process coarse coal particles while flotation was selected to process fine coal particles. HM consumption was kept around 4.5 kg/t on average before 2009. Magnetite consumption was reduced to 2.71 kg/t on average from January to June 2009 after efforts were made to improve magnetic separator operation, periodically flip HM drain sieve bends and enhance water spray pressure on D&R screens. Table 2 shows monthly magnetite consumption in Coal Preparation Plant H from January to June 2009.

**Table 2:** Monthly Magnetite Consumption in the Plant during Jan-Jun 2009

| Month | 1 | 2 | 3 | 4 | 5 | 6 | Average |
|---|---|---|---|---|---|---|---|
| **Magnetite Consumption (kg/t)** | 2.72 | 2.80 | 2.66 | 2.63 | 2.70 | 2.75 | 2.71 |

## 4. Statistics of medium magnetite consumption in coal preparation plants

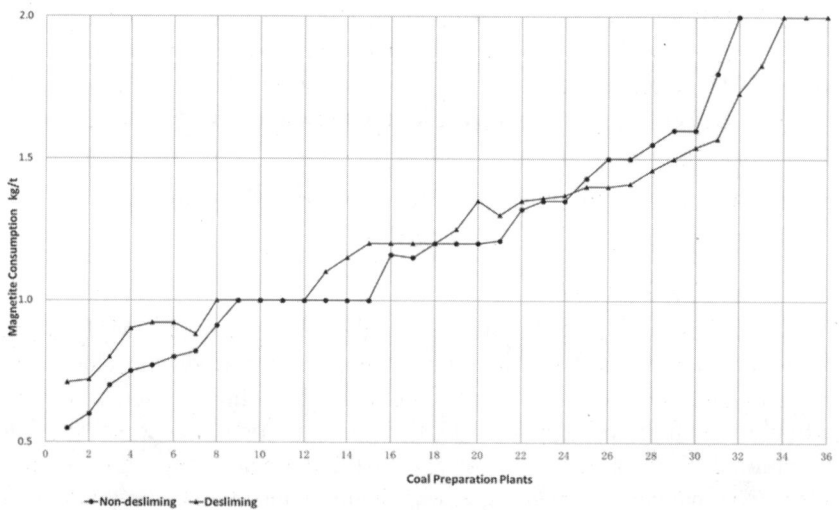

**Figure 1:** Magnetite consumption trends in the coal plants with and without desliming of raw coal feeds

Magnetite consumption data in a range of 0.51 to 2.0 kg/t were collected from coal preparation plants with and without desliming of raw coal feed. The magnetite consumption is 1.24 kg/t on average from 36 coal preparation plants equipped with HM cyclone process with desliming of raw coal feed and with their medium magnetite consumption ranged from 0.71 to 2.0 kg/t. Meanwhile, the magnetite consumption is 1.16 kg/t on average from 32 coal preparation plants equipped with HM cyclone process without desliming of raw coal feed and with their medium magnetite consumption ranged from 0.55 to 2.0 kg/t. Figure 1 shows trends of the HM consumption data collected from 36 coal plants equipped with HM cyclone process with desliming of raw coal feed and 32 coal plants equipped with HM cyclone process without desliming of raw coal feed.

## 5. Analysis of magnetite consumption data from hm cyclone processes with and without desliming of raw coal feed

There are no obvious differences of magnetite consumptions between desliming and non-desliming processes based on the data we could collect so far from the investigation of high, low and medium magnetite consumptions in many coal preparation plants equipped with HM cyclone processes using desliming and non-desliming of raw coal feed. It has been confirmed that the magnetite consumption of HM cyclone process is predominantly determined by selected coal process technologies, process flowsheets, equipment selection and plant operation management skill to reduce HM losses. Shichuan Dazhu Coal Power Ltd. Coal Preparation Plant is another example. It was designed with gravity-fed 3-product HM cyclone to process deslimed raw coal feed of 50-1.5 mm particles. The plant was designed at 3 MTPA of throughput of raw coal feed deslimed at 1.5 mm before fed to the cyclone. Magnetite consumption was as high as 3 kg/t when the plant was initially put into operation in October of 2007. The magnetite consumption was reduced to 1.54 kg/t on average during 2010-2012 due to the improvement of plant operation such as the way to add magnetite into HM circuit, stabilizing density of magnetite suspension, magnetic separator operation efficiency improvement, D&R screen dewatering enhancement, clarifying plant circulated washing water, etc. The magnetite consumption was further decreased to 0.92 kg/t on average of three years during 2014-2016. More and more similar examples can be given in terms of significant reduction of magnetite consumption resulting from the improvement of plant equipment and operation efficiency and plant management. Therefore, we do not need to use desliming of raw coal feed only in order to reach low HM consumption or reduce HM consumption. In other words, there are numerous coal plants with HM process and desliming of raw coal feed are still experiencing very high HM consumptions. Plant process circuit design, flowsheet & equipment selection, plant operation & management, etc. have significant effects on the HM consumption. Properties of plant raw coal feeds have significant effects on magnetite consumption in

HM process plants as well. Generally speaking, the magnetite consumption in US coal plants is lower than that in China coal plants, which primarily results from significant differences in raw coal feed characteristics. Percentage of fine particle fraction in raw coal feed to China plants is much higher than that to US coal plants.

## 6. Conclusion

The magnetite consumption mainly depends on selected process technologies, flowsheets, equipment selection, plant operation & management to reduce HM technical loss and management loss.

Currently the gravity-fed or pressure-fed 3-product HM cyclone can effectively process raw coal feed to 0.25 mm in particle size due to its super-high pressure feeding of HM suspension. Fine coal HM cyclone can effectively process fine coal particle size down to 0.1 mm. It is well recognized that HM cyclone process is the most efficient coal process in the world at moment. Therefore, it is not reasonable or inacceptable to use water-based separators (such as WO cyclone or TBS) with low separation accuracy to process $\leq 1.0$ (or 0.5) mm fraction of coal particles deslimed from the raw coal.

Compared to conventional technology of HM cyclone process with desliming of raw coal feed, the modern process circuit, using gravity-fed 3-product HM cyclone process without desliming of raw coal feed, has been considerably simplified without any increase of magnetite consumption, which has been well recognized by more and more coal preparation plants in China coal industry.

## 7. References

[1] Jiang, X. and et al., 2016, "Practice of super-low magnetite consumption at 0.45 kg/t in Xinyi Mine Coal Preparation Plant", Coal Quality Technology, No. 5, 2016, pp. 65-68.

[2] Liu, G. and et al., 2013, "Exploration and practice to reduce magnetite consumption in Matou Coal Preparation Plant", Proceedings of Symposium of National Coal Preparation Plant Energy Saving and Consumable Reduction, June 2013.

[3] Shu, H. and et al., 2015, "Comparison and application of desliming and non-desliming of raw coal feed in HM process in Gaozhuang Coal Preparation Plant", Coal Quality Technology, No. 6, 2015, pp. 37-44.

[4] Wang, Y., 2013, "Application and study of heavy medium process with magnetite consumption less than 0.50 kg/t", Proceedings of Symposium of National Coal Preparation Plant Energy Saving and Consumable Reduction, June 2013.

[5] Cao, H., 2014, "Exploration of reduction of magnetite consumption in Linxi Mine Coal Preparation Plant", Proceedings of Symposium of China Coal Preparation Development, May 2014.

[6] Li, H. and Chen, L., 2014, "Practice of heavy medium recovery in Tucheng Mine Coal Preparation Plant", Coal Quality Technology, No. 6, 2014, pp. 64-66.

[7] Qiao, J., 2015, "Optimization of process design and plant layout for Shaqiu Coal Preparation Plant", Coal Engineering, No. 12, 2015, pp. 22-24.

[8] Wang, W., 2012, "Practice of energy saving and consumable reduction in Xiqiu Mine Coal Preparation Plant", Shangxi Coking Coal Technology, No. 6, 2012, pp. 41-43.

# 40

# New rotary breaker device to improve its efficiency

D. Gnana Bharathi

*CSIR Central Leather Research Institute, Chennai, India*

**Abstract:** The run of mine (ROM) ores are crushed before being processed. There are various crushers used for this purpose. In jaw crusher, roller crusher, etc. the minerals are hit at both the ends. This leads to increased generation of fine particles. In the rotary breaker on the other hand breaks the mineral due to gravity fall which is single directional. As a result, breakage happens at points of weakness such as foliation, cracks, etc. Further, it generates less fine particles.

Rotary breaker is a horizontal cylindrical drum with series of holes in the drum. As the rotary breaker rotates at the optimum speed, the minerals are lifted up and fall within the rotary breaker due to gravity. Due to the impact, the minerals break into pieces and pass through the holes. The harder minerals do not break. The rotary breakers are positioned such a manner, the harder minerals move towards other end of the rotary breaker and disposed of.

There are various issues attached with rotary breakers. Diameter size of rotary breaker has to be sufficient enough for a good result. If the diameter of the rotary drum is less, the impact of falling minerals reduces. This results in poor crushing. Rotary breakers also serve as a screen. Only those particles that are reduced below the size of the holes can pass through it. Until then the particles get recirculate, automatically.

Paper discusses ways and means to overcome the issues related to the rotary breaker, a new design is developed which addresses these issues.

**Keywords:** Rotary breaker, coal crushing, gravity crushing, coal particles

## 1. Introduction

Coal is the most widely and easily available natural hydrocarbon resource. It is evacuated by underground mining and opencast mining. Due to the increased demand and mechanization, large quantities of coal are excavated from the mines. These run of mine (ROM) coals contain fines as small as a few microns to large pieces of more than 250 mm size.

The ROM ores contain coal as well as gangue. While hardness of coal particles ranges from 1.14 to 2.65 hardness of gangue ranges from 3.27 to 6.57 (Yang et al., 2018). In order to increase the liberation of mineral matter from the difficult-to-wash Indian coals, the coals are crushed finer before beneficiation leading to excessive generation of fines (Parthasarathy et al., 1995).

There are various crushers used for this purpose. In jaw crusher, roller crusher, etc. the minerals are hit at both the ends. This leads to increased generation of fine particles. The rotary breaker on the other hand breaks the mineral due to gravity fall which is single directional. As a result, breakage happens at points of weakness such as foliation, cracks, etc. Further, it generates less fine particles.

## 2. Rotary breakers design

Rotary breakers are used for over 100 years in the mineral processing industries. Rotary breaker is a horizontal cylindrical drum with series of holes in the drum. Minerals with low hardness such as coal are supplied into the rotary breaker. As the rotary breaker rotates at the optimum speed, the minerals are lifted up and fall within the rotary breaker due to gravity. There are lifting mechanisms to carry the minerals up. Due to the impact, the minerals break into pieces and pass through the holes. There are number of inventions that improve the efficiencies of the rotary breaker such as (Haltof, 1962; LIU, 2008).

The harder minerals do not break. The lifting mechanism is so arranged or the rotary breakers are positioned such a manner, the harder minerals move towards other end of the rotary breaker and disposed of.

Diameter size of rotary breaker has to be sufficient enough for a good result. If the diameter of the rotary drum is less, the impact of falling minerals reduces. This results in poor crushing.

Rotary breakers also serve as a screen. Only those particles that are reduced below the size of the holes can pass through it. Until then the particles get recirculate, automatically.

## 3. Issues related to rotary breakers

There are various issues attached with rotary breakers. Some of them are:

1. Size of rotary breaker cannot be increased significantly. If the diameter of the rotary drum is more, it occupies wider space and result in exponential increase in weight of the drum.

2. If the crushed particles do not pass through the holes, they are rotated to the top and fall again. This further reduces the size of the particles beyond the optimum size.

3. As the drum acts as a carrier as well as impact plate, many a times particles fall on the other. This result in double impact leading to fine grains.

4. Fewer the holes more recirculation resulting in more fine grain generation.

5. More holes result in poor impact.

## 4. New rotary breaker design

The design is basically elliptical or oval or rounded triangular shaped platform positioned by pillars on both the sides. Depending on the nature and type of coal, height of the platform can be fixed. The platform acts as rails to carry the interlinked carrier plates (CP). The carrier plate is constructed by joining of a shaft, rollers, cover sheet, and lifting plate. Two rollers are attached at through the shaft at both end and placed on the rail platform. Each carrier plate is connected with adjoining carrier plates through interlocking mechanism. The shaft, cover sheet, and the lifting plate extend from on one end of the platform to the other end.

The cover sheet (CS) is firmly fixed over the shaft. The gaps between the cover sheet act as a screen for the undersized particle to fall. The screen size between two cover sheets can be controlled by adjusting the width of the cover sheets. The lifting plate (LP) is positioned perpendicular or acute angle to the carrier plate and fixed with the shaft. A strap or a flap in the inner part of the lifting plate prevents the free fall of the coal particles while being lifted.

The lifting plate can be positioned in horizontally or in an inclined position crisscrossing the shaft. The inclination will ensure gravitational movement of the coal particles from one side rotary breaker to the other. The lifting plate also has check bars in-between to limit the downward slippage of the coal particles. Each cover sheet extends with wheels at both the ends through the shaft. The cover sheet is attached with lifting plate which has necessary knobs to prevent coal particles to slide away.

The device has a strong and impact plate (IP) placed horizontally just over the conveyor belt which drops the ROM coal. Impact plate extends from one end to the other end of the rotary breaker so as to ensure all coal particles fall on it from the top. The impact plate is tilted to some degrees to ensure the broken particles automatically slide down and fall on the carrier plates coming at the bottom.

Whatever be the height of the ellipse shaped rotary breaker, the impact plate should be free of any particle every time the coal particles fall from the lifting plates from the top. As a result the slope of the impact plate alone may not be sufficient. The impact plate may need horizontal shaking to ensure fall of coal particle from the impact plate.

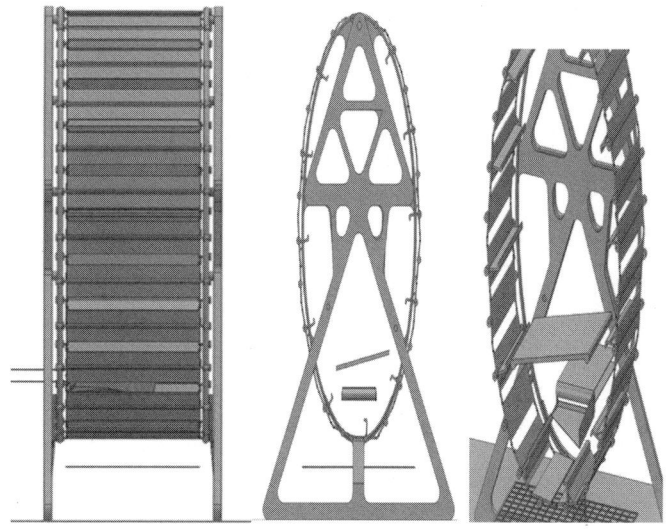

**Figure 1:** Three different views of a model design of a 2.0 m high ellipse shaped rotary breaker

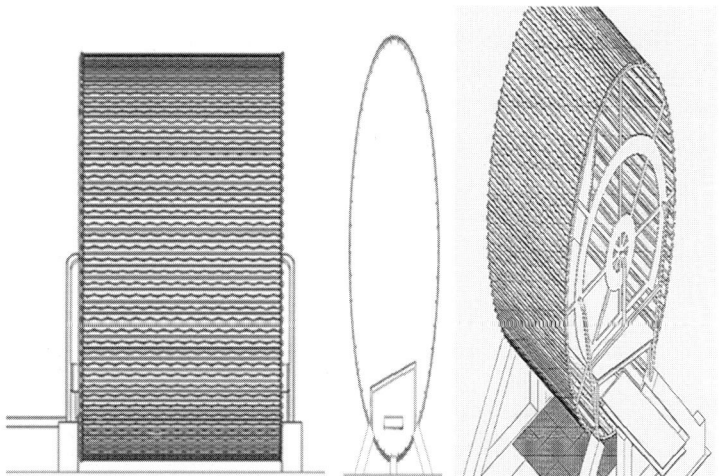

**Figure 2:** Three different views of a 10 m high ellipse shaped rotary breaker

The lift plate, conveyor belt and impact plate reduces the height of fall less than 10–20%. Yet for a 10 m tall ellipse shaped rotary breaker, direct fall of at least 9 m is ensured (Figs. 1 and 2).

The drive unit consisting of sprocket is positioned near top end of the ellipse shaped rotary breaker. The arc of the sprocket is same as the arc of the top end of the rotary breaker. The teeth of the sprocket are optimized to push and rotate the shaft of the carrier plates. The sprocket is rotated by a motor which in turn rotates the circular plates of the rotary breaker.

At the lower level of ellipse platform a screen or two can be placed to separate the crushed coal into different sizes.

## 5. Operation

The ROM coal transported in the conveyor belt is dropped at the one end of the rotary breaker. They are carried upward by the cover sheet and lifting plates that are more or less in horizontal and vertical positions, respectively, when the carrier plate is at the bottom of the rotary breaker. Many small coal particles will fall through the gap in between the cover sheets. As the carrier plates move up, the remaining small particles will pass through the gap in-between the cover sheets. As a result, only the larger particles are carried to the top.

On reaching or closer to the zenith, the coal particles fall due to gravity. The falling particle hit with the impact plate which is positioned just over the conveyor belt. As the impact plate is positioned standstill, the impact will be direct. The coal gets crushed due to gravity which is one directional. It effects in the breaking of over planes of weakness such as foliation, cracks, cleavages, etc. It results minimum production of fines in the product.

As the lifting plates are positioned in tilted position, the coal particles carried by them will slip towards other end of the rotary breaker. The check bars in the lifting plate act as barriers of slippage. The number of check bars determines how many times the oversized coal particles recirculate. The uncrushed particles will ultimately fall out of the other end of the rotary breaker.

The undersized crushed particles pass through the gap between the cover plates. Coal that is not product size continues to be lifted and dropped until it passes through the screen plate. The unbroken particles will be disposed through other end of the rotary breaker.

The impact plate has to be strong enough to withstand impact of the coal and stone particles. As the hit is borne by the impact plate, the cover plate can be made of lightweight material. This reduces overall weight and cost.

## 6. Conclusion

Coal fed into the rotary breaker can be crushed and sorted out within a few rotations. As the fall of the coal particles is due to gravity and hit directly on a stable impact plate, the particles crushed through the natural lines of weakness. This ensure low fine generation.

# 7. Acknowledgement

The author acknowledges the Director, CSIR Central Leather Research Institute for the permission to publish the article. The author also acknowledges the IPKP of CSIR-CLRI for processing the patent application.

# 8. References

[1] Haltof, H. (1962). US3186649 (A).

[2] Liu, X. (2008). CN201147708Y.

[3] Parthasarathy, L., Sharma, M. K., Choudhury, A. J., & Sharma, R. P. (1995). Preparation of high ash Indian coals for carbonization. ISIJ International, 35(7), 819–825. Retrieved from <Go to ISI>://WOS:A1995RP25800001. DOI: 10.2355/isijinternational.35.819.

[4] Yang, D. L., Li, J. P., Zheng, K. H., Du, C. L., & Liu, S. Y. (2018). Impact-crush separation characteristics of coal and gangue. *International Journal of Coal Preparation and Utilization, 38*(3), 127–134. Retrieved from <Go to ISI>://WOS:000424322700002. DOI: 10.1080/19392699.2016.1207634.

# 41

# Environmental aspects of decommissioning of a coal industry enterprises

O. Yehurnov[1], N. Kalugina[2], V. Krasnik[3]

*[1]ANA-TEMS LLC, Dnipropetrovsk, Ukraine*
*[2]Institute for Physics of Mining Processes of the National Akademy of Sciences*
*of Ukraine, Kyiv, Ukraine*
*[3]National Coal Company, Kyiv, Ukraine*

**Abstract:** The assessment of the situation associated with the decommissioning of coal industry enterprises in Ukraine has been carried out. The environmental aspects at the completion of mining operations in the Donetsk region have been analyzed.

The methods and means of economic and environmental feasibility study for the decommissioning of coal enterprises, developed by the Institute for Physics of Mining Processes of the National Academy of Sciences of Ukraine, have been suggested.

**Keywords:** Coal industry enterprise, environmental problems, decision making

High-intensive coal mining over the last century has led to the Donetsk region becoming one of the most environmentally neglected areas of Ukraine. Hundreds of mines have become the main source of pollution and environmental destruction. Over the past 20 years, numerous coal enterprises have ceased operations or are subject to closure. However, the issues of safety, environmental, and social consequences of this process remain unresolved in the country.

Since the beginning of the state Coal and Peat Industry Restructuring program in Ukraine, from 1996 to the present, 143 unprofitable enterprises have been prepared for closing down. To date, 61 enterprises have been completely dissolved, of which 34 were closed during the period of 2009–2013. Currently, 82 coal mines are in the stage of decommissioning, 10 mines are being prepared for the closing down, and another 29 mines are recognized as loss-making ones, and will also be closed in the future.

For the last 23 years of its existence, the Coal and Peat Industry Restructuring Program of Ukraine has never been financed at the level justified by the Ministry of Energy and Coal Industry in the budget request, which was drawn up in accordance with the project dissolution schedules. For example, the funds allocated in the budget amounted to 46% in 2012, 36% in 2013, and 26% in 2014 of the funds substantiated in the budget requests.

Constant underfunding of the restructuring program at a level lower than the semi-fixed costs of the restructured enterprises resulted in the following negative consequences: failure to meet the original project deadlines for mines closure from 4 to 15 years (some mines have been closing down since 1996, that is, more than 20 years, with the project closure deadlines of no more than 3–4 years), which in turn led to an increase in the cost of closing down 82 mines by 0.72 billion dollars – the accumulation of debts for consumed electricity and subsidized household fuel up to 124 and 40 million dollars, respectively; the impossibility of transferring the loss-making mines that are not extracting coal at the moment for preparation for closure, which in turn results in an increase in state subsidized funds for their maintenance. At the moment, the restructuring program finances: 12 mines, where preparations for the closing down are carried out; three regional directorates for the closure of loss-making coal enterprises; SE Ukrshahtgidrozaschita – an enterprise maintaining drainage complexes after their reconstruction, in order to ensure the hydro-safety of neighbouring operating mines and adjacent territories.

The total number of workers in these enterprises was 7041 people. Forty-six mines are in the mode of continuous drainage. According to clause 43 of the Mining Industry Law of Ukraine [1], only in 2014, it was necessary to provide 23.5 thousand people with subsidized household fuel.

To solve the set tasks, the corresponding Program was developed, the main theses of which are as follows:

- transfer within 2 years of 29 mines for preparation for closure, including 10 mines in 2014 and 19 mines in subsequent years;
- development of a mine closure preparation plan by 2020, completion of decommissioning projects and transfer of drainage complexes after reconstruction for their further permanent maintenance;
- justification of the schedule of financing all the activities of the program. Thus, to close down 122 mines, 2.5 billion dollars were required for 7 years.

The implementation of the program involved the implementation of all relevant hydrogeological environmental standards and the standards of reclamation of industrial sites, waste dumps, and sludge collectors. As of January 1, 2019, out of 30 coal-mining enterprises located on the territory controlled by the Ukrainian government, four mines and Alexandriyaugol open pits are now being closed, and four enterprises are being prepared for decommissioning (https://dn.gov.ua).

The armed conflicts in the east of Ukraine exacerbate the problems associated with the functioning of existing mines and with the state of the mines subject to closure. According to experts [2], in the conditions of military operations, the environmental conditions in the east can be catastrophic. Flooding of mines and adjacent areas as a result of de-energizing and damage to equipment of mining enterprises is one of the main causes of potential pollution of groundwater and surface water in contact with mine water contaminated with harmful substances. According to experts opinion, obtained as part of the study of the OSCE Project Coordinator in Ukraine, more than 35 mines are in the process of flooding on both sides of the contact line, 64 mines in the decommissioning stage are located in the territory beyond the Ukraine's control (Table 1, [2]).

**Table 1:** The state of mines in the east of Ukraine

| Coal mines | Controlled territory | Beyond control territory | Total |
|---|---|---|---|
| Mines in operation | 29 | 75 | 104 |
| In the drainage mode | 1 | 116 | 17 |
| In the process of flooding | 1 | *35* | *36* |
| At the decommissioning stage | 6 | 64 | 70 |
| Total | 37 | 190 | 227 |

*Note:* Estimates obtained by the method of expert rapid assessment, italics indicate approximate values.

Analysis of previous practical experience of partial and complete flooding of 310 unpromising mines within the period from 1960 to 2004 indicates an unpredictable deterioration of the ecological condition, the scale and nature of which lead to disastrous consequences. Neglect of the drainage regulations during the conservation, the full maintenance of which lacks funds, results in the opposite effect of environmental stabilization and provokes a significant change in groundwater and surface water, flooding of territories, salinization, and pollution of rivers, subsidence of the earth's surface, activation of karst and landslide processes.

It should be noted that the closure of mines is also associated with the need for post-decommissioning measures to ensure the hydro safety of existing coal mining enterprises. At the same time, the costs of such measures often exceed the costs of the mines decommissioning. Data analysis shows that in some coal areas the costs required to maintain the specified levels of groundwater due to the operation of powerful drainage systems exceed 90% of the cost of mine decommissioning.

After the mines closure, one of the most significant problems is the possibility of methane leakage from coal beds to the surface. A significant concentration of methane is characteristic of many coal beds developed in Donbas. Control of this highly flammable and explosive gas is one of the major safety issues in production management. Long-term studies of the features of physical processes occurring in coals and rocks in various geological conditions during underground mining of coal beds in Donbas, the study of physical properties and processes occurring in the coal–methane system, allow us to predict the dangerous phenomena that may occur during mine decommissioning [3–5]. Being a source of constant danger for miners, methane, even when removed to the daylight surface by ventilation systems or as a result of the free migration of gases through fractures and cracks from the beds to the surface, significantly pollutes the atmosphere, contributes to the greenhouse effect, can ignite and pose a threat to communication lines and structures as a result of explosions.

To study the resources of methane in closed mines, including the period of restructuring of the coal industry since 1996, the Institute for Physics of Mining Processes of the National Academy of Sciences of Ukraine processed geological information on 114 coal mines, studied and analysed the characteristics of resources and reserves of hydrocarbon gases and their capture in 27 exploration areas of five geological industrial areas. As a result of research, the method for predicting methane migration from the mined-out spaces of closed mines was improved, an algorithm was developed for calculating the amount of methane gas that migrates to the daylight surface and the period of its migration [6].

At the same time, if in the context of mine closure, the issue of methane leakage must always be under control, the scientifically based methods for estimating residual gas in closed mines may allow identifying promising areas for its use as an energy resource.

In addition, a comprehensive solution to the problem of the coal-mining enterprises decommissioning must take into account their residual energy resource. These are residual and substandard reserves of coal left after mine-field development, underground waters in the flooded mine openings, a powerful reservoir potential of abandoned coal deposits. The analysis of the global scientific and practical experience in the development of the natural and industrial resource of coal deposits indicates its high profitability in a number of countries around the world (for example, in the old coal regions of Poland, Germany, and the USA), as well as the inefficiency of the existing geotechnologies in Ukraine and its underachievement in this area. Geotechnological solutions aimed at stabilizing the ecological and energy environment in Donbas may include geo-circulation systems for storing summer heat and winter cold in water-bearing formations, creating gas storages in water-bearing formations [7], underground burning of residual coal reserves [8,9], and geomodules for development of the thermal resource of the flooded mine field [10].

Subject to the restoration of Donbas infrastructure, it is necessary to make a balanced and informed decision on the resumption of work or the closure of coal mining enterprises. Making such a decision is only possible with the availability of all the modern information characterizing mining facilities. This is the information of mines and ministries, scientific research of dozens of institutions that have been dealing with the problems of the mining industry for a long time. Such information is very important for risk assessment, operation, and closing down of coal mining enterprises.

The Institute for Physics of Mining Processes of the National Academy of Sciences of Ukraine has accumulated a vast experience in risk assessment – environmental, organizational, technological, and geological. There is an opportunity to analyze mining development programs, as well as to assess the budget overruns and project implementation time. There is also an experience in solving complex applied multi-parameter problems using the discrete mathematical methods, which miners, power engineers and metallurgists often face in the process of manufacturing end-use products. If such a task is represented in the form of alternative graphs (Fig. 1), it is possible to optimize them using classical algorithms, transforming them into appropriate systems. For example, the choice of the optimal technology allows taking into account the quality of the mineral, the conditions of its occurrence and extraction, the technological cycle of mining, transportation, processing, the impact of production on the environment, the possibility of restoring the natural state after man-induced impact [11].

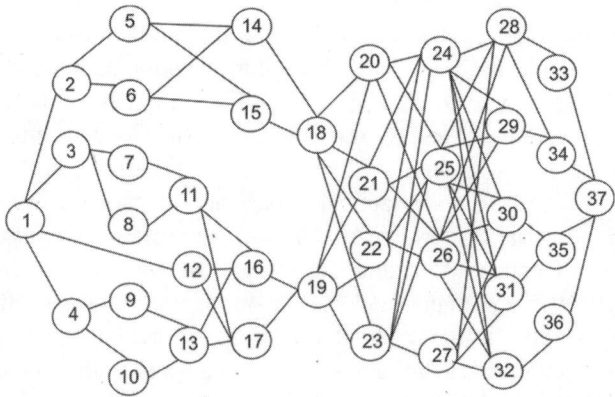

**Figure 1:** Alternative graph to determine the field development strategy

Equipment types, technology options, etc. are taken as vertices, and possible relationships between the objects are taken as edges. Each level in the alternative graph (2–4, 5–10, ..., 33–36) corresponds to the solution of a specific problem. For example, to obtain end-use products in the form of coal, it is necessary to determine the technological parameters of the production face (place length, working width of the bed), choose equipment, choose means of transporting coal from the production face to the shaft bottom, arrange coal transportation to the processing plant, choose a method of coal preparation, to resolve issues related to the tails disposal. Naturally, at each stage there will be alternative options to equipment, technology, etc., which can be expressed as a price $P_c$. Point 1 corresponds to the beginning of the production cycle, and point 37 corresponds to the obtainment of finished products. Points 2–4 correspond to a specific solution (in our case, the technological parameters of the production face). The distance from point 1 to points 2–4 corresponds to the price $P_c$ (the cost of development and maintaining the openings). Then, to make the best decision, it is necessary to find the shortest route between points 1 and 37. To achieve this, the optimization algorithms can be used on networks and columns.

To automate the selection process, the staff of the Institute for Physics of Mining Processes of the National Academy of Sciences of Ukraine developed appropriate software [13,14]. The program contains a database and provides for various options for data input, and moreover, is integrated with office applications.

To automate the selection process, the staff of the Institute for Physics of Mining Processes of the National Academy of Sciences of Ukraine developed appropriate software [4]. The program contains a database and provides for various options for data input, and moreover, is integrated with office applications. To optimize, let's plot points on the effective range and set the distance between the points to which the vertices correspond. After that, the program will automatically find the shortest route, which will correspond to the production cycle with maximum productivity and minimum cost.

In developed countries of today, the environmental aspect becomes the issue of the highest priority as the main criterion in decision-making during the development of resources. The research development potential in the field of optimization of network models is sufficient to carry out an economic assessment of the environmental aspect of the mineral deposits development [12]. The procedure for such an assessment involves the financial comparison of two development scenarios – the most economically feasible and environmentally preferable ones.

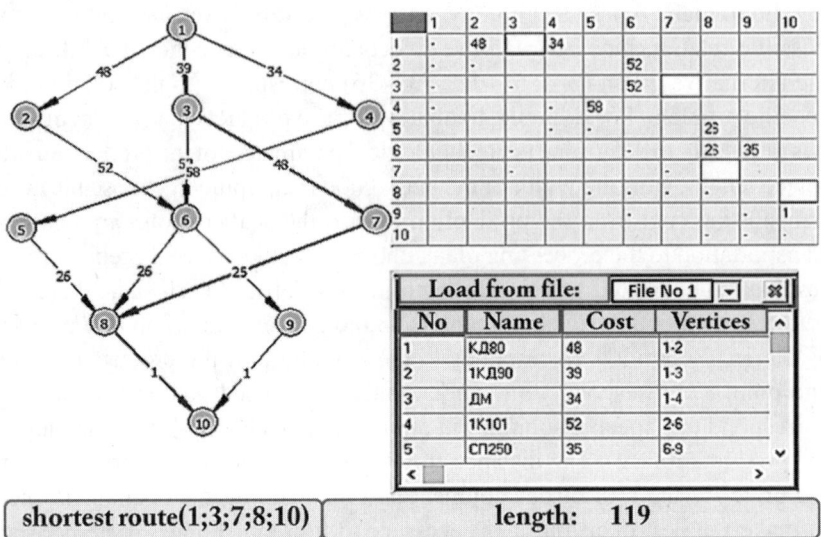

**Figure 2:** Optimization of the process flow (for example, the choice of equipment)

The economic analysis, using the developed program, allows choosing a cycle process flow with the maximum productivity and minimum cost. For environmental assessment, each of the stages can be evaluated from the standpoint of the man-induced impact of the enterprise on the environment. In this case, the assessment can be carried out comprehensively and take into account not only obvious environmental consequences, but also assess the dynamics of ecosystem rehabilitation. For example, in the development of coal deposits, one may take into account the nature of landscape changes, the amount of methane emitted to the surface, changes in the level of aquifers, the environmental consequences of flooding of workings, and the number of areas allocated for waste dumps.

Such approaches to the methodology and software for the economic assessment of the environmental aspect, if necessary, can be applied in restoring the Donbas infrastructure after the liberation of the occupied territories in eastern Ukraine by a long-term perspective planning, taking into account strategic directions.

Resource, technical, and many social and economic conditions of Ukraine have no equivalents, allowing the application of the best experience in solving such issues in world practice. The joint work of foreign representatives with such experience and the Ukrainian structures for the adaptation and implementation of the best global practices of operation and closure, is very desirable.

## Conclusion

1. Analysis of the few studies carried out in Donetsk and Luhansk region, torn by the military conflict, shows that the operation of the coal-mining complex in "extreme" conditions unbalances the ecological situation of Donbas, which was maintained by the mandatory implementation of design ecological and hydrogeological solutions.

2. Environmental problems require special attention during the completion of mining operations: huge areas of imbalance of the subsoil resources, large accumulations of coal mining waste and waste coal in the form of dry dumps and sludge reservoirs in a small densely populated area, neglect of drainage schedule during the conservation, flooding of mines and adjacent areas due to the de-energizing and damage to equipment of enterprises as a result of armed clashes, leakage of explosive methane, etc.

3. A comprehensive solution of the issues of coal industry enterprises decommissioning requires scientifically-based methods of studying residual gas in closed mines and its cost-effective use, as well as geotechnological solutions aimed at stabilizing the environmental and energy situation in Donbas.

4. The Institute for Physics of Mining Processes of the National Academy of Sciences of Ukraine has patented the methodology and software allowing economic and environmental assessment to be carried out for both explored fields and operating coal-mining enterprises or the enterprises being decommissioned. The goal to be achieved is to make weighted and informed decisions regarding the coal industry enterprise decommissioning or its restoration during the period of rehabilitation of the environmentally unstable territories.

## References

[1] Mining Industry Law of Ukraine. Bulletin of the Verkhovna Rada of Ukraine, 1999, No. 50, art. 433. (Гірничий закон України. Відомості Верховної Ради України (ВВР), 1999, No 50, ст. 433.)

[2] Denisov, N., & Averin, D. (2017). Ocinka ekologichnoi shkodi ta prioriteti vidnovlennya dovkillya na skhodi Ukraini. Kyiv, VAITE. 88 p.

[3] Alexeev A. D. Phase states of methane in fossil coals, A.D. Alexeev, T.A. Vasilenko, E.V. Ulyanova, Solid State Communication. 2004, Vol. 130, No 10, P. 669–673.

[4] Diffusion-filtration model of methane escape from a coal seam, A.D. Alexeev, T.A. Vasilenko, K.V. Gumennik, N.A. Kalugina, E.P. Feldman, Technical Physics. 2007, Vol. 52, No 4. P. 456–465.

[5] Fel'dman E.P. Physical kinetics of coal–methane system: Mass transfer, pre-outburst events, E.P. Fel'dman, T.A. Vasilenko, N.A. Kalugina, Journal of Mining Science. 2014, Vol. 50, No 3. P. 448–464.

[6] Grinev V.G. Study of the process of methane migration from the mined-out area of closed mines, V.G. Grinev, A.I. Sergiyenko, A.A. Podrukhin, Physical and Technical Problems of Mining Industry: Collection of Studies, 2009. Edition 12, P. 74–79. (Гринев В.Г. Исследования процесса миграции метана из выработанного пространства закрытых шахт / В.Г. Гринев, А.И. Сергиенко, А.А. Подрухин // Физико-технические проблемы горного производства: Зб. наук. пр. 2009. Вип. 12. С. 74–79).

[7] Inkin O.V. Hydrogeomechanical assessment for parameters of underground gas storage in the aquifer, I.O. Sadovenko, D.V. Rydakov O.V. Inkin, Scientific Reports on Resource Issues (Rock Strength, Rock Fragmentation and Effective Use of Energy Potential of Geotechnical systems), International University of Resources. TU Bergakademie Freiberg, Germany, 2012. Vol. 2, P. 196–208.

[8] Inkin O. Influence of coal layers gasification on bearing rocks, O. Inkin, V. Timoshuk, V. Tishkov, Geomechanikal Processes during Underground Mining. Proceedings of the school of underground mining. Dnipropetrovsk, Yalta, 2012. P. 109–113.

[9] Inkin O. Integrated analysis of geofiltrational parameters in the context of underground coal gasification relying upon calculations and modeling, O. Inkin, V. Tishkov, N. Dereviahina, E3S Web of Conferences Volume 60, 00035 (2018) Ukrainian School of Mining Engineering. Pp 1–9.

[10] Inkin O. Geotechnikal schemes to the multi-purpose use of geothermal energy and resources of abandoned mines, O. Inkin, I. Sadovenko, D. Rydakov, Progressive Technologies of Coal, Coalbed Methane, and Ores Mining. Taylor & Francis Group, London. 2014. P. 443–450.

[11] Hrinov V., Khorolskyi A. (2018). Improving the process of coal extraction based on the parameter optimization of mining equipment. In E3S Web of Conferences (Vol. 60, p. 00017). EDP Sciences. https://doi.org/10.1051/e3sconf/20186000017.

[12] Khorolskiy A.A., Grinev V.G. Dynamic Programming Application for Mining Industry Designing at Scarce Resources. Innovative Technologies in Education, Science and Production: Materials of the International Conference, November 17–18, 2018. Minsk: BNTU–[Electronic source]–Access mode: http://www.bntu.by/images/stories/mido/ntik6/horol.pdf (Хорольский А.А., Гринев В.Г. Использование динамического программирования для проектирования горного производства при ограниченных ресурсах. Инновационныетехнологии в образовании, науке и производстве: Материалы международной конференции, 17–18ноября 2018г. Минск: БНТУ. [Электронный ресурс].– Режим доступа: http://www.bntu.by/images/stories/mido/ntik6/horol.pdf).

[13] Grinev V.G., Khorolskiy A.O. (2018). Kompiuterna prohrama «Prohrama vyboru optymalnykh komplektatsii ochysnoho obladnannia na osnovi universalnykh hrafiv» («CountsCEM.v1.p2.6_c25»). Patent No. 74856, Ukraine.

[14] Grinev V.G., Khorolskiy A.O. (2018). Kompiuterna prohrama «Prohrama znakhodzhennia naikorotshykh vidstanei mizh usima vershynamy merezhevoi modeli» («GraphON.v1.2017»). Patent No. 75055, Ukraine.

# 42

# Development of an energy-efficient and wear-reduced hydraulic piston press for coal briquetting

Dr.-Ing. Franz Fehse[1,2], Dipl.-Ing. André Schmidt[1,2], Thomas Müller, M.Sc.[1,2], Dr.-Ing. Hans-Werner Schröder[1]

[1]*TU Bergakademie Freiberg, Institute of Thermal-, Environmental- and Resources' Process Engineering (ITUN), Freiberg, Germany*
[2]*ATNA Industrial Solutions GmbH, Leipzig, Germany*

**Abstract:** For several coal-upgrading processes the appropriate agglomeration process is an important step for the applications that follow. Because of technical constraints of certain agglomeration technologies the range of feedstock materials treatable is restricted. This circumstance was the starting point for the development of a new agglomeration technology, which is characterised by a wide range of applicable feedstocks and which is suited to high quality briquette production. By performing a hydraulic-driven two-stage press process, the operational expenses can be significantly reduced and the operating principle offers the opportunity to reduce the relative movement between the emerging briquette and the forming tool at high pressure. This leads to a significant wear reduction.

In laboratory-scale tests the compression characteristics of different coals were systematically investigated. As a result, parameter sets for the maximisation of the briquette quality were determined. With these parameter sets the briquetting behaviour of different coals was investigated at the demonstrator (throughput 50 kg h−1) in semi-industrial scale. As a result the quality of semi-industrial scale matched with those of industrial scale and can also compete with the quality of commercial products. In the next step a prototype of the new press with a throughput of 3 t h−1 is assembled and tested at the ITUN technical centre. After successful commissioning the prototype will be tested under industrial conditions.

**Keywords:** Briquetting press, hydraulic piston press, coal, press agglomeration, briquetting, energy efficiency, wear reduction

## 1. Introduction

The agglomeration of finely-dispersed materials is one of the major applications in process industry. For several coal-upgrading processes it is an essential step for the following applications, e.g. lump-coke production using non-baking coals [1], coal gasification in fixed-bed gasifiers [2] as well as the production of design fuels [3]. The evaluation of technical opportunities for the production of high-quality briquettes highly depends on the particular physical and chemical characteristics of the feedstock. Furthermore, certain press technologies are limited to selected feedstock materials.

With a new agglomeration technology a wide range of feedstocks should be agglomerated to high-quality briquettes.

A major focus of the technical development was the task to find a briquetting process with a minimum specific power consumption. This can be achieved by performing a hydraulic driven two-stage-press-process, with an optimisation of the length of hydraulic cylinder extension. Thus a significant reduction of operational expenses can be achieved, as well as an increase of the die profitability of industrial scale coal upgrading plants. This new operating principle opens the opportunity to reduce the relative movement between the emerging briquette and the forming tool at high pressure and results in a significant wear reduction between the forming tools. Hence a second major factor for the reduction of operational expenses can be derived. By extending the operating time of the forming tools the machine downtime for maintenance can be decreased and higher annual throughputs can be achieved. Furthermore, an easily maintainable machine design allows fast forming tool replacement and short maintenance periods.

Another major motivation for this new technology development was the situation at briquetting machine market for coal applications. Since decades the dominating briquetting technology in Germany has been the Ram Extrusion Press. There are industrial coal processing plants where presses manufactured in the 1910s are still in operation. Hence to the operating principle of generating the counterforce for briquetting by friction and narrowing of the forming channel a massive wear occurs at the forming tools [4,5]. Depending on feedstocks ash content, the operating time of a forming tool can go down to just some days. Thus the operational expenses and annual operating time of the single machine have a negative impact on plant profitability. Besides this technology is connected with huge capital expenses.

The only currently available substitutions are roller presses for the production of egg and pillow-shaped briquettes, as well as coal scabs. The main disadvantage of these machines is a limited applicability for different coal types and fluctuations in feedstock quality. Huge briquette quality problems occurred when trying to adjust the roller press operation to feedstocks like soft-, subbituminous- and hard lignite. It has been shown that an addition of binders such as starch, molasses or tar pitch is often inevitable for a stable briquetting process with high quality products.

By performing a two-stage press process with hydraulic cylinders it is easily possible to adjust the effective press force to the feed-stock conditions.

This enables the new universal high-performance press to adjust itself in operation and to generate a stable high quality product output.

In a first step the demonstrator with a throughput of 50 kg h$^{-1}$ was constructed, built and tested to provide the proof of principle of the technology and gives the opportunity to investigate the briquetting behaviour of various feedstocks.

For the present investigations the compression behaviour of different coals is determined on a hydraulic stamp press in laboratory scale to obtain the parameter set for the maximisation of the briquette quality of each coal. Using those parameter sets, the briquetting behaviour of these coals is investigated at the demonstrator at ITUN technical centre.

## 2. Materials

For the following investigations two German brown coals, and Indonesian and a Chinese brown coal were used.

The proximate analysis of the coals are given in Table 1.

**Table 1:** Moisture, proximate analysis and lower heating value (LHV) of the used coals

| Origin | M | A (d) (wt.%) | VM (d) (kJ kg$^{-1}$) | FC (d) (kJ kg$^{-1}$) | LHV (d) (kJ kg$^{-1}$) |
|---|---|---|---|---|---|
| Germany 1 | 56.2 | 5.60 | 48.90 | 45.50 | 25.028 |
| Germany 2 | 51.7 | 4.07 | 48.96 | 46.97 | 25.967 |
| Indonesia | 47.8 | 2.96 | 48.99 | 44.29 | 24.931 |
| China | 47.0 | 5.34 | 48.55 | 46.20 | 26.073 |

The German brown coals offer the highest moisture content ($M$) with 51.7% and 56.2%. The Indonesian brown coal offers the lowest ash content ($A$) (2.96% (d)). The ash content of the German brown coal 2 reaches 4.07%. The volatile matter (VM) of the coals is on the same level.

## 3. Methods

### 3.1 Processing and agglomeration of the coal

The methods used in these investigations are shown in Fig. 1.

For laboratory-scale tests the coal samples were comminuted in a hammer mill to a grain size of $\Delta d = 4/0$ mm. The crushed coal was dried in a laboratory draying cabinet at $T = 105\,°C$ to different moisture contents depending on the coal sample. After drying the coal is fine comminuted to $\Delta d = 2/0$ mm in a fine impact mill without any discharge screen and to $\Delta d = 1/0$ mm using a 4 mm discharge screen. In the laboratory tests the coal was briquetted at a hydraulic stamp press (Zeulenroda Presstechnik, Type PYE) with various briquetting-parameter sets (briquetting pressure $p$, briquetting temperature $T_p$ and briquetting time at maximum pressure $t_p$).

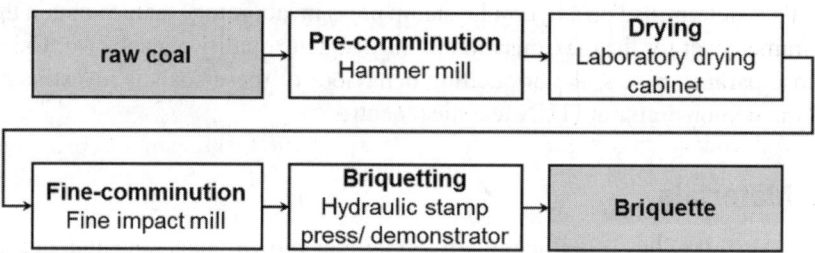

**Figure 1:** Procedure for the processing and briquetting of the coals in laboratory and technical centre

For investigations in technical centre the coal was processed as in laboratory-scale investigations. Briquetting was done using the demonstrator with a throughput of 50 kg h$^{-1}$ (Fig. 2, left).

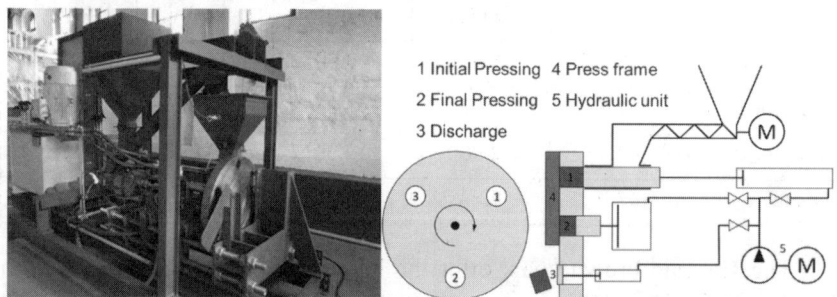

**Figure 2:** Demonstrator of the new high-performance press (left) and working principle of the demonstrator (right)

This demonstrator of the new high-performance press performs a two-stage press process. As shown in Fig. 2 (right) the press consists of a forming tool disk with cylindrical forming tools. At the beginning the coal is charged to a bunker. With a screw conveyor the coal is transported into a smaller feed tank. When the piston of the initial-pressing-step cylinder drives back, the coal falls into the press channel of the initial pressing step. In the initial pressing the piston pushes the coal into the forming tool at position 1 and forms a stable pre-agglomerate with maximum pressure under 5 MPa. Then the forming-tool disk rotates to position 2 and the pre-agglomerate is densified to the final briquette by the main compaction hydraulic cylinder which can reach a maximum pressure of 160 MPa. After one more rotation of the forming tool disk to position 3 the briquette is pushed out of the forming tool by the piston of the ejection cylinder.

The briquetting parameter sets were chosen according to the investigations in laboratory scale.

## 3.2 Scope of experiments

In preliminary investigations the parameter sets for maximum briquette quality were determined for $\Delta d = 2/0$ mm by variation of the briquetting pressure and moisture content of the coal according to [6]. As a result of these investigations the quality of the briquettes of the different coal samples is discussed at a constant parameter set to underline the influence of the raw material base.

To increase the briquette quality, the grain size of the coal was reduced to $\Delta d = 1/0$ mm and the briquetting pressure was increased to 140 MPa. The other briquetting parameters $(T_p, t_p)$ remained constant according to the previous investigations.

With those results from laboratory-scale investigations the briquetting behaviour of the coals were determined at the demonstrator using the parameter sets of maximum briquetting pressure in laboratory scale.

## 3.3 Parameters of evaluation

The quality of the briquettes was determined by the following parameters: Compressive strength (according to former TGL 9491). The briquettes are loaded with a force between two stamps of 30 mm diameter until breakage. The maximum pressure is defined as compressive strength σpB. Raw density: After briquetting, the agglomerates were measured with a calliper gauge and weighted to calculate the raw density $\rho_{raw}$. Abrasion resistance: Five briquettes are set into a drum of 500 mm diameter and length with four blades of 80 mm displaced by 90. The briquettes are loaded for 100 revolutions at 25 min$^{-1}$. The briquette abrasion resistance is indicated as the residue of the briquettes on the 30 mm sieve (Eq. 1) after the test.

$$R30(100) = \frac{m_{30}}{m_{tot}} 100\% \quad (1)$$

## 4. Results

### 4.1 Investigations in laboratory scale

In first investigations the briquette quality was determined for the four different brown coals at $\Delta d = 2/0$ mm under constant briquetting conditions ($p = 120$ MPa, $w = 11\%$, $t_p = 3$ s, $T_p = 70°C$). Figure 3 shows the raw density, compressive strength and abrasion resistance of the briquettes.

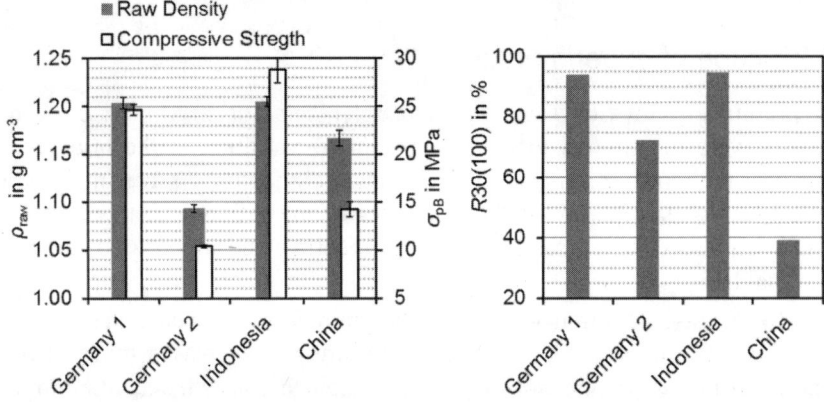

**Figure 3:** Raw density, compressive strength and abrasion resistance of the briquettes at the parameter sets of maximum briquette quality for four different coals for the grain size $\Delta d = 2/0$ mm ($p = 120$ MPa, $w = 11\%$, $t_p = 3$ s, $T_p = 70°C$)

As shown in Fig. 3, the provenience of the coal has a high effect on the quality of the briquettes. The briquettes of German brown coal 1 offer high raw density and compressive strength. The abrasion resistance exceeds 90%. The briquettes from German brown coal 2 show a lower quality. The raw density, compressive strength and abrasion resistance are on a significantly lower level. Since the coal shows a higher brittleness than the German brown coal 1 which leads to a more difficult briquetting behaviour. The briquettes from Indonesian Coal offer a comparable quality as the briquettes from German brown coal 1. The compressive strength even exceeds 28 MPa. The raw density and compressive strength of the briquettes from the Chinese brown coal are located between German brown coal 1 or Indonesian brown coal and German brown coal 2. This coal also shows a high brittleness since it shows a slightly higher coal rank. The higher brittleness causes the very low abrasion resistance ($R30\ (100) < 40\%$).

To increase the briquette quality the grain size of the brown coal was reduced to $\Delta d = 1/0$ mm and the briquetting pressure was increased to 140

MPa. Raw density, compressive strength and abrasion resistance are given in Fig. 4.

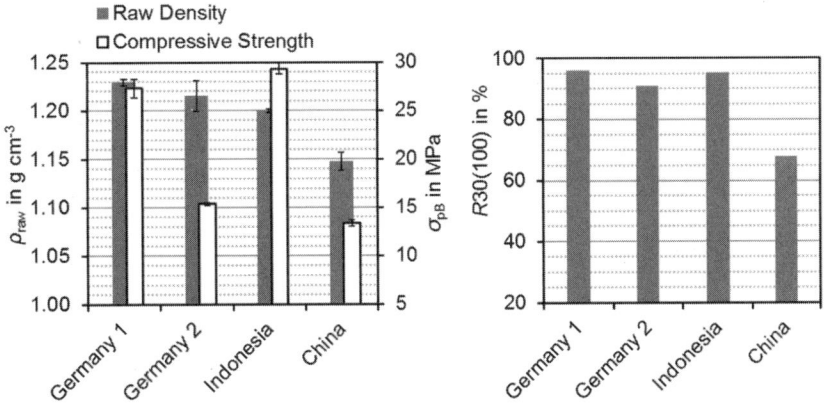

**Figure 4:** Raw density, compressive strength and abrasion resistance of the briquettes at the parameter sets of maximum briquette quality for four different coals for the grain size $\Delta d$ = 1/0 mm ($p$ = 140 MPa, $w$ = 11%, $t_p$ = 3 s, $T_p$ = 70°C)

The intense comminution of the coal leads to an increasing briquette quality especially for the German brown coal 2 and the Chinese brown coal. For the German brown coal 2 a slight increase of the raw density was determined. The compressive strength increases from 10.4 MPa to 15.4 MPa and the abrasion resistance exceeds 90%. For the Chinese brown coal the raw density remains on the same level as before, the compressive strength decreases slightly but the abrasion resistance significantly increases from 39.0% to 67.7%.

In the preceding investigations the coal was briquetted in a single compression step. To investigate the effect of the two-stage press process the German brown coal 2 was briquetted with pre-compaction and main-compaction step at the hydraulic stamp press. The coal was pre-compacted with a pressure of 5 MPa, a pre-compaction temperature of 70 °C and a pre-compaction time of 10 s. Afterwards the compaction was done with a briquetting pressure of 120 MPa, briquetting temperature of 70 °C and a briquetting time of 3s. Figure 5 shows the raw density, compression strength and abrasion resistance of briquettes produced with the one-stage and two-stage briquetting regime.

The two-stage briquetting process leads to an increasing raw density. In the pre-compaction step the raw material is rearranged and densified by the low pressure. It is also possible that elastic and plastic deformation of the coal particles may occur. In the step of main compaction the volume between the coal particles is reduced to a minimum and binding mechanisms like van der Waals forces and hydrogen bonds as well as interlocking bonds lead to the creation of a denser and stronger briquette.

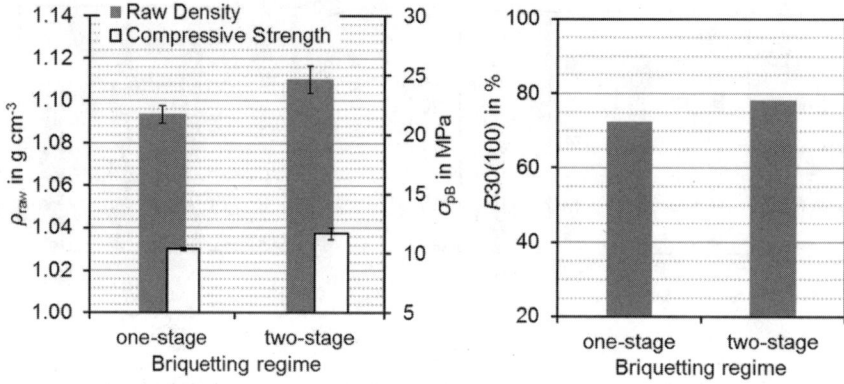

**Figure 5:** Raw density, compressive strength and abrasion resistance of the briquettes for one-stage and two-stage briquetting regime for the grain size $\Delta d$ = 2/0 mm of brown coal Germany 2

## 4.2. Investigations with the demonstrator

For the investigations at technical centre the coals were pre-comminuted, dried to a moisture content of 15% and fine-comminuted to the grain size of $\Delta d$ = 2/0 mm. For the tests the German brown coal 2 and the Indonesian brown coal were briquetted on the demonstrator with the briquetting parameters shown in Table 2.

**Table 2:** Briquetting parameters for the German brown coal 2 and the Indonesian brown coal at the demonstrator

|  | Germany 2 | Indonesia |
|---|---|---|
| $p$pre-compatcion (MPa) | 5 | 5 |
| $t$pre-compaction (s) | 10 | 5 |
| $p$main-compaction (MPa) | 140 | 120 |
| $t$main-compaction(s) | 3 | 3 |
| $T_p$ (°C) | 20 | 20 |

The raw density, compressive strength and abrasion resistance of the produced briquettes are shown in Fig. 6.

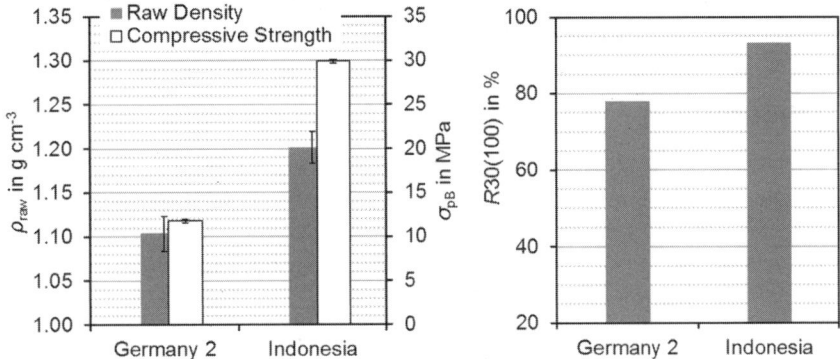

**Figure 6:** Raw density, compressive strength and abrasion resistance of briquettes produced on the demonstrator with German brown coal 2 and Indonesian brown coal

As shown in Fig. 5 the briquettes produced on the demonstrator show quality comparable to the laboratory scale. Analogue to the laboratory-scale investigations the quality of the briquettes of the Indonesian coal is higher than the briquette quality of German brown coal 2. This means, a scale-up of the briquetting behaviour from laboratory scale to semi-industrial scale is possible using the new press technology.

## 5. Summary

Within this work the briquetting of four coals of different provenience was investigated in laboratory scale and at the demonstrator of a new high-performance briquetting press. The work showed that the briquette quality highly depends on the provenience of the coal as well as the briquetting parameters. By using a finer grain size of the coal or a two-stage briquetting the briquette quality can be increased significantly. The investigations on the demonstrator showed that a scale up of the briquetting approach is possible from laboratory to semi-industrial scale with nearly the same briquetting parameters. The briquettes produced at the demonstrator showed a comparable quality to the laboratory scale. In the next step a prototype of the new press with a throughput of 3 t h$^{-1}$ is assembled and tested at the ITUN technical centre. After successful commissioning the prototype will be tested under industrial conditions.

## 6. Acknowledgement

The investigations are part of the research transfer project "EVA – Development of an energy-efficient and wear-reduced agglomeration technology" and were supported by the Federal Ministry for Economic Affairs and Energy of the Federal Republic of Germany, support code 03EFKSN123.

## 7. References

[1] Fehse, F., Rosin, K., Schröder, H.-W., Kim, R., Spöttle, M., Repke, J.-U.: Influence of briquetting and coking parameters on the lump coke production using non-caking coals. Fuel, 203 (2017), p. 915–923.

[2] Schmidt, A.: BGL Application & ZEMAG Slagging Gasification Technology, Gasification India 2015, New Delhi, 13–14.02.2015.

[3] Naundorf, W., Wollenberg, R., Trommer, D.: Verringerung der $SO_2$-Emission bei der Verbrennung von Briketts aus mitteldeutscher Braunkohle. Braunkohle, 6 (1994), p. 11–18.

[4] Krug, H., Naundorf, W.: Braunkohlenbrikettierung, Deutscher Verlag für Grundstoffindustrie.

[5] Rosin, A., Schröder, H.-W., Repke, J.-U.: Briquetting press as lock-free continuous feeding system for pressurized gasifiers. Fuel, 116 (2014), p. 871–878.

[6] Fehse, F., Schröder, H.-W., Kim, R., Spöttle, M., Repke, J.-U.: A new approach for processing and agglomeration of low-rank coals for material usage, in: Litvinenko et al. (eds.) 18th International Coal Preparation Congress, St. Petersburg, 28.06.2016–01.07.2016.

# 43

# Mechanism of drying coal fines by means of contact sorption

Marco le Roux[1], Quentin Campbell[1], Jakobus Hoffman[2]

[1]*North-West University, Potchefstroom, South Africa*
[2]*South African Nuclear Energy Corporation (NECSA), Pretoria, South Africa*

**Abstract:** Contact sorption drying has been proofed as a possible alternative drying method to the ones currently in use. It comprises of mixing a sorbent, having a large surface area, with wet coal particles to aid in moisture transport from the coal to the sorbent. It was found that commercially available sorbents have the ability to dry a fine coal filter cake, typically starting at about 0.30 g(moisture)/g(coal and moisture), to well-below 0.08 g(moisture)/g(coal and moisture) within 10 min. In addition, it was proved that these sorbents had extended process life and can be regenerated in a packed column using ambient airflow. Energy balances on this process setup in a laboratory showed it to be energy positive and therefore a viable option for larger scale testing.

The aim of this paper was to quantify the moisture transportation mechanism. This was done using a series of small scale setups which isolated the moisture transport mechanism and showed it to occur predominantly via the liquid phase. X-ray computed tomography verified this by tracing barium nitrate rich water from a coal filter cake to sorbent particles over the cause of three hours, showing the sorption thereof into the alumina rich ceramic beads.

**Keywords:** Activated alumina, fine coal dewatering, contact sorption drying, moisture transport

## 1. Introduction

Coal fines, defined here as +0.15 mm −0.5 mm, are steadily on the increase due to mechanised mining and processing methods, which has the disadvantage of containing added moisture (De Korte and Mangena, 2004). It is not uncommon for the size fractions to retain up to 25 %wt after dewatering, and even above 30 %wt if the particle size drops below 0.15 mm (Bourgeois and Barton, 1998). Since these numbers exceed coal moisture limits, plants have to mix the wet fines into dryer coarser material up to the point where contractual limits are reached, and then discard the remaining fines (Hand, 2000). In South Africa alone, approximately one billion tons of wet, valuable fine coal is currently lying in ponds (Mining Review Africa, 2016), increasing by 60 Mtons each year. Attempts to dry these coals via traditional mechanical or thermal methods proved fruitless due to the large retention forces or high capital and operational costs, respectively. Alternative methods such as contact sorption drying need to be understood and developed in an attempt to solve this age old dilemma.

This paper stems from previous work done on contact sorption drying, aiming to put forward this drying technique as a viable alternative or add-on for current drying processes. In doing so, this paper will discuss the process and a summary of results found, while identifying the main moisture transfer mechanism during drying by using both experimental work and computed X-ray tomography. A true understanding of the transfer mechanism will enable future development and implementation of this technology on a larger scale.

## 2. Background

### 2.1 Moisture in coal

Coal is porous by nature which results in coal particles having large surface areas for moisture to adsorb onto, or be retained within the crevasses of these particles. Depending on where the moisture is located in the filter cake, it can be classified to be either inter-particulate water, intra-particulate water (both being retained via capillary forces) or surface water (adhering to the surfaces of the particle (Asmatulu and Yoon, 2012). Fine and ultra-fine coal tend to contain more moisture compared to larger coal particles. The physical properties of smaller particle sizes are the main contributor to the high retention of moisture. These smaller particles have a larger total surface area, increasing its surface tension and in return its ability to attract and retain water. When lumped together, the smaller inter-particulate pore radii associated with small particles increase the capillary forces between the particles, for example a filter cake, which in turn increases the moisture holding capacity of the total coal sample (Tao et al., 2006).

### 2.2 Contact sorption drying

Contact sorption is not a new technology and has been used in the chemical and minerals industries with great success over an extended period of time. In short, it refers to the transfer of a phase (liquid or gaseous) from the surface of one substance to that of another (Parashar, 2015). When applied as a drying mechanism, moisture is transported (in either liquid or vapour phase) from the inner and outer surfaces of a coal particle, to the surface of an adjacent sorbent particle until equilibrium between the coal and sorbent attraction forces are reached. This equilibrium point is mainly a function of available moisture in both the coal and sorbent particle, inter and intra-capillary forces within both substances as well as the binding forces of moisture on the particle surfaces. This process can schematically be represented by the simultaneous occurrence of two mechanisms, A and B (Dąbrowski, 2001). These mechanism depicts the movement of moisture from the inner particle, diffusing through the boundary layer to reach the free moisture regime (A) from where it is adsorbed in a reverse manner into the sorbent (B).

## 2.3 X-ray computed tomography

X-ray computed tomography is a non-destructive process by which the internal structures of various objects and materials get mapped and drawn into a virtual three-dimensional image (Schabowicz et al., 2018). It has long been used in the medical field to view the skeletal structure of patients without having to operate on the person. There are different ways that X-rays can be implemented to image the internal structure of objects, but this investigation exploited the density difference of materials to map the location of water during its diffusion from coal particles to the sorbents. Figure 1 depicts the setup that was used at the South African Nuclear Energy Corporation's (NECSA) micro-focus X-ray computer tomography facility.

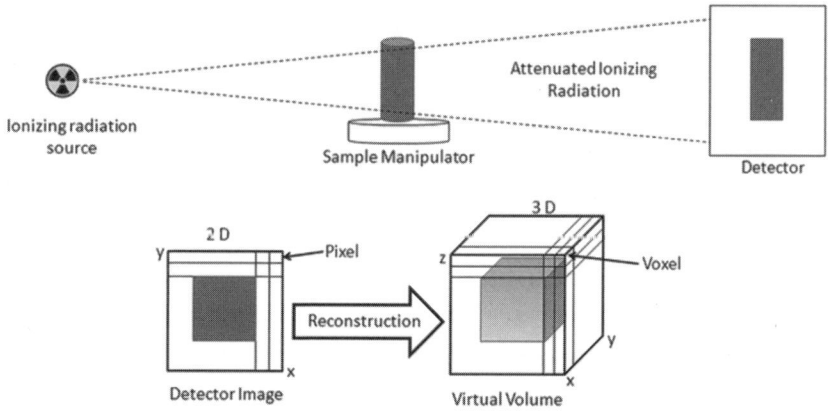

**Figure 1:** Schematic illustration of the SANRAD facility at NECSA

Van der Merwe et al. (2006) have already illustrated that the setup depicted in Fig. 1 can be used to image the location of water inside alumina ceramic spheres by using a solution of barium nitrate as dye in order to get better contrast between the solid and liquid phases. X-rays are produced when current flows at a specific frequency through a tungsten-based filament. The rays loose energy as it passes through the sample, which is placed on a rotating platform. Once a ray has passed through the object, it travel on towards an array of sensors that measure the intensity (energy loss) of X-rays across a flat two dimensional plane. All the measurements are sent to a computer which reconstructs the two-dimensional image. After one image of the sample has been taken, the rotating platform gets adjusted to make a different part of the sample face directly towards the X-ray source. This rotation and imaging process is repeated until the platform has turned a full 360°. All the two-dimensional images are then collectively transformed into one three dimensional virtual object which can be analysed using dedicated computer software (Viljoen et al., 2015).

## 3. Experimental

### 3.1 Materials used

Medium-rank C bituminous coal from the Highveld coal field in South Africa was used in this study. The proximate analysis and the sulphur content of the coal are given in Table 1.

**Table 1:** Proximate analysis (air dry basis) and sulphur content

| Test type | Composition (%wt) | Test method standard |
|---|---|---|
| Inherent moisture | 3.6 | ACT-TPM-010 rich on ISO11722: 1999 |
| Ash yield | 14.9 | ACT-TPM-011 rich on ISO 1171: 2010 |
| Volatile matter | 29.6 | ACT-TPM-012 rich on ISO 562: 2010 |
| Fixed carbon | 51.9 | (By difference) |
| Total sulphur | 0.97 | ACT-TPM-013 rich on ISO 19579: 2006 |

As received coal samples were left to dry under ambient conditions and later crushed and sieved to produce samples of −2 mm +1 mm, −1 mm +0.5 mm and −0.5 mm +0.25 mm. The samples were then submerged in water for 24 h and dewatered in a positive pressure filter to obtain a minimum moisture content which ranged between 15 %wt and 30 %wt depending on the size fraction. Alumina rich 3 mm and 5 mm spherical beads were bought and used as sorbent as received. The composition of the beads are given in Table 2 whilst the physical properties as shown in Table 3.

**Table 2:** Compositions of ceramic sorbent

| Mineral content | Composition (%wt) |
|---|---|
| Aluminium oxide (Al2O3) | 92.7 |
| Silicon oxide (SiO2) | 0.02 |
| Iron (iii) oxide (Fe2O3) | 0.02 |
| Sodium oxide (Na2O) | 0.3 |

**Table 3:** Physical properties of ceramic sorbent

| Physical properties | Measurement |
|---|---|
| Surface area (m²/g) | 350 |
| Pore volume (cm³/g) | 0.5 |
| Bulk density (kg/m³) | 769 |
| Moisture (%wt) | 1.8 |

## 3.2 Experimental setup

The experimental work was done on a laboratory scale unit as shown in Fig. 2.

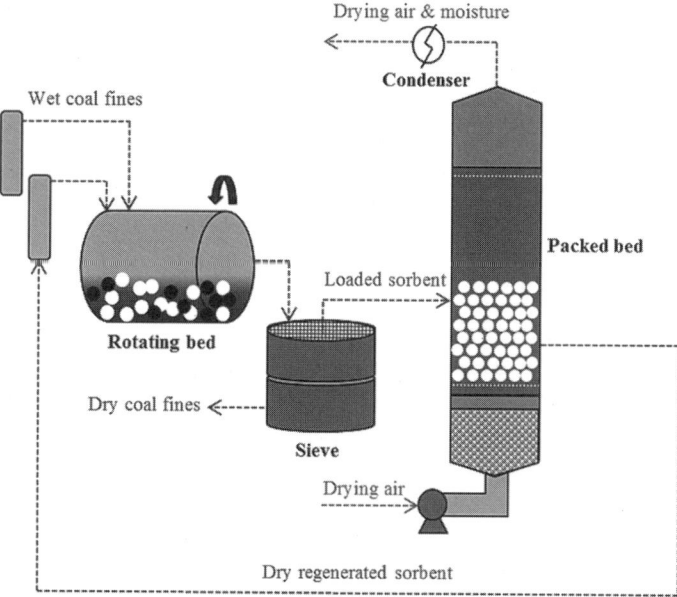

**Figure 2:** Diagram of the laboratory setup for contact sorption drying and regeneration

A 5.5 cm inner diameter vessel with a length of 5 cm were filled with a mixture of the wet coal filter cake and sorbents, and placed on rollers to ensure a rotating speed of 3 revolutions per minute. The rotation was used to warrant proper mixing between the coal and sorbent. The conditions inside the vessels were kept constant at 22°C and 40% relative humidity. Experimental conditions were as follows:

- *Motion*: Stationary and rotating bed.
- *Sorbent type*: Alumina rich sorbent.
- *Sorbent to coal mass ratio*: 0.5:1, 0.75:1, 1:1, 2:1 and 3:1.
- *Sorbent particle size*: 3 mm and 5 mm.
- *Coal size fraction*: +1 mm −2 mm, +0.5 mm −1 mm and +0.25 mm −0.5 mm.

The rotating bed setup consisted of 10 cylinders each containing an identical coal and sorbent mixture at the same conditions. At a predetermined time interval as single cylinder was removed, sampled and moisture analysis were done on the coal and sorbent. Hereafter the sorbents were placed in an 8 cm inner diameter packed bed for regeneration using ambient air flow at 22 °C and 40% relative humidity at upwards velocities ranging between 1.5 m/s and 1.7 m/s.

## 3.3 Supplementary testing

A set of nine focused experiments were completed to determine which of vapour or liquid phase transfer dominates the transport of moisture from the coal to the sorbent. A schematic representation of the focussed experimental setups for the first eight can be seen in Fig. 3, with the ninth test being the X-Ray computed tomography scans. A description of each is as follows:

(1) A vessel was completely filled with pure liquid water, a known amount of sorbents were added and the vessel was enclosed. Timeous sampling allowed for the determination of the maximum liquid adsorption and capacity within the sorbents. (2) An empty vessel was filled with a known amount of the conditioned dry sorbent material and was left open and exposed to controlled atmospheric conditions of 22°C and 40% relative humidity. The setup made it possible to determine the rate of vapour adsorption. (3) A single sorbent bead was placed in a water solution containing the colour pigment gentian violet. The colouring was used to examine the diffusion path of liquid water into the sorbent. (4) This setup was similar to the first set-up with the exception of using only one sorbent bead, allowing to determine single particle capacity. Setup (5) is a repeat of Setup (2) but for coal instead of sorbents. It allows for

the determination of total moisture vaporization from a coal filter cake at the indicated conditions. (6) The previous setup was repeated at higher relative humidity to test the influence of saturation levels on vaporization of moisture from the coal. (7) In the seventh setup, mixtures of wet coal filter cake and conditioned sorbents were tightly packed in a vessel and filled to capacity. The enclosed vessel limited air contact and allowed for predominantly liquid water transfer to be studied. (8) In the last setup in Fig. 3, two open vessels, one filled with sorbent beads and the other with wet coal, were sealed off inside a third container with initial air conditions at 22°C and 40% RH. Predominant vapour transfers were studied here. Lastly (9), X-ray computed tomography was used to trace barium enriched water transfer from coal to the sorbents.

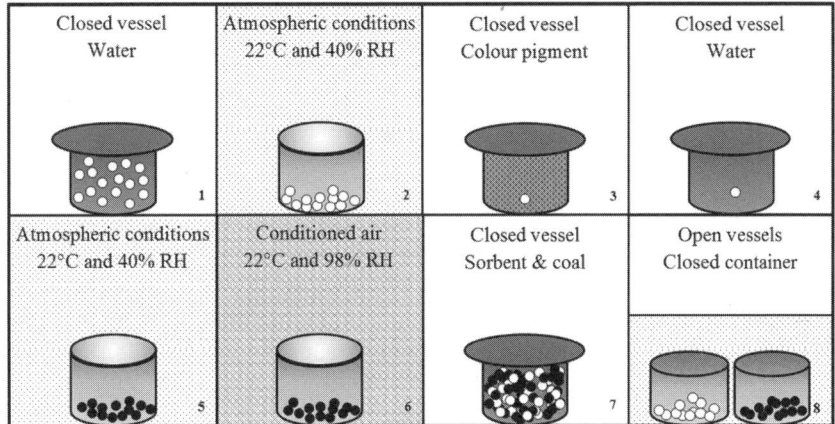

**Figure 3:** Experimental setups; focussed experiments

# 4. Results and discussions

## 4.1 Contact sorption drying

The base experiment chosen for discussion was the dewatering of +0.5 mm −1 mm coal fines in the rotating vessels with 5 mm alumina rich sorbent in a 1:1 mass ratio. Repeatability of the results showed a variation between 0.6% and 2.8%. The dewatering results are shown in Fig. 4.

Free moisture and inter-particulate moisture are initially transferred during the first stage of dewatering as indicated by stage A, in Fig. 4. According to Klaewkla et al. (2011), this stage usually happens very fast with most moisture transfer happening during this time. This is due to the abundance of open pores available for adsorption and a greater concentration gradient. After 2.5 min, the rate of drying decreases, as indicated by stage B. Here, the sorbent pores become saturated to a point where the concentration gradient between the coal and sorbent is less and the rate of moisture transfer slows down. Equilibrium is reached in stage C. This typical run proved that it was possible to dewater coal fines with an initial moisture content of 0.25 g(water)/g(coal and water) to 0.13 g(water)/g(coal and water) after 2.5 min, and to 0.05 g(water)/g(coal and water) after 10 min.

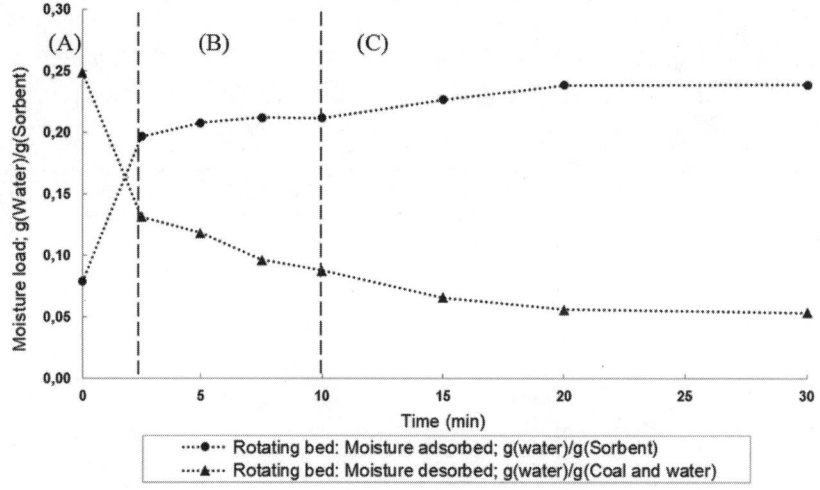

**Figure 4:** General moisture transfer; +0.5 mm −1 mm coal fines in a 1:1 ratio with 5 mm alumina rich sorbent

## 4.2 Feasibility

To evaluate the feasibility of this process, spent sorbents were regenerated in a packed bed using ambient air before recycling it to be mixed as feed to the drying vessels. Moisture desorption from the sorbents took less than 10 min, rendering a dry enough particle for use, which compared remarkable well with fresh sorbent feed when used in the drying vessel, as indicated by Fig. 5.

There is a remarkable similarity between the performance of unused and regenerated sorbents during all three moisture transfer phases. It supports the notion that it is possible to regenerate the ceramic spheres in a packed bed at ambient conditions. By regenerating and recycling the sorbents in this manner in a closed loop during large operations, significant reduction in capital and operational expenditures can be achieved. Bratton et al. (2012) reported similar findings, observing no substantial difference in the effectiveness of sorbents in coal drying after regeneration at high temperatures.

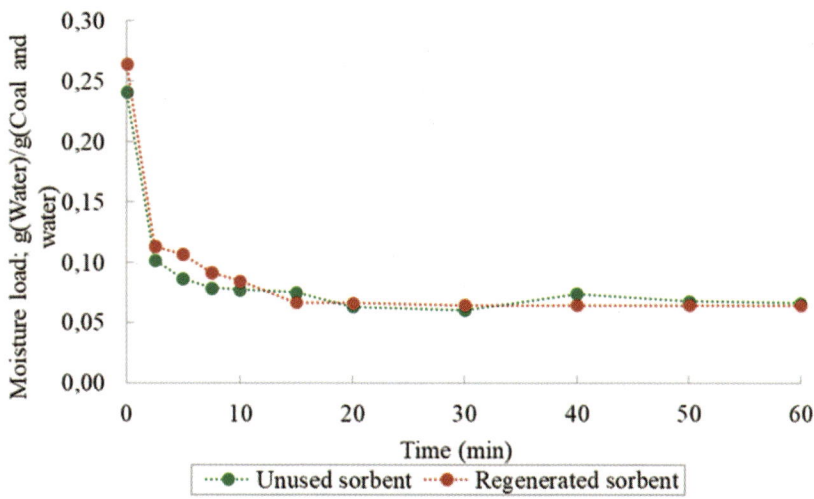

**Figure 5:** Unused and used sorbent; +0.5 mm −1 mm coal fines in a 1:1 ratio

## 4.3 Supplementary results

Setups 1 and 2 from Fig. 3 compared water only and vapour only adsorption onto the sorbents. The large difference in adsorption capabilities are shown in Fig. 6(A). After 40 min have elapsed, the submerged sorbents had a moisture content of 0.43 g(water)/g(sorbent) as appose to 0.14 g(water)/g(sorbent) for the air exposed ones. Exposure to available liquid will clearly dominate sorbent adsorption behaviour.

Similar behaviour was observed when comparing Setups 5–7 from Fig. 3, as shown in Fig. 6(B). The desorption of vapour from coal into different atmospheric conditions were shown to be very similar, and much less than the desorption of liquid moisture from coal into ceramic beads. This was confirmed with Setup 8 from Fig. 3 where virtually no moisture transfer was recorded. Therefore, when combining both the adsorption mechanism into the ceramic beads, as well as the desorption of moisture from coal, the predominant mechanism was shown to be liquid transfer.

In a final attempt to quantify the transport mechanism, micro-focus X-ray tomography scans were done on a coal filter cake, containing barium enriched water, which had sorbent beads placed on the inside. Figure 7 shows the progression of moisture into the sorbent as if diffuses to the centre of the bead over time. Each one of the tomopraphs are of the same bead, showing the moisture progression into the sorbent, which are mapped out by the blue lines. Using software, the volume moisture adsorbed by the bead was calculated over time, to give an estimate of the adsorption rate. It was calculated as 0.22 mm$^3$/min, which correlated well with work done by Le Roux et al. (2014) on static drying of fine coal using ceramic drying agents.

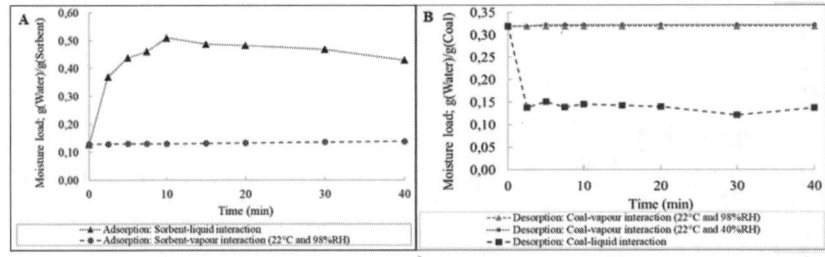

**Figure 6:** (A) Adsorption: sorbent–liquid vs. sorbent–vapour interaction and (B) Desorption: coal–liquid vs. coal–vapour interaction

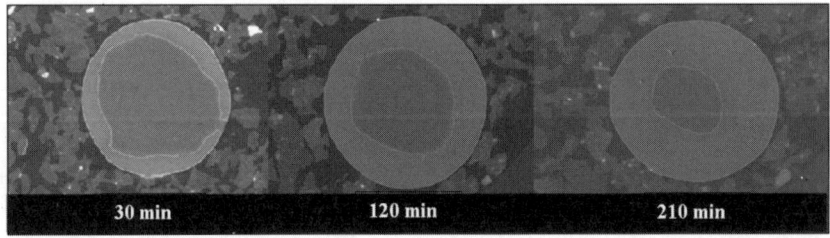

**Figure 7:** Microfocus X-ray tomograph show moisture adsorption onto a ceramic bead

## 5. Conclusion

Sorbent assisted drying of fine coal was proven to be successful and feasible when targeting a coal product with less than 10% moisture. In addition, it showed that ceramic particles can be regenerated at ambient conditions, using high airflow and reused as drying agent to achieve similar results. In an attempt to isolate the transport mechanism, a series of supplementary tests indicated that liquid moisture transport dominate vapour transport to the extent that it can be regarded as the sole transport mechanism. Resistances for liquid flow from coal to the sorbents were identified and contributed to available contact area, diffusion time and intra-particulate capillary forces.

## 6. Acknowledgement

The authors would like to thank Coaltech, the Southern African Coal Processing Society, the Centre of Excellence in Carbon Based Fuels, North-

West University, South Africa and NECSA, for their contribution to this project, financial and/or otherwise.

This work is based on the research supported by the South African Research Chairs Initiative of the Department of Science and Technology and National Research Foundation of South Africa (Chair Grant No: 86880, UID 85643, Grant No: 85632 and Grant UID 72310). Any opinion, finding or conclusion or recommendation expressed in this material is that of the author(s) and the NRF does not accept any liability in this regard.

## 7. References

[1] Asmatulu, R. and Yoon, R.H., 2012. Effects of surface forces on dewatering of fine particles. *Separation Technologies for Minerals, Coal and Earth Resources*, pp. 95–102.

[2] Bratton, R., Ali, Z., Luttrell, G., Bland, R. and McDaniel, B., 2012. Fine coal: Nano-drying technology. Coal Age, June. p. 50–57.

[3] Bourgeois, F.S. and Barton, W.A., 1998. Advances in the fundamentals of fine coal filtration. *Coal Preparation*, *19*(1–2), pp. 9–31.

[4] Dąbrowski, A., 2001. Adsorption – from theory to practice. Advances in colloid and interface science. Maria Curie-Sklodowska University: Faculty of chemistry. Lublin, Poland. p. 135–224.

[5] De Korte, G.J. and Mangena, S.J., 2004. Thermal Drying of Fine and Ultra-fine Coal. Report No. 2004 – 0255, Division of Mining Technology, CSIR. p. 5–24.

[6] Hand, P.E., 2000. Dewatering and drying of fine coal to a saleable product. Coaltech 2020, Task 4.8.1. p. 8–100.

[7] Karthikeyan, M., Zhonghua, W. and Mujumdar, A.S., 2009. Low-rank coal drying technologies – current status and new developments. *Drying Technology, 27*(3), pp. 403–415.

[8] Klaewkla, R., Arend, M. and Hoelderich, W.F., 2011. A review of mass transfer controlling the reaction rate in heterogeneous catalytic systems. Mass transfer: Advanced aspects. https://www.researchgate.net/profile/Raweewan_Klaewkla, p. 68–684. Date of access: 20 Jan. 2017.

[9] Le Roux, M., Campbell, Q.P. & Van Rensburg, M.J., 2014. Fine coal dewatering using high airflow. *International Journal of Coal Preparation and Utilization, 34*:220–227.

[10] Mining Review Africa, 2016. Coal waste material: An untapped power generation solution. Issue 4. p. 50–53.

[11] Parashar, K., 2015. Adsorption. University of Johannesburg: Department of Applied Chemistry. http://www.slideshare.net/Kamyaparashar/adsorption-presentation-44669901. Date of access: 12 Feb. 2017 (PowerPoint presentation).

[12] Schabowicz, K., Jóźwiak-Niedźwiedzka, D., Ranachowski, Z., Kudela, S. and Dvorak, T., 2018. Microstructural characterization of cellulose fibres in reinforced cement boards. *Archives of Civil and Mechanical Engineering, 18*(4):1068–1078.

[13] Tao D., Yu, S. and Parekh, B.K., 2006. Picobubble enhanced fine coal flotation. In *XVI International Coal Preparation Congress & Exhibition* (pp. 470–471). China.

[14] Van der Merwe, W., Nicol, W. and De Beer, F., 2006. Internal wetting dynamics of alpha- and gamma-alumina catalyst spheres using X-ray computed tomography. *South African Journal of Science, 102*:585–588.

[15] Viljoen, J., Campbell, Q.P., Le Roux, M. and De Beer, F., 2015. An analysis of the slow compression breakage of coal using microfocus X-ray computed tomography. *International Journal of Coal Preparation and Utilization, 35*:1–13.

# 44

# CHRONOS short-cycle non-thermal drying process

Elena N. Chernysheva and Vadim A. Kozlov

*Coralina Engineering LLC, Moskva, Russia*

**Abstract:** Application of coal slurry wet preparation processes at coal-preparation plants results in producing high-moisture products that does not allow them to be added to the total concentrate without additional drying. Thus, most of coal-preparation plants are forced to build drying units. However, the traditional thermal drying of high-volatile coal is fire-hazardous and explosion-hazardous. Strict control of oxygen content in the fuel gas is necessary. The drying units should not allow air ingress from the environment and should be manufactured with maximum tightness of joints, connections, and gas pipelines. This paper discusses the new safe process of short-cycle non-thermal drying "CHRONOS" for fine coal of any rank and minerals. The CHRONOS process enables removing moisture from coal without direct heating it, but with mechanical contact with special type sorbents. The sorbent adsorbing required amount of moisture from coal is regenerated and returned to the process. The desorption of moisture from the sorbent pore volume is performed in a special regeneration unit. A pilot packaged drying unit with the capacity of up to 2 tph was built to study the effectiveness of different types of sorbents for drying coal of different ranks, fineness, ash content, and initial moisture. Coraline Engineering implements projects for the installation of equipment for non-thermal dewatering coal by sorbents using the CHRONOS process at operating coal-preparation plants.

**Keywords:** Sorbent, moisture adsorption, moisture desorption, coal slurry dewatering, drying process

## 1. Introduction

Most of the currently used equipment for mechanical dewatering of 0–3 mm fine coal, such as screw, vibrating, screen bowl centrifuges, filter presses, vacuum filters, and hyperbaric filters, in most cases, do not provide enough product moisture to obtain standard concentrate. Fine particles of coal slurry have developed surface and are capable to retain a significant amount of moisture. Removing this bound moisture requires extra effort. Moisture in the slurry is retained mainly by sorption and capillary forces. The amount of bound moisture is governed by the coal chemical composition, degree of metamorphism, which determines the presence of chemically active groups on the surface of the coal particles, and the particle size distribution of the coal slurry. The bound moisture in a coal occurs in different states: held by capillary forces in fractures and pores, and in the form of films adsorbed at the surface of the coal particles. A monomolecular water layer is most strongly held on solid surface of coal by hydrogen bonds. The subsequent layers of water molecules are bound with coal weaker depending on the distance from the coal surface according to the van der Waals bonding law (Chernysheva, 2016).

Coal slurry of 0–0.2 (0.5) mm fineness introduces the largest share of moisture in the total concentrate of a coal-preparation plant. The greater slurry share in the final product, the higher the saleable product moisture. The finer the slurry, the greater amount of moisture is retained on its surface due to intermolecular forces (adsorption) (Novak, 2016).

The energy required to remove the adsorbed moisture from coal depends on the strength of the physical bonds of water molecules with the coal surface. Until recently, it was possible to remove such moisture from coal only by thermal drying at very high temperatures.

Most thermal dryers are based on feeding hot air to a drying device with direct contact of hot fuel gas and the material to be dried. Large particles of coal (more than 1 mm) and small ones (less than 1 mm, especially fine particles of 0–100 microns), are unevenly heated in the air flow. While the larger particles are dried, the small particles overheat and begin to soar in the dryer volume. This causes the danger of the overheated dust explosion in the drying chamber. Thermal drying of high-volatile coal is especially dangerous. Even modern thermal dryers, in which the risk of explosion or inflammation is mitigated by reducing the duration of coal stay in the drying chamber, bear a certain degree of explosion risk due to possible oxygen ingress into the dryer, and also imply large capital and operating costs.

In the last decade, innovative methods of drying fine coal – infrared dryers, microwave dryers, as well as drying coal using sorbents – have become increasingly popular. The idea of drying wet viscous and loose materials using sorbents was born over 50 years ago. However, the commercial-scale implementation of this method has not yet taken place and remains at the stage of experiments and bench tests.

The essence of the process lies in the fact that when a moist material comes into contact with a sorbent, dewatering is more effective compared to mechanical methods. This enables the adsorbed moisture to be removed from the coal without heating the drying agent. When the sorbent comes into contact with a moist material, due to the presence of active centers and developed pore space on the surface of the sorbent, water molecules having the form of dipole are attracted to these centers, penetrate into the structured pore space due to capillary suction, are fixed on the surface of the pores, and fill large volume of the pore space. The strength of the sorption activity of the sorbents is such that the sorbent sucks out free, film, and even a part of the internal strongly-bound moisture from the material to be dried in relatively short period of time.

The simplicity and cost-effectiveness of the process lies in the fact that, after the contact, the sorbents, which absorbed moisture, and the dried material are separated by simple screening on a traditional sieve with mesh larger than the size of the largest coal particle. The sorbents are sent to the regeneration unit and reused to dry the next batch of material. This method allows dewatering not only coal, but also minerals, as well as any loose materials, which cannot be heated for various reasons. The main issue in the development of the process for material dewatering by sorbents is the method for the effective non-destructive regeneration of the sorbents in moisture sorption–desorption cycles.

## 2. History of development of sorbent-based drying process

The idea of using sorbents for drying wet and viscous materials arose in the 1960s. One of the well-known patents is the patent of Nelson Severinghaus, who proposed the idea of drying wet dusty materials in a traffic flow (on conveyor belt and transporting screw conveyor) (Severinghaus, 1971). In 2012, the American company Nano Drying Technologies LLC proposed theoretical solutions for the continuous drying mineral and coal slurry using molecular sieves (sorbents). The proposed process (Bratton et al., 2012) has not yet been implemented on commercial scale until recently. The probable reasons for this delay, in our opinion, were the lack of suitable sorbents on the market, insufficiently studied features of the sorbents applied for drying coal, and the lack of efficient and cost-effective methods to regenerate the sorbents (Kirillov et al., 2016). However, many experts in coal preparation methods managed to get acquainted with their publications on the laboratory research.

In 2012, Coraline Engineering performed a laboratory research to objectively assess the capabilities of sorbents to remove adsorbed moisture and the influence of the dried material properties on the drying process. Based on the research findings a patent was obtained (Novak et al., 2012). In continuation of this work, an unique laboratory cyclic test method has been developed to date. Comprehensive research on drying coal of different ranks, including high-volatile coal from coal-preparation plants of the Kuzbass, Vorkuta, Yakutia, and Donbass, were performed. Concentrates from screen bowl centrifuges, cake from belt and chamber filter presses, raw coal, and minerals were used as the material to be dried. The research performed in the company's laboratory (Kirillov et al., 2016) allowed collecting the sufficient amount of experimental findings and developing the new method of short-cycle incomplete sorbent regeneration, enabling reducing the regeneration time and increasing the sorbent granule integrity in the sorption–desorption cycles. On the basis of the data obtained, the Coraline Engineering has developed and implemented on commercial scale the CHRONOS short-cycle non-thermal drying process (CHRONOS).

## 3. Solution to the problem of sorbent application to drying coal

Sorbents regenerative capability makes their commercial-scale application rather cost-effective. Depending on the scope of use, the process of sorbent regeneration can be based on the thermal effect of a heated gas, the application of high pressure, or vacuum generation in the regeneration chamber, and also be combined. At present time, attention is focused on developing efficient and cost-effective methods of regenerating sorbents to enable their commercial-scale application. An interesting and promising method in terms of simplicity of implementation is regeneration of sorbents by exposure to microwave electromagnetic radiation in resonant chamber (Muller et al., 2012). However, this regeneration method requires rather long time (about 10–15 min). Most of the existing traditional regeneration methods are focused on sorbents working with water vapor in gaseous environment. The number of sorbent types capable to effectively remove condensed moisture without subsequent destruction of the granule structure and losing sorption properties turned out to be very limited. Regeneration of sorbents by the existing traditional methods accelerates the process of destruction of the granule structure and reduces their sorption capacity, significantly decreasing the sorbent performance life in the sorption–desorption cycles, as it is aimed at the almost complete removal of absorbed moisture at high energy impacts, in particular, when exposed to high temperatures. For example, complete regeneration of NaX zeolite takes 70–75 min at temperature of 300°C, whereas, when exposed to microwave radiation, it takes 10–15 min at temperature of 80–100°C only (Muller et al., 2012). Insufficient knowledge of the dynamics of moisture adsorption and desorption by different types of sorbents depending on the physical, physico-chemical, and structural characteristics of the dried materials is a deterrent for the development of the commercial-scale process for loose material dewatering by sorbents.

To date, a new method of short-cycle incomplete sorbent regeneration on the basis of a laboratory research complex has been developed in the Coraline Engineering Research Center. The method enables reducing the regeneration time and increasing integrity of sorbent granules in the sorption–desorption cycles (the method is at the stage of patent formalities). This regeneration method is implemented in the commercial-scale process of short-cycle non-thermal drying, refer to as CHRONOS.

## 3.1 The essence of the proposed process

The kinetics of the condensed moisture desorption process at incomplete sorbent regeneration is characterized by the absence of linear dependence of the amount of water removed from the sorbent on the time of the exposure to the drying factor at its constant intensity. During the thermal exposure of the sorbent granules the total amount of moisture removed unequally increases with time of the exposure to high temperature. In the beginning, the moisture evaporation rate increases, and then decreases; the dependence is exponential. This is due to two different types of moisture removed from the sorbent capillaries. The first type is strongly bound moisture, adsorbed on the surface of solids in the pore channels, and another type is weaker bound capillary moisture. When water molecules get into the sorbent pores, they are firmly fixed on the pore walls at the active centers of the interatomic bonds of the sorbent material atoms. Then water molecules are drawn into the pores by capillary forces and are bound in the pore space by much less strong intermolecular forces. The capillary moisture inside the pores and channels is removed relatively readily, whereas the moisture adsorbed on the solid surface requires additional energy to be removed. The moment of transition, in the process of evaporation, from the capillary moisture to the adsorbed bed removal is characterized by dropping the evaporation rate and expressed by an inflection of the evaporation kinetics curves.

At dynamic capacity of sorbents for water vapor up to 25–28%, the dynamic sorption capacity of some sorbents with pore size of up to 1 nm can reach a value of 35.5% when in contact with condensed and film moisture, that corresponds to 0.55 g of water per 1 g of sorbent. Most of the moisture can be attributed to the capillary type. These types of sorbents are capable of repeatedly adsorbing and desorbing moisture in the amount of up to a half or more of their dry weight due to capillary forces. When applying the sorbents in a gaseous environment, the concentration of water vapor is insufficient for the active accumulation of water in the internal pore channels. Most of the moisture in the pores is represented by adsorbed moisture, which, in the course of the sorbent regeneration, requires long-term high-temperature heating to be removed. The adsorption bed of water molecules is readily and firmly fixed on the pore walls and hardly removed from them. While the capillary moisture, filling the remaining pore volume following the adsorption moisture, can relatively easily be removed from the pore volume due to weak forces of interaction between water molecules.

This feature of moisture behavior in sorbents allows saving time for moisture desorption. The desorption may be limited to evaporation of capillary moisture in the sorbent pores only, even if internal pore channels of the sorbents remain permanently filled with adsorption moisture (up to 60% of the total pore volume).

Demonstrative evidences of the described process of moisture evaporation from the sorbent pores are the findings of evaporation kinetics research, performed by the authors on samples of two sorbents: (1) synthetic zeolite NaX (Sorbent-1) and (2) active oxide of aluminum AOA (Sorbent-2), subjected to heating in hot air flow at fixed temperature for periods of time from 1 to 5 min. The size of the synthetic zeolite NaX pore aperture is 0.9 nm. The sorbent is in form of granules up to 6 mm in diameter. Pore aperture of the aluminum oxide is 0.5 nm. The sorbent is in form of granules up to 5 mm in diameter.

Before testing, each sample of each sorbent was saturated to the maximum possible value of moisture $W_{t\,max}$ (for all types of the sorbents under consideration, the maximum sorption capacity is different). Kinetics curves of the moisture evaporation from the sorbent pores are presented in Fig. 1. It is seen from the curves that for the first 2 min the evaporation rate falls almost linearly. This segment of the curve corresponds to the capillary moisture evaporation. After 2 min, the moisture evaporation rate decreases, that is, evaporation of adsorbed moisture begins, which requires more intense and prolonged exposure to be removed. Longer heating becomes ineffective for evaporation of residual moisture in terms of the ratio of energy consumption for heating the heat medium to the degree of the sorbent moisture decreasing.

For example, when heating Sorbent-1 granules, with increase of the temperature exposure time, the intensity of moisture removal from the sorbent begins to decrease after the second minute of heating, when 52.85% of the initial moisture has been evaporated from the sorbent. After heating for 3 min, 63.12% of the initial moisture has been removed from the sorbent, and the difference in water weight between the second and the third minutes was 10.27% only. Then this difference decreases further with time.

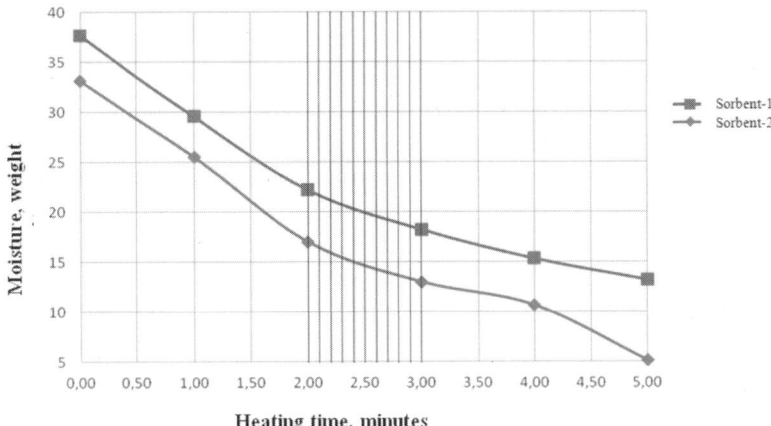

**Figure 1:** The dependence of the final moisture of the sorbent
from the time of its heating

The similar process is observed in the tests of the second type of sorbent (active alumina), designated as Sorbent-2. In both cases, after 2 min of heating, the evaporation rate decreased, that corresponded to decreasing the amount of capillary moisture in the sorbent pores and starting evaporation of the adsorption moisture, which requires further additional energy. This range on the graph is highlighted by vertical hatching. The heating time of 2–3 min is chosen as the optimal one.

The value of the optimal temperature for performing incomplete regeneration of the sorbent was determined in the course of testing to study the dependence of the weight of residual moisture in the pores of the sorbents on the heating temperature of the sorbent granules in hot air flow. The tests were performed on similar samples of the sorbents: synthetic zeolite NaX (Sorbent-1) and aluminum oxide (Sorbent-2) The test findings are shown in the graph (Fig. 2).

For the tests, sub-samples of sorbents weighing 50 g with fixed initial moisture of 15% were selected. This means that the weight of water in the sorbent pores was 7.5 g. The amount of moisture corresponding to this weight in the pores of the sorbent was taken as 100%. The heating time in each test was equal to 5 min. Temperature and moisture content readings during the tests were taken automatically and continuously.

As seen from the graph, an abrupt jump in decreasing the amount of residual moisture in the pores of the sorbent is observed in the temperature interval from 95°C to 105°C. This, in our opinion, corresponds to the stage of evaporation of capillary moisture from the pore volume of the sorbents.

Further, in the interval from 105 °C to 120 °C, the decrease in the rate of the evaporation process is observed. This characterizes the almost complete evaporation of capillary moisture residues, heating of the adsorption moisture bed, and accumulation of energy for the subsequent evaporation of strongly bound moisture. Starting from the point of 120 °C, the curve begins to slowly decrease that corresponds to the process of gradual evaporation of the adsorption moisture beds.

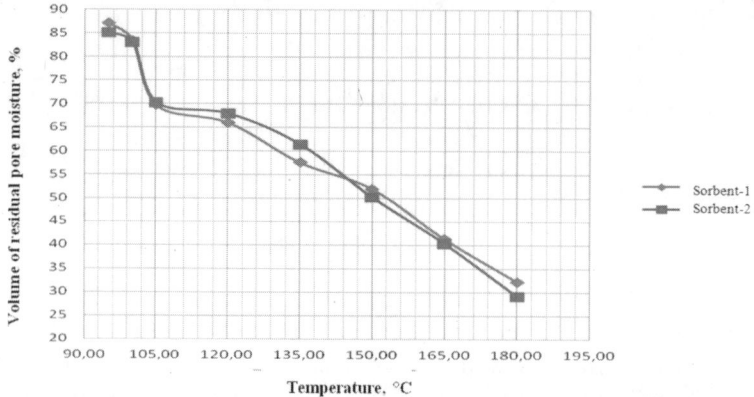

**Figure 2:** The dependence of the amount of residual moisture in the sorbent pores from the temperature of its heating

The parallel tests of the sorbents for the moisture absorption from being dried materials showed that, for efficient drying moist materials by the sorbents, it is enough to free the sorbent pores from capillary moisture without affecting the adsorption one and not spending additional energy for its evaporation. Thus, we accepted the temperature in the interval from 105°C to 120 °C as optimum, above which the costs for heating the sorbents would be excessive and uneconomic.

## 4. Conclusion

Based on the research conducted in 2015, the CHRONOS pilot unit with capacity of up to 2 tph of initial moist coal was developed, designed, and put into operation (Fig. 3). The patent for the unit was obtained in 2015 (Kirillov et al., 2015).

At this unit, preliminary tests of various products are carried out on request of coal-preparation plants. This allows to select the coal drying operating modes, calculate parameters of the process and equipment for drying, and develop design solutions for a specific customer. Currently, this unit is being tested in the operating production conditions at one of the plants in the Kuzbass.

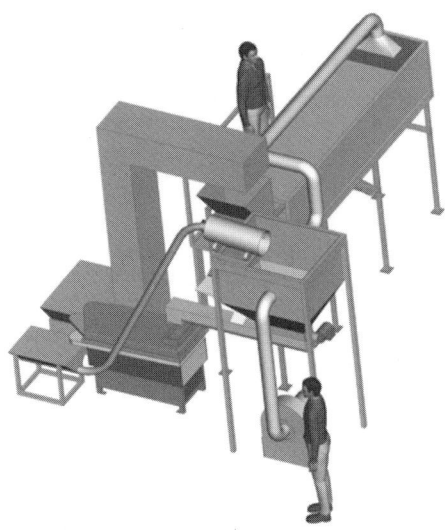

**Figure 3:** 3D model of the CHRONOS pilot unit

To date, Coraline Engineering offers for customers the safe non-thermal process for drying coal of particle size of 0.05–3 mm, ash content up to 20%, with initial moisture up to 30%, implemented in the package plant with capacity of up to 70 tph of the initial moist product. The schematic diagram of the commercial plant is shown in Fig. 4.

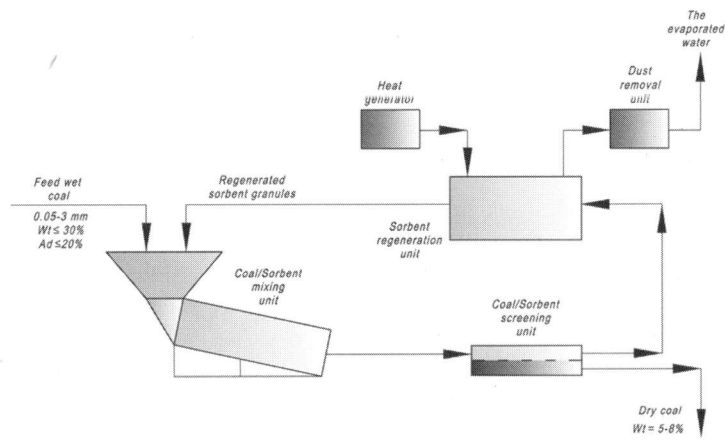

**Figure 4:** Schematic diagram of the CHRONOS process-based commercial plant

The proposed commercial-scale CHRONOS process of short-cycle non-thermal drying enables to safely dry high-volatile coal with volatile content above 30%, typical of coal ranks B, D, G, and Zh. The cost of the plant in terms of the dried saleable concentrate is comparable to the cost of operating high-performance thermal dryers. At the same time, this plant requires significantly lower capital costs and ensures full explosion and fire safety of the production.

## 5. References

[1] *R. Bratton, Z. Ali, G. Luttrell, R. Bland, B. McDaniel.* Nano Drying Technology, Coal Age. June. 2012. C. 50–55.

[2] *E. Chernysheva.* Coal moisture content as the product quality indicator, Ugol' – Russian Coal Journal, 2016, No. 8, pp. 125–128.

[3] *K. Kirillov, V. Kozlov, V. Novak, E. Chernysheva.* Installation for dehydration of fine classes of ore and nonmetallic materials. Patent No. 2588529 of March 26, 2015.

[4] *K. Kirillov, E. Chernysheva, V. Kozlov.* Innovative Drying Technology "Chronos". Deep Non-Thermal Dewatering of Coal and Mineral Fines. XVIII International Coal Preparation Congress: 28 June–01 July. 2016. Saint-Petersburg, Russia Vladimir Litvinenko (ed.). p. 695–700.

[5] *R. Muller F., V. Olshanskaya, A. Rumyantsev.* The method of regeneration of sorbents by non-thermal exposure to microwave electromagnetic radiation. Patent of the Russian Federation No. 2438774 of January 10, 2012.

[6] *V. Novak.* Coal-preparation plant cake issue. Who is at fault, and what is to be done? Ugol' – Russian Coal Journal, 2016, No. 10, pp. 70–73.

[7] *V. Novak, M. Pikalov, V. Kozlov.* Installation for coal slurry dewatering. Patent No. 130235 of December 21, 2012.

[8] *N. Severinghaus.* Patent #US3623233, 1971. Method and apparatus for drying damp pulverant materials by adsorption.

# 45

# How far do you have to go – Filtration makes the difference pressure filtration and steam pressure filtration of coal flotation concentrates

Juergen Hahn

*BOKELA GmbH, Karlsruhe, Germany*

**Abstract:** HiBar steam pressure filtration is capable to produce extremely dry ultrafines below 10% w/w moisture which now offers new options in coal ultrafines treatment. To turn coal ultrafines from a waste into product the cake moisture must not exceed 9–10 weight %. HiBar steam pressure filtration produces ultrafines at this target moisture and can be either marketed as own product or admixed to the coarse and fine fraction in any wanted amount.

Pilot plant operations with the BOKELA HiBar pilot plant took place from 2014 to 2017 and proved for different concentrates from three coal washeries that steam pressure filtration produces dry coal ultrafines. In these pilot operations moisture contents from 7% w/w to 10% w/w were achieved depending from the concentrate. Meanwhile, a first HiBar Disc Filter has been sold into the Australian Coal Industry and is expected to be operational by Q1 2019. The paper reports on results of the steam pressure filtration pilot plant operations and will present operating results of the industrial scale HiBar Filtration plant.

**Keywords:** HiBar, pressure filtration, steam pressure filtration, coal flotation, ultrafines

## 1. Introduction

Depending on the coal deposit the amount of ultrafines can sum up to some 10–40% w/w. If the free moisture content of the combined coarse and fine coal fraction is below 10% w/w, then filtration of ultrafines (particle size $x <$ 250 µm) with modern rotary vacuum disc filters is a profitable way to produce a dewatered ultrafine product with 20–30% w/w free moisture, allowing a mixture of a considerable amount of ultrafines in to the end product. Results of filtration and dewatering of coal ultrafines (Australian flotation concentrates, South African coking coal flotation cell overflow and underflow) with vacuum disc filters have been reported by Hahn and Essack (2011). If the combined coarse and fine coal fractions have free moisture content close to 10% w/w or greater, then the continuous HiBar steam pressure filtration is capable to produce extremely dry ultrafines below 10% w/w free moisture which now offers new options in coal ultrafines treatment. Ultrafines dewatered with steam pressure filtration can be either marketed as own product or admixed to the coarse and fine fraction in any wanted amount.

## 2. Ultrafines dewatering with continuous HiBar steam pressure filters

### 2.1 Plant and process design of HiBar filtration technology

HiBar filtration technology uses a rotary filter – for coal dewatering a disc filter is used – which is installed inside a pressure vessel (Fig. 1) filled with compressed air (7 bar(a)). The filtrate pipes are connected to the environment and the suspension is pumped by an appropriate pump into a pressurized vessel. The filter cake is removed from the filter cloth by compressed air blowback and discharged from the pressurized zone through a sluice system. The vacuum pumps used with a conventional vacuum filter are replaced by a compressor that supplies the necessary compressed air to the vessel and for compressed air blowback. The compressed air from cake blowback also serves as process air to maintain the overpressure in the vessel for the filtration process. Inside the vessel, the filter runs with a differential pressure of up to $Dp = 6$ bar.

An industrial HiBar team pressure filtration plant with a 70 m² disc filter is shown in Fig. 2.

**Figure 1:** HiBar disc filters (70 m² each) with steam cabins for steam pressure filtration in pressure vessel (left) and filter building with two HiBar disc filters 70 m² each (right)

### 2.2 HiBar steam pressure filtration

For steam pressure filtration the filter discs are equipped with steam cabins and with feed pipes for steam supply. The use of steam at continuous HiBar steam pressure filtration leads to a combined thermal/mechanical phenomenon during filter cake dewatering followed by a subsequent convective drying with pressurized gas (Gerl and Stahl, 1996).

At HiBar steam pressure filtration the filter cake is only partially exposed to steam, which accelerates and intensifies the dewatering. This process can be explained as follows:

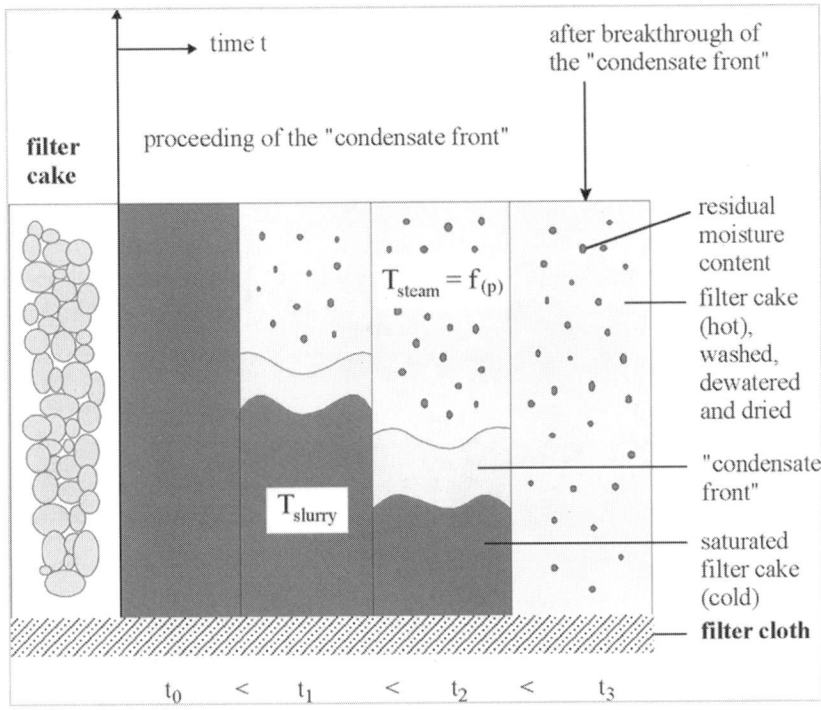

**Figure 2:** Steam pressure filtration: model of the "condensate front"

A filter cake which is formed at the low temperature of the feed slurry enters a specially designed steam cabin immediately after emerging from the slurry in the filter trough. Here, a superheated steam atmosphere exists and the following phenomena take place which can be described by the model of the "condensate front" (Fig. 2):

- The steam condenses on the cold cake surface and a homogeneous condensate layer is formed moves through the cake in a piston like flow ("condensate front").
- The moving "condensate front" replaces nearly 100% of the 'mother liquor'.

- When the "condensate front" reaches the filter cloth, the filter cake is 'heated up' completely to the steam temperature. At this point the cake leaves the steam cabin.
- Then compressed air passes the pre-dewatered and hot filter cake causing a very effective thermal drying which leads to extremely low cake moisture contents.

These thermal/mechanical processes inside the cake lead to a nearly homogeneous and highly intensive cake demoisturing without pressure and energy loss by a "fingering".

## 2.3 Results of ultrafines dewatering with HiBar steam pressure filtration

Steam pressure filtration achieves extremely dry ultrafines in a new dimension that eliminates former limits in ultrafines treatment. Typical values for solids throughput and moisture content are below 10% w/w as shown in Fig. 3 and Table 1.

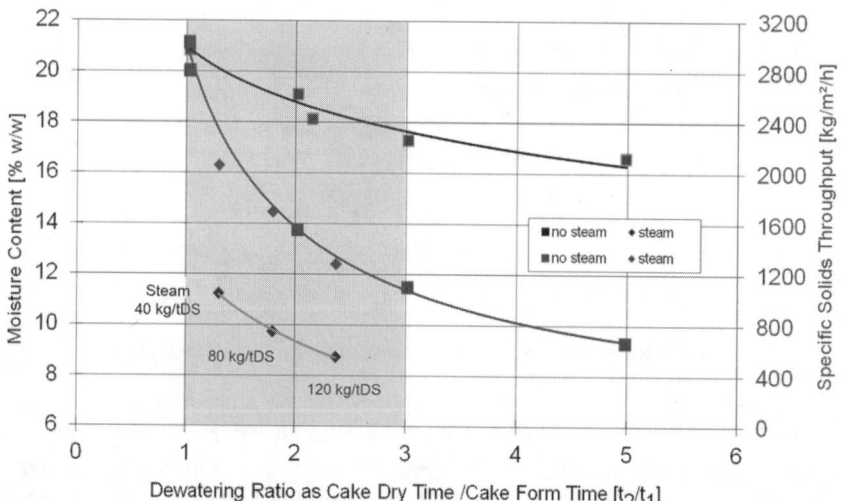

**Figure 3:** Free moisture content and specific solids throughput for flotation coal with pressure filtration and steam pressure filtration versus dewatering ratio (dewatering ratios on rotary filters range to 1–3)

In Fig. 3 moisture content for pressure filtration (black line) and steam pressure filtration (blue line) for a flotation coal ($x_{50}$ < 45 μm) is shown versus the so called dewatering ratio of dry time $t_2$ to form time $t_1$ which is equivalent to the ratio $a_2/a_1$ ($a_2$ = dewatering angle, $a_1$ = cake formation angle). Results shown have been attained from lab tests with a pressure difference of 5.5 bar, feed solids of 350 g/l and flocculant dosage of 13 g per 1000 kg. The red line shows the corresponding specific solids throughput. Typical dewatering ratios on rotary filters range to 1–3 as indicated in Fig. 3. Within this range pressure filtration achieves free moisture contents from 21 to about 17.5% w/w while steam pressure filtration achieves free moisture contents from 11% to 9% w/w depending on dewatering ratio and amount of steam. As a rule of thumb 10 kg steam/t (d) are necessary to reduce moisture by 1%-point compared to pressure filtration.

Since steam pressure filtration accelerates and intensifies the dewatering not only lower moisture contents are achieved compared to pressure filtration but also higher specific solids throughputs are attained since the filter can be operated with a bigger cake formation angle $a_1$ and higher filter speed.

**Table 1:** Typical moisture contents of coal ultrafines with vacuum filtration, pressure filtration and HiBar steam pressure filtration for −250 μm

| Method | Free moisture % w/w |
|---|---|
| Vacuum | 23–28 |
| Continuous pressure filtration | 16–19 |
| HiBar steam pressure filtration | 8–12 |

## 3. HiBar pilot plant operations for steam pressure filtration of coal ultrafines

In July 2014 the BOKELA HiBar pilot plant was operated at the coal washery Auguste Victoria (RAG), Germany, for steam pressure filtration of coal ultrafines (Fig. 6). Experts from the international coal industry travelled to Ruhrcoal to witness this event. It was a world premiere and for the first time filter cakes of dry coal ultrafines below 9% w/w free moisture were produced in a semi-industrial scale.

Figure 4 shows steam consumption versus moisture content for HiBar steam pressure filtration of coal ultrafines as achieved during these test trials. Pressure difference for cake dewatering was D$p$ = 3 bar (vessel pressure = 4 bar). As can be seen cake moistures significantly below 10% w/w were achieved with steam consumptions of some 120 kg/t (d). As mentioned above it can be stated as a rule of thumb that with coal ultrafines 10 kg steam per kg (d) effect a moisture reduction of 1% w/w which is valid in the range of 20–8% w/w moisture. Slight deviations from this correlation in the graph of

Fig. 4 result from edge effects which falsify steam consumption to somewhat higher values due to the small filter area of the pilot filter of only 1 m².

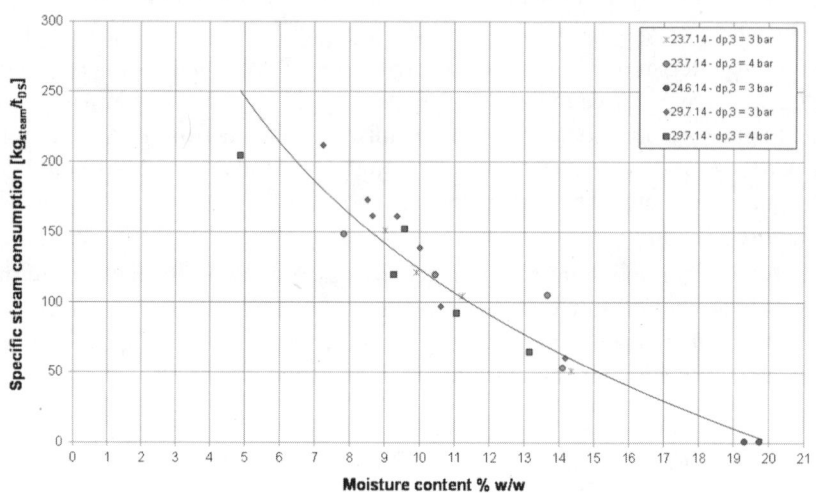

**Figure 4:** Steam consumption versus moisture content for HiBar steam pressure filtration of coal ultrafines; results of HiBar pilot plant operation (1 m² disc filter) at the Coal Mine Auguste Victorica/Germany

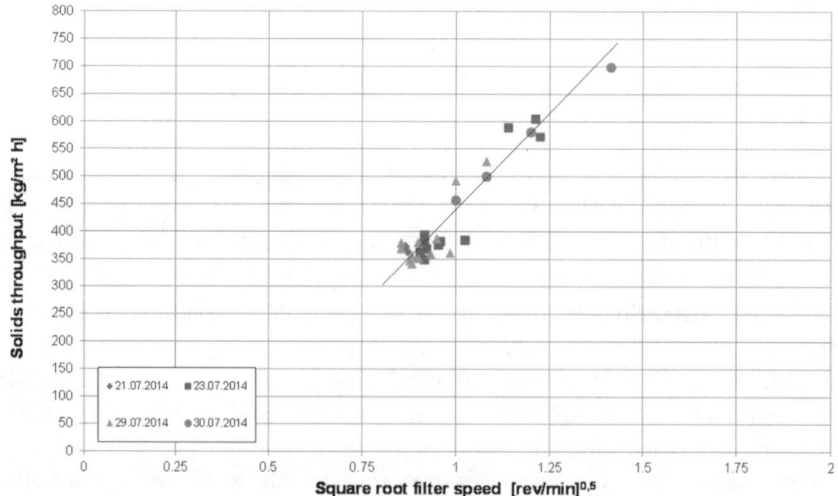

**Figure 5:** Results of HiBar pilot plant operation (1 m² disc filter) at the Coal Mine Auguste Victorica/Germany; solids throughput versus square root of filter speed for HiBar steam pressure filtration of coal ultrafines

Figure 5 shows solids throughput versus square root of filter speed for values of filter speed from 0.7 to 2 rev/min. According to filtration theory (Ehrfeld et al., 2008) the solids throughput increases linearly with square root of filter speed from about 350 kg/m² h to 750 kg/m² h. While pressure difference for cake dewatering was $Dp = 3$ bar the pressure difference for cake formation was adjusted via throttle valve to only $Dp = 1$ bar. This limitation in pressure difference for cake formation was necessitated by the discharge sluice of the pilot plant which was not able to handle all solids which would have been produced with full pressure difference $Dp = 3$ bar in cake formation zone. With this full pressure difference and with use of flocculant, solids throughput for these coal ultrafines would increase – with cake moisture unchanged – by a factor of 3–4 to values which would be comparable to solids throughput values (red curve) in Fig. 3.

**Figure 6:** HiBar pilot plant operation (1 m² disc filter) at the coal washery Auguste Victoria (RAG), free moisture content of discharged filter cake: 8% w/w

In March 2016 the HiBar pilot plant was operated together with the Australian Coal Association at Tahmoor coal mine in Australia, NSW. The pilot plant was operated with two different flotation concentrates and could demonstrate the feasibility of steam pressure filtration also for Australian coal flotation concentrates. Moisture contents below 10% w/w were achieved.

In October 2017 the pilot plant was operated at the Pniowek Mine in Poland. With the slightly coarser concentrates of the Pniowek Mine moisture contents down to 7% w/w were achieved.

## 4. Waste to product: Advantages of dry ultrafines

The continuous HiBar steam pressure filter is able to operate at extraordinarily high throughputs in continuous filtration and also to achieve dry filter cakes below 10% w/w moisture, which eliminates former limits and offers new options in the treatment of coal ultrafines, such as:

- Allowing mixtures of coarse and fine fractions in any given amount;
- marketing as a separate product;
- reduced transport costs through reduced water content;
- improved bulk flow behaviour for discharge of railway wagons;
- improved transport in cold regions with long frost periods (prevention of freezing solids); and
- lower or even no energy cost for thermal drying.

To turn coal ultrafines from a waste into product the cake moisture must not exceed 9–10% w/w. BOKELA HiBar Steam Pressure Filtration produces ultrafines at this target moisture. For this some 10 kg steam/t (d) are necessary to reduce moisture by 1%. Regarding a filter cake of 17% w/w free moisture achieved with pressure filtration steam pressure filtration requires 80 kg steam/t (d) to reduce free moisture to 9% w/w ($Dmc = 8\%$ w/w). Total operational cost is approximately US$5 per 1000 kg solids. Considering current market prices, this represents an increased revenue of about US$50 per 1000 kg solids, if the waste is now product coal. As a result, the return of investment of a large turnkey HiBar steam pressure filtration plant, with a throughput of 300,000 t/a, is less than one year.

## 5. Waste to product: Advantages of dry ultrafines

A first HiBar disc filter for pressure filtration of coal ultrafines has been sold into the Australian Coal Industry. The project is a culmination of extensive work undertaken by Bokela in conjunction with the Australian Coal Association (see Pos. 3). The HiBar filter has a maximum design capacity of 135 tph and will operate at 6.5 bar gauge producing a filter cake with 13–17% moisture. The unit will be operational by Q2 2019.

## 6. Conclusion

HiBar filtration technology for continuous pressure and steam pressure filtration enables profitable solutions and new options in filtration of coal ultrafines and iron ore concentrates. The continuous HiBar steam pressure filtration is capable of producing extremely dry coal ultrafines (i.e. below 10% w/w free moisture) and so offers new options in coal ultrafines treatment e.g. can turn waste into a saleable product. Ultrafines dewatered with steam pressure filtration can be either marketed as own product or admixed to the coarse and fine fraction in any wanted amount. Iron ore concentrates can be dewatered to lowest moisture contents of 3% w/w. Comparable to coal ultrafines this improves transport in cold regions with long frost periods or makes transport possible at all.

## 7. References

[1] Ehrfeld, E., Bott, R. and Langeloh, Th., 2008, "Layout of rotary filters on the basis of laboratory results", 10th World Filtration Congress, Leipzig, Germany, 14–18 April.

[2] Gerl, S. and Stahl, W., 1996 "Improved dewatering of coal by steam pressure filtration", Coal Preparation, Volume 17, Issue 1–2, 137–146.

[3] Hahn, J. and Essack, H., 2011, "Performance, operation and maintenance experience of coal ultrafines filtration with modern high speed disc filters", South African Coal Preparation Society Conference, Secunda.

[4] Hahn, J., Bott, R. and Langeloh, Th., 2016, "HiBar steam pressure filtration of coal ultrafines – New developments and results", ICPC 2016, XVIII. International Coal Preparation Congress & Exhibition 2016, St. Petersburg.

# 46

# The process flowsheet design to maximize pyrite rejection and enhance coal heating value using dry beneficiation technology

Yunkai Xia and Gongmin Li

*Tangshan Shenzhou Manufacturing Co., Ltd*

**Abstract:** Coal sulfur content is a critical property affecting thermal coal marketability. Sulfur level reduction greatly enhances coal value. In the Ordos coal basin region of the China sulfur levels in some low rank coals are often marginal (1.0-4.0 percent). On many occasions, much of the sulfur is concentrated in the form of pyrite in the -13mm fine coal fractions. Traditional wet process circuitry (heavy medium separation and jigging) tends to produce low levels of product sulfur at cost of increased clean coal product moisture, lower heating value and lower clean coal yield. Being able to run at higher densities in the 13-1mm size fraction reduces sulfur level considerably, removal of high-density pyrite by a combination of primary and secondary circuits allows low sulfur froth concentrates to be produced while maintaining higher clean coal yield. Circuit design and plant results for a couple of recently commissioned dry coal preparation plants in China using this technology are discussed.

**Keywords:** Sulfur Reduction, Fine Coal Cleaning, Dry Beneficiation Technology, Clean Coal Moisture, Lignite, ZM separator

## 1. Introduction

In general, China's coal-fired power plants require fuel coal has sulfur content below 0.8%. For high sulfur coal-producing enterprises, adopting coal desulfurization technology is one of the main steps to improve clean coal product quality. Sulfur content in coal is usually divided into organic sulfur and inorganic sulfur content, inorganic sulfur component are mainly made of pyrite minerals. Inorganic sulfur desulfurization methods mainly include gravity separation, the froth flotation, the selective flocculation, granulation separation, high gradient magnetic separation, microwave desulfurization [1-5], electric separation method and X-ray sorting method [6], etc. Gravity separation is the main method to remove pyrite sulfur from thermal coal. Gravity separation processes can be divided into wet separation process and dry separation process. The wet process includes heavy medium separation, jigging and spiral separation, etc [7-8]. With the development of coal mining mechanization and geographical condition changes, the fine coal content in ROM coal keeps increasing. At some coal mines, fine coal rate reaches over 70% and the sulfur content in some coal seams is above 3%. During fine thermal coal processing, although clean coal ash and sulfur content decrease after washing, the product moisture increases and the calorific value of coal is not obvious or even declines due to the negative effect of moisture. Therefore, most of thermal coal preparation plant operators are forced to bypass mot of fine coal, which sulfur content cannot be controlled. Coal slurry treatment also has been a problem in plant operation. When there is higher desulfurization requirements, the more raw coal have to be washed, the more the slime problem processing problems. Fortunately, and the advantages of dry coal preparation just can overcome the shortcoming of wet coal preparation.

Dry coal desulphurization process mainly made of compound air separation desulfurization and air dense medium desulphurization process [9-11]. In China, R&D of dry coal desulfurization technology and equipment should be a priority to thermal coal preparation industry. Dry separation has advantages of no moisture increase in clean coal product, saving product dewatering and coal slime water treatment so that the coal preparation procedures can be greatly simplified and therefore cost of capital investment and production operations. After nearly 30 years, dry coal desulfurization technology is well developed in China and this technology is currently at the door of large-scale promotion and application. It is of important practical significance to choose an efficient, economical, and reasonable dry coal desulfurization process according to their specific coalmine condition and raw coal wash abilities.

## 2. Structures of compound coal dry separator and dry cleaning system

The compound dry separator is composed of one separation deck, vibrators, air chambers, frame and hanging mechanism and other parts. Two vibration electric motors are fixed on the separation deck electric motor frame and the separation deck with vibrators is suspended on the machine frame by steel wire ropes with damping springs. As shown in Fig. 1, the separation deck is composed of right angle trapezium deck surface with evenly distributed air passing holes, riffles, back plate, discharging dampers and refuse product door. Below the deck surface, there are several compressed air chambers linked to the centrifugal fan, each compressed air chamber is controlled by compressed air valve, and the compressed air chambers are fixed with the deck surface in the same body with the separation deck.

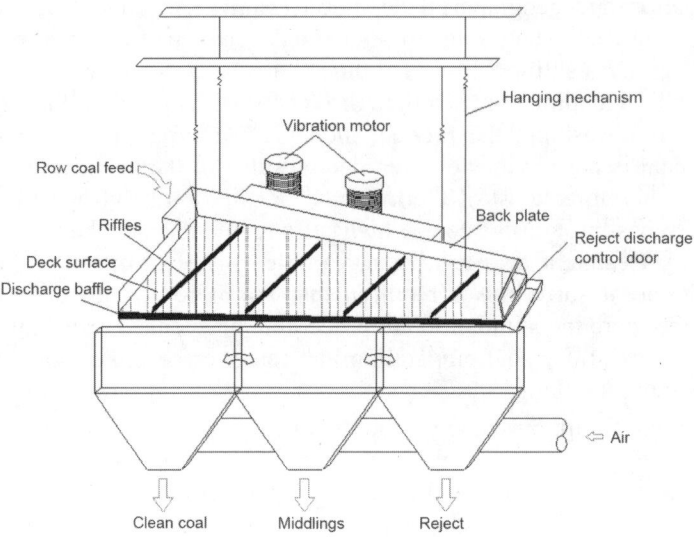

**Figure 1:** Compound dry separator structure drawing

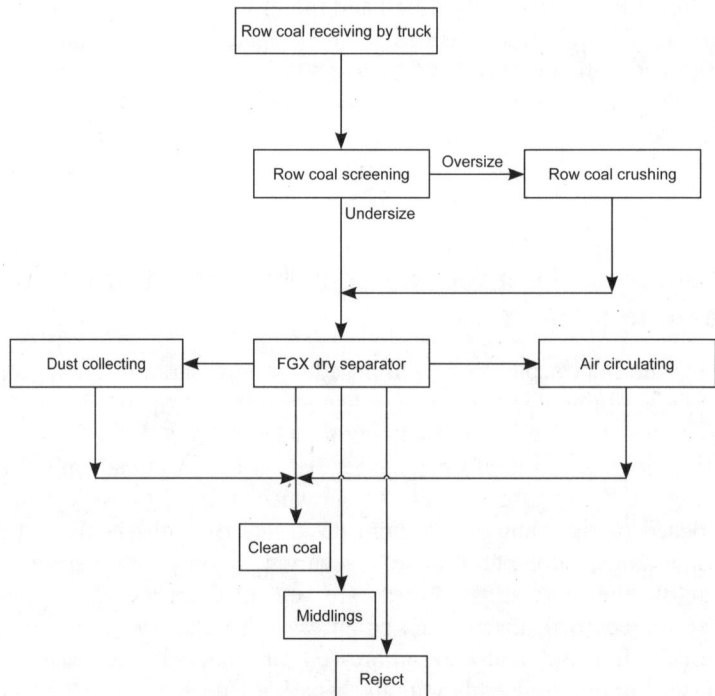

**Figure 2:** Process flow block diagram

The general process flow block diagram of the dry cleaning system is shown in Figures 2. Run-of-Mine (ROM) coal is delivered to the ROM pad or directly to ROM dump bunkers by haul trucks. The top size of raw coal is assumed to be 300 mm; raw coal is transported to a raw coal classifying screen via a belt conveyor. In general, raw coal screen has a screen panel aperture of 80 mm; the oversize product is crushed down to -80mm and blended with screen undersize product. Raw coal is then fed to a dry separator to get clean coal, middlings and reject. One dust collection system is used to clean the recycled airflow; the cyclone dust collector and bag filter dust collector are used to prevent dust accumulation in recycled air and to remove the dust from the air being emitted into the atmosphere.

The density range of coal particle, reject gangue and pyrite are 1.3 ~ 1.8 g/cm$^3$, 1.8 ~ 2.2 g/cm$^3$ and 3 to 4 g/cm$^3$ respectively. Gravity separation utilizes the density difference among these three group minerals. In compound air dry separator, powered by upward air flow and mechanical vibration forces, the bulk coal is loosened and separated based on density difference. The cutting density of dry separator is generally above 2.0 g/cm$^3$, which is especially suitable for high-density deshaling and sulfur reduction. The new generation of a dry separator, ZM type series dry separation equipment, have the higher unit capacity and sharper separation than conventional FGX type separator. In addition, the newly designed dust collecting system is in line with new strict environmental protection regulations of the state.

## 3. Typical dry coal desulfurization process

Three desulfurization process flowsheet utilizing dry coal beneficiation technology are introduced below:

## 3.1 Mixed coal desulfurization process

Mixed coal desulfurization is suitable for processing of easy-to-wash raw coal which has high ash and sulfur content, the flowsheet is shown in Figure 3. The top size limit of coal feed is about 80mm for a plant with capacity above 1.0 Mtpa, the small plant will have a feed top size limit in a range of 50-60 mm. The dry separation system will produce three products: clean coal, middlings and reject. Whether middlings product will be re-processed depends on clean coal product requirements. If clean coal sulfur content requirements are not high, the middlings will simply blend with clean coal product otherwise the middlings will be directed back to raw coal feed to improve the overall separation efficiency of the dry separation system. The middlings might be crushed before recirculation cleaning according to the liberation conditions between coal and pyrite.

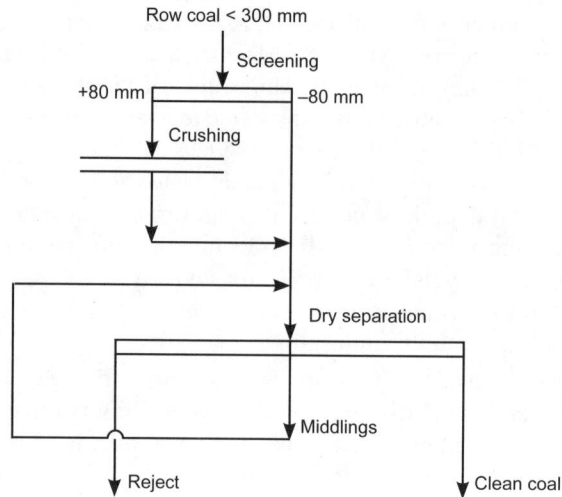

**Figure 3:** Simple rough separation process flowsheet

Application example: The Fengshenkui coal mine in Ordos city of Inner Mongolia in China will process No. 5 high sulfur bituminous coal and the clean coal sulfur content target is 0.8%. The size distribution analysis of raw coal sampled from the coal yard is shown in Table 1. The waste rock and pyrite mineral are mainly concentrated in +3 mm size fraction. The sulfur content of 25-13mm size fraction is 2.76%, the float-sink analysis of this size fraction with the highest sulfur content is listed in Table 2. The < 1.6 g/cm³ density component has low sulfur and the intermediate density material content is small, this size fraction shows easy washability at high cutting density. In this plant, one set of 120 TPH dry separator is selected, the top size of raw coal feeding the plant is controlled below 80mm, and the separation performance of the system is shown in Table 3. The clean coal has a yield of 87.07% and its sulfur content is 0.6%, the desulfurization rate reaches 63.8%, which meets the clean coal quality requirement.

**Table 1:** Size distribution analysis of No. 5 raw coal

| Size, mm | Wt % | $M_t$, % | $A_d$, % | $S_{t,d}$, % | Qnet.ar/(MJ·kg⁻¹) |
|----------|------|------|------|------|------|
| >25 | 28.39 | 22.3 | 32.13 | 1.68 | 14.77 |
| 25~13 | 17.20 | 21.8 | 29.37 | 2.76 | 15.75 |
| 13~6 | 19.41 | 23.3 | 23.78 | 2.57 | 16.90 |
| 6~3 | 11.44 | 24.4 | 20.4 | 1.29 | 17.38 |
| 3~1 | 13.14 | 24.6 | 20.43 | 0.92 | 17.43 |
| <1 | 10.42 | 24 | 27.36 | 0.57 | 15.54 |
| Total | 100 | 100 | 23.1 | 26.66 | 1.78 |

**Table 2:** Float-sink testing analysis of 25~13mm raw coal

| Density (g·cm³) | Wt, % | $S_{t,d}$, % |
|---|---|---|
| <1.6 | 84.14 | 0.6 |
| 1.6~1.8 | 1.84 | 8.43 |
| >1.8 | 14.02 | 17.05 |
| Total | 100.00 | 3.05 |

**Table 3:** Dry separation performance of No.5raw coal at Fengshenkui coal mine

| Product | Wt,% | Ad, % | $S_{t,d,}$% |
|---|---|---|---|
| Clean Coal | 79.35 | 8.21 | 0.52 |
| Middlings | 7.72 | 11.81 | 1.46 |
| Reject | 12.93 | 35.34 | 7.16 |
| Total | 100 | 12.00 | 1.45 |

Table 4 shows quality analysis of clean coal from dry separation system, the sulfur content of each size fraction meets clean coal requirement except -3mm coal. If clean coal sulfur cannot satisfy the demand, the -3mm fine coal could be screened out from clean coal product in rough separation in order to meet clean coal quality requirement.

**Table 4:** Clean coal size distribution analysis

| Size, mm | Wt, % | Ad, % | $S_{t,d}$, % |
|---|---|---|---|
| +25 | 28.42 | 6.15 | 0.31 |
| 25-13 | 23.86 | 6.73 | 0.38 |
| 13-6 | 23.59 | 7.66 | 0.59 |
| 6-3 | 9.38 | 8.62 | 0.67 |
| 3-0 | 14.75 | 14.26 | 0.85 |
| Total | 100 | 8.07 | 0.51 |

## 3.2 Fine coal desulfurization process

The majority of Chinese thermal coal preparation plants adopt part washing process, i.e. only +25mm lump coal is separated in the heavy medium vessel and minus 25mm fine is simply bypassed and unwashed. For high sulfur coal, quality of -25mm fine coal is unstable and its sulfur content often cannot meet the demand of power plant. Although it is possible to reduce ash and sulfur content if fine coal is handled by wet process, there are associated problems such as clean product moisture increase, high amount of coal slime produced and high operation cost. The coal dry separation desulfurization can avoid these problems.

In dry separation, if the feed top size is 25mm, the efficient separation size range is 25~1mm. The fine coal desulfurization process flowsheet is shown in Figure 4. This flowsheet is suitable for desulfurization of -25mm high sulfur raw coal which has low sulfur content in -1mm size faction. It also can be used for the desulfurization of high sulfur low rank coal such as lignite. In this flowsheet, +25mm lump coal is fed to wet process, -25mm fine coal is handled by a dry separator which will produce clean coal, middlings and rejects. The middlings can be recycled for re-cleaning to improve overall separation efficiency.

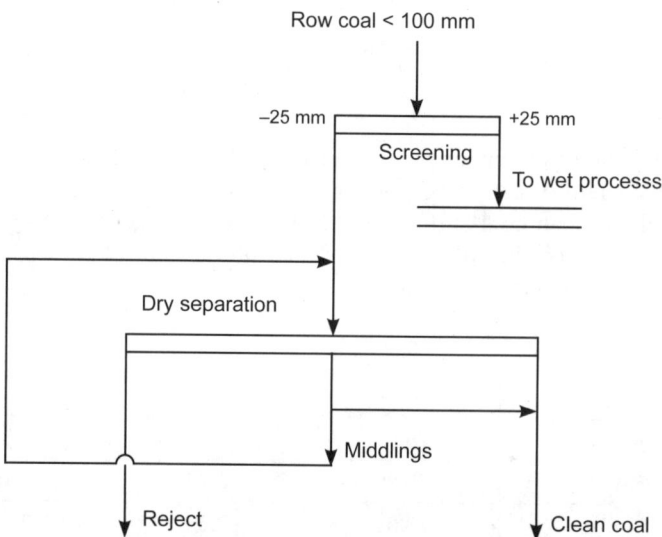

**Figure 4:** Fine coal desulfurization process flowsheet

Application example: Selian No.1 coal mine of Datong coal industry group company produces long flame coal, the sulfur component is mainly made of sulfide minerals (Sp), the organic sulfur (So) and sulfate sulfur (Ss) contents are very low. In this plant, only >25mm coal is washed by heavy medium vessel, the <25mm fine coal is of high ash and intermediate sulfur content. The size distribution analysis of <25mm raw coal is shown in Table 5. The >6mm coal has high sulfur content (0.94%) while <3mm size fraction is of low sulfur coal. The product balance of separation of <25mm fine coal is list in Table 6.

**Table 5:** Size distribution of <25mm raw coal

| Size, mm | Wt,% | Index | | | | |
|---|---|---|---|---|---|---|
| | | Mt% | Mad% | Ad% | St.d% | Qnet,ar MJ/kg |
| 25-13 | 22.911 | 23.9 | 7.70 | 27.54 | 0.77 | 15.95 |
| 13-6 | 20.27 | 24.3 | 7.83 | 27.54 | 1.14 | 15.70 |
| 6-3 | 27.22 | 25.3 | 8.01 | 24.46 | 0.58 | 16.03 |
| 3-0 | 29.60 | 25.6 | 8.33 | 33.04 | 0.48 | 13.48 |
| 合计 | 100.00 | 24.87 | 8.00 | 28.33 | 0.71 | 15.19 |

**Table 6:** Product balance of dry separation of <25mm raw coal

| Product | Wt,% | Index | | | | |
|---|---|---|---|---|---|---|
| | | Mt% | Mad% | Ad% | St.d% | Qnet,ar MJ/kg |
| Clean Coal | 79.97 | 25.99 | 8.85 | 21.24 | 0.48 | 16.41 |
| Middlings | 12.72 | 23.61 | 7.99 | 30.99 | 0.63 | 14.57 |
| Reject | 7.31 | 10.90 | 2.55 | 76.39 | 5.56 | 3.76 |
| Total | 100.00 | 24.59 | 8.28 | 26.51 | 0.87 | 15.25 |

After dry separation of -25mm fine coal, the clean coal yield reaches 79.97% with a sulfur content of 48%，the heating value is increased to 16.41MJ/kg. Compared to raw coal feed, the ash content is lowered by 9.19 percentage points and heating value increased by 1.54MJ/kg，the overall desulfurization rate is 65.73%. The dry separation shows good performance in deshaling and sulfur removal. If same raw coal is handled by a wet process, such as fine coal heavy medium cyclones, the clean coal surface moisture will be increased at least by 5 percentage points and clean coal heating value will be lowered by 1.46MJ/kg.

## 3.3 Rough and re-cleaning separation process

The grain-size has a significant influence on the dry separation performance, the fined the particle size, the lower the separation efficiency. The separation precision will also decrease with the increase of the cutting density. In the case of difficult-to-wash high sulfur coal, high content of fine coal and very low clean coal sulfur content requirement, the rough separation might not meet clean coal sulfur target. The -13mm size fraction of rough clean coal product and middlings need to be re-cleaned. The rough and re-cleaning separation process flowsheet is shown in Figure 5. Rough clean coal is classified at 13mm, the crushed middlings and -13mm rough clean coal are mixed and fed to the re-cleaning dry separator. The clean coal from the re-cleaning operation and +13mm clean coal from the rough operation are blended as a final clean coal product. This flowsheet is suitable for cleaning of raw coal made of multiple coal seams and raw coal having severe sulfur content fluctuation. If raw coal sulfur content becomes low, the re-cleaning operation can be stopped and bypassed.

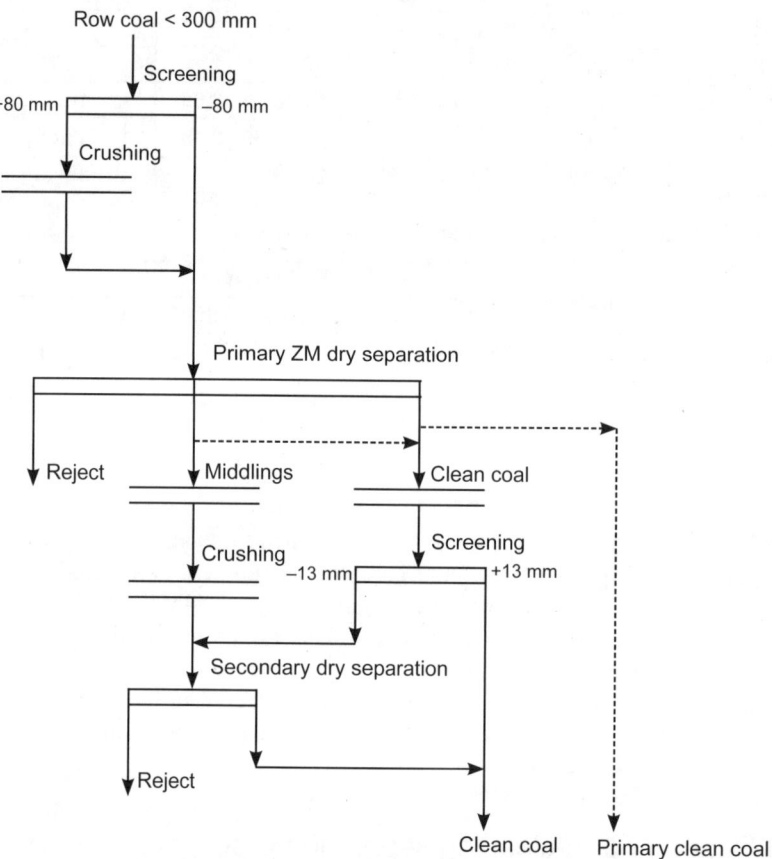

**Figure 5:** Rough and re-cleaning separation process

Application example: Yushujin coal preparation plant is designed to clean high sulfur lignite which is easy to degrade. In this plant, all <80mm raw coal is processed by dry separator and clean coal product is classified at 30mm. The raw coal generally contains 2% sulfur and heating value is above 13.40MJ/kg, the clean cal sulfur content target is 1.2%. After the plant was put into operation, raw coal feed has a heating value lower than 13.4MJ/kg. Statistic analysis of plant operation data of one week (2018/4/13 ~ 2018/ 4/21) shows that the calorific value of clean coal product will gradually increase with raw coal heating value, the higher of raw coal heating value, the lower the rate of increase. When the raw coal heating value is at 11.72MJ/kg and 13.9MJ/

kg respectively, the net heating value of clean coal product will be increased by 3.77 and 2.09MJ/kg respectively; When the sulfur content of raw coal is increased from 1.1% to 2.4%, the sulfur content of +30mm lump clean coal and -30mm fine clean coal are stabilized at 1.1% and 1.3% (Figure 6 and 7). The fine coal sulfur content cannot meet the clean coal product requirement. For example, On April 6, 2018, the production data (Table 7) shows that raw coal low calorific value is only 10.97MJ/kg and sulfur content is 1.87% in raw coal. After dry separation, the +30mm lump clean coal yield is 10.52% and its heating value is 16.79MJ/kg. The -30mm fine clean coal yield is about 52.96% and its heating value is 15.19MJ/kg. The sulfur content of fine clean coal is very close to the sulfur target. The overall desulfurization rate is 60.54%. Although the dry separation shows good deshaling and desulfurization performance, recently the fine clean coal sulfur content shows severe fluctuations.

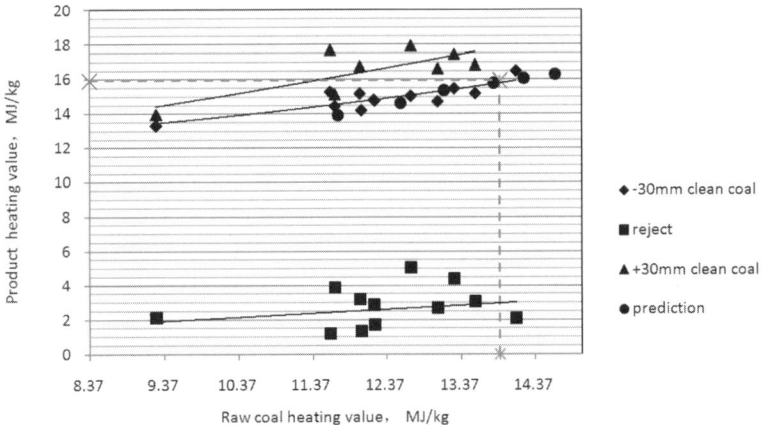

**Figure 6:** Heating value changes of product with raw coal quality

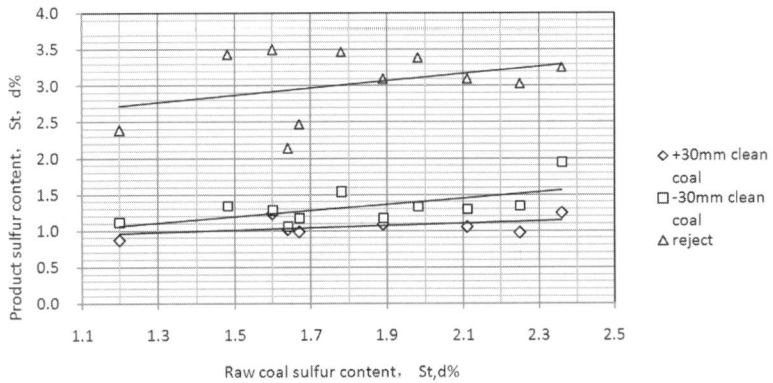

**Figure 7:** Sulfur content changes with raw coal quality

**Table 7:** Product balance of dry separation (Date: 2018.04.06)

| Product | Wt,% | Mt,% | Ad,% | St.d,% | Qnet,ar, MJ/kg |
|---|---|---|---|---|---|
| Raw coal | 100.00 | 20.17 | 43.32 | 1.87 | 10.97 |
| +30mm clean coal | 10.52 | 26.5 | 15.84 | 1.09 | 16.79 |
| -30mm clean coal | 52.96 | 24.9 | 24.97 | 1.18 | 15.19 |
| reject | 36.52 | 11.5 | 77.86 | 3.10 | 3.19 |

The size distribution analysis of one sample of -30mm fine clean coal taken from rough separation is shown in table 8. The ash content and sulfur content shows an increasing trend with decreasing particle size. The -13mm fine coal shows higher sulfur content that sulfur target, it is should be re-cleaned to control sulfur content. It should be noted that the content of -3mm coal powder is 35.73% and its ash content and sulfur content are high, this size fraction cannot be efficiently in rough operation and it has a significant influence on clean coal quality. The -13mm fine coal rough separation is processed again in a dry separator. The re-cleaning results are shown in Table 9.

In re-cleaning operation, the coal feed has a sulfur content of 1.63%, the clean coal yield is 87.88% and its sulfur content is successfully lowered to 1.14%, the desulfurization rate is 38.16%. Both middlings and reject amount are small, therefore there is no significant heating value increase after re-cleaning.

**Table 8:** Size distribution of -30mm clean coal from rough separation

| Size, mm | Wt% | Mt% | Ad% | St.d% | Qnet,ar MJ/kg |
|---|---|---|---|---|---|
| +25 | 8.82 | 22.9 | 20.21 | 0.75 | 16.15 |
| 25-13 | 16.35 | 22.5 | 20.17 | 1.02 | 16.14 |
| 13-6 | 20.16 | 22.2 | 19.81 | 1.29 | 16.10 |
| 6-3 | 18.94 | 22.1 | 22.30 | 1.68 | 15.40 |
| 3-0 | 35.73 | 22.3 | 36.02 | 1.46 | 12.79 |
| Total | 100 | 22.3 | 26.17 | 1.33 | 14.80 |

**Table 9:** Product balance of re-cleaning of -13mm fine clean coal from rough separation

| Product | Wt,% | Mt% | Ad% | St.d% | Qnet,ar, MJ/kg |
|---|---|---|---|---|---|
| Clean coal | 87.88 | 24.81 | 25.93 | 1.14 | 14.57 |
| middlings | 4.45 | 22.90 | 33.14 | 1.98 | 13.27 |
| reject | 7.67 | 13.20 | 68.18 | 6.87 | 5.19 |
| Total | 100 | 23.83 | 29.49 | 1.62 | 13.79 |

## 4. Conclusion

The choose of the desulphurization process will depend on raw coal washability and product sulfur content requirement. Three typical dry coal beneficiation processes for pyritic sulfur removal are introduced. When clean coal sulfur requirement is not high, the mixed coal desulfurization is suitable for processing of easy-to-wash raw coal which has high ash and sulfur content; Fine coal desulfurization process can be used to clean bypassed fine raw coal in majority thermal coal preparation plants, it shows good deshaling and desulfurization effect and avoid some problems associated with process; In case of difficult-to-wash high sulfur coal, high content of fine coal and high clean coal sulfur content requirement, two stage separation, i.e. rough separation of -80mm raw coal and re-cleaning of -13mm rough clean coal, can be used to control fine clean coal sulfur content. The dry beneficiation technology has advantages of no water using, simple procedures, high capacity, low capital and operation costs. It has been successfully applied in coal desulfurization in China in large scale, this technology shows good potential in the area of utilizing high sulfur coals and environment protection.

## 5. References

[1] Biao Li and Sha Jie, 2010, The before-combustion desulfurization method and its application status ,Mining machinery,38(20),9-14.

[2] Jianhua Qin,2000, Coal preparation is the preferred method of coal desulfurization in China, Coal Preparation Technology，1,10-11.

[3] Xiangkuan Huang, 2015, Desulfurization of high sulfur coal before mining operation,Clean Coal Technology,6, 26-29.

[4] Liu Haiyin, 2016, Application of fine coal desulfurization technology at Nantong coal preparation plant, Coal Processing and Comprehensive Utilization,3,9-12.

[5] Zhou Ming, Yangxu and Song Liqiang, 2013, Advance of microwave coal desulfurization technology research, Journal of Henan Polytechnic University (natural science)，6, 760-763，767.

[6] Lŭtke Von Ketelhodt, 2010, The dry separation by dual energy X ray projection method, XVI International coal preparation congress, Translated and published by Coal Preparation Society of China,305-311.

[7] P. Bethell, B. Watters and E. Wolfe, 2013, Optimizing fine circuit design to maximize pyrite rejection and enhance coal marketability[C]. XVII International coal preparation congress, Istanbul, Turkey.:395-400.

[8] Shi Bin, 2011, Desulfurization of high sulfur coal before combustion and strengthened desulfurization method Coal Preparation Technology,2, 68-71.

[9]  Zhang Sanshan, 2002,The status analysis of the compound dry separator application , Mechanical Management and Development, 4： 53-54.

[10]  Chen Qinru, Jiao Hongguan and Pan Lanyin, 2007, Two stage high efficient dry coal cleaning technology before coal combustion, Journal of Henan Polytechnic University (natural science)，4： 345-352.

[11]  Zuo wei, Luo Zhenfu and Wu Wangcan, 2009, Dry coal beneficiation technology for coal desulfurization，Coal Processing and Comprehensive Utilization,6,17-20.

# 47

# Upgradation of coal by dry deshaling

Pradip Kr Baranwal and S.K. Jayswal

*CMPDI, Ranchi, India*

**Abstract:** Coal is a gift of nature to the mankind and is a prime fuel for generation of commercial energy in India. Coal India Limited (CIL) shares the major responsibility to cater the need of coal to the wide spread thermal power plants throughout the nation. The coal in this part of the continent are sourced from drift origin leading to lot of inter-grown impurities, usually referred to as 'ash', which can be handled through suitably selected coal beneficiation technology. However, during mining of coal, some quality dilutions have been observed in form of the external shales and stones that get mixed with the run-of-mine coal, mostly inadvertently. Sometimes, this leads to slippage of grade from expected ones.

The conventional wet coal beneficiation technologies are effective way of reducing ash content for quality improvement, however, are usually cumbersome and require lot of essential paraphernalia. Dry technology may establish itself as good candidate for deshaling the free stone/shale. Some initiatives have been taken in this field with the assistance of indigenous institute/company embracing research activity in the field of coal beneficiation. Dense medium fluidization, X-ray based sorting and winnowing are few basic concepts, which have been selected for pilot/demonstration scale study to assess the efficacy for dry deshaling. This paper is an attempt to present summary of efforts made with dry technology of deshaling coal with their performance.

**Keywords:** Washery, dry beneficiation, winnowing, air-jig, X-ray

## 1. Introduction

Coal is a gift of nature to the mankind, which is used as prime fuel for generation of commercial energy in India. Coal India Limited (CIL) shares the major responsibility to cater the need of coal to the wide spread thermal power plants throughout the nation.

The coal in this part of the continent is sourced from drift origin leading to lot of inter-grown impurities, usually referred to as 'ash', which can be handled through suitably selected coal beneficiation technology. However, during mining of coal, some quality dilutions have been observed in form of external shales and stones that get mixed with the run-of-mine coal, mostly inadvertently.

Such coal, if supplied to the consumers directly, leads to slippage of grade commitments and is a matter of concern. Grade slippage with presence of shale and stones found in the coal has been usual complaint reported by consumers. Beneficiation before dispatch to consumer is a solution but processing cost usually acts as deterrent.

Therefore, if only grade quality is to be maintained by removing the external contaminants present in form of shale or stones, cumbersome beneficiation process may be sidestepped and simple deshaling or destoning technologies may be adopted at pit head itself before dispatch of coal to the customers.

## 2. Deshaling or destoning technology

The condition for application of deshaling or destoning technology is that the input coal should have free shales or stones, else it is difficult to improve the quality of input coal with simple flow scheme without employing commdnition circuit. Deshaling technologies are, however, way away from expectation to deliver upgraded coal with a targeted ash content. Deshaled coal usually is the residual coal obtained after removal of shale present in the input side.

It is almost certain that any industrial processing for value addition essentially attracts cost. The challenge is striving a balance between quality and economics.

Conventional gravity separation wet technologies may be adopted but usually have complicated circuits and have limitation at high end of cut-gravity at which the deshalers have to operate. Wet jigs are simple in operation but have challenges with regular variation in feed coal quality. Heavy media bath and heavy media cyclone are very effective coal beneficiation technologies but have limitation at high cut-gravity and media stability may also get disturbed at very high cut-gravity. Moreover, heavy medium systems do have complicated circuits. The inadvertent addition of moisture in wet methods offset the benefit of deshaling to some extent.

Dry technology may be a sensible alternative and is likely to circumvent these problems.

## 3. Exploring dry beneficiation

Based on the differential physical properties of the particles (i.e., coal and ash) such as density, size, shape, magnetic susceptibility, electrical conductivity, etc. many technological developments have been targeted and different technological equipment have been developed for dry beneficiation, that may be broadly classified as air jigs, optical sorters, radiometric sorters, magnetic separators, fluidized bed separator, air-dense medium fluidized bed separators, electrostatic separators, etc.

Coal India Limited and Ministry of Coal have given due consideration to assess the dry technology and a number of projects have been approved for research supports. The approach is oriented towards development of new concept as well as embracing the existing technology. Dry technology plant usually has simpler circuit and the plant set up time is also less compared to wet washing plants.

## 3.1 New development

Indigenous efforts by government agencies as well as independent private investigators have been given due regards for development in India with lab-scale design, modelling and testing, along with simultaneous demonstration of available technology at pilot/plant level.

### 3.1.1 Coal winnowing

The winnowing technology has been in use in agriculture sector for sorting of the chaff or husk and rice from rice stalk after harvesting and threshing since time immemorial. Separation of particles occurs in horizontal air stream due to their differential specific gravity. Such machines are easy to operate and maintain (Figs. 1 and 2).

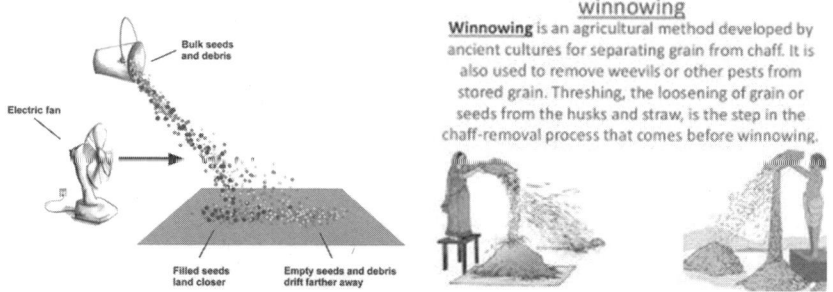

**Figure 1:** Winnowing application for separating chaff from grains

Efforts have been made to sort coal and ash particles by application of winnowing technique in association with Central Institute for Mining & Fuel Research, Nagpur. The basics was first established with the outcome that beneficiation of coal is possible with the application of winnowing technology after due understanding of governing parameters, of which the dynamics of air flow around coal particles has been observed to be very prominent.

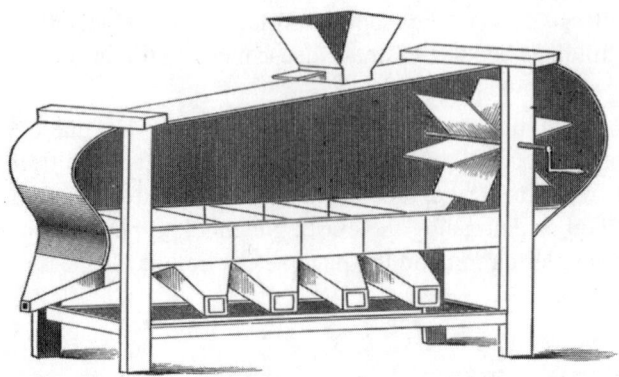

**Figure 2:** Winnowing machine used to separate chaff from tea leaves.
*Source:* https://etc.usf.edu/clipart/86600/86686/86686-winnowing-machine.htm

If two grains of different sizes are introduced with the same initial velocity into air stream, the acceleration imparted to the particle is dependent on their drag coefficient $(C_d)$ and the velocity $(V_r)$ of the particle relative to the air stream and inversely on their diameters and their densities. The aerodynamic drag force $(F_d)$ exerted upon a particle by a stream of air is a function of the frontal area A of the particle, the particle density and the relative velocity between the air and the particle $(V_r)$.

In the winnowing type air stream separator, air is blown horizontally or at an inclined angle to the horizontal against mixed feed (coal) injected along the vertical plane. The aerodynamic drag force $(F_d)$ exerted upon a particle by a stream of air is a function of the frontal area $(A)$ of the particle, the particle density $\rho_a$ and the relative velocity between the air and the particle $(V_r)$, and is governed by the formula (Fig. 3):

$$F_d = \frac{1}{2} C_d \rho_a A V_r^2$$

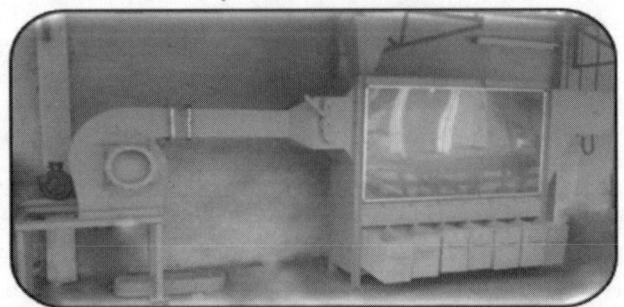

**Figure 3:** Coal winnowing laboratory set-up

The drag force $(F_d)$ accelerates the particle until it acquires the velocity of air stream. The heterogeneous materials are displaced along the horizontal plane at various distances based on their aerodynamic properties. This has advantage of producing two or more products based on gravity within a short time. There were seven boxes placed one after another in a row to collect particles sorted based on density. Experiments in coal winnowing machine were carried out at different coal size fractions such as 50–25 mm and 25–13 mm on different coal samples. One of the preliminary test data is presented in Table 1 below.

**Table 1:** Test data with laboratory set-up, CIMFR-CSIR

|  | Box (from feed end) | | | | | | | Total |
|---|---|---|---|---|---|---|---|---|
|  | 1 | 2 | 3 | 4 | 5 | 6 | 7 |  |
| Wt. of coal (g) | 1151 | 3160 | 2407 | 1689 | 789 | 446 | 386 | 10,000 |
| Moisture (%) | 6.4 | 6.2 | 7.0 | 7.4 | 7.8 | 8.4 | 8.4 |  |
| Ash (%) | 38.7 | 37.1 | 28.3 | 25.7 | 21.9 | 16.6 | 18.8 | 30.5 |

Further work is in progress for scale up design.

### 3.1.2 Air fluidized vibrating deck separator technology

Fluidization and de-fluidization of particles in cyclic manner allow the particles to stratify in layers according to increasing specific gravity from top to bottom. Particles are removed by some specific technique. Work is in progress in association with CSIR-NML, Jamshedpur on air-fluidized-vibrating-deck separator that works on the same principle with integration of the separation principles of vibrating deck most close to the conventional tableconcentrator and is meant for treating coarser fraction (50–6 mm).

Tests with non-coking coal are being conducted and effect of equipment variables such as effect of longitudinal angle, transverse angle, and air-flow rate are being studied. Preliminary studies with Hingula non-coking coal (Ash ~ 42%) of Mahanadi Coalfields Ltd. has indicated potential to reduce the ash level up to 34%. However, more tests are planned to be conducted to arrive at conclusion.

### 3.1.3 X-ray based radiometric technology

Detection of coal and mineral laden ash particles with the application of dual energy X-ray technique is another area where CIL has recently extended its support for indigenous development in association with M/s Ardee Hi Tech (P) Ltd. The sorter has been given a name of 'ArdeeSort'. Multi-energy radiometry has been established as an ideal system for rapid quality determination of coal particles in close size range (preferably, top-size:bottom-size, 2:1) that has been used for subsequent acceptance or rejection based on pre-set parameters. The know-how encompasses – coal chemistry, radiation physics, sensor technology, electronic hardware and software, mechanical handling systems, pneumatic sorting technology, interface systems and final integration of all these know-how into one composite equipment.

Attenuation of X-rays by particles depends primarily on their atomic numbers. The primary combustibles in coal, viz., carbon, oxygen and hydrogen, and derived elements, have relatively low atomic numbers. The incombustibles like silica, alumina, iron, calcium, etc. have higher atomic numbers. Therefore, the attenuation of a particle will depend primarily on the constituents – higher the ash content, higher the attenuation. Dual energy sensor technology is utilized to neutralize the effects of size of particle on attenuation. The mathematical algorithm that analyses the dual energy signal automatically computes the composite ratio of organics to inorganic in the particle. The operator has the freedom to set the limits of inorganic (ash) percentage in particles that can be accepted, which is akin to setting the specific gravity in a dense media bath. In fact, the distinct advantage of such a radiometric technology is that the target for clean coal or the threshold value for rejection can be planned and set as per need. This equipment is under different stages of testing at existing Madhuband washery.

## 3.2 Demonstration of existing technology

Development of a technology from fundamental and subsequent validation for confidant industrial application is resource intensive; especially significant time as well as substantial money is required. Coal India Limited as a commercial organization is simultaneously striving for identification of established dry coal beneficiation technology with potential for operation with Indian non-coking coal round the year.

## 4. Conclusion

The coal washery business in India is passing through sea change in almost all aspects, viz. business growth, open market scenario, technology freedom, and research support. Dry technology for deshaling or beneficiation is the need of posterity for enabling more effective resource management. Involvement of all stakeholders will boost the efficient utilization of large resource of low grade coal in India. There is a vast opportunity for the globally established players in the coal washing sector as quality conscious customers need to be fed with right quality coal for satisfaction.

## 5. Acknowledgement

The authors like to show their gratitude to the management of CMPDI, the company they are serving to, for allowing them to share their wisdom on the subject gathered during various research projects.

## 6. References

[1] Bartram K, Zyl J V, Viti E, 2013, 'Dual Energy X-ray Transmission Sorting of Coal for Deshaling and Ash Content'; ICPC, Istanbul.

[2] Robben C, Korte J de, Wotruba H, Robben M, 2013, 'Experiences in Dry Coarse Coal Separation Using X-Ray-Transmission-Based Sorting'; ICPC, Istanbul.

[3] Sakhre D K, 2014, 'Design and Development of Coal Winnowing System for Dry Beneficiation of Coal based on CFD modeling & Simulation' (Completion Report: Project Code: CP/45) by *CIMFR*-CSIR, Nagpur, India.

[4] Sakhre D K, Singh S P, Mehatre R S, Acharya R K, 2013, 'Dry coal Beneficiation: A future of Coal Washing' in Coal Preparation Technology-2013 – Recent Trends & Future Needs in Coal Preparation, at CIMFR Digwadih Campus, Dhanbad, India.

[5] Website Reference: 'Report of the Commissioner of Agriculture for the year 1877 (Washington, District of Columbia: Government Printing Office, 1878)'; https://etc. usf.edu/clipart/86600/86686/86686-winnowing-machine.htm, (2019, Jan 3).

# 48

# Dry deshaling: Expanding options for cleaning coal

Dr. G.V. Ramana (Managing Director, Ph.D.)

*Ardee Hi-Tech Private Limited, Plot No. 1, Ardee Building,*
*Balaji Nagar, Visakhapatnam, India*

**Abstract:** ArdeeSort dry deshaling technology tested at Madhuban washery of BCCL is a path-breaker. It offers solutions for improving coal quality not available earlier to coal industry or coal user. What successful testing of ArdeeSort coal dry beneficiation at Madhuban washery does is to enhance basket of options to coal industry for improving quality of coal.

ArdeeSort dry deshaling can help prioritization of quality improvement for coal industry. Mines where quality is slipping due to external contamination, causing revenue losses, can invest and start reaping the benefits within a short time. This is irrespective of whether they would like to go in for full-fledged washing facilities since the output of the deshaling process can serve as input to the washery which will sharply reduce washing costs and improve yields.

Only option previously was jig or heavy media wet technology. With dry beneficiation or deshaling, a viable option emerges for low cost beneficiation without use of water or chemicals. It becomes a powerful tool in hands of coal industry to leverage quality for better price realization. Many options can be explored to reduce washery size and/or water consumption.

**Keywords:** Dry beneficiation, deshaling, coal sorting, ArdeeSort, X-ray sorting, radiometric sorting, hybrid washer

## 1. Introduction

Most countries wash a substantial proportion of their coal before dispatch from the mine due to strict standards on use of cleans coals. India is trying to improve its act. Since it has to depend on low quality coal as its prime source of energy but at the same time, pay attention to environment and costs. Environment means not only emissions but also conserving water which in most areas of the country is in acute short supply.

## 2. Trials of Ardee sort with Indian coals

ArdeeSort dry deshaling technology was successfully tested at Madhuban washery of BCCL and is a path-breaker in that respect. A retrofitting exercise of accommodating a dry deshaling circuit in an existing washery was a technical challenge. Several compromises had to be made including accommodating deshaling and pre-screening facilities in a very compact space. Since it was a demonstration scale unit, adequate care had to be taken to ensure that it could be put into commercial operation once the trials are over.

Trials were conducted over six months and the results represent are encouraging. For washery operators, it enhances the bouquet of options available to improve washery performance, lower costs and be environment-friendly. The technology has demonstrated improvement in ash in 25–50 mm and 13–25 mm by 14–9%. Input coal in Madhuban washery fluctuated with ash range varying between 45% and 55% in size 13–50 mm with a median value of 50% ash. Reduction of ash by deshaling to 39% represents an improvement in yield in second stage washing by 25–27%. Main advantage in pre-treating by deshaling is removal of external contaminants, freeing up space to feed more coal, and reduce crowding out effect in washery. This effectively facilitates much better separation efficiency. In case of a Greenfield washery, the wet circuit can be designed for a smaller capacity which reduces capex and opex. It enables better yield and lowers cost of clean coal since washing costs are attributed only on clean coal portion obtained from washing.

The tables below gives results of experiments conducted in two phases. In final phase, results were exciting with several glitches reduced and settings fine-tuned. More needs to be done to improve consistency especially feed size, quality, and sampling arrangements. Data base needs to be built up to convert average dual energy readings into accurate ash and weight readings (Tables 1A and B and 2A and B).

**Table 1A:** Effect of deshaling coking coal in size range of 13–25 mm (Phase I)

| Date | Raw RBC2 | 25–13 mm | | | Reduction in ash |
|---|---|---|---|---|---|
| | 50–0 mm | Raw coal | Product | Reject | |
| 24-9-2010 | 53.2 | 51.6 | 42.6 | 59.2 | 9.0 |
| 28-9-2016 | N/A | 61.2 | 52.8 | 70.4 | 8.4 |
| 1-11-2016 | 45.3 | 55.8 | 51.8 | 64.1 | 4.0 |
| 10-11-2016 | N/A | 54.9 | 51.7 | 58.0 | 3.2 |
| 11-11-2016 | 53.1 | 46.7 | 36.1 | 47.9 | 10.6 |
| 14-11-2016 | 41.7 | 39.5 | 35.4 | 61.3 | 4.1 |
| 15-11-2016 | 42.1 | 58.9 | 37.0 | 49.6 | 21.9 |
| 15-11-2016 | 39.0 | 41.7 | 38.2 | 48.6 | 3.5 |
| *Average* | *45.7* | *51.3* | *43.2* | *57.4* | *8.1* |

**Table 1B:** Effect of deshaling coking coal in size range of 13–25 mm (Phase II)

| Date | Raw RBC2 | 25–13 mm | | | Reduction in ash |
|---|---|---|---|---|---|
| | 50–0 mm | Raw coal | Product | Reject | |
| 9-1-2017 | 48.5 | 51.4 | 35.8 | 54.8 | 15.6 |
| 10-1-2017 | 61.4 | 52.4 | 35.5 | 58.3 | 16.9 |
| 12-1-2017 | 39 | 49.5 | 31.7 | 58.1 | 17.8 |
| 20-1-2017 | 57.4 | 52.4 | 40.7 | 65.4 | 11.7 |
| 21-1-2017 | 54.3 | 47.1 | 33.4 | 51.9 | 13.7 |
| 21-1-2017 | 53.6 | 54.7 | 43.3 | 58.6 | 11.4 |
| 21-1-2017 | 49.1 | 50.1 | 38.3 | 59.7 | 11.8 |
| 28-1-2017 | 49.8 | 46.4 | 40.4 | 50.3 | 6 |
| 28-1-2017 | 43.3 | 46.6 | 36.6 | 60.4 | 10 |
| 30-1-2017 | 61 | 56.2 | 41 | | 15.2 |
| 30-1-2017 | 62.6 | 50.6 | 34.1 | 62.4 | 16.5 |
| 31-1-2017 | | 53.6 | 45.9 | 50 | 7.7 |
| *Average Jan-17* | *52.7* | *50.9* | *38.1* | *57.3* | *12.9* |
| *Average overall* | *49.2* | *51.1* | *40.6* | *57.3* | *10.5* |

*Note:* Several perverse results and results where more than two sets of information were not available due to sampling errors were not considered in the analysis.

**Table 2A:** Effect of deshaling coking coal in size range of 25–50 mm (Phase I)

| Date | Raw RBC2 | 50–25 mm | | | Reduction in ash |
|---|---|---|---|---|---|
| | 50–0 mm | Raw coal | Product | Reject | |
| 16-9-2016 | 53.3 | 56.1 | 46.2 | | 9.9 |
| 20-9-2016 | 55.5 | 59.1 | 53.8 | 64.8 | 5.3 |
| 24-9-2016 | 53.2 | 58.4 | 52.9 | 68.1 | 5.5 |
| 27-9-2016 | 61.8 | 56.2 | 51.4 | 62.3 | 4.8 |
| 31-10-2016 | 53.8 | 54.6 | 48.7 | 58.7 | 5.9 |
| 1-11-2016 | | 61.0 | 42.9 | 67.0 | 18.1 |
| 7-11-2016 | 54.1 | 48.3 | 40.9 | 58.6 | 7.4 |
| 9-11-2016 | 39.4 | 45.9 | 40.8 | 57.6 | 5.1 |
| 9-11-2016 | 62.7 | 59.8 | 46.2 | 64.8 | 13.6 |
| 10-11-2016 | | 63.5 | 51.3 | 65.5 | 12.2 |
| 14-11-2016 | 41.7 | 41.0 | 35.8 | 45.7 | 5.2 |
| 15-11-2016 | 39.0 | 51.2 | 40.8 | 60.2 | 10.4 |
| *Average** | *51.5* | *54.6* | *46.0* | *61.2* | *8.6* |

**Table 2B:** Effect of deshaling coking coal in size range of 25–50 mm (Phase II)

| Date | Raw RBC2 | 50 - 25 mm | | | Reduction in ash |
|---|---|---|---|---|---|
| | 50–0 mm | Raw coal | Product | Reject | |
| 9-1-2017 | 48.5 | 56.2 | 47.6 | 63.0 | 8.6 |
| 10-1-2017 | 61.4 | 56.3 | 35.6 | 67.3 | 20.7 |
| 11-1-2017 | 45.2 | 61.9 | 55.5 | 77.2 | 6.4 |
| 12-1-2017 | 39.0 | 49.8 | 30.6 | 64.6 | 19.2 |
| 13-1-2017 | 53.8 | 57.3 | 38.7 | 66.1 | 18.6 |
| 20-1-2017 | 57.8 | 47.2 | 39.9 | 69.2 | 7.3 |
| 21-1-2017 | 54.3 | 53.9 | 33.1 | 68.1 | 20.8 |
| 21-1-2017 | 53.6 | 50.6 | 36.9 | 64.4 | 13.7 |
| 21-1-2017 | 50.1 | 50.8 | 44.9 | 67.9 | 5.9 |
| 23-1-2017 | 54.7 | 59.7 | 40.2 | 62.7 | 19.5 |
| 23-1-2017 | | 57.3 | 37.7 | 65.8 | 19.6 |
| 27-1-2017 | 48.3 | 51.7 | 43.3 | 58.0 | 8.4 |
| 27-1-2017 | 43.9 | 51.4 | 37.9 | 56.3 | 13.5 |
| 27-1-2017 | 50.2 | 42.8 | 25.4 | 61.7 | 17.4 |
| 28-1-2017 | 49.3 | 41.9 | 36.0 | 59.2 | 5.9 |
| 28-1-2017 | 49.8 | 51.6 | 34.6 | 75.3 | 17 |
| 28-1-2017 | 55.1 | 54.5 | 34.3 | 61.5 | 20.2 |
| 30-1-2017 | 59.1 | 63.4 | 36.1 | 53.4 | 27.3 |
| 20-1-2017 | 51.5 | 45.1 | 36.5 | 72.0 | 8.6 |
| 31-1-2017 | 45.3 | 57.6 | 37.0 | 67.5 | 20.6 |
| *Average Jan-17* | *51.1* | *53.1* | *38.1* | *65.1* | *15.0* |
| *Overall Average* | *51.4* | *53.7* | *42.4* | *63.6* | *11.3* |

## 3. Ardee sort installation at Madhuban washery

While the second phase gave excellent results from deshaling coking coal, the conditions were much better including better screening due to lower moisture, better size control with some degree of manual intervention in feed, etc. The difference between the two phases is also on account of a better understanding of the quality of incoming coal and appropriate parametric settings to prevent rejection of usable coals.

### 3.1 Expanding beneficiation options

ArdeeSort dry deshaling can help prioritization of quality improvement for coal industry. Mines where quality is slipping due to external contamination, causing revenue losses, can opt for dry beneficiation and start reaping benefits within a short time. Whether they would like to go in for full-fledged washing facilities later is immaterial since output of deshaling plant at mine can serve as input to washery. This will reduce washing costs and improve yields while reducing consumption of magnetite and water. Capex and opex of washery will come down, water requirements and addition of moisture to clean coal will be lower.

Emergence of dual energy X-ray transmission technology (DE-XRT) deshaling options is a rather recent phenomenon and a major game changer. Dry deshaling techniques like air jigs or air tables depend upon existence of large density differential between contaminants and coal. If near gravity material (NGM) is high, cross-migration is likely to be high. This is where ArdeeSort steps in and opens the deshaling route. Charts below illustrate the difference:

**Chart 1:** Performance of other dry benefication technologies

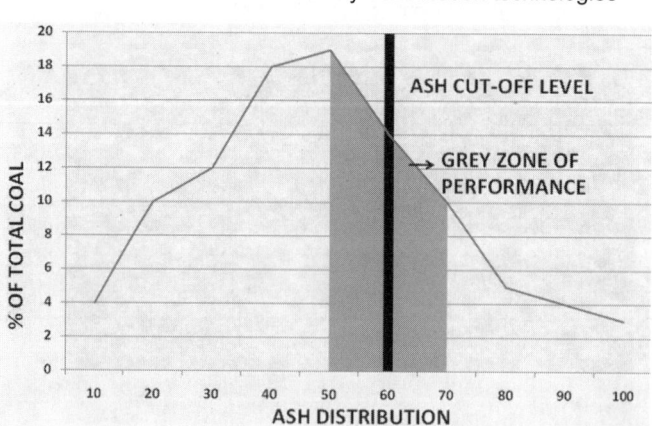

As can be observed from above graph, there is a grey zone around ash cut-off point (specific gravity cut-point is proxy normally used) traversing 20% ash points where particles are likely to encounter misplacement. This account for nearly 28% of total coal in above example where cross migration can affect yield and/or ash in cleans by as much as 5% points. In several mines in India, where coals are classified as "difficult to wash", this percentage can be much more.

With high degrees of cross migration in coals having high, several industries, especially sponge iron industries, and cement industries who invested in air jig technologies are in the hunt for more optimal solutions. This is despite sponge iron industries using both cleans and rejects from dry beneficiation air jig technology in the kilns and power plants respectively. Hence, no part of input coal is wasted. Cement plants having captive generation use entire product since they can also use ash generated to mix with cement up to 7–10%.

Advantage with ArdeeSort is that it can be programmed to set rejection level unlike mechanical logic systems like air jig or air table. With accurate calibration, results can be quite precise. The grey zone around ash cut-off point is narrower in ArdeeSort and hence efficiencies will be much higher. These testing procedures need to be conducted in a more rigorous set up than what is feasible at Madhuban. Following graph gives projected narrow grey zone band in ArdeeSort which opens tremendous opportunities for coal washery operators.

**Chart 2:** Performance of Ardeesort

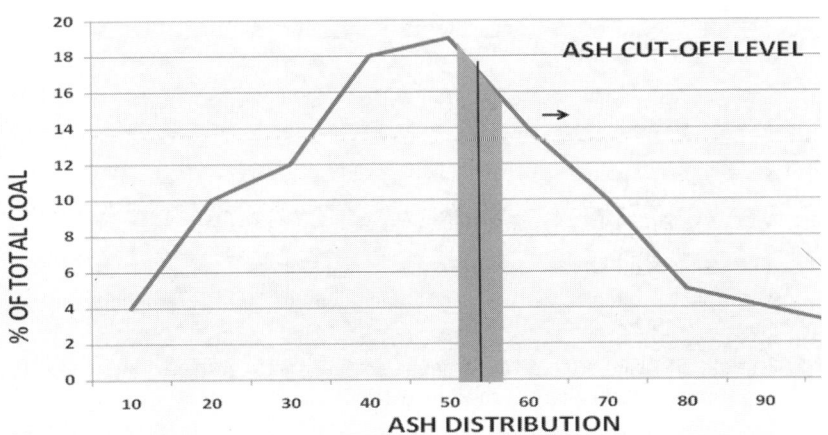

**Option 1:** In two-stage dry beneficiation, usable coals are first retrieved by rejecting those particles that are beyond a lower cut-off limit. Rejects from

this stage are re-processessed at a higher ash limit. In our example, rejection limit at 50% and then 70% ash respectively gets three products from dry beneficiation route. All particles below 50% limit will be clean coals. This equals 60% of coal above 13 mm size. Re-processing higher ash material to eliminate particles having >70% ash in above example is 20% by weight of +13 mm size coals. Cleans in this stage are re-sized to liberate ash and washed in conventional washery improve yield at desired ash level. Washery size comes down to 1/5th of conventional route, moisture addition to clean coal is minimized since overwhelming portion of coal is dry beneficiated, overall cost and water requirement, chemicals and flocculants and other consumables also reduced substantially.

**Option 1A:** Single stage dry beneficiation removing usable coal from washing circuit will reduce volume of coal required to be washed by 60%. Only 40% of capex and opex will be required for conventional washing apart from the general coal handling and dry beneficiation costs.

**Option 2:** Single stage deshaling by which only high ash particles, say beyond 65–70% ash will be removed. This accounts for 30% of the total weight of coal. Washing rest of the coal reduces load on washery by 30%, improves yield and opens up the possibility to have better quality middlings for use in power generation if the end use is not for power.

## 4. Conclusion

Every washery will have to evolve their own flow sheets depending on the quality and washability of coals that they will handle. ArdeeSort empowers operators' with additional tools.

## 5. References

[1] Biswal, S.K., Sahu, A.K., Parida, A., Reddy, P.S.R., and Misra, V.N., 2002, "Prospects of dry beneficiation of Indian high ash non-coking coal – A review." Journal of Mines Metals and Fuels, 51, pp. 53–57.

[2] Frankland, S.C., 2000, "Review of the Technology Requirement of the Coal Preparation Sector of India." DTI Cleaner Coal Technology Programme, Project Summary PS 263, 2000.

[3] Kamall, Roshan, 2001, "Coal Preparation." DTI Cleaner Coal Technology Programme." Technology Status Report 15, London.

[4] Mitra, S.K., 2004, "Effective and economic coal preparation technologies available for cleaning of high ash Indian thermal coal." In Proceedings of Mineral Processing Technology, (G.V. Rao and V.N. Misra, Eds.), Bhubaneswar, India: Allied, pp. 575–578.

[5] Ramana, G.V., 2009, "Dry Beneficiation of Coal – Relevance for Indian Coals." CPSI Journal, Vol. I, No. 1, New Delhi, India, pp. 35–39.

[6] Ramana, G.V., 2013, "Technology for Affordable Coal Quality – Emergence of X-ray On-Line Solutions." Coal Preparation Technology 2013, Proceedings of National Seminar on Recent Trends and Future Needs in Coal Preparation, CSIR-Central Institute of Mining & Fuel Research, Dhanbad, India.

[7] Ramana, G.V., and Ramakrishna, V., "Upgrading Low Rank Coals – The Dry Method." CPSI Journal, Vol. VI, No. 17, New Delhi, India, pp. 137–139.

[8] Riedel, F., and Wotruba, H., 2004, "Pre-concentration by sensor-based sorting devices in mineral processing." In Proceedings of Mineral Processing Technology, (G.V. Rao and V.N. Misra, Eds.), Bhubaneswar, India: Allied, pp. 37–44.

[9] Sakhre, D.K., Singh, S.P., Mehatre, R.S., and Acharya, R.K., 2013, "Dry Coal Beneficiation : A Future of Coal Washing." Coal Preparation Technology 2013, Proceedings of National Seminar on Recent Trends and Future Needs in Coal Preparation, CSIR-Central Institute of Mining & Fuel Research, Dhanbad, India, p. 7.

[10] Singh, S., and Masand, N.K., 2011, "RAMDARS Technology for Sustainable Energy – Dry Beneficiation of Lignite." Mining Engineers' Journal, Vol. 13, No. 5, December, 2011, pp. 23–26.

[11] Sinha, V.K., Dey, B.K., and Baranwal, P.K., 2016, "Coal Preparation in India: New Business Opportunities & Need for Dry Separation Technology" Proceedings of XVIII International Coal Preparation Congress, 28 June–01 July, 2016, (Ed. Vladimir Litnivenko), pp. 1167–1170, Springer.

# 49

# Study of dry coal beneficiation in an air pulsated stratifier

Abhishek Kumar, Ganesh Chalavadi, Kalicharan Hembrom, Snehashish Tripathy

*CSIR-National Metallurgical laboratory, Jamshedpur, India*

**Abstract:** Coal keeps on doing a noteworthy job in the economic development of a country, particularly in metallurgical industries and power generation sectors. Coal is currently beneficiated predominantly in wet condition. The conventional methods for processing of coarse coal like heavy media bath, jigging, etc. utilize water. In the near future, coal resource and water resource is going to have a reverse distribution thus posing several hurdles to wet beneficiation technologies in coal cleaning. Thus, it is exigent to develop efficient dry beneficiation technology for coal. This paper presents a summarized assessment of performance of an air pulsated stratifier with particular reference to Indian coal. The study was carried out on coal samples having ash in the range of 30–33%. Response parameters considered for this study were ash reduction and yield of clean coal. Results are found to be encouraging as the experimental trials resulted in absolute reduction of ash percentage in the range of 7–9% in single stage.

**Keywords:** Coal, heavy media bath, jigging, coal cleaning, dry beneficiation, air pulsated stratifier, ash reduction, yield

## 1. Introduction

Coal is an organic sedimentary rock, a natural fossil fuel formed over millions of years from the remains of decaying trees and vegetation (Subba Rao and Gouricharan, 2016). Water is the richest asset on earth. India is among one of those countries with vast coal reserves where water is in scarcity. Water Resources Information System of India revealed that in the year 2010, 1588 m³ of fresh water is available per capita per year. However, it has been projected that per capita surface water availability is likely to be reduced to 1401 m³ and 1191 m³ by the end of year 2025 and 2050, respectively (Web 1). Due to rapid rise in population and growing economy of the country, there will be a continuous increase in demand for water, and it will become scarce in coming decades (Web 1). Presently, most of the coal preparation plants are based on wet beneficiation techniques that records huge volume of water consumption. So, there is an alternative to wet beneficiation technique called dry beneficiation and it has an evident lead over wet beneficiation (Lockhart, 1984). Dry beneficiation has gained significant attention in the recent years because of dearth of fresh water and economic benefit over wet beneficiation as it circumvents the treatment and storage of wastewater (Honaker et al., 2008; Sahu et al., 2009). For treatment of 50 mm × 6 mm coal feed FGX separator is used with limitation of −6 mm maintaining in the range of 10–20% in order to ensure a fluidized bed (Honaker and Luttrell, 2007), for 3 mm × 1 mm coal particles AKAFLOW is being explored and for fine coal i.e. −1 mm coal particle Airtable is being used (Chalavadi et al., 2016). However, the treatment of coal fraction in the size range −6 mm to 3 mm is a bit less explored. In order to explore the dry coal beneficiation in this size range, authors have fabricated in house air pulsated stratifier unit and this paper discus about it in detail.

## 2. Experimental

### 2.1 Equipment details

For processing of coal samples of size range −6 mm to 3 mm, air pulsated stratifier (APS) is designed and fabricated in-house. APS comprises of following components (a) blower for air supply, (b) a rotary valve for pulsating airflow, (c) a stratifying chamber with provision of removing products layer wise, and (d) a perforated sheet of 3 mm opening placed at the bottom of the stratification chamber through which air is distributed uniformly in to the chamber. This equipment works on batch scale (100 kg). Particle separation results with the help of pulsating airflow in order to produce the stratification. Since minimum fluidization velocity is a function of density and size, the particles will be distributed layer wise i.e. denser material will be at the bottom and lighter material will be at the top. Forces acting on a particle in APS are drag force, gravity force, and buoyancy. The separation process generated three product streams i.e. clean coal product, middling, and reject stream (Fig. 1).

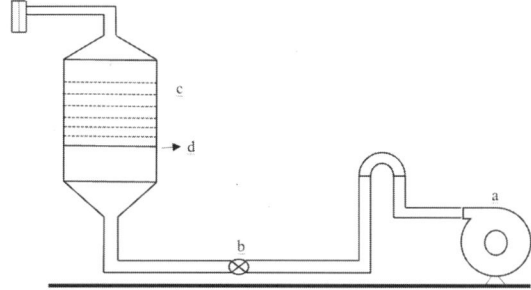

**Figure 1:** Schematic diagram of air pulsated stratifier. (a) Blower, (b) rotary valve, (c) a stratifying chamber, and (d) perforated sheet

## 2.2 Feed characterization

Coal feed sample has been subjected to Bomb calorimeter for estimation of gross calorific value (GCV), which yields a GCV of 4090.3 kcal/kg. Further, proximate analysis of the feed sample was done to estimate moisture, volatile matter (VM), ash, and fixed carbon (FC) and results are tabulated below in Table 1:

**Table 1:** Proximate analysis of feed sample

| Moisture% | VM% | Ash% | FC% |
|-----------|-----|------|-----|
| 7.56 | 25.88 | 31.02 | 35.54 |

The representative feed sample was subjected to sink-float tests by using the liquids with relative density in the range of 1.30–2.00 prepared with the help of bromoform and benzene. Washability data is shown graphically in Fig. 2.

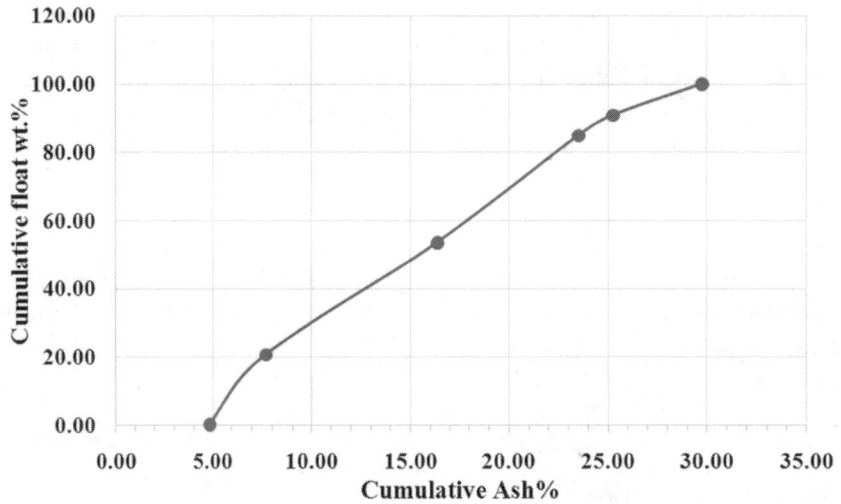

**Figure 2:** Washability curve of coal feed sample

Coal feed sample was analyzed with the help of electron probe micro-analyzer (EPMA). EPMA results were presented pictorially and graphically in Fig. 3.

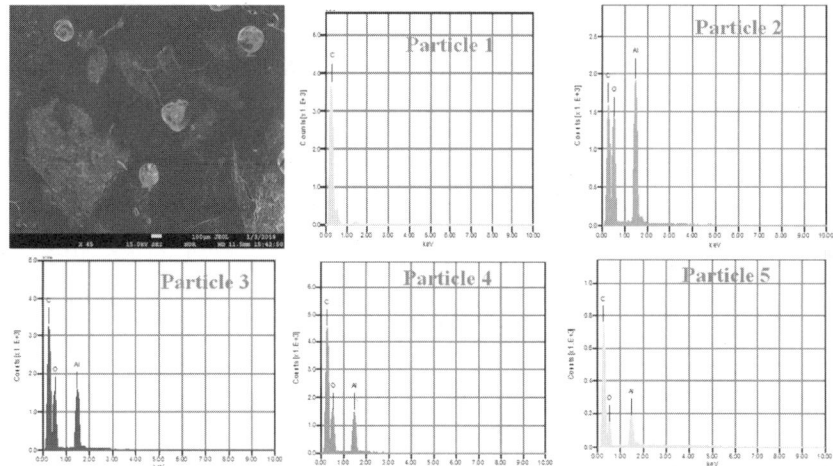

**Figure 3:** EPMA analysis results of coal feed sample

From EPMA results it is observed that particle 1 is showing peaks of coal particles and particles 2–5 are showing the presence of aluminosilicates.

## 2.3 Campaign description

Coal samples were prepared for processing in APS in the size range of −6 mm to 3 mm fractions. Samples were prepared by crushing them in a jaw crusher followed by a roll crusher. Two stage crushing was done in a jaw crusher to bring down the size to around 13 mm. Further reduction was carried out in roll crusher in closed circuit in order to minimize the generation of fines. After screening, the required size class of −6 mm to 3 mm was obtained for carrying out campaign in APS. Effect of operating variables i.e. blower frequency (BF) and pulsation frequency (PF) were studied in terms clean coal grade and recovery. Optimum test performances and conditions were identified by varying the operating parameters systematically. After each campaigns each product streams were collected and weighed and representative samples are withdrawn for ash analysis.

## 3. Results and discussion

The values of two critical operating parameters (blower frequency and pulsation frequency) were varied over six test campaigns. Campaign results are tabulated in Table 2 below:

**Table 2:** Results of the campaign with varying parameters

| Operating parameters | Streams | Wt.% | Ash% |
|---|---|---|---|
| BF = 40 Hz<br>PF = 60 Hz | Clean coal | 50.04 | 26.93 |
| | Middling | 30.22 | 28.44 |
| | Reject | 19.73 | 41.48 |
| BF = 40 Hz<br>PF = 50 Hz | Clean coal | 55.88 | 29.10 |
| | Middling | 34.36 | 32.04 |
| | Reject | 9.76 | 56.60 |
| BF = 40 Hz<br>PF = 40 Hz | Clean coal | 54.59 | 22.92 |
| | Middling | 35.51 | 34.43 |
| | Reject | 9.90 | 56.97 |
| BF = 50 Hz<br>PF = 40 Hz | Clean coal | 50.56 | 22.91 |
| | Middling | 39.55 | 37.29 |
| | Reject | 9.89 | 57.04 |
| BF = 50 Hz<br>PF = 50 Hz | Clean coal | 52.81 | 26.22 |
| | Middling | 35.75 | 35.76 |
| | Reject | 11.44 | 56.56 |
| BF = 50 Hz<br>PF = 60 Hz | Clean coal | 53.26 | 24.10 |
| | Middling | 36.88 | 36.61 |
| | Reject | 9.87 | 56.20 |

From the results tabulated above it can be inferred that, pulsation frequency (PF) is playing important role in separation as compared to blower frequency (BF). At higher levels of PF ash content is on the higher side corresponding to each level of BF. Best results of clean coal yield of 54.5% with 22.9% ash obtained at BF = 40 Hz and PF = 40 Hz.

## 3.1 Product characterization

Clean coal products obtained at best operating conditions were analyzed with the help of EPMA. Results are presented in Fig. 4 pictorially and graphically.

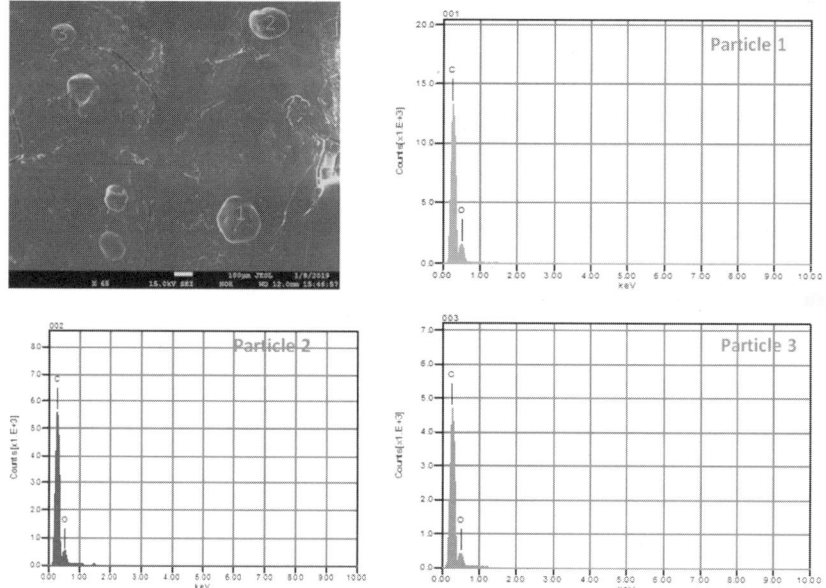

**Figure 4:** EPMA analysis of clean coal products

## 4. Conclusion

The observation from the campaign reflects ability of air pulsated stratifier to efficiently reduce down the ash content. The yield was found to be around 54% at 22.9% ash, reducing the ash of coal feed sample by 8 units in single stage processing.

## 5. Acknowledgement

Authors are thankful for the support provided by in-house project support group (i-psg) committee, CSIR-National Metallurgical Laboratory, Jamshedpur.

## 6. References

[1] G. Chalavadi, R.K. Singh, and A. Das, 2016. "Processing of coal fines using air fluidization in an air table." *International Journal of Mineral Processing* 149: 9–17.

[2] R.Q. Honaker, and G.H. Lutrell, 2007. Final Technical Report "Development of an advanced deshaling technology to improve the energy efficiency of coal handling, processing, and utilization operations".

[3] R.Q. Honaker, M. Saracoglu, E. Thompson, R. Bratton, G.H. Luttrell, and V. Richardson, 2008. Upgrading coal using a pneumatic density based separator. *International Journal of Coal Preparation and Utilization* 28: 51–67.

[4] N.C. Lockhart, 1984. Dry Beneficiation of coal, *Powder Technology* 40(1–3): 17–42.

[5] A.K. Sahu, S.K. Biswal, and A. Parida, 2009. Development of air dense medium fluidized bed technology for dry beneficiation of coal – a review, *International Journal of Coal Preparation and Utilization* 29(2009): 216–241.

[6] D.V. Subba Rao, and T. Gouricharan, Coal Processing and Utilization, 2016, CRC press, ISBN no. 9781138029590-CAT # K29610.

## Referencing websites

Web 1: www.india-wris.nrsc.gov.in/wrpinfo/index.php?title=India%27s_Water_Wealth.

# 50

# Dry cleaning of high ash Indian thermal coal using fluidization based vibrating deck

Shobhana Dey and Ratnakar Singh

*CSIR-National Metallurgical Laboratory, Jamshedpur, India*

**Abstract:** In India, reserve of non-coking coal is 86.6% and contents high ash due to its inherent lithological association, intercalated shale of varied thickness and out of seam dilution through open cast mining. About 20% non-coking coal is processed for ash reduction. Ministry of Environment and Forest, Govt. of India made a regulation that if coal is transported more than 1000 km from pit-heads, or if burnt in urban areas, ash content in the coal should be less than 34%. The coal beneficiated by wet method retains moisture (6–15%) depending upon the feed size. Dry beneficiation of coal has several advantages over the wet beneficiation and is also cost effective.

In the present study, air fluidized vibrating deck separator was used for dry beneficiation of high ash non-coking coal. Coal characterization including washability studies were carried out to assess the cleaning performance. The critical process variables like frequency of deck vibration, air flow rate, longitudinal, and transverse angle of the deck were studied. The responses were measured in terms of yield of clean coal, recovery of combustibles in clean coal and organic efficiency. The cleaning performance was described by probable error (Ep). It has been noted that Ep is 0.10 for coarser size and it increases to 0.24 as the feed size decreases. The studies reveal encouraging results in the reduction of ash, improvement in the recovery of combustibles and calorific value of clean coal.

**Keywords:** Dry processing, washability study, partition co-efficient, air fluidization, differential acceleration

## 1. Introduction

The resources of coal in India are about 301.56 billion tonnes. Out of which 88% is non-coking coal. The quality of the Indian coal is inferior because of drift origin and causes dissemination of the mineral matter. Presently, washing of thermal coal in India is typically targeted to yield less than 34% ash. Ministry of Environment and Forest promulgated regulations, amended in 2014, mandating that coals must be cleaned to less than 34% ash content if transported more than 1000 km from pit-heads, or if burnt in urban areas, environmentally sensitive or critically polluted areas irrespective of their distance from the pit-head (MoEF, 2014).

Coal is currently beneficiated by wet method through gravity and flotation techniques. Dry cleaning of coal has attracted the attention of the researchers worldwide because of the obvious economic advantages it offers. Studies have been carried out in most of the major coal producing countries to evaluate the performance of the dry coal cleaning units. In early 1916, dry beneficiation of coal fines was introduced in United States (Lockhart, 1984). The air density separators were widely practiced between 1930 and 1960. Several researchers found that dry beneficiation of coarse coals are found to be effective in reducing the ash in clean coal by different techniques (Honaker et al., 2008; Yang et al., 2013; Gupta et al., 2012). The autogeneous medium produced by the mixture of air and fine coal powder behaves as a separating density in air fluidization technique. The degree of segregation of fine lignite in a vibrated gas-fluidized bed was studied by Zhao et al. (2015). Theoretical considerations show that effective separation in air table can be achieved when the time of free fall could be minimized (Osborne, 1988; Haider and Levenspiel, 1989; Zhao et al. 2002). The separation of fine particles in air is difficult due to low differences in settling velocity. However, the modern techniques assist to increase the efficiency of separation by introducing other parameters, like transverse oscillation, slanting the deck in longitudinal and transverse direction which increases the flow of the particles according to the desired grade.

Therefore, efforts have been made to produce a product suitable for thermal power plant by dry processing of non-coking coal. In the present investigation, high ash non-coking coal has been studied for reduction of ash to 34% and also for understanding the separation behavior.

## 2. Materials and methods

A bulk coal sample in the form of lumps was collected from the Dakra coal block of the North Karanpura coalfields which belongs to the Central Coalfields Limited (CCL). Detailed characterization includes proximate analysis, size and size-wise ash analysis, determination of calorific value, coal petrography, and washability study are discussed below.

## 2.1 Coal characterization

Petrographic studies of the coal sample (Fig. 1) were carried out following BIS procedures (IS 9127-5, 2004) on polished mounts. It was observed that volume of vitrinite macerals dominate (35%) and occur with plain surface and grey color. Inertinites (19.8%) are mainly represented by semifusinites with small cellular cavities and inertodetrinites. Liptinite macerals contribute only 3% and occur with dark thread-like appearance. Mineral matter (mainly argillites and some carbonate minerals) occur as cavity fillings or in disseminated form which accounts for about 42.2 vol.% (Table 1).

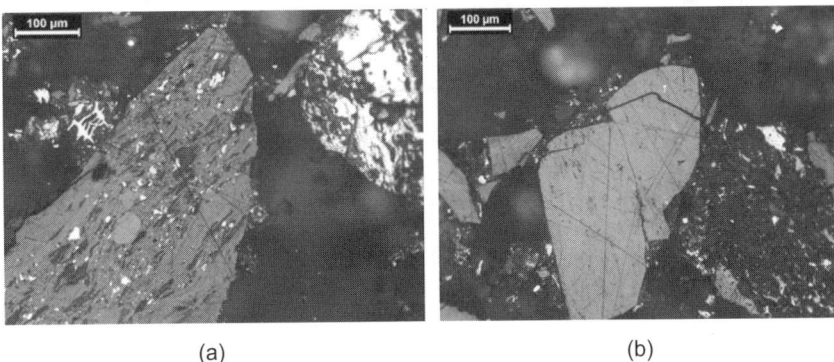

(a)                                      (b)

**Figure 1:** Photographs showing petrographic characters under DM4500 polarizing coal petrological microscope with oil immersion and using 20× objective. (a) Grains of impure vitrinite (greyish), liptinite (whitish), and mineral matter (smoky-black). (b) A vitrinite maceral (grey) surrounded by mainly mineral matter (smoky-black)

**Table 1:** Maceral composition in coal sample

| Petrographic constituents | Vitrinite | Inertinite | Liptinite | Mineral matter |
|---|---|---|---|---|
| Vol.% | 35 | 19.8 | 3 | 42.2 |

**Table 2:** Proximate analysis of coal sample

| Moisture (%) | Ash (%) | VM (%) | FC (%) | Heat value (kcal/kg) |
|---|---|---|---|---|
| 5.5 | 43.1 | 21.5 | 29.9 | 3420 |

**Table 3:** Size and ash distribution in feed coal sample

| Size (mm) | Wt (%) | Ash (%) | Moisture (%) |
|---|---|---|---|
| −50 + 25 | 40.0 | 42.2 | 5.8 |
| −25 + 13 | 17.3 | 42.9 | 6.0 |
| −13 + 6 | 15.3 | 40.0 | 5.6 |
| −6 + 3 | 8.3 | 40.7 | 5.3 |
| −3 | 19.1 | 45.3 | 5.9 |
| Total | 100 | 42.45 | 5.3 |

The proximate analysis of as-received coal sample (Table 2) indicates that ash is 43.1% which is quite high. The moisture is 5.5%, volatile matter

(VM) and fixed carbon (FC) are 21.5% and 29.9%, respectively. The size and size-wise ash distribution of the coal crushed to 50 mm given in Table 3 shows that ash content in the finest fraction of −3 mm is about 48.9% which is quite significant. The coarser fraction of −50+13 mm contains 42.5% ash and the size fraction of −13+3 mm contains about 40% ash.

## 2.2 Washing characteristics

The washability study of coal sample crushed at −50 mm was carried out to ascertain the washing characteristics of the coal for achieving the maximum yield at targeted ash level. The study was conducted at varied density of the medium from 1.4 to 2.2. It has been found from Fig. 2 that recoverable clean coal could be 79% at 34% ash level at relative density of 2.0.

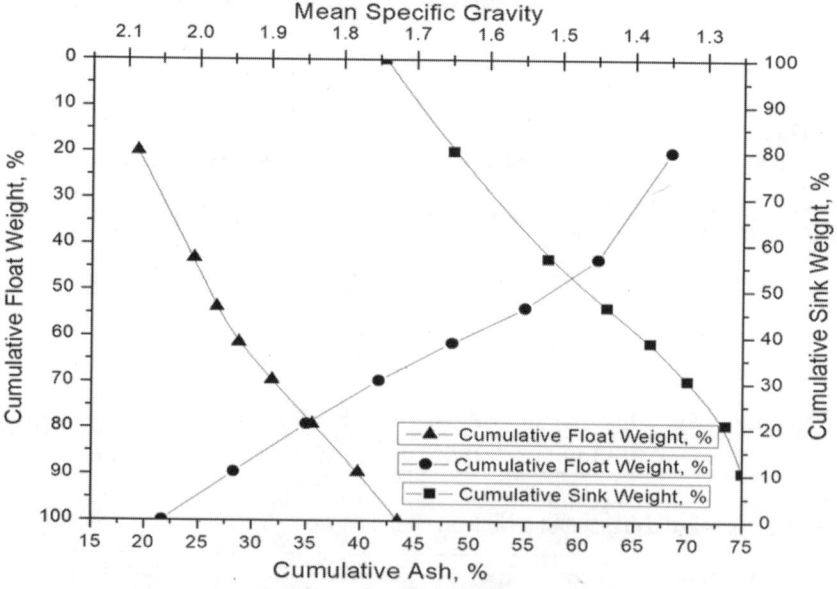

**Figure 2:** Washability characteristics of −50 mm coal

## 3. Results and discussion

The air fluidized vibrating deck separator was used for dry processing of coal. It consists of a perforated separating deck, three air chambers, and vibrating and hanging support mechanism. The longitudinal and transverse angle of the separating deck can be changed by the supporting mechanism. The fine coal particles and air makes the autogeneous medium due to the mix of air and fine coal and behaves like a separating medium within it. The separation takes place based on Archimedes principle. Airflow supplied from the draft fan fluidizes the bed material and the vibration mechanism imparts a helical turning motion to particles as they slide towards and the refuse end. Particle stratification on the separating deck takes place under the action of the vibration mechanism and the fluidizing force of the air flow. With a suitable combination of vibration force and the upward pressure of airflow, stratification of solids particles can be achieved mainly based on their differences in density. As a result, high density particles remained on the bottom-most layer, i.e. the layer closest to the deck surface. The buoyancy effect produced by the interaction of heavier particles can effectively control the misplacement of low density coal particles into the refuge bed, thus ensuring the purity of the refuse stream. The density of autogeneous medium during separation could be controlled by air flow rate and the fines in the feed sample. In the present study, process variables considered are air flow rate, deck vibration frequency, transverse angle, longitudinal angle, baffle plate height, and splitter position.

The air flow rate, oscillation assisted by the deck vibration, deck inclination (transverse and longitudinal angle), and splitter position play a significant role in air fluidized vibrating deck separation and they essentially controlled the yield and ash of clean coal. The transverse angle of the deck and air flow rate decides the product yield and grade of the clean coal. The increased transverse angle results to move the material towards the lower side of the deck and increases the ash of lighter fraction (clean coal) as the high density ash bearing mineral has a propensity to report towards the lower end of the deck. Thus, the effective reduction of ash does not occur. When the transverse angle is reduced, more material moves towards the higher end of the deck. The deck frequency and transverse angle are inversely inter-related and needs to be synchronized for effective separation.

The results obtained in single stage dry processing under different conditions are given in Table 4 and compared with the washability results in Fig. 3. The yield varies from 43% to 62% with variation of ash level from 29% to 35.76%. An attempt has also been made to improve the total yield of the concentrate by reducing the size of the middling to −13 mm for better liberation and to recover the clean coal with 34% ash. The results of two stage dry processing are given in Table 5. It has been found that two stage dry processing increases the yield of clean coal to 71.0% with 34% ash.

**Table 4:** Results of single stage processing of −50 mm using air fluidized vibrating deck separator

| CC, Yd% | CC, ash% | Mid, Yd% | Mid, ash% | Reject, Yd% | Reject, ash% | Organic efficiency |
|---|---|---|---|---|---|---|
| **Single stage processing** | | | | | | |
| 62.3 | 35.76 | 34.2 | 52.38 | 3.6 | 67.32 | 77.9 |
| 58.8 | 33.14 | 33.5 | 49.86 | 7.7 | 63.15 | 74.4 |
| 61.7 | 31.29 | 26.1 | 48.07 | 12.2 | 75.47 | 77.1 |
| 57.5 | 34.8 | 30.88 | 51.7 | 11.6 | 79.71 | 72.8 |
| 47.7 | 30.2 | 34.0 | 49.57 | 18.3 | 75.49 | 60.3 |
| 44.8 | 29.0 | 48.3 | 47.4 | 6.9 | 77.22 | 56.7 |
| 43.1 | 31.35 | 38.7 | 43.07 | 17.5 | 62.27 | 40.0 |

CC, clean coal; Yd, yield; and RC, recovery of combustible.

**Table 5:** Results of two stages processing of −50 mm using air fluidized vibrating deck separator

| CC, Yd% | CC, ash% | Mid, Yd% | Mid, ash% | Reject, Yd% | Reject ash% | Organic efficiency |
|---|---|---|---|---|---|---|
| **Two stage processing** | | | | | | |
| 71.0 | 34.0 | 10.5 | 55.36 | 18.5 | 70.21 | 90.0 |
| 67.8 | 33.18 | 14.7 | 53.82 | 17.5 | 69.42 | 85.82 |
| 69.0 | 34.09 | 11.2 | 52.91 | 19.8 | 67.73 | 87.34 |

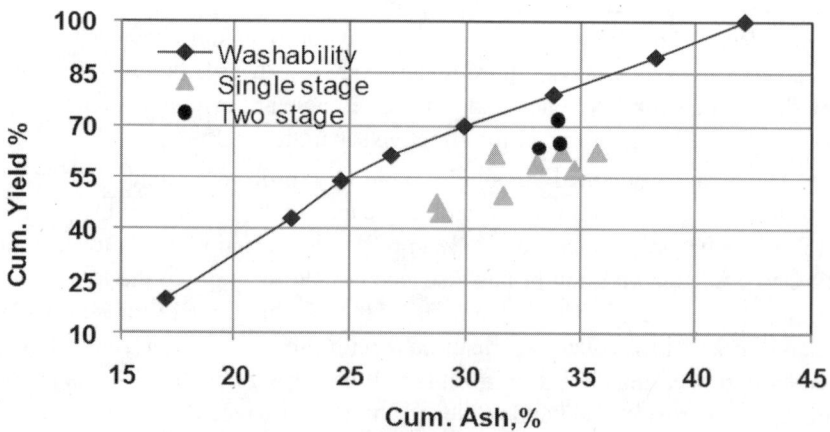

**Figure 3:** Results of air fluidized vibrating deck separator and washability of −50 mm coal

# 4. Conclusion

The findings of the investigation carried out for high ash Indian non-coking coal are enumerated as below:

- The drag force due to air velocity and vibratory motion of the deck eccentric play significant role in segregation and stratification of the particles and reducing the ash in clean coal.
- Increased transverse angle had negative effect on the grade of the clean coal comparing with other variable parameters.
- In a single stage processing, yield of the product at 34.14% ash is 58.8% with 66.7% recovery of combustibles; however, there is a loss of combustibles in middling stream. The process efficiency is measured by organic efficiency which is 77.9%.
- Two stage processing brought a significant improvement in separation. It improves the clean coal yield to 71% and the organic efficiency increases to 90%.
- A reject containing 70.21% ash and recovery of combustible 9.7% could be generated which is a great achievement in minimizing the loss of carbonaceous matter.

# 5. Acknowledgement

The authors would like to express their sincere gratitude to Council of Scientific and Industrial Research, Delhi, India for supporting the Mission mode net work project on dry beneficiation of Indian thermal coal (Project: ESC 0109).

# 6. References

[1] Gupta N, Bratton R, Luttrell GH, Ghosh T, Honakar RQ (2012) Application of air-table technologies for cleaning Indian coals. In: Young CA, Luttrell GH (eds) Separation of technologies for minerals, coal, and earth resources. Society for Mining, Metallurgy, & Exploration, Englewood, pp 199–209.

[2] Honaker RQ, Sracoglu M, Thompson E, Bratton R, Luttrell GH, Richardson V (2008) Upgrading coal using a pneumatic density based separator. Int J Coal Prep Utili 28(1):51–67.

[3] Haider A, Levenspiel O (1989) Drag co-efficient and terminal velocity, of spherical and non-spherical particles. Powder Technol 58(1):63–70.

[4] IS 9127-5: Methods for the petrographic. Analysis of bituminous coal and anthracite, Part-5, 2004.

[5] Lockhart NC (1984) Review paper: dry beneficiation of coal. Powder Technol 40:17–42.

[6] Li HB, Luo ZF, Zhao YM, Wu WC, Zhang CY, Dai NN (2011) Cleaning of South African coal using a compound dry cleaning apparatus. Min Sci Technol 21(1):117–121.

[7] Ministry of Environment and Forest (MoEF), Gazette Notification GSR 02(E), 2014.

[8] Osborne DG (1988) Pneumatic separation, coal preparation technology. Graham and Trotman, Norwell, pp 373–386.

[9] Yang Xuliang, Zhao Yuemin, Luo Zhenfu, Song Shulei, Duan Chenlong, Dong Liang (2013) Fine coal dry cleaning using a vibrated gas-fluidized bed. Fuel Process Technol 106:338–343.

[10] Zhao ZF, Chen YM, Chen QR, Fan MM, Tao XX (2002) Separation characteristics for fine coal of the magnetically fluidized bed. Fuel Process Technol 79(1): 63–69.

[11] Zhao PF, Zhao YM, Chen ZQ, Luo ZF, (2015), Dry cleaning of fine lignite in a vibrated gas-fluidized bed: segregation characteristics. Fuel 142:274–82.